EXS 99

Molecular, Clinical and Environmental Toxicology

Volume 1: Molecular Toxicology

Edited by Andreas Luch

Birkhäuser Verlag
Basel · Boston · Berlin

Editor

Andreas Luch
Federal Institute for Risk Assessment
Thielallee 88-92
14195 Berlin
Germany

Library of Congress Control Number: 2008938291

Bibliographic information published by Die Deutsche Bibliothek
Die Deutsche Bibliothek lists this publication in the Deutsche Nationalbibliografie;
detailed bibliographic data is available in the Internet at http://dnb.ddb.de

ISBN 978-3-7643-8335-0   Birkhäuser Verlag AG, Basel – Boston – Berlin

The publisher and editor can give no guarantee for the information on drug dosage and administration contained in this publication. The respective user must check its accuracy by consulting other sources of reference in each individual case.
The use of registered names, trademarks etc. in this publication, even if not identified as such, does not imply that they are exempt from the relevant protective laws and regulations or free for general use.

© 2009 Birkhäuser Verlag AG
Basel – Boston – Berlin
P.O. Box 133, CH-4010 Basel, Switzerland
Part of Springer Science+Business Media
Printed on acid-free paper produced from chlorine-free pulp. TFC ∞
Cover illustration: with friendly permission of Andreas Luch
Cover design: Benjamin Blankenburg, Basel, Switzerland

Printed in Germany
ISBN 978-3-7643-8335-0   e-ISBN 978-3-7643-8336-7

9 8 7 6 5 4 3 2 1   www. birkhauser.ch

Contents

List of contributors

Volker M. Arlt, Institute of Cancer Research, Section of Molecular Carcinogenesis, Brookes Lawley Building, 15 Cotswold Road, Sutton, Surrey SM2 5NG, United Kingdom; e-mail: volker.arlt@icr.ac.uk

Mónica R. Calera, Instituto de Física, Universidad Autónoma de San Luis Potosí, México; e-mail: mealera@ifisica.uaslp.mx

Ulrike Camenisch, Institute of Pharmacology and Toxicology, University of Zürich-Vetsuisse, Winterthurerstrasse 260, 8057 Zürich, Switzerland; e-mail: u.camenisch@access.uzh.ch

Salah-Dine Chibout, Investigative Toxicology, Preclinical Safety, Novartis Pharma AG, 4002 Basel, Switzerland; e-mail: salahdine.chibout@novartis.com

Jeff Chou, Microarray Group, National Institute for Environmental Health Sciences, Research Triangle Park, NC, USA; e-mail: chou@niehs.nih.gov

Alexei Degterev, Department of Biochemistry, Tufts University School of Medicine, Boston, MA 02111, USA; e-mail: alexei.degterev@tufts.edu

Wolfgang Dekant, Department of Toxicology, University of Würzburg, Versbacher Straße 9, 97078 Würzburg, Germany; e-mail: dekant@toxi.uni-wuerzburg.de

Steven G. Gilbert, Institute of Neurotoxicology & Neurological Disorders, 8232 14th Ave NE, Seattle, WA 98115, USA; e-mail: sgilbert@innd.org

Jan-Åke Gustafsson, Department of Biosciences and Nutrition, Karolinska Institutet, Hälsovägen 7, 14157 Huddinge, Sweden; e-mail: jan-ake.gustafsson@ki.se

Antoinette N. Hayes, Wyeth Research, One Burtt Road, Andover, MA 01810, USA; e-mail: hayesa@wyeth.com

Johan Högberg, Institute of Environmental Medicine, Karolinska Institutet, 17177 Stockholm, Sweden; e-mail: johan.hogberg@ki.se

G. Jean Horbach, Department of Pharmacology, NV Organon (Schering-Plough), Molenstraat 110, 5340 BH Oss, The Netherlands; e-mail: sjeng.horbach@spcorp.com

William K. Kaufmann, Department of Pathology and Laboratory Medicine, Center for Environmental Health and Susceptibility, University of North Carolina at Chapel Hill, Chapel Hill, NC 27599, USA; e-mail: wkarlk@med.unc.edu

David Kim, Department of Environmental Health, School of Public Health, Harvard University, 401 Park Drive, Boston, MA 02215, USA; e-mail: kbkim@hsph.harvard.edu

Andreas Luch, Department for Product Safety and ZEBET, Federal Institute for Risk Assessment, Thielallee 88-92, 14195 Berlin, Germany; e-mail: Andreas.Luch@bfr.bund.de

Ci Ma, Department of Environmental Health and Center for Environmental Genetics, University of Cincinnati College of Medicine, 3223 Eden Ave, Cincinnati, OH 45220, USA; e-mail: maci@email.uc.edu

Sari Mäkelä, Functional Foods Forum & The Department of Biochemistry and Food Chemistry, University of Turku, Finland; e-mail: sarmak@utu.fi

Jennifer L. Marlowe, Investigative Toxicology, Preclinical Safety, Novartis Pharma AG, 4132 Muttenz, Switzerland; e-mail: jennifer.marlowe@novartis.com

B. Alex Merrick, Laboratory of Respiratory Biology, National Institute for Environmental Health Sciences (NIEHS), 111 Alexander Drive, Research Triangle Park, Durham, NC 27709, USA; e-mail: merrick@niehs.nih.gov

Jonathan Moggs, Investigative Toxicology, Preclinical Safety, Novartis Pharma AG, 4132 Muttenz, Switzerland; e-mail: Jonathan.Moggs@novartis.com

Hanspeter Naegeli, Institute of Pharmacology and Toxicology, University of Zürich-Vetsuisse, Winterthurerstrasse 260, 8057 Zürich, Switzerland; e-mail: naegelih@vetpharm.uzh.ch

Leena A. Nylander-French, Department of Environmental Sciences and Engineering, School of Public Health, University of North Carolina at Chapel Hill, Chapel Hill, NC 27599, USA; e-mail: leena_french@unc.edu

Pauliina Penttinen-Damdimopoulou, Functional Foods Forum & The Department of Biochemistry and Food Chemistry, University of Turku, Finland; e-mail: paelpe@utu.fi

David H. Phillips, Institute of Cancer Research, Section of Molecular Carcinogenesis, Brookes Lawley Building, 15 Cotswold Road, Sutton, Surrey SM2 5NG, United Kingdom; e-mail: david.phillips@icr.ac.uk

François Pognan, Investigative Toxicology, Preclinical Safety, Novartis Pharma AG, 4132 Muttenz, Switzerland; e-mail: francoise.pognan@novartis.com

Alvaro Puga, Department of Environmental Health and Center for Environmental Genetics, University of Cincinnati College of Medicine, 3223 Eden Ave, Cincinnati, OH 45220, USA; e-mail: Alvaro.Puga@uc.edu

Ingemar Pongratz, Department of Biosciences and Nutrition, Karolinska Institutet, Hälsovägen 7, 14157 Huddinge, Sweden; e-mail: ingemar.pongratz@ki.se

Joëlle Rüegg, Department of Biosciences and Nutrition, Karolinska Institutet, Hälsovägen 7, 14157 Huddinge, Sweden; e-mail: joelle.ruegg@ki.se

Roberto Sánchez-Olea, Instituto de Física, Universidad Autónoma de San Luis Potosí, México; e-mail: rsanchez@ifisica.uaslp.mx

Willem G.E.J. Schoonen, Department of Pharmacology, NV Organon (Schering-Plough), Molenstraat 110, 5340 BH Oss, The Netherlands; e-mail: willem.schoonen@spcorp.com

Ilona Silins, Institute of Environmental Medicine, Karolinska Institutet, 17177

Stockholm, Sweden; e-mail: ilona.silins@ki.se

Ulla Stenius, Institute of Environmental Medicine, Karolinska Institutet, 17177 Stockholm, Sweden; e-mail: ulla.stenius@ki.se

Soon-Siong Teo, Investigative Toxicology, Preclinical Safety, Novartis Pharma AG, 4132 Muttenz, Switzerland; e-mail: soon-siong.teo@novartis.com

Paul B. Watkins, Center for Drug Safety Sciences, The Hamner Institutes for Health Sciences, University of North Carolina at Chapel Hill, Research Triangle Park, NC, USA; e-mail: pbwatkins@med.unc.edu

Walter M.A. Westerink, Department of Pharmacology, NV Organon (Schering-Plough), Molenstraat 110, 5340 BH Oss, The Netherlands; e-mail: walter.westerink@spcorp.com

Frank A. Witzmann, Department of Cellular and Integrative Physiology, Indiana University School of Medicine, Indianapolis, IN 46202, USA; e-mail: fwitzmann@iupui.edu

Tong Zhou, Center for Drug Safety Sciences, The Hamner Institutes for Health Sciences, University of North Carolina at Chapel Hill, Research Triangle Park, NC, USA; e-mail: TZhou@thehamner.org

Preface

Toxicology is the science of poisons and the study of adverse effects of chemical agents upon living organisms. The root of the word "toxic" is taken from the Latin word *toxicus* (meaning poisonous), which is itself derived from the ancient Greek word *toxikon*, a term used to describe poisons into which arrows were dipped (arrow poisons). Thus, assessing the biological effects of toxic chemical species (*i.e.* toxins) is the central theme and focus in toxicology, whether these are rather "small" (*small molecules*) or "macro" (*macromolecules* such as proteins and other biomolecules), or being synthesized in a chemical laboratory or in "mother nature's lab".

As a medical discipline, toxicology developed over millennia from the very early observations and the reclamation of both healing and adverse properties of plants, animals and corresponding extracts or preparations made thereof. During the last decades, toxicology has made rapid and great strides in moving forward from more or less descriptive ground fields towards a molecular, mechanism-based scientific discipline. While toxicology was considered the smaller sibling of an unlike pair for long periods in more recent times in modern ages till at least the mid-20th century, these recent developments fostered the emancipation from its greater sibling, pharmacology, and led to the rise of a highly interactive and independent field of research stimulated and strengthened by all natural sciences and human medicine. This interplay of many disciplines is what makes the work on biological effects of toxic substances interesting and exciting but – at the same time – also rather difficult and complex.

Recent advances in synthetic and analytical chemistry, in biophysics and structural biology, or in biochemistry and molecular cell biology offered great and valuable tools for deciphering the fate and effects of chemical compounds in biological systems across the entire range of complexity represented by the living world, *i.e.* from single macromolecules such as receptor proteins, structural proteins, nucleic acids to subcellular structures such as membranes and organelles, up to individual organs or whole organisms towards human-related environments and local or world-wide ecosystems. One important consequence of this development is that toxicology has moved towards a great variety of subdisciplines such as molecular toxicology, analytical toxicology, clinical (medical) toxicology, forensic toxicology, military toxicology, occupational (industrial) toxicology, regulatory toxicology (risk assessment), food toxicology, ecotoxicology (environmental toxicology), and so forth.

While all of these directions certainly have their eligibility and value, overall the body of knowledge acquired during the last decades in these fields ultimately and primarily will be applied for the sake of protecting human health and well-being. This is where medical aspects come into focus again, thereby closing the circle between recent efforts and the very early beginnings.

For most periods in history, toxicology has focused on deadly toxic chemicals. However, it was already recognized and recorded in the sixteenth century by Phillipus Theophrastus Aureolus Bombastus von Hohenheim, better known by the name Paracelsus, that *"In all things there is a poison, and there is nothing without poison. It depends only upon the dose whether a poison is poison or not ..."*. Of course, there are extremely potent toxins on one hand, and much less toxic compounds that require high concentrations at target tissue sites to convey adverse effects on the other. Moreover, in consideration of the importance of timing in producing adverse effects in living organisms, toxicologists roughly distinguish between acute toxicity and effects emerging only after long-term and chronic exposures. It was within the scope and one of the goals of this book project to address this broad range in toxicities, from extremely potent, deadly toxic compounds through environmental agents (xenobiotics) that exert toxicity rather in the wake of long-term accumulation and affect within certain tissues of living organisms. This concept led to the proposal of three individual volumes in which acute, clinically relevant toxins (2nd volume) and environmental compounds (3rd volume) would be handled separately in affiliation to an introducing volume that provides insights on chemical induced effects at the molecular level in sub-cellular compartments. To this end, the complete text would cover all aspects of modern toxicology with special emphasis on recent developments and achievements.

The first volume in its final composition presented today encompasses a very exciting journey from the fields of toxicokinetics, genotoxicity and DNA repair, cellular signaling and death, and receptor-mediated toxicities towards most recent developments in epigenetic profiling, toxicogenomics, toxicoproteomics, and high-throughput screening analysis of *in vitro* toxicity. While all of these chapters focus on molecular issues important in the understanding of the biological effects specific compounds would induce, the introducing chapter provides a historical account on toxicology in its broadest sense, from the very early beginnings until modern ages. This perspective is meant to remember and honor the great work and huge efforts of past generations of researchers and physicians in our field. As said above, toxicology developed as a medical discipline and, over hundreds of years was mostly focused on dramatic and fatal effects of deadly-toxic compounds. Work in this field was much more dangerous and health-consuming in former times compared to our days. At the same time, the knowledge acquired had great influence on many practical decisions in societies and thus shaped daily lives of people decisively. Until today, this aspect did not change very much although the focus now shifted more towards environmental compounds and the adverse health effects they might exert on individuals or the entire human population and ecological systems in a long term perspective.

Major parts of our societies, politicians, decision-makers, and sometimes even educated scientists do not know much about the role, the benefits and the great contributions toxicology has made over time in the development of an understanding of how hazardous chemicals would work in the body and on how this understanding can be used in a practical sense to save lives and to prevent chronic diseases by means of risk assessment and regulation of chemical exposures. That is one of the reasons why the value of toxicology as a scientific discipline is grossly underappreciated and why the field of toxicology faces major challenges these days. No matter which country in the western world we are looking at, toxicology is in deep trouble to survive as a scientific discipline due to an extreme shift in public funding towards basic science fields such as genetics, molecular biology and cell biology. In my mind, toxicology only will have a chance to survive when it saves its roots and keeps its interdisciplinary character as a highly interactive field of research along with practical applications stimulated and further strengthened by all natural sciences *and* human medicine. The acquisition of live-saving knowledge on the acute effects of potent toxins in the organism and the applicability of specific antitoxins and inhibitors in clinical treatment on one hand, and the growth in understanding of long-term mediated toxicity exerted by xenobiotics on the other, is unique and unmatched by any other academic discipline. It is essential to emphasize and to stress this distinctiveness in the context of discussions (and decisions) related to the positioning and preservation of our discipline. Toxicology has a very practical mission: research and gain of knowledge and understanding for the sake of human health and the saving of human lives. This goal is very real and concrete. To give up this mandate and to try to compete with basic research disciplines in natural sciences as those mentioned above by focusing on fundamental scientific issues only, would result in the sustained elimination of toxicology in academia for decades. Sadly, it looks as if we are already in the middle of this process. The main incentive and motive force driving my decision to present this work to the scientific community can thus be summarized by simply a few words: For the Unity and Future of Toxicology!

Last but not least, I would like to thank all of the authors who contributed to this volume with their great work and the time they invested to nicely and comprehensively cover a specific topic in the field of molecular toxicology. The main goal of this volume is to outline and to communicate the progress that has been achieved in molecular toxicology during the past decade. I deeply hope that it will find its place and acceptance in the scientific community and that it provides a guide for students and professionals in medicine, sciences, public health or engineering who are demanding reliable information and knowledge on toxic or potentially harmful agents and their effects in the human body.

Andreas Luch
Berlin, October 2008

Molecular, Clinical and Environmental Toxicology. Volume 1: Molecular Toxicology
Edited by A. Luch

Historical milestones and discoveries that shaped the toxicology sciences

Antoinette N. Hayes[1] and Steven G. Gilbert[2]

[1] *Wyeth Research, Andover, MA, USA*
[2] *Institute of Neurotoxicology & Neurological Disorders, Seattle, WA, USA*

Abstract. Knowledge of the toxic and healing properties of plants, animals, and minerals has shaped civilization for millennia. The foundations of modern toxicology are built upon the significant milestones and discoveries of serendipity and crude experimentation. Throughout the ages, toxicological science has provided information that has shaped and guided society. This chapter examines the development of the discipline of toxicology and its influence on civilization by highlighting significant milestones and discoveries related to toxicology. The examples shed light on the beginnings of toxicology, as well as examine lessons learned and re-learned. This chapter also examines how toxicology and the toxicologist have interacted with other scientific and cultural disciplines, including religion, politics, and the government. Toxicology has evolved to a true scientific discipline with its own dedicated scientists, educational institutes, sub-disciplines, professional societies, and journals. It now stands as its own entity while traversing such fields as chemistry, physiology, pharmacology, and molecular biology. We invite you to join us on a path of discovery and to offer our suggestions as to what are the most significant milestones and discoveries in toxicology. Additional information is available on the history section of Toxipedia (www.toxipedia.org).

Introduction

The history of toxicology provides a fascinating perspective on not only the development of the science of toxicology but also society's changing approach to preventing disease. Popular history focuses mostly on the deadly uses of various poisons. For example, Greeks used aconite as an arrow poison, and hemlock as a method of execution (e.g., Socrates). As society developed so did the more varied uses of poisons, such as the use of arsenic to achieve the "milk and roses" complexion many women envied and mercury to treat syphilis. There was also recognition of the occupational hazards related to the use of lead and mining of mercury. With the widely adopted use of the scientific method came more sophisticated understanding of the principles of toxicology and the dose-response relationship. This was followed by the recognition that specific agents could cause cancer, birth defects, or defects of learning and development. Certain agents were discovered to be environmentally persistent and their accumulation in biological organisms can cause inter-generational effects. Advances in the biological sciences, accompanied by technological advances, have placed a greater emphasis on understanding the biological mechanism of specific agents. There is also a far greater emphasis on the subtle sub-clinical effects of an agent and the need to account for individual sensitivity.

Despite our vast amount of toxicological knowledge we still repeat lessons learned. Much of our knowledge came from trial and error. For example, we learned about harm that can be caused by persistent chemicals such as DDT, PCBs, and lead, yet repeated this mistake with brominated flame retardants (e.g., PBDEs). On the other hand we did learn the lesson of thalidomide exposure and enacted efficacy and safety testing standards for new drugs. An important part of examining the milestones and history of toxicology is to evaluate lessons learned. Another important aspect is to examine how society has responded to advances in the toxicological sciences. There are several good general resources in recent toxicology textbooks on the history of toxicology, which also include good lists of references [1–3]. An excellent and wide-ranging list of references for the history of toxicology was developed by Stirling [4].

Toxicology etymology and definition

The word "toxicology" may have multiple origins. The word "toxic" was used before the word "toxicology" [5]. The best assumptions, based on available sources suggest it was derived from ancient Greek. The Greeks defined all drugs or potions as "pharmaka" or "pharmakon" giving no distinction between those causing harm or those used for treating disease [6]. Some time later the term pharmakon took on the meaning of poison. The Greek word "toxicos" or "toxikos" was an adjective of the noun "toxon" which meant "bow". Other Greek derivations of toxicology were: "toxicon", "toxikon", or "toxicos," meaning poison or poisons into which arrows were dipped [7]. Herodotus (484–425 B.C.E.), the Greek historian, wrote of the Scythians and their archery skills [8]. The Greek philosopher Aristotle wrote about the "scyticon" or "toxikon" as the poison or poison-tipped arrows the Scythians used in warfare [9]. The Greek word for poison "pharmacon" or "pharmakon" was compounded with the word for bow, "toxicon", to derive the phrase for arrow poison, "toxicon pharmacon" [10]. This phrase specifically meant "arrow poison" and was distinct from toxicon, which referred to any poison. The Romans shortened this phrase in their Latin derivation of the word for poison to "toxicum". They derived the word for arrow instead of the word for poison [10]. Therefore, toxicology's true origins were in the Greek word for bow, "toxon", which, initially, really had nothing to do with poisons. Interestingly, later derivations of the word toxon in the English language revert back to the original Greek meaning of the word bow; toxophily (archery), toxoplasma (bow shaped organism), toxocara (nematode with a head shaped like a bow) [11].

From Persia, the English word "toxin" was derived from the word "tekw" or "toxsa", meaning to run or flee [1]. A toxin was defined as a poison used on the tips of arrows, which were often fashioned from the branches of the yew tree (*Taxus baccata* derived from Latin). The modern use of the word "toxin" refers to poisons derived from biological sources, mainly those derived from animals [12].

The definition for toxicology has gone through many transformations over the years. Our knowledge of how poisons and drugs work has necessitated revisions that incorporate these changes. The various definitions of toxicology are defined in Table 1. These definitions demonstrate the increasing complexity of the field over time.

The last definition offers the most accurate description to date. It covers the widest range of toxic assaults and includes those that originate within the living system as a result of an exposure to a substance. It also covers the study of diseases using compounds that are known to be harmful to living systems. An example of this is 1-methyl-4-phenyl-1,2,3,6-tetrahydropyridine (MPTP), a neurotoxic compound first synthesized and used as an analgesic in 1947 by Ziering [18]. It was re-discovered in the 1980s when illicit drug synthesis led to the poisoning of several people who demonstrated symptoms that resembled Parkinson's disease. MPTP is now used in animal models to

Table 1. Definition of toxicology

Date	Definition of toxicology
1875	Toxicology is the branch of medical science that relates to the history and properties of poisons, and their effects upon the living body [7].
1891	The science of toxicology treats of the nature, symptoms, effects, doses, and modes of detection of poisons [13].
1911	The name of that branch of science that deals with poisons, their effects, and antidotes [14].
1939	The science of poisons; the department of pathology or medicine that deals with the nature and effects of poisons [10].
1959	The branch of medical science that deals with the nature, properties, effects, and the detection of poisons [15].
1975	The first edition of Casarett and Doull's Toxicology: The Basic Science of Poisons offers the broadest definition of toxicology. This edition describes the field as a complex amalgam of many fields that employs scientists from a variety of backgrounds to study the effects of drugs and chemicals for occupational health and drug safety [16].
1986	Toxicology is described as both a science and an art. "Toxicology is the study of adverse effects of chemicals on biological systems." The science of toxicology involves careful observation and the collection of raw data, whereas the "art" of toxicology involves the prediction of adverse outcomes based on these data [17].
1998	The organized study of the nature, effects, and mechanisms of action of toxic substances on living systems as well as the application of the acquired body of knowledge to human interests [11].
2007	The study of adverse effects of xenobiotics and further, the study of molecular biology, using toxicants as a tool. The study of the mechanisms of endogenous compounds such as oxygen radicals and reactive intermediates generated from xenobiotics and endobiotics [2].

study Parkinson's disease. The definition can be taken even further by including the effects of compound deficiencies on the body such as those created by vitamin A, vitamin C, vitamin E, and antioxidant proteins such as glutathione whose depletion results in increased oxidative stress on the cellular environment.

Antiquity: 3000 B.C.E.–90 C.E.

The earliest record of poisons involves a god or goddess called Gula or Ninisina. She was associated with poisons and references to god appear as early as 4000 B.C.E. This female deity was revered by the Sumerians and known as the "Goddess of Healing", "Mistress of Charms and Spells", "Controller of Noxious Poisons", and the "Terrible Goddess". She was described in detail on a cuneiform tablet written around 1400 B.C.E. The inscription read: *"Gula, the woman, the mighty one, the prince of all women. His seed with a poison not curable, without issue in his body may she place all the days of his life, blood and pus like water may he pour forth"* [19]. Gula was often depicted as a canine and the ancient Gula temples in the region known as Mesopotamia were believed to be places where medical diagnoses occurred and where illnesses were treated. Within these temples, ancient medical texts were housed and used as reference texts. Shen Nung (Shennong, 2737 B.C.E.) or the Yan Emperor brought agriculture to China by inventing the plow and the hoe and teaching the Chinese people how to cultivate the soil for the production of grain [20]. He was the father of the Chinese pharmacopoeia (Shen Nung called the Divine Farmer) and, according to legend, tasted three hundred or more herbs and plants in order to elucidate their medicinal properties. It was believed that he discovered numerous toxic substances by this method of tasting. One source stated that Shen Nung became extremely ill on at least seventy different occasions after having tasted certain substances. Shen Nung was credited with being the source for the work titled "The Divine Farmers Herb-Root Classic" (Divine Husbandman's Materia Medica or Shen Nung Pen Ts'ao Ching), written centuries later [21]. The true authorship is unknown. Within this work are listings of over 300 medicines derived from plants, minerals, and animals [22]. The work of Shen Nung took place before Chinese written history and much of what we know of him was the result of oral history, myth, and legend. He was considered to be the founder of medicine and the discoverer of natural drugs in China.

The Ebers Papyrus (1500 B.C.E.) is among the oldest written documentation of medicines in the ancient world. It is the oldest known ancient Egyptian medical scroll, dating back to a period between 1550 and 1500 B.C.E. [23]. It was allegedly discovered between the legs of a mummy in the district of Assassif in Thebes. Somehow it made its way to the city of Luxor in the mid 1800s and was subsequently sold to Edwin Smith (1822–1906) in 1862. The papyrus remained in Smith's collection until 1869 when it appeared in the catalog of an antiquities dealer. Georg Moritz Ebers (1837–1898) purchased the document in 1872 (other dates listed are 1873 and 1874) (the papyrus is named after this owner). Georg Moritz Ebers was a German Egyptologist and historical fiction novelist. He published a facsimile of the papyrus with an English-Latin vocabulary and introduction [23]. In the year 1890, Joachim translated the papyrus to German. Cyril Bryan translated Joachim's version from German to English. The papyrus is now located at the University of Leipzig, Germany, which is where Georg Ebers held a chair on the University's faculty until his retirement in 1889. The Ebers Papyrus mentions over 700 substances and over 300 recipes detailing incantations and concoctions using poisonous or medicinal minerals and plants. Among those listed are opium, hemlock, lead, antimony, aconite, and cannabis. The classes of substances mentioned include: stimulants, sedatives, motor excitants and depressants, hypnotics, narcotics, analgesics, antispasmodics, expectorants, myotics, mydriatics, tonics, sialogogues (increases flow of saliva), emetics, anti-emetics, cathartics, astringents, choleretics, anthelmintics, restoratives, antipyretics, anti-inflammatory agents, diuretics, labor inducers, ecbolics, demulcents, antizymotics, disinfectants, deodorants, antidotes, styptics, caustics, hemostatics, and parasiticides. Numerous delivery systems are mentioned as well and include: infusions, injections, pills, capsules, powders, potions, lotions, ointments, plasters, and mastications [24].

In the epic tales of Homer (The Odyssey and The Iliad), which were written about 850 B.C.E., events surrounding the Trojan War, believed to have taken place about 1200 B.C.E. are described in detail. The hero, Achilles, was struck down by Paris when he was shot with an arrow that pierced his ankle. The arrow was presumably poisoned since one would not expect for a superficial ankle wound to cause the death of a person, even in antiquity. There were several versions of this story of the death of Achilles. One told of how, after being struck, Achilles stumbled about in abject pain from the sudden envenomation. According to legend, he called out to his well-hidden executioner, to face and fight him man to man. Philoctetes, a Greek hero and famed archer, then drew back his own arrow, also dipped in poison, and killed Paris. The arrows of Philoctetes were, according to myth, given to him by Hercules and turned the tide of the war to the Greeks [10]. Poison arrows were prevalent throughout antiquity in both fact and folklore. During the Trojan War (1200 B.C.E.), Homer writes of Odysseus poisoning arrows (800 B.C.E.) in preparation for war. The poem speaks of Odysseus traveling to Ephyera to learn how to prepare poison arrows for his journey. *"For voyaging to learn the direful art to taint with deadly drugs the barbed dart."* There are legends of Hercules dipping his arrows into the blood of the Hydra (a multiple-headed poisonous serpent) and subsequently using these poisoned arrows to defeat his enemies. Ironically, it would be a poison arrow that would be the death of Hercules. Ovid writes (in Metamorphoses): *"The poison, heated by fire, coursed through his limbs. His blood, saturated by the burning poison, hissed and boiled. There was no limit to his agony as flames attacked his heart and the hidden pestilence melted his bones."* The word hydra in ancient Greek means water snake and some believe that the Hercules legend is believed to actually be a fictionalized account of the practice of using venom of a water snake to poison the tips of arrows. Other substances that were believed to have been used besides snake venom included certain plants such as wolfbane, hellebore and aconite. Aristotle (384–322 B.C.E.) described the use and preparation of arrow poisons. As mentioned above, the word for poison in ancient Greek was "toxicon" and it was derived from the word "toxon" that meant bow. Toxicology owes its nominal origins to the poison-dipped arrows used in ancient warfare. In antiquity, a poison or "toxicon" referred only to a deadly substance that was applied to an arrow or shield. Literary sources indicate that the Greeks, Scythians, and Nubians all used poisoned-dipped arrows, javelins, and other weapons in battle, indicating that there was, at least in some parts of the ancient world, both a vast knowledge of animal and plant poisons and some understanding of their effects on living organisms [8]. Theophrastus (371–287 B.C.E.) was a pupil of Aristotle and wrote extensively on the subject of botany. He wrote "De Historia Plantarum" and a variety of poisonous plants within this work are mentioned. In 399 B.C.E., Socrates (470–399 B.C.E.) was charged with religious heresy, and the corruption of Athenian citizens. He was tried, found guilty, and subsequently sentenced to death by hemlock. In ancient Greece, a man of Socrates status was given the opportunity to choose his own punishment. It was clear that Socrates chose death, as he could have escaped or been exiled had he wanted. Plato described his final moments after the poison began to take effect: *"… and he walked about until, as he said, his legs began to fail, and then he lay on his back, according to the directions, and the man who gave him the poison now and then looked at his feet and legs; and after a while he pressed his foot hard, and asked him if he could feel; and he said, "No"; and then his leg, and so upwards and upwards, and showed us that he was cold and stiff. And he felt them himself, and said: "When the poison reaches the heart, that will be the end." He was beginning to grow cold about the groin, when he uncovered his face, for he had covered himself up, and said – they were his last words – he said: "Crito, I owe a cock to Asclepius; will you remember to pay the debt?" "The debt shall be paid", said Crito; "is there anything else?" There was no answer to this question; but in a minute or two a movement was heard, and the attendants uncovered him; his eyes were set, and Crito closed his eyes and mouth."* [25]. There is speculation that the concoction of hemlock that Socrates ingested was mixed with a very large dose of opium [8]. This is likely the case since Socrates died a rather painless, quiet death and hemlock or the active ingredient coniine causes a more severe and spastic reaction before the body succumbs and death results. Coniine induces an ascending muscle and respiratory paralysis, which eventually leads to death due to a lack of oxygen supply to the heart, brain and tissues. These symptoms are more or less consistent with Plato's account of Socrates' death; however, there is some debate surrounding his account of the events of that day and whether the actual poison was, in fact, hemlock, a mixture of hemlock and opium, or something else entirely.

Hippocrates (460–370 B.C.E.) and Democritus (460–370 B.C.E.) were contemporaries and studied, extensively, the effects of animal, plant, and mineral poisons on the body. Hippocrates, for example, made observations on the toxic effects of copper and hellebore. He also used hellebore, in small doses, for the treatment of various diseases. Alexander the Great (356–323 B.C.E.) was possibly poisoned with hellebore or strychnine in 323 B.C.E. [26, 27]. Mithridates VI (Mithridates the Great) of Pontus (132–63 B.C.E.) tested hundreds of substances (antidotes to poisons) on himself, animals and prisoners in order to inoculate himself against any and all poisons. In Appian's "Roman History" there is an account of the death of Mithridates: "*Mithridates then took out some poison that he always carried next to his sword, and mixed it. There two of his daughters, who were still girls growing up together, named Mithridates and Nyssa, who had been betrothed to the kings of [Ptolemaic] Egypt and of Cyprus, asked him to let them have some of the poison first, and insisted strenuously and prevented him from drinking it until they had taken some and swallowed it. The drug took effect on them at once; but upon Mithridates, although he walked around rapidly to hasten its action, it had no effect, because he had accustomed himself to other drugs by continually trying them as a means of protection against poisoners. These are still called the Mithridatic drugs. Seeing a certain Bituitus there, an officer of the Gauls, he said to him, "I have profited much from your right arm against my enemies. I shall profit from it most of all if you will kill me, and save from the danger of being led in a Roman triumph one who has been an autocrat so many years, and the ruler of so great a kingdom, but who is now unable to die by poison because, like a fool, he has fortified himself against the poison of others. Although I have kept watch and ward against all the poisons that one takes with his food, I have not provided against that domestic poison, always the most dangerous to kings, the treachery of army, children, and friends." Bituitus, thus appealed to, rendered the king the service that he desire"* [28]. The term "mithridatum" means antidote. Lucius Cornelius Sulla (138–78 B.C.E.) was a Roman general and dictator who preceded the rule of Ceasar. Around 81 B.C.E., he issued the law, Lex Cornelia de Sicariis et Veneficiis (Lex de Sicariis, or Lex Cornelia), that protects public safety and punishes the act of poisoning. The law clearly states that any person who makes, sells, possesses, or purchases a "venenum malum" (poisonous or noctious substance) with the intention of committing murder may be tried by a commission for that act (Cicero).

The Hellenistic queen, Cleopatra VII (69–30 B.C.E.), was descended from Ptolemy I, a general who served under Alexander the Great and the first of a long line of successive Hellenistic rulers (male and female) in Egypt. Cleopatra VII is perhaps the most famous of the Ptolemaic rulers because of the intrigues that surrounded her court. She was, at different points during her rule, involved with Julius Ceasar and Mark Antony and caused political tension in Rome as a result (with Ceasar and Pompey, and Antony and Octavian). Antony and Cleopatra were defeated by Octavian in Actium in September of 31 B.C.E. Plutarch writes that Cleopatra tested numerous poisons on prisoners and animals in an effort to discover the least painful means of invoking death. She ended her own life by allowing an Egyptian asp snake to bite her (August of 30 B.C.E.).

Demosthenes (384–322 B.C.E.) was a Greek orator, speech writer and statesman. He strongly opposed Macedon's expansion into Greece and organized failed revolts against both Phillip II and Alexander the Great. When Antipater came into power after the death of Alexander, he sent his guard Archias to arrest Demosthenes who fled to the temple of Poseidon on the island of Calauria. Antipater knew that Demosthenes could incite the Athenians to another revolt and condemned him to death along with many others. Archias found Demosthenes at the temple and allegedly waited outside while he wrote a final letter to his family. Apparently Demosthenes who only pretended to write was actually sucking poison from the hollow tip of his pen. When Archias grew tired of waiting he went into the temple to arrest Demosthenes who stood as if to walk out but, instead, fell dead from the poison (Plutarch).

Pedanius Dioscorides (40–90 C.E.) was a Greek physician during the time of Nero who wrote "De Materia Medica", an extensive pharmacopoeia organized by drug classes (animal, vegetable, mineral) and physiological effect on the body. His five-volume book remained in print and in use until the 1600s and became the basis for the modern pharmacopoeia. His book was reprinted for nearly 1500 years and many editorial liberties were taken with each printing (some organized the contents in alphabetical order). The books mention over 600 medicinal and toxic plants with full illustrations, uses, preparations, and dosages. Nicander of Colophon (130 B.C.E.) was a physician and poet who wrote two surviving poems about poisons, Theriaca and Alexipharmaca. Theriaca is a 958-verse poem detailing the venomous bites of animals and how they should be treated and the Alexipharmaca contains 630 verses with details on poisons and their antidotes.

Titus Livius (59–17 C.E.) also called Livy, was a Roman historian who wrote an extensive history of Rome dating back to about 750 B.C.E. (possible date of the founding of Rome). He wrote of a period in Roman history where poisoning was rampant. Titus Livius writes that in 331 B.C.E. there was some type of epidemic in Rome that took many lives [29]. The cause was unknown but many speculate that a group of women were conspiring to poison the citizens of Rome. There was never any real proof of this and the cause of the many deaths was never resolved. According to Livy, in 331 B.C.E. there were a group of married women who were accused of the crime of poisoning. A slave girl confessed on the grounds of immunity, that she knew of a group of women who were mixing a poison in a large cauldron. The women were discovered but denied they were making poisons. They were made to drink their concoction and many of them died immediately afterwards. At this time, Livy writes (in Book 8) that a law was enacted to make poisoning a crime. If this is the case, this law pre-dates the Lex Cornelia passed by Sulla in 81 B.C.E. (Book 8, Titus Livius). In 184 and 180 B.C.E. there were allegedly mass poisonings in Rome and many were accused and convicted of the crime, although there is some doubt as to their actual guilt [6]. During the time of Nero (37–68 C.E.) a woman named Locusta was a notorious poisoner and poisoned many people in Rome. It is believed that she may have been hired by Agrippina to kill Claudius and was arrested for doing so. It is believed that Claudius was killed with the poisonous "death cap" mushroom, *Amanita phalloides* or an edible mushroom that was tainted with poison. Locusta was rescued from her conviction by the Praetorian Guard so that she could use her knowledge of poisons to kill for the Roman court of Nero. Nero allegedly had Locusta kill Britannicus, the son of Claudius and Nero's competition for the throne. Britannicus was allegedly poisoned with arsenic or a cyanic substance. Nero attempted to cover up the crime by powdering Britannicus's corpse with white chalk to conceal the *post mortem* lividity that spread about the victim's face hours after his death but his crime was revealed when rain washed away the chalk. Locusta tested poisons on prisoners and animals while in the court of Nero, and opened schools that taught the dark art of poisoning so that her knowledge could be passed on. When Nero was assassinated, Locusta was arrested shortly afterward and convicted and sentenced to death for her poisoning crimes. According to legend, Locusta only succeeded in killing Britannicus after her second attempt at poisoning [29].

The volcanic eruption of Mount Vesuvius in 79 C.E. killed Pliny the Elder and thousands of others who lived in or near the cities of Pompeii and Herculaneum. The eruption sent pyroclastic flows raging through the city and buried its inhabitants. The weight of the volcanic ash caused the roofs of homes to collapse and the suffocating gases killed thousands. It is believed that these toxic gases are what killed Pliny the Elder; however, this is debated since others who were with him survived. Pliny was on the city's border when he collapsed and there is some speculation that the gases did not reach his party. It is possible that Pliny's age made him more susceptible to the gas-filled atmosphere. Toxic gases typically emitted from volcanoes are carbon dioxide, carbon monoxide, and sulfur dioxide. The eruption of 79 C.E. was not the first time Vesuvius erupted. There is evidence that the volcano erupted several times earlier, three of which were significant in magnitude. The eruption in 1800 B.C.E. (The Avellino eruption) leveled Bronze Age settlements nearby. The volcano is still active and has erupted many times since the 79 C.E. eruption. The last eruption occurred in the 20th Century.

Cornelius Celsus (25–50 C.E.) was a Roman encyclopedist who wrote a large volume on medicine. Only one small section of this work remains, a work titled "De Medecina" and in it Celsus details the benefits of human and animal experimentation and on the preparation of drugs such as opium. He describes the symptoms or signs of inflammation (calor meaning warmth, dolor meaning pain, tumor meaning swelling, and rubor meaning redness) and treatments for various diseases.

A historical time line of chemicals isolated from plants is summarized in Table 2 and a toxicology time line of antiquity is summarized in Table 3.

Table 2. Chemical isolation from plants

Year	Compound	Plant source (Latin)	Scientists
1800	Curare	*Strychnos toxifera*	Humboldt
1805	Morphine	*Papaver somniferum*	Sertürner
1817	Narcotine	*Papaver somniferum*	Robiquet
1818	Strychnine	*Strychnos nux vomica*	Caventou and Pelletier
1818	Veratrine	*Veratrum album*	Caventou and Meissner
1819	Brucine	*Strychnos nux vomica*	Caventou and Pelletier
1819	Colchicine	*Colchicaceae colchicum*	Caventou and Meissner
1820	Caffeine	*Rubiaceae coffea*	Caventou, Pelletier and Runge,
1820	Quinine	*Rubiaceae cinchona*	Caventou and Pelletier
1812	Picrotoxin	*Anamirta paniculata*	Boullay
1822	Emetine	*Psychotria ipecacuanha*	Magendie and Pelletier
1827	Coniine	*Apiaceae conium*	Giesecke, Geiger and Hess
1828	Nicotine	*Nicotiana tabacum*	Posselt and Reimann
1828	Salicin	*Salix alba*	Leroux and Piria
1831	Aconitine	*Aconitum napellus*	Mein, Geiger and Hess
1832	Codeine	*Papaver somniferum*	Robiquet
1833	Atropine	*Atropa belladonna*	Geiger and Hess
1833	Hyoscine	*Atropa belladonna*	Geiger
1833	Thebaine	*Papaver somniferum*	Dumas and Pelletier
1842	Theobromine	*Theobroma cacao*	Woskresenky
1848	Papaverine	*Papaver somniferum*	Merck
1851	Choline	*Amanita muscaria*	Babo and Hirschbrunn
1860	Cocaine	*Erythroxylum coca*	Niemann
1864	Physostigmine	*Physostigma venenosum balfour*	Jobst and Hesse
1870	Muscarine	*Amanita muscaria*	Koppe and Schmiedeberg
1875	Digitoxin	*Digitalis lanata*	Schmiedeberg
1887	Ephedrine	*Ephedra sinica*	Nagai
1889	Ricin	*Ricinus communis*	Hermann and Kobert
1903	Camphor	*Cinnamomun camphora*	Komppa
1939	Monocrotaline	*Crotolaria spectabilis*	Adams and Rogers
1964	Cannabinoids	*Cannabis sativa*	Ederly, Gaoni, and Mechoulam
1967	Taxol	*Taxus brevifolia*	Wall and Wani

Note: The species listed above are a common source of the substance and not necessarily the source for the first isolation. For example, choline was first produced in 1851 by the decomposition of sinapine with barium hydroxide (and was called sincaline instead of choline). In 1875 Schmiedeberg and Harnack isolated choline from *Amanita muscaria* and called it amanitine.

Table 3. Toxicology history time line: Antiquity

Year	Event
2696 B.C.E.	Shen Nung: The Father of Chinese medicine, noted for tasting 365 herbs and said to have died of a toxic overdose.
1500 B.C.E.	Ebers Papyrus: Egyptian records contain 110 pages on anatomy and physiology, toxicology, spells, and treatment, recorded on papyrus.
1400 B.C.E.	Gula: Sumerian texts refer to a female deity, Gula. This mythological figure was associated with charms, spells and poisons.
850 B.C.E.	Homer: Wrote of the use of arrows poisoned with venom in the epic tale of *The Odyssey and The Iliad*. From Greek toxikon arrow poison.
399 B.C.E.	Socrates: Charged with religious heresy and corrupting the morals of local youth. Death by hemlock – active chemical alkaloid coniine.
377 B.C.E.	Hippocrates: Greek physician, observational approach to human disease and treatment, founder of modern medicine, named cancer after creeping crab.
356 B.C.E.	Alexander the Great: Born in the year 356 B.C.E. in Macedonia, died in 323 B.C.E. possibly poisoned with aconite or strychnine.
131 B.C.E.	Mithridates VI of Pontus: Tested antidotes to poisons on himself and used prisoners as guinea pigs. Created mixtures of substances leading to term mithridatic.
82 B.C.E.	L. Cornelius Sulla: *Lex Cornelia de sicariis et veneficis* – law against poisoning people or prisoners; could not buy, sell or possess poisons.
69 B.C.E.	Cleopatra: Experimented with strychnine and other poisons on prisoners and poor. Committed suicide with Egyptian Asp.
40 C.E.	Pedanius Dioscorides: Greek pharmacologist and physician, wrote *De Materia Medica* basis for the modern pharmacopeia.
79 C.E.	Mount Vesuvius: Erupted August 24th – Cities of Pompeii and Herculaneum destroyed and buried by ash. Pliny the Elder suffocated by volcanic gases.
100 C.E.	Aulus Cornelius Celsus: Roman encyclopedist and possibly a physician. His only surviving work *De Medicina* is only a small part of his larger encyclopedia but is one of the best sources of Alexandrian medical knowledge.

The Medieval and Renaissance period of toxicology: 476–1699 C.E.

The Middle Ages were filled with scientific discovery and gave rise to the Renaissance period, even though generally it has been considered to be a time of little accomplishment. For medicine and toxicology it was mostly a period filled with a belief in folklore, superstitions, and religion. Alchemy, the prelude to chemistry, was widely practiced and through this many discoveries were happened upon by experimentation. Charlemagne (742/747–814 C.E.) was King of the Franks from 768 C.E. until his death in 814 C.E. Generally considered to be the founder of the French and German monarchs, Charlemagne ruled over the Holy Roman Empire with his brother Carlomann, under the Pope (Leo III). During this period of history poisoning was common. The brothers were not amenable to each other and shortly after they were put into power, Carlomann suddenly died. There was some suspicion that Charlemagne may have hastened his brother's death by poisoning him. At least nine of Charlemagne's successors were murdered by poisoning. Charlemagne reportedly owned a tablecloth made of asbestos and demonstrated its fire resistant properties to his guests. Interestingly, the properties of asbestos were known to Pliny the Elder (23–79 C.E.) who described the effects of asbestos on the lungs of slave weavers who made garments from the substance.

Ancient napalm or "Greek fire" was described by the Crusaders in 1248 C.E. as consisting of naphtha, quicklime, sulfur, and saltpeter or potassium nitrate. In 994 C.E., an outbreak of ergot killed nearly 40 000 people who ingested contaminated wheat/rye bread. The resulting symptoms, known as St. Anthony's Fire, consisted of painful peripheral gangrene, convulsions, hallucinations, and nausea. Moses Maimonides (1135–1204) or Moses Ben Maimon, was a Jewish philosopher and physician who wrote "Treatise on Poisons and Their Antidotes". The book was divided into two parts. The first part covered all manner of bites including those of mad dogs, snakes, scorpions, bees and wasps. The second part covered poisons in food, minerals, and plants. It also covered antidotes and efforts that could be undertaken to prevent absorption of poisons such as emesis. Albertus Magnus (1193–1280) was a Dominican friar who wrote extensively on the compatibility of religion and science. He believed that metals had magical or occult properties and reported that he discovered the "philosopher's stone". He is credited with the discovery and isolation of arsenic in 1250. Raymundus Lullius (1232–1315), a Spanish chemist, discovered ether in 1275 and called his discovery "sweet vitriol". Paracelsus knew of the hypnotic effects of ether but it was not used as an anesthetic until 1842 when Crawford Long (1815–1878), a physician, used it when he removed a tumor from the neck of one of his patients. The philosopher's stone was a mythical substance that could, allegedly, turn ordinary metals into gold. The whole of the Middle Ages was filled with alchemists experimenting with minerals and fire in search of the philosopher's stone. A book titled "Mutus Libris" or "wordless book" of illustrations outlined the method for creating the philosopher's stone. The Knights Templars (1118–1307) were a Christian military order said to be experts with poisons. They searched for the "Elixir of Life", or potion that would grant anyone who drank it, eternal life. Later, this "Elixir of Life" was interpreted as being the "holy grail". The Templars were disbanded in 1307 by the Roman Catholic church. Petrus de Abano (1250–1315) was an Italian scholar who translated the works of Hippocrates and Galen into Latin and wrote "De Venenis" (first print in 1473). The Black Death (1347–1351), probably bubonic or pneumonic plague, spread throughout Europe causing the death of the highest proportion of the population in recorded history.

The Venetian Council of Ten (1310–1791) was a governing body made up of ten people (law makers or "governors") who were put into power to keep order and prevent crimes against the state; however, records indicate they routinely accepted payments for murder contracts carried out by poisoning. The ten voted for or against the proposed murder contract. They kept a written record of the reason for the murder contract, the amount paid, and when the deed was done the word "factum" (translated "done deed") was written next to the record. In 1543, a Franciscan brother by the name of John of Ragusa offered his poisoning services to the Council. He offered a price list for specific poisoning jobs; 500 ducats for the Sultan Mahomet II, 150 ducats for the Duke of Milan, and 100 ducats for the Pope. One recorded document indicated a plot to poison soldiers by contaminating a well from which they all drank. Pope Clement VII and Pope Clement XIV were likely poisoned. Clement VII may have been murdered with either poisoned edible mushrooms or with the poisonous death cap mushroom, *Amanita phalloides*. Clement XIV may have had poison slipped into his food. Pope Benedict XI may have eaten poisoned figs in 1304 C.E. Pope John VIII was allegedly poisoned and clubbed to death in 882 C.E. Pope Pius III was also allegedly poisoned in 1503 C.E. The Borgia family was suspected of being notorious poisoners and was the ruling family in Rome from the early 1400s to the mid 1500s. The members of the Borgia family who took center stage in the malevolent dealings of their time in power were Alonso, the father who became Pope Calixtus III, his son, Roderigo who became Pope Alexander the VI, and his children, Cesare, and Lucrezia. Lucrezia was used as a pawn in the Borgia's power struggle and, though it is doubtful that she actively participated in poison plots, she is often associated with poison and intrigue in the Borgia court. She was married off and divorced at the behest of her father several times, though divorce was generally not allowed in the Roman Catholic church. These arranged marriages were politically motivated. The Borgia's used a poison called "Cantarella" or "La Cantrella" believed to be a concoction of arsenic and phosphorous usually dispensed in wine to unsuspecting victims. Preparation of the poison involved killing a hog with arsenic and adding more poison to the animal's abdominal cavity. The animal was allowed to putrefy and the liquid, which oozed from the animal's carcass, was collected and evaporated to obtain a powder. Leonardo Da Vinci (1452–1519 C.E.) experimented with poisons in Renaissance Italy. He studied with the bioaccumulation of poisons using a procedure he called "passages" that involved dosing an animal with a lethal dose of poison, preparing an extract of the animal's organs and feeding the extract to yet another animal. The process would be repeated several times over in order to increase the strength of the poison.

Perhaps the most popular doctrine in the field of toxicology was proposed by Paracelsus (Philipus Theophrastus Aureolus Bombastus, von Hohenheim, 1493–1541 C.E.). Paracelsus was a physician, alchemist, and philosopher of the late Medieval, early Renaissance period. His contribution of the dose-response concept of toxicology is well documented in nearly every textbook on the subject. Paracelsus was born in Einsiedeln in 1493 and later moved to Austria where there were many metal smelting industries and miners. His father, Wilhelm, taught mineralogy in Austria and it is likely that Paracelsus learned the art of manipulating metals here. Paracelsus wrote extensively and in one document he wrote of humoral pathology or disease in terms of "five principles" or Entia. The first principle or "Ens Astri" related diseases under the influence of the stars and meteorology. The second principle or "Ens Veneni" dealt with poisoning and disturbances of metabolism. The third principle or "Ens Naturale" was concerned with the constitution of the patients and his "humors" and hereditary traits that influenced his condition. The fourth principle or "Ens Spirituale" dealt with diseases that originated from mental instability and the fifth principle or "Ens Dei" was concerned with diseases sent by God or incurable diseases. Paracelsus also believed that there was a point of disease progression beyond which no man could be cured and death was imminent. His five principles were accompanied by five types of doctors or healers that could treat them. There were "Naturales" or doctors who treated contrary with contrary, "Specifici" who used specific remedies to treat disease, "Characterales" or magi who treated "like with like", "Spirituales" who used chemical drugs and the last was "Christ and the Apostles" for treating those diseases sent by God. Paracelsus taught that all bodies were composed of three basic elements, sulfur, mercury, and salt. Sulfur, he taught, was the soul or life force. Mercury was the force of the mind or intellect and salt was the principle of matter. Paracelsus infuriated many scholars of his time by going against the teachings of Galen and the teachings of other accepted scholars and concepts. When Paracelsus returned to Carinthia to take possession of his father property after he died, the leaders of the community protested his presence. Paracelsus responded by writing his famous "apologia" titled "Septem Defensiones". The document defended the seven charges laid before him; the charges being, that his theories were new, that he described diseases that did not exist or were not accepted in the medical community and gave them "made up" names, that his prescriptions were poisonous, that he was a vagrant, that he unjustly attacked the medical profession and his colleagues, that his manners were course resulting in his students abandoning him and his teachings, and that he, not unlike any doctor, could not cure all diseases. In his famous defense of the third charge, his "Third Defense", he argued the point that *"… the dose makes the poison"*. He wrote: *"In all things there is a poison, and there is nothing without poison. It depends only upon the dose whether a poison is poison or not… I separate that which does not belong to the arcanum from that which is effective as the arcanum and I prescribe it in the right dose… then the recipe is correctly made… That which redounds to the benefit of man is not poison; only that which is not of service to him, but which injures him, is poison."* This was in response to his critics who disputed his use of inorganic compounds to treat disease, deeming them too toxic. Paracelsus also believed that diseases were primarily associated with a specific organ in the body, giving rise to the idea of "target organ toxicity". The contributions Paracelsus made to science were notable if not profound, but were shadowed by an equal distribution of the absurd, as was the tradition of many alchemists whose practical study of the properties of matter yielded unexplainable phenomenon. Perhaps in his darkest period, Paracelsus claimed the ability to create a diminutive man called a "homonculus" from a mixture of semen, horse manure and human blood. Paracelsus questioned the long-standing ideas of his time and challenged those around him to do the same. He tried to rationalize the treatment of disease and rationalize all aspects medicine, always relating back to the nature of things. He authored many important medical and philosophical works including: Greater Surgery, Astronomia Magna, Labyrinth, Defensiones, Chronicle of Carinthia, De Natura Rerum, and Archidoxa, which was an alchemy book dedicated to the production of medicine (not gold as was the custom of alchemy books of the time). Paracelsus died in 1541.

Giulia Toffana (1650 C.E.) was credited with poisoning over 600 people in Italy with her concoction known as "Aqua Toffana". The poison was believed to consist of arsenic trioxide in solution. Giulia Toffana or Tophania (1635–1719) undoubtedly poisoned many people. Several sources suggest that she also poisoned two Popes, Clement IV and Pius III; however, inconsistent dates lend some doubt to these claims. The Middle Ages and the first part of the Renaissance were times where poisoning was not only rampant, it was accepted. There were schools of where one could learn the dark art of poisoning and payment for services rendered was quite common. It was not exceptional for guests to bring their own wine to dinners and gatherings to ensure that they were poison free, as it was the custom of the time to administer poison in the drink. This act was not considered an insult to the host. As early as 400 B.C.E., poison was administered in wine. Xenophon (400 B.C.E.) wrote that poisoning was so frequent among the Medes that tasters were employed to sample the wine first, before the king or nobleman, to ensure it was safe and unadulterated [30].

A historical time line of toxicology events in the medieval and renaissance period is summarized in Table 4.

The Chemical Age of toxicology: 1700–1899

In 1700, Bernardino Ramazzini's (1633–1714) book titled "Discourse on the Diseases of Workers" or "De Morbis Artificum Diatriba" was published. This book brought much attention to the burgeoning field of industrial hygiene and occupational toxicology in that it related some 50 or more occupations with its associated disease. Ramazzini included specific methods of analysis for disease detection and a methodical approach to disease prevention, which became the foundation for factory safety. Duke Francesco II assigned Ramazzini to the University of Modena in 1682, giving him the title of "Medicinae Theoricae", and it was here (in the 1690s) that he observed and studied the diseases of man and animals, in the rural areas surrounding Modena. His seminal work, "Discourse on the Diseases of Workers" outlined particular health hazards of metals, dust, chemicals and chemical abrasives, associated with occupations such as weaving, farming, mining, masonry, and nursing. Ramazzini discussed causal factors of occupational diseases, treatment, and preventative measures

Table 4. Toxicology history time line: Middle Ages

Year	Event
673 C.E.	Greek Fire: Ancient "napalm" described by the Crusaders as consisting of naphtha, quicklime, sulfur, and saltpeter.
994 C.E.	Ergot Outbreak: 40 000 died from eating contaminated wheat/rye causing gangrene – known as St. Anthony's Fire.
1135 C.E.	Moses Maimonides: Jewish philosopher and physician wrote *Treatise on Poisons and Their Antidotes.*
1193 C.E.	Albertus Magnus: Dominican friar wrote extensively on compatibility of religion and science and isolated arsenic in 1250.
1275 C.E.	Raymundus Lullius: Ether discovered by Spanish chemist and later called "sweet vitriol".
1300 C.E.	Knights Templar (1118–1307): Christian military order alleged to be experts with poisons. They searched for the "Elixir of Life".
1315 C.E.	Petrus de Abano (1250–1315): Italian scholar, translated works of Hippocrates and Galen into Latin; wrote *De Venenis.*
1347 C.E.	The Black Death (1347–1351): Bubonic and pneumonic plague ravaged Europe leaving the highest number of casualties in history until the flu pandemic of 1918.
1386 C.E.	Arsenic is mentioned in *Chaucer's Canterbury Tales* in the context of poisoning someone.
1419 C.E.	Venetian Council of Ten: Regulatory body in Venice whose members carried out murders with poison for a fee and kept detailed records of the transactions. When the murder was committed they marked the record, "factum" meaning "done deed".
1423 C.E.	Zhou Man: Chinese explorer lost thousands of crew members from uranium exposure while mining lead in Jabiru, Australia.
1450 C.E.	Rodrigo and Cesare Borgia (1400–1500): Poisoned many people in Italy for political and monetary gain. Used arsenic in a concoction called "La Cantrella".
1452 C.E.	Leonardo da Vinci (1452–1519): Experimented with bioaccumulation of poisons in animals and called the procedure "passages".
1478 C.E.	Pope Clement VII (1478–1534): Died (possibly murdered) after eating *Amanita phalloides* (death cap mushroom).
1493 C.E.	Paracelsus (1493–1541): Dose-response concept in toxicology. "All substances are poisons; there is none which is not a poison. The right dose differentiates a poison from a remedy."
1494 C.E.	Georgius Agricola (1494–1555): Wrote *De Re Metallica* published 1556. The most comprehensive book on mining and metallurgy.

in detailed summaries that accentuated the classical ideals of Galen, Hippocrates, and Celsus with a closing summation of his own observations. His advice to physicians: *"When you come to a patient's house, you should ask him what sort of pains he has, what caused them, how many days has he been ill, whether the bowels are working and what sort of food he eats ... I may add one more question: what occupation does he follow?"* The last question was Ramazzini's personal addition to the previous questions, which were often suggested by the classics such as Hippocrates and Galen. Ramazzini was not the first to suggest that some occupations are more hazardous than others or the first to associate diseases of workers with exposures to certain agents. Ulrich Ellenbog (1440–1499) published in 1480 on the effects of mercury and lead exposure in goldsmithing, and Georgius Agricola (1494–1555) discussed the diseases of miners in his book titled "On the Nature of Metals" or "De Re metallica" published in 1556.

Percivall Pott (1714–1788) was a surgeon at London's Bartholomew's Hospital from 1744–1787. He is well known in the field of toxicology (especially occupational toxicology) for the observation that squamous cell carcinoma of the scrotum was a hazard associated with chimney sweeps due to their repeated exposure to soot. This was the first reported example of polyaromatic hydrocarbon carcinogenicity. Others before Pott made observations on the toxicity of smoke and soot, including Paracelsus who described the conditions of miners continuously exposed to soot and heavy metals. Richard Mead (1673–1754) published "A Mechanical Account of Poisons in Several Essays", the first book written in the English language devoted entirely to the subject of poisons. John Jones was an English physician who studied, extensively, the effects of opium and wrote "Mysteries of Opium Reveal'd" in 1701. The book describes the history of opium and how it was used in different cultures. It describes preparations of opium and the effects of opium use and withdrawal. Thomas De Quincey (1785–1859) wrote an extensive account of his own experiences with opium in "Confessions of an English Opium Eater" (published in 1823). The book details his use of opium, experiences and thoughts while under its influence, and his inability to stop using the drug. The field of chemistry grew and the isolation of pure drug substances led to more potent medicines and addiction was rampant. The concept of addiction or addiction to a particular substance was not apparent in the 18th and even the early 19th centuries. Addiction was merely thought of in the context of being a compulsive, yet voluntary, behavior. It was only recently (in the past hundred or so years) that we have an understanding of the mechanisms behind addiction and some insight into how it should be treated. The concept of treating the effects of one drug with another drug was fairly new.

Antoine Lavoisier (1743–1794), Joseph Priestley (1733–1804), and Carl Wilhelm Scheele (1742–1786) were pivotal in the advancement of chemical knowledge in the 18th century. They made extraordinary discoveries and observations that led to a reform of the ideas held at the time. The phlogiston theory, for example (theory that a "fire-like" element was contained within combustible bodies and released in the combustion process) was completely discredited. Scheele and his contemporaries discovered a number of important elements including oxygen, hydrogen, chlorine, and sulfur dioxide. Carl (Karl) Wilhelm Scheele was a German-Swedish chemist and contemporary of Lavoisier. He discovered oxygen, chlorine, hydrogen cyanide and many other gases and compounds (Scheele, Lavoisier and Priestley are often attributed with independently making the same discoveries). Like many chemists of his time, he worked as a pharmacist and conducted experiments in the pharmacy where chemicals were readily available and exploration of their properties could be undertaken without difficulty. Scheele believed that a complete description of a compound meant uncovering all of its attributes, including taste. He believed this to be a necessary evil if one wanted to elucidate every property of substance and a full characterization was not complete without taste. He routinely tasted his chemical discoveries (reported the taste of cyanide) and eventually became gravely ill, possibly from a self-imposed poisoning from his years of exposure to toxic chemicals and heavy metals. Scheele knew of the risks involved with his work and referred to his illness as being *"the trouble of all apothecaries"*. Scheele died on 26 May 1786.

In the 1700s and 1800s, many chemists, pharmacists, and doctors were inspired to adapt methods for the isolation of drugs from natural products such as plants and minerals. As the field developed and knowledge expanded as a result of experimentation, a greater understanding of chemical properties allowed for the synthesis of new compounds from basic starting materials. Chemistry, pharmacy, medicine, and toxicology evolved collectively, from the observations of the actions of substances on living things, to the identification of plants, animal, and mineral substances responsible for these actions. This evolution involved the qualitative and quantitative chemical detection of substances, to their pure isolation and, finally, to their complete synthesis from very different starting materials. These fields evolved with each century or new era, building on the knowledge acquired from the preceding one. During this period, the number of chemicals and medicinal products grew exponentially with little oversight or care of possible negative human or environmental exposure effects. The period 1700–1900 was a time of rapid growth around the world as the industrial revolution was the impetus for inventions in mining, medicine, building, and transportation. No field was left untouched by its approach. The field of toxicology was taking shape on the backdrop of the chemical and industrial revolution. The new industries that sprouted up all over the world brought new-found wealth to many, and the age-old tradition of poisoning, while ever present, became a new target for the emergent scientific and medical community. The process of detecting poisons in biological material became an important endeavor. Toxicology as it exists today owes its humble beginnings to the field of medical jurisprudence and forensics. Medical and criminal cases where poisonings were suspected could not be proven prior to the 18th century. The court relied on full confessions, witnesses, or circumstantial evidence to prove a poisoning had occurred. Richard Weston with co-conspirators was tried and convicted of poisoning Thomas Overbury with arsenic and mercury in 1616. During his trial, the jury was warned not to expect any proof that the poison used was arsenic or mercury. Weston confessed after his conviction that the poison he used was white arsenic. In 1752, Mary Blandy went on trial for murdering her father 1 year earlier in 1751 by poisoning him with arsenic. This was the first legal trial in England where chemical tests to detect arsenic were allowed. Surprisingly, the tests that were conducted were more or less qualitative and not without ambiguity. They were conducted by Dr. Anthony Addington who also medically treated Mr. Blandy for the alleged poisoning, and who also conducted the autopsy. He accepted the "white powder" from someone in the household who told him that this was the same powder mixed with Mr. Blandy's food and taking them on their word, he conducted tests on this powder and determined that it was white arsenic. Interestingly, Dr. King who did not testify at the trial, also conducted tests. The arsenic tests were not done under the observation of witnesses so that the results could not be confirmed in real time. Such evidence in a trial conducted today would be considered insufficient but since this was the best available test at the time, it was accepted as irrefutable. Johann Daniel Metzger (1739–1805) discovered that if a substance containing arsenious oxide is heated over charcoal, it will form a black "mirror"-like deposit on cold plate that is held over the coals. He made his discovery of the "arsenic mirror" in 1787. James Marsh (1794–1846) developed the Marsh test for arsenic in 1836 and published his findings in the Edinburgh Philosophical Journal. The test involved adding sulfuric acid and arsenic-free zinc to a substance suspected of containing arsenic. The reaction produced arsine gas, which was ignited to decompose the gas and pass it through a cold tube. If arsenic was present in the material a black "mirror"-like deposit formed on the surface of the tube. Marsh's test for arsenic was highly successful but had limitations. The test could only be performed in a controlled laboratory environment by an experienced chemist. Hugo Reinsh also developed a test for arsenic in 1841. Reinsh's test required that a substance suspected of containing arsenic be dissolved with hydrochloric acid and boiled with copper. If arsenic was present, a gray metallic material (arsenic) formed on the surface of the copper. Carl Fresenius (1818–1897) and August von Babo (1827–1894) also developed screening methods for arsenic and other poisons. In 1851, the Arsenic Act was passed in England. This law stated that arsenic sold in quantities of 10 pounds or more had to be colored with a dye of indigo or soot, that a record of the sale had to be kept in a book on the premises, and that the purchaser and seller had to sign this book, and when the buyer was unknown to the seller, there had to be a third party present to witness the sale. Arsenic adulterated candies killed 20 people and poisoned over 200 people in the town of Bradford, England in 1858. The confectioner thought he was mixing in a filler substance called "daft" which was often used to adulterate foods because it was a lot cheaper than sugar. He was actually adding arsenic trioxide.

In the 1800s, the field of organic chemistry was on the rise due to the discovery that inorganic compounds could be turned into organic compounds by simple chemical transformations. In 1812, a British doctor and amateur chemist synthesized phosgene (John Davy, 1790–1868) by exposing carbon monoxide and chlorine to sunlight. The complete history of the first synthesis of mustard gas is somewhat clouded. Several scientists claimed to have synthesized the gas; however, their claim to having been the first is questionable since they did not describe the very toxic blistering properties, which would have been obvious if they had actually come into physical contact with the compound.

Friedrich Sertürner (1783–1855) was a German pharmacist who discovered and isolated morphine from opium in 1804. He named the compound morphium, after Morpheus, the Greek god of dreams. Morphium or morphine is generally accepted as being the first medicinal alkaloid isolated from any plant. Joseph Caventou (1795–1877) and Pierre Pelletier (1788–1842) were French pharmacists who applied the methods of Friedrich Wilhelm Sertürner and were among the first to isolate a number of extremely toxic, yet medicinally important alkaloids including emetine in 1817, strychnine in 1818, quinine in 1820, and veratrine in 1821. They isolated quinine from the cinchona bark in their pharmacy and sold their preparations to the public for the treatment of malaria. They also made their process of isolation public so that others could repeat their process and produce quinine to treat the disease that had already claimed so many lives. Friedlieb Ferdinand Runge (1794–1867) first isolated caffeine and quinine in 1819, phenol in 1834 (simultaneously with Eilhard Mitscherlich, 1794–1863), and he studied the mydriatic effects of belladonna using cats for *in vivo* toxicity studies. Pelletier and Caventou are generally given credit for first isolating caffeine and quinine. Runge lived in poverty and lacked the funding to commercialize his discoveries so his efforts were not acknowledged in his lifetime. The German chemist Friedrich Gaedcke (1828–1890) isolated cocaine in 1855 (he called it erythroxyline). Albert Niemann (1834–1861) of Göttingen University improved the process in 1859 and called the compound "cocaine". The Stas-Otto process for the extraction and isolation of alkaloids was published in 1851 by the Belgian analytical chemist, Jean Servais Stas (1813–1891) and German chemist Friedrich Julius Otto (1809–1870). Hermann Emil Fischer (1852–1919), who was awarded the Nobel Prize in 1902 for his work on purines and sugars, was the first to synthesize caffeine in 1895. Gustaf Komppa (1867–1949), a Finnish chemist, synthesized camphor in 1903. Constantin Fahlberg (1850–1910) and Ira Remsen (1846–1927) discovered saccharin in 1879. Lewis Lewin (1854–1929) studied and classified many psychoactive plants and was the first to classify them according to their effects (euphorics, hypnotics, inebriants, excitants, and hallucinogens). By the year 1880 chloroform, carbon tetrachloride, diethylether, carbonic acid, and petroleum had all been synthesized and put to some use. At this time little attention was devoted to the caustic, or toxic properties of any of the hundreds of thousands of chemicals that were made and produced in unrecorded quantities. There was very little, if any, legislation to govern the production or distribution of chemicals, hazardous or not.

Francoise Magendie (1783–1885) studied the toxic effects of strychnine, emetine and arrow poisons (curare and tubocurare). Claude Bernard (1813–1878) was a French physiologist who studied the effects of carbon monoxide and curare. Magendie heavily influenced his work. Bernard published many of his observations in "Leçons sur les effets des substances toxiques et medicamenteuses", a book of lectures published in 1857. Ascanio Sobrero (1812–1888) was an Italian chemist who discovered nitroglycerine in 1847, by adding glycerol to concentrated sulfuric and nitric acids. After a laboratory explosion injured Sobrero, scarring his face permanently, he abandoned his work with nitroglycerine deeming it too dangerous to handle. He was quoted: *"When I think of all the victims killed during nitroglycerine explosions, and the terrible havoc that has been wreaked, which in all probability will continue to occur in the future, I am almost ashamed to admit to be its discoverer"*. Sobrero knew of the mortal dangers associated with handling this new chemical, which could not be handled except with extreme caution. Alfred Nobel (1833–1896) was Sobrero's student and he conducted experiments with nitroglycerine to stabilize it for commercial use with the invention of dynamite (nitroglycerine stabilized in diatomaceous earth). Nitroglycerine is still used, medicinally, as a potent vasodilator to treat certain heart conditions.

Theodore Wormley (1826–1897) wrote the first American text dedicated to the study of poisons. The book was titled "The Microchemistry of Poisons Including Their Physiological, Pathological and Legal Relations" and was published in 1869. Oswald Schmiedeberg (1838–1921), a pioneer in drug metabolism, studied the relationship between the chemical structure of narcotics and their effects. He studied substances such as digitalis, muscarine, nicotine, camphor, and various heavy metals. Mathieu Bonaventure Orfila (1787–1853) is considered to be the father or founder of modern toxicology. He was born in Minorca, Spain. He practiced and studied medicine in Spain and France, settling later in Paris when a suitable professorship did not turn up in his home country. He became the pupil of Louis Nicolas Vauquelin (1763–1829), a French pharmacist and chemist. Orfila became a professor at the University of Paris and began conducting lectures. Orfila was given many honors under France's last king, King Louis-Philippe I (1773–1850), who ruled from 1830–1848. He opened the Museum of Pathological Anatomy of the Medicine Faculty of the University of Paris in 1835. Orfila's definitive work "Traité des poisons tirés des règnes minéral, végétal et animal; ou, Toxicologie générale" was first published around 1814 and the last edition was published in 1826. Orfila published "Traite de Toxicologie" in 1843 and the last edition was published in 1852 just 1 year before his death. Orfila dedicated his life to the scientific detection of poisons in organic materials. He testified as an expert witness in countless trials, the most famous of which was the trial of Marie Lafarge, accused and convicted of poisoning her husband in 1841. Poisons and poisonings were widespread throughout the western world and through the 18th century very little had changed to dissuade the use of poisons to murder people. Poisons, most notably arsenic, were very easily obtained by anyone as they were used to kill vermin or used as additives in industrial products. For these reasons, many well-known poisonous compounds could be purchased from drugstores and were found in practically every home in some form and their purchase or presence in the home would not arouse any suspicion. Robert Christison (1797–1882) wrote his "Treatise on Poisons in Relation to Medical Jurisprudence, Physiology, and the Practice of Physics" in 1829. Christison also created the first poisoned harpoon in 1831 for the whaling industry. Christison's design incorporated a special indentation that contained prussic acid (hydrogen cyanide) in a glass vial. Upon harpooning, the prussic acid vial was broken and the poison was released into the bloodstream of the whale, killing it instantly.

Charles John Samuel Thompson (1862–1943) or C.J.S. Thompson was a physician by trade and a writer by leisure pursuit. At the age of about 36, already a published author ("The Mystery and Romance of Alchemy and Pharmacy", 1897), he began working for Wellcome Burroughs and Company in London. He became the head curator for Henry Wellcome's library in 1897 (later the Wellcome Institute) and began collecting a huge assortment of medical and science instruments and ephemera. As the libraries collection of books and medical history grew, its head curator used the collection as a resource for more than forty books on the subject of poisoning, most notably "Poison Romances and Poison Mysteries" in 1899, "Poison Mysteries in History Romance and Crime" in 1923, and "Poisons and Poisoners, With Historical Accounts of Some Famous Mysteries in Ancient and Modern Times" in 1931.

Felice Fontana (1730–1805) was an Italian physiologist and physicist; however, he made significant contributions to toxicology (considered the father of toxinology) with his study of the toxins in various snake venoms. Fontana performed thousands of experiments with various species to elucidate the dosage, symptoms, and effects of viper toxins and proved, experimentally, that ligatures could delay the onset of poisoning (Fontana also recognized the limitations of the ligature). The Devonshire colic epidemic of the mid to late 1700s was responsible for the death of hundreds of people. It was eventually attrib-

Table 5. Toxicology history time line: 16th and 17th centuries

Year	Event
1519	Catherine de Medici (1519–1589): Queen of France, expert assassin, tested poisons on the poor and the sick.
1530	Jean Nicot (1530–1600): French diplomat and scholar, brought the tobacco plant to Europe and it was named after him as well as nicotine.
1600	William Shakespeare (1564–1616): From Romeo & Juliet – act 5 "Here's to my love! O true apothecary! Thy drugs are quick. Thus with a kiss I die."
1633	Bernardino Ramazzini: Italian physician, was one of the first to note the link between worker occupation and health, which he documented in his book *De Morbis Artificum Diatriba* (Diseases of Workers).
1640	William Piso: in Brazil, studied effects of *Cephaelis ipecacuanha*, an emetic; treat dysentery.
1659	Hieronyma Spara: ~1659 Roman women & fortune teller organized wealthy wives and sold them an arsenic elixir to murder their husbands.
1660	Guilia Tophania (1635–1719): Tophania was a supplier of poison to wives who wanted to be widows.
1680	Catherine DeShayes Monvoisin (1640–1680): Accused sorcerer and convicted poisoner in France. She was burned at the stake.
1682	King Louis XIV (1682): Passed royal decree forbidding apothecaries to sell arsenic or poisonous substances except to persons known to them.

uted to apple cider tainted with lead that was used to line cauldrons holding the cider. George Baker (1722–1809), a physician in the area, studied the clinical presentations of the sick and conducted experiments that clearly proved the deaths were caused by drinking of lead-tainted cider.

An historical time line of the chemical age of toxicology in the 16th and 17th centuries is provided in Table 5, and the 18th century in Table 6.

Table 6. Toxicology history time line: 18th century

Year	Event
1700	Devonshire Colic: 1700s Devonshire, England. High incidence of lead colic from drinking contaminated cider.
1701	John Jones: English doctor wrote *The Mysteries of Opium Reveal'd* described many treatments of opium, but also withdrawal and addiction.
1702	Richard Mead (1673–1754): Wrote *A Mechanical Account of Poisons* dedicated to poisons snakes, animals and plants.
1742	Carl Wilhelm Scheele (1742–1760): Swedish apothecary and chemist, discovered oxygen, barium, chlorine, manganese, and hydrogen cyanide.
1745	Benjamin Rush (1745–1813): Was a Founding Father of the United States and published first American chemistry textbook.
1755	Christian Friedrich Samuel Hahnemann (April 10, 1755 in Meiben, Saxony – July 2, 1843 in Paris, France): German physician who founded homoeopathic medicine.
1760	Percivall Pott (1714–1788): British physician who recognized that coal-tar caused cancer of the scrotum in chimney sweeps. Chimney Sweepers Act of 1788.
1767	Felice Fontana: Italian chemist and physiologist who was the first to study venomous snakes. Discovered that viper venom affects blood.
1777	Bernard Courtois: French chemist who discovered iodine in 1811 and later isolated morphine.
1783	Friedrich Sertürner (1783–1841): Isolated an alkaloid from opium poppy in 1803. He named it morphine after Morpheus, the Greek god of dreams.
1783	Francoise Magendie (1783–1855) Discovered emetine and studied effects of strychnine and cyanide. Called the father of experimental pharmacology.
1786	Fowler's Solution (1786–1936): Potassium arsenite solution prescribed as a general tonic and used from about 1786 to 1936. Used by Charles Darwin?
1797	Pierre Ordinaire (1797–1915): Created elixir using absinthe popularized and sold by Henry Pernod. Absinthe was used by Vincent Van Gogh; banned in 1915.

Chemical elixirs, patent medicines and death in a bottle

Quack cures in the form of elixirs, tonics, potions, liniments, and bitters were rampant throughout the 17th, 18th, and 19th centuries. The poor and disabled were most susceptible to these "nostrums" as they were known, which had so many indications they became a panacea for a host of seemingly unrelated ailments. The public, who often had no knowledge of the ingredients, effectiveness or safety, used these remedies faithfully according to the indiscriminate dosage prescribed. The manufacturers, distributors, and sellers were often not doctors or pharmacists and had little, if any, training in the medical profession. They usually managed to escape liability because it was impossible to ascertain if the cure made someone ill or if the ailment for which they sought out the cure had merely gotten worse. Elixirs of opium were marketed under hundreds of different trade names and for a host of, often unrelated, indications. Two particularly shocking elixirs of opium were marketed under the names "Godfrey's Cordials" and "Atkinson's Royal Infants' Preservative" and were indicated for the hyperactive child, or the teething or colicky baby. The infamous Dr. Hawley Harvey Crippen, hanged in 1910 for murdering his wife with hyoscine, made his living by selling quack cures as did Dr. H.H. Holmes, a notorious serial killer who stalked his victims at the 1893 World's Fair in Chicago, Illinois [31, 32].

Many of the elixirs contained known poisons such as arsenic and strychnine; however, some were not without some medicinal value. Thomas Fowler (1736–1801) was an English doctor who instituted the use of his own tonic remedy under his own name. The elixir, known as Fowler's Solution contained a 1% solution of potassium arsenite. It was used from 1786 to the mid 1900s for the treatment of a number of ailments including malaria and syphilis. Fowler's Solution, although never officially approved or withdrawn by the U.S. Food and Drug Administration (FDA) was revived for use as a therapy for promyelocytic leukemia in 2000. Dover's powder was a mixture of ipecacuanha, opium, and potassium sulfate and was used to induce sweating (sudoforic) to prevent the onset of fever. The mixture is named after Dr. Thomas Dover (1662–1742), an English physician and adventurer. Pierre Ordinaire (1741–1821) was a French doctor living and working in Couvet, Switzerland. He created an elixir for use as a digestive. The elixir was primarily composed of locally grown herbs and wormwood and was later made into a distilled drink. Wormwood had a long history of use prior to its popularization in the early 19th century. Galen (131–201 C.E.) wrote that an infusion of absinthia with wine could be given as a type of digestive to relax the stomach if the recommended therapy of olive oil was not effective. Hippocrates recommended the use of the absinthia plant for jaundice and rheumatism.

A time line of toxicology in the 19th century is summarized in Table 7 and of chemical elixirs in Table 8.

The Poison Squad of 1883, Harvey Wiley and the FDA

Prior to the formation of the FDA there was another organization that protected the food supply. In 1862, under President Lincoln, the Department of Agriculture was formed with the chemist Charles M. Wetherill heading up the Division of Chemistry. This department was formed to investigate the adulteration of agricultural products and was the precursor to the FDA [33]. Harvey Washington Wiley took over as Chief Chemist at the then Bureau of Chemistry in 1883 and formed the "Poison Squad", a group of men who volunteered to taste food additives to determine if they were toxic [33].

Table 7. Toxicology history time line: 19th century

Year	Event
1813	Mateu J. B. Orfila (1787–1853): Considered the father of modern toxicology. In 1813 he published *Traite des Poisons*, which described the symptoms of poisons.
1820	Joseph Caventou and Pierre Pelletier: French pharmacists isolated quinine from bark of Cinchona tree in back of their pharmacy.
1821	Thomas de Quincey (1785–1859): English writer became addicted to opium in early 1800s and published *Confessions of an Opium Eater* in 1821.
1821	Napoleon Bonaparte: Died 1821 and was suspected of being poisoned with arsenic.
1822	Edward Jukes and F. Bush: Simultaneously invented gastric lavage. They both experimented with the removal of opium and other poisons from the stomach using plastic tubing and a syringe.
1829	Robert Christison (1797–1882): Toxicologist at University of Edinburgh wrote *Treatise on Poisons* in 1829. Invented poison harpoon for whaling that contained prussic acid.
1830	Claude Bernard (1813–1878): French physiologist studied the effects of carbon monoxide and curare. Influenced by Francois Magendie.
1833	Alfred Nobel (1853–1896): Swedish chemist, engineer, innovator, armaments manufacturer and the inventor of dynamite.
1840	James Marsh (1794–1846): Chemist developed and perfected the Marsh test for arsenic. The improved Marsh test used forensically for the first during the trial of Marie Lafarge.
1847	Ascansio Sobrero (1812–1888): Italian chemist, discovered nitroglycerine in 1847, a powerful explosive and vasodilator. Alfred Nobel was his student.
1847	James Young Simpson (1811–1870): Discovered the anesthetic properties of chloroform.
1851	Arsenic Act: Required arsenic to be colored with soot or indigo to prevent "accidental" poisoning.
1854	Lewis Lewin (1854–1929): German pharmacologist studied and classified hallucinogenic plants, alcohols and other psychoactive compounds.
1855	Friedrich Gaedcke: Isolated cocaine from *Erythroxylon coca*.
1860	William Jennings Bryan (1860–1925): Was the chief proponent of the Harrison Narcotics Tax Act of 1914, which regulated and taxed the production, importation, distribution and use of opiates.
1861	Thallium discovered by Sir William Crookes (1832–1919).
1874	Charles Wright (1844–1894): Synthesized heroin at St. Mary's Hospital in London.
1869	Theodore Wormley (1826–1897): Wrote the first American book dedicated to poisons in 1869 entitled *Microchemistry of Poisons*.
1884	Robert Koch: German physician, who described numerous microbes and developed Koch's postulates that define causation.
1888	Rudolf Arndt (1835–1900) and Hugo Schulz (1853–1932) propose the idea of hormesis.
1895	Emil Fischer (1852–1919) synthesized the stimulant caffeine.
1896	Henri Becquerel: French physicist, Nobel laureate, discovered radioactivity in 1896 while investigating phosphorescence in [uranium] salts.
1897	Felix Hoffmann (1868–1946) synthesized Aspirin at Bayer Laboratories.

Table 8. Toxic chemical elixirs

Name	Compound Ingredients	Indication(s)
Fowler's Solution	Potassium arsenite	General tonic or remedy
Dover's Powder	Ipecacuanha and opium	Analgesic and purgative
James's Powder	Antimony oxide, sulfuric acid, lime phosphate and ammonia	Antipyretic
Elixir of Vitriol	Sulfuric acid, alcohol and cinnamon or ginger	Digestive disorders
Hiera Picra (Piera)	Aloe and cinnamon powder	Purgative
Black Drop	Vinegar, opium, sugar and various spices	Analgesic
Laudanum	Opium and alcohol	Analgesic
Paregoric	Opium and camphor	Analgesic
Donovan's Solution	Arsenic triiodide	General tonic or remedy
De Valagin's Solution	Arsenious acid and hydrochloric acid	General tonic or remedy
Tartar emetic	Antimony and potassium	Emetic and expectorant
Storey's Worm Cakes	Mercurous chloride	Antihelmintic
Cling's Worm Lozenges	Mercurous chloride	Antihelmintic
Santonin Worm Lozenges	Santonin from the flowering part of the toxic species *Artemisia maritima*	Antihelmintic
Strychnine Tonic or Tincture	Strychnine	General tonic or heart and respiratory stimulant
Caper Spurge Tonic	Caper Spurge seeds or oil (ingenol esters)	Violent purgative
Croton Oil Tonic	Croton oil	Violent purgative and anti-epileptic
Mrs. Winslow's Soothing Syrup	Opium	Teething Baby
Godfrey's Cordials	Opium	Infant soothing or sleep agent
Quietness	Opium	Infant soothing or sleep agent
Mother's Friend	Opium	Infant soothing or sleep agent

Note: The ingredients listed above are the primary ingredients in the elixir owing to its effects or they are used in the production process as in the case of James's Powder. According to one formulary for James's Powder, ammonia is added to "alkalinize" the water after the addition of sulfuric acid, antimony oxide, and lime phosphate, after which a precipitate forms and this precipitate is dried and the powder is collected and used for the remedy.

Chemical warfare toxicology

Chemical agents such as vesicants (blistering agents) and gases are often thought of as relatively new forms of terrorism or war strategy tools; however, the use of these materials in various forms can be traced to antiquity. In the year 600 B.C.E., Solon, an Athenian dictator ordered the poisoning of the water supply in the battle at Kirrha [34]. The water was contaminated with hellebore root, which contained poisonous saponines, protoanemonin and bufadienolides. The Scythians also made use of weaponized poisons. The Scythians were well known in antiquity as expert archers and sometimes poisoned the tips of their arrows [35]. Aristotle (384–322 B.C.E.) wrote that the Scythians used snake venom or putrefied blood to poison their arrowheads [8]. Long before the invention of modern napalm (naphthenic and palmitic acids mixed with fuel substances), similar incendiary devices were used in antiquity. The emperor Constantine used burning liquid projectiles, which were difficult, if not impossible to extinguish. The Byzantine Greeks of the Middle Ages mastered a similar formulation, the so-called "Greek Fire", that may have consisted of naphtha, quicklime (calcium oxide), sulfur, or saltpeter (potassium nitrate). The exact formulation of "Greek Fire" is unknown, possibly because the armies who used it heavily guarded its manufacturing process. During World War I (1914–1918) chemical warfare agents were re-introduced on the field of battle in the form of irritant gases. Chlorine and phosgene (carbonyl chloride) gases, and the sulfur mustard vesicants were responsible for thousands of deaths and by the end of the war the number of casualties was high on all sides. These agents were discovered in the early 18th and 19th centuries. Carl Wilhelm Scheele, mentioned earlier in this chapter, discovered chlorine in 1774 and John Davy (1790–1868), the brother of Humphrey Davy, discovered phosgene in 1812, although neither were used in warfare until much later.

Insecticide research led to some of the most dangerous and deadly chemicals known to man. Prior to the 1900s the primary means of reducing insect populations was nicotine. Nicotine was first used as an insecticide in 1763 [36]. It was first isolated in pure form in 1828 by Wilhelm Heinrich Posselt and Karl Ludwig Reimann, and first synthesized by Posselt and Arnold Rotschy in 1904 [36]. The first organophosphate insecticide compound able to block the cholinesterase enzyme was tetraethyl pyrophosphate (TEPP). TEPP was synthesized in 1854 by Wladimir Petrovich Moshnin (died 1899 or 1900) and Philippe de Clermont (1831–1921) while they were students under Adolphe Wurtz (1817–1884) in France [37]. Paul Hermann Müller (1899–1965) developed DDT as an insecticide in 1939, although it had been discovered 65 years earlier in 1874 [38]. The spread of vector borne pathogens such as typhus and malaria necessitated a rapid and inexpensive means for reducing or eradicating harmful insect populations. The wide usage of DDT, distributed almost exclusively by Geigy, eventually led to resistance in some insect populations and many chemical companies seized upon the opportunity to develop new insecticides. Pesticide research led to the synthesis of malathion, parathion, and other organophosphate derivatives with anti-cholinesterase activity. Ranajit Ghosh of the Plant Protection Laboratories of the British firm Imperial Chemical Industries (ICI) and Lars Erik Tammelin (1923–1991) of the Swedish Institute of Defense Research discovered the VX series of gases independently in 1952.

Toxicology events from 1900–1930 are summarized in Table 9.

Table 9. Toxicology history time line: 1900–1930s

Year	Event
1902	Emil Fischer (1852–1919) and Joseph von Mering (1849–1908) synthesized diethylbarbituric acid (first active barbiturate synthesized).
1903	Pierre and Marie Curie: Marie Curie, and Henri Becquerel won the Nobel Prize in Physics, "… in recognition of the extraordinary services they have rendered by their joint researches on the radiation phenomena".
1905	Upton Sinclair (1878–1968): Published *The Jungle* in 1905. Chronicled the unsanitary conditions in meat packing industry in Chicago.
1906	Pure Food and Drugs Act: Harvey Washington Wiley (1844–1930). Law prevents production or trafficking of mislabeled, adulterated or poisonous foods, drugs, medicines, and liquors.
1910	Arsphenamine (Salvarsan) discovered as therapeutic against syphilis by Paul Ehrlich (1854–1915) and Sahachiro Hata (1873–1938).
1912	Sir William Richard Shaboe Doll (1912–2005): Was a British epidemiologist, physiologist, and a pioneer in the research linking smoking to health problems.
1914	Harrison Narcotics Act: William Jennings Bryan was the chief proponent of the Harrison Narcotics Tax Act, which regulated and taxed the production, importation, distribution and use of opiates.
1915	Fritz Haber (1868–1934): German chemist developed blistering agents used in World War I; chlorine and cyanide gases.
1918	Roger Adams (1889–1971): Synthesized "adamsite" or chloroarsine.
1919	U.S. Prohibition (1919–1933): Law that made the production and sale of alcoholic beverages illegal but very profitable.
1923	Wilhelm Conrad Röntgen (1845–1923): Was a German physicist who produced and detected electromagnetic radiation in a wavelength range today known as X rays or Röntgen rays, an achievement that earned him the first Nobel Prize in Physics in 1901.
1925	Geneva Protocol: Banned use of chemical weapons. Updated in 1993 as the "Chemical Weapons Convention" to include banning production.
1929	Ginger Jake: Alcoholic tonic produced illegally during prohibition adulterated with TOCP (triorthocresyl phosphate) that caused OPIDN (organophosphorous induced delayed neurotoxicity, Jake Leg), affecting 50 000 adults.
1930	Hawks Nest Incident (1927–1935): Hundreds of black workers die from acute silicosis while digging tunnel for a hydroelectric project for Union Carbide.
1930	Gerhard Schrader (1903–1990): German chemist who, while working on developing insecticides, assembled nerve agents.
1930	U.S. Food and Drug Administration: Formed to regulate the content and safety of consumer drugs and food. The FDA was established as a government agency.
1930	Percy Julian (1899–1975): Synthesized physostigmine in 11 steps.
1931	Rolla Harger (1890–1983): Invented the "drunkometer", a colorimetric test for alcohol.
1937	Elixir Sulfonilamide: Over 100 people, many children died when Elixir Sulfanilamide was distributed without testing and contained diethylene glycol as a vehicle.
1937	Marijuana Tax Act: Federal criminal offense to possess, produce, or dispense. Non-medical use prohibited in California (1915) and Texas (1919).
1938	Food Drug and Cosmetics Act.
1938	Albert Hofmann: Synthesized LSD in the Sandoz Laboratory (now Novartis). In 1943 Hoffmann tested LSD on himself and recorded his infamous "trip".
1939	Ernest Volwiler (1852–1992) and Donalee Tabern (1900–1974) synthesized sodium pentathol.
1939	DDT: Recognized as insecticide by the Swiss scientist Paul Hermann Müller, who was awarded the 1948 Nobel Prize in Physiology and Medicine. Banned in 1972.

Toxicology impacts society: 1950–present

In the post-war era, toxicology increased in importance. War research led to the invention of new drugs and chemicals with toxic properties, and their uncensored use had repercussions that needed to be addressed. The burgeoning field of toxicology was the answer. Chemicals invented for industrial uses were improvised for household applications and, as a result, more poisonings occurred in the home and more often in small children. Louis Gdalman, a pharmacist at the Chicago Hospital began a poison call service with a self-compiled, handwritten, card index of poisons and antidotes. He accepted calls both at home and at the hospital to address acute poisonings and was, perhaps, the first *defacto* poison control center in the United States. The center was formalized and made an official part of the Chicago hospital in 1953 and soon others followed [39]. The poison control center in North Carolina was formed at Duke University in 1954 and a third opened in Boston in 1955.

In the late 1950s, the drug thalidomide was prescribed to pregnant women in the United Kingdom to treat morning sickness. Frances Kelsey of the FDA blocked U.S. approval after observing an increase in a specific type of birth defect induced in women who were taking the drug. Thalidomide was later withdrawn from the market but not before thousands of children were affected by its use ("thalidomide babies"). Thalidomide has since re-emerged as a cancer therapy. The fallout from the thalidomide disaster was a reform of the FDA, which now requires that all drugs be tested for teratogenicity [40].

The Chisso Corporation in Kumamoto Japan manufactured fertilizers, carbide, calcium nitrate, petrochemicals, plastics, drugs and perfumes over a period spanning nearly a century beginning in the early 1900s. The chemical factory expanded its operations to the Kumamoto area and adopted the practice of discarding chemical waste into the Minamata Bay and later, the Minamata River. The production of acetaldehyde, which involved the use of organic mercury, led to the accumulation of mercury waste, which was also discarded in the Minamata Bay and River beginning around 1932. Years after the dumping started, it was apparent to many that the people living around the Minamata Bay were developing neurological symptoms and becoming gravely ill. Doctors Hajime Hosokawa and Kaneyoshi Noda of the Chisso Corporation first recognized an increase in patients with neurological symptoms of unknown origin and set out to conduct experiments to determine the cause. Hosokawa and Ichikawa conducted experiments with cats using the waste water generated at the acetaldehyde plant and concluded, from these experiments, that the cause of the neurological syndromes observed in the cats was due to the organic mercury present in the waste water. They surmised that the organic mercury waste that was dumped into the water supply made its way into the food chain of the local population and caused the symptoms observed in so many inhabitants. Their findings were not published and only came to light some years later (1970) after subsequent studies confirmed organic mercury as the cause of the, then-coined, "Minamata disease" observed in the inhabitants surrounding the Minamata Bay and River.

More recent toxicology events are summarized in Tables 10 and 11 and incidents of drug recalls are summarized in Table 12. Additional information on many of these people, events, or incidences are available at Toxipedia (www. toxipedia.org).

Table 10. Toxicology history time line: 1940–1970s

Year	Event
1940	Juda Quastel: British Canadian biochemist who discovered one of the first hormone herbicides known as 2,4-D (2,4-dichlorophenoxy acetic acid).
1944	Thomas Midgley Jr (1889–1944): American chemist who developed both the tetra-ethyl lead (TEL), additive to gasoline, and chlorofluorocarbons (CFCs).
1944	John Henry Draize (1900–1992): Developed the Draize Test for chemical eye irritation.
1946	2,4-D: Developed during World War II at British Rothamsted Experimental Station, by Juda Hirsch Quastal and sold commercially in 1946. Used to control broadleaf plants.
1950	Sir Austin Bradford Hill: English epidemiologist and statistician, pioneered the randomized clinical trial and, together with Richard Doll, was the first to demonstrate the connection between cigarette smoking and lung cancer in 1950 papers.
1950	Minamata (1950s): Minamata Bay contaminated with mercury by chemical industry. Thousands adults and children were poisoned from eating fish contaminated with methyl mercury.
1952	London Great Smog: London was overwhelmed by The Great Smog, also referred to as the Big Smoke, starting on December 5, 1952, and lasted until December 9, 1952. This catastrophic event caused or advanced the death of as many as 12 000 people and formed an important impetus to the modern environmental movement.
1953	Poison Control Centers: First, Chicago 1953 under Louis Gdalman (1910–1995), second at Duke University, NC in 1954, and third opened in Boston 1955.
1954	Robert Borkenstein (1812–2002): Invented the "breathalyzer" test for alcohol.
1954	Alan Turing (1912–1954): Mathematician and logician committed suicide with a cyanide laced apple which was found near his body with a single bite taken
1959	Thalidomide (1959–1960s): Drug prescribed to pregnant women for morning sickness induced birth defects. Frances Oldham Kelsey of the Food and Drug Administration (FDA) blocked approval in U.S.
1961	Arnold J. Lehman: He collaborated with other scientists in his field to produce the first large compilation of toxicology named *Procedures for the Appraisal of the Toxicity of Chemicals in Foods*. Lehman also was a co-founder of the Society of Toxicology and its journal *Toxicology and Applied Pharmacology*
1961	Society of Toxicology (SOT) founded March 4, 1961 and held the first meeting in 1962.
1962	Alice Hamilton (1869–1970): Pathologist who associated worksite chemical hazards with disease. She studied the effects of lead and rubber on workers.
1962	Rachel Carson (1907–1964): Scientist who led the crusade against the use of DDT. Published *Silent Spring* in 1962.
1970	Arsenic poisoning in Bangladesh from contaminated tubewells.
1970	Earth Day held for the first time on April 22nd.
1970	Occupational Safety and Health Act passed to ensure workers have a safe workplace.
1970	Environmental Protection Agency (EPA) established.
1970	The Society of Forensic Toxicology (SOFT) founded.

(continued on next page)

Table 10. (continued)

Year	Event
1971	Mr. Yuk symbol adopted by the Pittsburgh Poison Center as an education tool.
1971	Over 40 000 people poisoned in Iraq after ingesting mercury contaminated seed grain.
1975	Modern toxicology textbook published. Louis J. Casarett and John Doull edited Toxicology: *The Basic Science of Poisons*, now in its 7th printing.
1978	Love Canal, August 7, 1978: U.S. President Jimmy Carter declared Love Canal a federal emergency. 42 million pounds of over 200 chemicals contaminated Love Canal, disrupting many lives.
1978	Georgi Ivanov Markov: Bulgarian dissident assassinated by poisoning when stabbed with a ricin tipped umbrella.
1979	Church Rock Dam: On July 16, Church Rock dam burst, spilling more than 1100 tons of radioactive mill waste and 90 million gallons of contaminated liquid making it the worst uranium accident in U.S. history.
1979	American Board of Toxicology (ABT) formed in 1979, first exam in 1980.

Table 11. Toxicology history time line: 1980–2000

Year	Event
1980	International Union of Toxicology (IUTOX) established.
1981	International Society for the Study of Xenobiotics (ISSX) established.
1981	Academy of Toxicological Sciences (ATS) established.
1982	Tylenol Tampering: Seven people were killed with cyanide-laced tylenol pills.
1983	Times Beach: Community in Missouri is evacuated after dioxin is discovered in the soil.
1984	Bhopal Disaster: Release of 40 tons of methyl isocyanate from a Union Carbide plant in India killed thousands and injured hundreds of people.
1986	Excedrin Tampering: Stella Nickell poisoned her husband with cyanide-laced Excedrin and poisoned several others (killing one) when she laced other bottles to cover her crime.
1986	Lake Nyos: The lake erupted a carbon dioxide gas bubble that killed over one thousand people in the surrounding village. Many animals were killed as well.
1986	Rhine Valley: A chemical spill on November 1, caused a fire at the Sandoz laboratory that led to tons of chemicals spilling into the Rhine River in Basel, Switzerland.
1986	Chernobyl accident at the nuclear power plant meltdown released radioactive waste into the atmosphere poisoning hundreds of thousands of people.
1989	Exxon Valdez Oil Spill: On March 24, the oil tanker Exxon Valdez hit Bligh Reef spilling millions of gallons of oil into the enclosed Prince William Sound.
1990	Libby Montana: The Environmental Protection Agency warned that anyone living in the northwest Montana town for 6 months anytime prior to January 1991 was most likely exposed to harmful levels of asbestos.
1992	Pitohui: The *Pitohui* is the only poisonous bird known to date. An account of it was first published in *Science* in October 1992.
1995	Tokyo subway sarin gas attack: Members of religious group Aum Shinrikyo released sarin gas in five places in Tokyo subway, killing 12 and injuring 6000.

(continued on next page)

Table 11. (continued)

Year	Event
1997	Mercury tragedy at Dartmouth: Dartmouth College professor and toxic chemical specialist Dr. Karen E. Wetterhahn was fatally poisoned by dimethylmercury.
1998	A Civil Action (1999–2004): Jonathon Harr novel based on the Woburn, Massachusetts class action lawsuit in 1985 where W.R. Grace and Beatrice Foods were accused of contaminating the water supply with perchloroethylene and trichloroethylene.
1999	Vioxx (1999–2004): A nonsteroidal anti-inflammatory, COX-2 selective inhibitor for treatment of osteoarthritis, produced by Merck & Co., was voluntarily withdrawn because of risk of heart attack and stroke.
2000	Tons of cyanide spill into the Yangtze River in China after a truck overturns.
2002	Russian movie theatre hostage crisis leads to hundreds of deaths by fentanyl overdose.
2004	Viktor Yushchenko poisoned with a single massive dose of dioxin (1000 times higher than the normal background concentration)
2001	Anthrax terrorists send poisoned letters that cripple the U.S. postal service and poisoned many.
2004	Ephedrine and ephedrine alkaloids banned by the FDA.
2006	Alexander Litvinenko: Former KGB/FSB assassinated after being poisoned with polonium 210.
2007	Casarett and Doull's Toxicology: *The Basic Science of Poisons* 7th edition published.

Table 12. Drug recalls

Drug name	Year withdrawn	Reason for withdrawal
Thalidomide (Neurosedyn)	1950	Teratogenic effects (approved for restricted use)
Lysergic acid diethylamide (LSD)	1950	Recreational use initiated a ban
Diethylstilbestrol (DES)	1970	Teratogenic effects
Phenformin	1978	Lactic acidosis
Azaribine (Triazure)	1978	Thromboembolism
Buformin	1978	Lactic acidosis
Tienilic acid (Ticrynafen)	1982	Hepatitis
Benoxaprofen (Oraflex)	1982	Jaundice
Zimelidine (Normud)	1983	Guillain-Barré syndrome
Zomepirac (Zomax)	1983	Anaphylactoid reactions
Methaqualone (Quaalude or Doriden)	1984	Addiction and overdose
Nomifensine maleate	1985	Hemolytic anemia
Suprofen tablets (Profenal)	1985	Flank pain syndrome
Encainide	1986	Increased mortality in patients with pre-existing arrhythmias
Glafenine	1990	

(continued on next page)

Table 12. (continued)

Drug name	Year withdrawn	Reason for withdrawal
Triazolam (Halcion)	1991	Adverse psychiatric effects
Temafloxacin (Omniflox)	1992	Hepatotoxicity, renal toxicity, hematopoietic toxicity
Flosequinan	1994	Increased mortality
Chlormezanone	1996	
Fenfluramin/phentermine (Fen-phen)	1997	Heart valve disorder
Bromfenac Sodium (Duract)	1998	Hepatotoxicity
Terfenadine (Seldane)	1998	Cardiac arrhythmias
Tolcapone (Tasmar)	1998	
Grepafloxacin HCl (Raxar)	1999	Torsades de Pointes
Astemizole (Hismanal)	1999	Torsades de Pointes
Amineptine	1999	
Aminophenazon	1999	
Trovafloxacin (Trovan)	1999	
Mibefradil (Posicor)	2000	Drug interactions
Troglitazone (Rezulin)	2000	Hepatotoxicity
Alosetron (Lotronex)	2000	Complications of constipation (approved for restricted use)
Cisapride (Propulsid)	2000	Cardiac arrhythmias
Ephedrine	2000	Stroke
Fowler's Solution (Trisenox)	2000	Intrinsically toxic (approved for restricted use)
Phenylpropylamine (Proin)	2000	
Sertindole (Serdolect)	2000	
Rapacuronium bromide (Raplon)	2001	Fatal bronchospasm
Cerivastatin (Baycol)	2001	Rhabdomyolysis
Levacetylmethadol	2003	
Rofecoxib (Vioxx)	2004	Myocardial infarction
Hydromorphone (Palladone)	2005	Overdose (when mixed with alcohol)
Pemoline (Cylert)	2005	Hepatotoxicity
Natalizumab (Tysabri)	2005	Progressive multifocal leukoencephalo-pathy
Amphetamine salts (Adderall XR)	2005	Stroke
Ximelagatran	2006	

Note: The drugs included on this list are those that have been withdrawn since the inception of the 1938 Food, Drugs, and Cosmetics Act requiring toxicity testing of all drugs prior to prescription to the general public. Drugs were either withdrawn or voluntarily removed from the market. Drugs often have more than one common name or trade name and the index above was not intended to include all known trade names. Drug withdrawals listed are general and may not apply to one specific country. Interestingly, Trisenox was unofficially banned but approved by the FDA in 2000 for use in the treatment of promyelocytic leukemia.

References

1 Borzelleca JF (2007) The art, the science, and the seduction of toxicology: An evolutionary development. In: AW Hayes (ed.): *Principles and Methods of Toxicology*, 5th edn. Informa Healthcare, New York, 1–22

2 Gallo MA (2008) History and scope of toxicology. In: CD Klaassen (ed.): *Casarett & Doull's Toxicology – The Basic Science of Poisons*, 7th edn. McGraw-Hill Company, New York, 3–10

3 Watson KD, Wexler P, Everitt JM (2000) Highlights in the history of toxicology. In: P Wexler, PJ Hakkinen, G Kennedy, FW Stoss (eds): *Information Resources in Toxicology*, 3rd edn. Academic Press, New York

4 Stirling DA (2006) History of toxicology and allied sciences: A bibliographical review and guide to suggested readings. *Int J Toxicol* 25: 261–268

5 Ramoutsaki IA, Ramoutsakis YA, Tsikritzis MD, Tsatsakis AM (2000) The roots of toxicology: An etymology approach. *Vet Hum Toxicol* 42: 111

6 Bailey MD (2007) *Magic and Superstition in Europe: A Concise History from Antiquity to the Present*. Rowman and Littlefield Publishers, New York

7 Taylor AS (1875) *Medical Jurisprudence and Medicine*. Henry C. Lea, Philadelphia

8 Mayor A (2003) *Greek Fire Poison Arrows and Scorpion Bombs: Biological and Chemical Warfare in the Ancient World*. Overlook Duckworth, New York

9 Askitopoulou H, Ramoutsaki IA, Konsolaki E (2000) Analgesia and anesthesia: Etymology and literary history of related Greek words. *Anesth Analg* 91: 486–491

10 Hutchinson J (1997) Words to the wise: Poison arrows. *Br Med J* 314: 7082

11 Onions CT (1939) *The Shorter Oxford English Dictionary: On Historical Principles*. Oxford University Press, London

12 Milles D (1999) History of toxicology. In: H Marquardt, SG Schäfer, RO McClellan, F Welsch (eds): *Toxicology*. California Academic Press, San Diego, 11–23

13 Reese JJ (1891) *Medical Jurisprudence and Toxicology*. WMF. Fell & Co., Philadelphia

14 Chisholm H (1911) *Encyclopedia Britannica*, Vol. 27, Horace Everett Hooper, UK

15 Dubois KP, Geiling EMK (1959) *Textbook of Toxicology*. Oxford University Press, New York

16 Casarett LJ (1975) Origin and scope of toxicology. In: LJ Casarett, J Doull (eds): *Casarett & Doull's Toxicology – The Basic Science of Poisons*. Macmillan Publishing Company, New York

17 Doull J, Bruce MC (1986) Origin and scope of toxicology. In: CD Klaasen, MO Amdur, J Doull (eds): *Casarett & Doull's Toxicology – The Basic Science of Poisons*. Macmillan Publishing Company, New York

18 Ziering A, Malatestinic N, Williams T, Brossi A (1970) 3'-Methyl, 8-methyl, and 8-phenyl derivatives of 5,9-dimethyl-6,7-benzomorphans. *J Med Chem* 13: 9–13

19 Thompson CJS (1924) *Poison Mysteries in History, Romance and Crime*. Lippincott, Williams & Wilkins, Philadelphia

20 Hucker CO (1975) *China's Imperial Past: An Introduction to Chinese History and Culture*. Stanford University Press, Stanford

21 Shang Z (1999) [Discussion on the date of appearance of the title Shen nong ben caojing (Shennong's Herbal Classic).] *Zhonghua Yi Shi Za Zhi* 29: 135–138

22 Hamada T (1980) [On the arrangement of the drugs contained in "shen nong ben cao jing" (2): Botanical drugs.] *Yakushigaku Zasshi* 15: 26–38

23 Haas LF (1999) Papyrus of Ebers and Smith. *J Neurol Neurosurg Psychiatry* 67: 578

24 Hood LJ, Leddy S, Pepper JM (2006) *Leddy's and Peppers Conceptual Basis of Professional Nursing*. Lippincott, Williams & Wilkins, Philadelphia

25 Jowett B (1900) *Dialogues of Plato*. Collier and Son Colonial Press, New York

26 Ashraf M (2007) *Top Ten: Lives of the Greatest Monarchs of History*. Lulu.com

27 Saunders NJ (2006) *Alexander's Tomb: The Two Thousand Year Obsession to Find the Lost Conqueror*. Basic Books, New York

28 White H, Denniston JD, Robson EL (1912) *Appian's Roman History*. Macmillan Company, New York

29 NYT (1884) *Some Famous Poisoners*. New York Times, New York

30 Decker WJ (1987) Introduction and history. In: TJ Haley, WO Berndt (eds): *Handbook of Toxicology*. Hemisphere Publishing Corporation, New York, 1–19

31 Larson E (2003) *Devil in the White City: Murder Magic, and Madness at the Fair that Changed America*. Random House, New York

32 Larson E (2006) *Thunderstruck.* Crown Publishers, New York
33 Kurian G (1998) *A Historical Guide to the U.S. Government.* Oxford University Press, New York
34 Szinicz L (2005) History of chemical and biological warfare agents. *Toxicology* 214: 167–181
35 Ekino S, Susa M, Ninomiya T, Imamura K, Kitamura T (2007) Minamata disease revisited: An update on the acute and chronic manifestations of methyl mercury poisoning. *J Neurol Sci* 262: 131–144
36 Haley TJ, Berndt WO (1987) *Toxicology.* CRC Press, Boca Raton
37 Petroianu GA (2008) The history of cholinesterase inhibitors: Who was Moschnin(e)? *Pharmazie* 63: 325–327
38 Smith AG, Gangolli SD (2002) Organochlorine chemicals in seafood: Occurrence and health concerns. *Food Chem Toxicol* 40: 767–779
39 Botticelli JT, Pierpaoli PG (1992) Louis Gdalman, pioneer in hospital pharmacy poison information services. *Am J Hosp Pharm* 49: 1445–1450
40 Botting J (2002) The history of thalidomide. *Drug News Perspect* 15: 604–611

Molecular, Clinical and Environmental Toxicology. Volume 1: Molecular Toxicology
Edited by A. Luch
© 2009 Birkhäuser Verlag/Switzerland

Physiologically based toxicokinetic models and their application in human exposure and internal dose assessment

David Kim[1] and Leena A. Nylander-French[2]

[1] *Harvard University, Boston, MA, USA*
[2] *University of North Carolina at Chapel Hill, Chapel Hill, NC, USA*

Abstract. Human populations may exhibit large interindividual variation in toxicokinetic response to chemical exposures. Rapid developments in dosimetry research have brought medicine and public health closer to understanding the biological basis of this heterogeneity. The toxicokinetic behavior of chemicals is, in part, controlled by the properties of the epithelium surrounding organs, some of which are effective barriers to penetration into the systemic circulation. Physiologically based toxicokinetic (PBTK) models have been developed and used to simulate the mechanism of uptake into the systemic circulation, to extrapolate between doses and exposure routes, and to estimate internal dosimetry and sources of heterogeneity in animals and humans. Recent improvements to PBTK models include descriptions of active transport across biological membranes, carrier-mediated clearance, and fractal kinetics. The expanding area of toxicogenetics has provided valuable insight for delineating toxicokinetic differences between individuals; genetic differences include inherited single nucleotide polymorphisms, copy number variants, and dynamic changes in the methylation pattern of imprinted genes. This chapter discusses the structure of PBTK models and how toxicogenetic information and newer biological descriptions have improved our understanding of variability in response to toxicant exposures.

Introduction

A question that continues to baffle researchers in the fields of medicine and public health is "Why do some people get sick and others do not?". Rapid developments in genomic analysis have brought medicine and public health closer to an answer. Also, developments in dosimetry research have revealed many reasons for interindividual variation in the bioavailability of drugs and environmental toxicants. The bioavailability of an environmental toxicant depends on the level of exposure, the rate of uptake into the systemic circulation, retention in storage tissues, biotransformation into water-soluble compounds, and clearance. An accurate assessment of these toxicokinetic properties is a key part of the risk assessment process.

Toxicokinetic studies have provided valuable insight into the uptake and distribution of toxicants in the body [1, 2]. The initial step in the uptake of toxicants into the systemic circulation is controlled by the epithelium surrounding organs, some of which are efficient barriers to penetration (e.g., epithelial cells of the skin and gastrointestinal tract). The extent of membrane permeability is governed by the diffusion of chemicals across the epithelium. Diffusion itself is determined by the lipophilicity of chemicals; however, lipophilicity alone is not predictive of bioavailability because of membrane proteins that are actively involved in the uptake and efflux of chemicals. A clear understanding of uptake across membranes is important for accurately predicting the extent of absorption and bioavailability of toxicants.

Physiologically based toxicokinetic (PBTK) models are helpful for simulating the uptake of chemicals into the systemic circulation. The majority of published PBTK models are based on the Ramsey-Andersen model for inhalation exposure to styrene [3]. These models have been used for improving our understanding of absorption, distribution, metabolism, and excretion (ADME) processes for a variety of chemicals by incorporating more complete descriptions of human biology than data-based compartmental models. Numerous PBTK models have been used in the last decade to estimate internal dose from environmental exposures.

The objective of this review is to illustrate how newer biological descriptions are incorporated into PBTK models. This is not meant to be an exhaustive review of all PBTK models published in the literature; for that, we refer the reader to Reddy et al. [1]. Rather, our intent is to highlight some of the ways in which PBTK models have been used to describe uptake of toxicants into the systemic circulation. We begin with a brief review of the structure and function of biological membranes, and then focus our discussion on various quantitative models for describing how chemicals are taken up into the systemic circulation. We conclude with a brief discussion of how PBTK models can be used with toxicogenetic information to elucidate interindividual differences in response to toxicant exposures.

Structure and function of biological membranes

Lungs

The lungs are the largest organ of the respiratory system (about 80 m^2) and they have three important roles: (1) to supply oxygen, (2) to remove wastes and toxins (e.g., carbon dioxide), and (3) to defend against harmful substances (e.g., dirt, viruses, and bacteria). Approximately 10% of the lungs are occupied by solid tissue, whereas the remainder is filled with air and blood. The functional structure of the lung can be divided into the conducting airways (dead air space) and the gas exchange region. Gas exchange occurs in the alveoli, which have a very large surface area and thin tissue layer to allow for rapid and efficient gas exchange. During inspiration, oxygen diffuses through the thin alveoli walls and the interstitial space into the blood. Carbon dioxide and other waste products diffuse in the opposite direction during exhalation. Gas exchange takes approximately 0.25 s or 1/3 of the total transit time of a red cell. The entire blood volume of the body passes through the lungs each minute (approximately 5 L/min) in the resting state.

Skin

The skin is a large complex organ that contains xenobiotic metabolism and dynamic immune response systems. The skin is generally considered to be a good barrier to the systemic absorption of chemicals. However, many chemicals can partially or fully breach the skin where they may be metabolized and interact with dermal macromolecules or components of the skin immune system. Therefore, skin is also a target for environmental and occupational chemical exposures and associated disease. Although an extensive literature exists on dermal exposure and percutaneous absorption, little is known about the action of xenobiotics (e.g., allergenic and carcinogenic agents) that may interact directly with the skin.

The skin is composed of two layers: (1) the epidermis, which provides much of the barrier function, and (2) the dermis, which is a highly vascularized layer containing protein filaments, blood vessels, nerves, lymphatics, and epidermal appendages [4]. The epidermis, which consists of two distinct layers, the *stratum corneum* (*sc*, outer layer) and the viable epidermis (*ve*, inner layer), is a complex and highly metabolic organ that has many functions. These include the ability to regenerate itself from stem cells, constant replacement of the protective *sc* by the basal keratinocytes, and the system of mixed-function oxygenases and glutathione-*S*-transferases (GST) that metabolize or conjugate for removal of both endogenous ligands and xenobiotics that penetrate the nonviable *sc* into the *ve*.

Chemicals deposited onto the skin can be absorbed into the skin to cause local effects, pass through the skin to cause systemic effects, or evoke allergic reactions both locally and systemically. As part of a dermal exposure assessment plan, it is important to understand the extent to which chemicals absorb into the skin and become systemically available.

Gastrointestinal tract

The gastrointestinal (GI) tract is an important site of absorption for numerous environmental pollutants. The GI membrane forms a first-line defense system against harmful substances. Absorption of xenobiotics takes place thought the entire GI tract but the majority of ingested compounds are absorbed in the stomach and small intestine mucosa. The stomach lining consists of four sections: mucosa, submucosa, muscle layers, and serosa. The primary barrier function of the stomach is performed by the mucosa, which consists of a superficial lining of epithelium, a secondary layer called the lamina propria, which contains ducts that secrete mucus, and a tertiary layer of smooth muscle. The total mucosa thickness varies from 530 to 900 μm [5]. This histological pattern is similar for the small intestine, the primary function of which is to absorb nutrients and water. The small intestine is approximately 5 m in length and presents a larger surface area for entry of xenobiotics into the systemic circulation.

Non-ionized (i.e., lipophilic) compounds tend to be readily absorbed across from the stomach lumen, while the ionized forms of the compounds are absorbed predominantly in the small intestine [6]. The vehicle carrying the compound is also an important factor that influences where the compound is absorbed [7]. All compounds, whether absorbed in the stomach or small intestine, are taken up into the liver and undergo first-pass metabolism.

Quantitative descriptions of absorption, distribution, metabolism, and excretion

A brief history of toxicokinetic modeling

The origins of applying a mathematical approach to physiological research can be traced back to the pioneering work by mathematically minded physiologists. Adolf Fick was a German physiologist well known for introducing laws of diffusion for gases across a fluid membrane [8]. Jean Poiseuille was a French physiologist known for developing the Poiseuille's law, which describes the flow of an incompressible liquid through a cylindrical tube [9]. In 1936, Torsten Teorell (who some consider the father of toxicokinetic modeling) published a mathematical approach to study the distribution of drugs from different routes of administration [10]. In later years, with the rapid improvements to computational resources, more complex PBTK models could be developed and solved routinely. The PBTK model developed for styrene [3] is an often-referenced example (Fig. 1). In all cases, the models put forward were constructed within a framework obedient to fundamental physiological, mechanical, and physicochemical principles.

The ADME information plays an important role in the characterization of the risk to human health associated with exposure to occupational and environmental pollutants. Modern-day PBTK models are used for organizing and deciphering ADME datasets generated from *in vitro* and *in vivo* experiments, and for predicting the time course of chemicals in blood, organs, and tissues. Each compartment in a PBTK model is described by a mass-balance differential equation (MBDE) that is solved by numerical integration. PBTK models are increasingly being used in chemical risk assessment in Asia, North America, and the European Union.

Inhalation exposure

The Ramsey-Andersen model for styrene has been used in the development of present-day PBTK models for inhalation exposure to volatile organic compounds (VOCs) [3]; a schematic of this model is shown in Figure 1. The Ramsey-Andersen model consists of four tissues groups representing: (1) rapidly perfused tissue (RPT), (2) muscle tissue, (3) fat tissue, and (4) metaboliz-

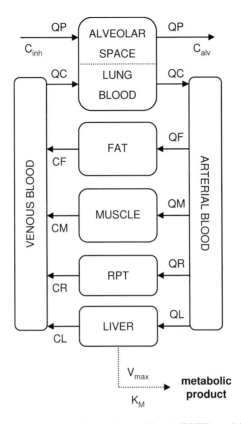

Figure 1. Diagram of the physiologically based toxicokinetic (PBTK) model used by Ramsey and Andersen [3] to simulate the toxicokinetic behavior of styrene in rats and humans following inhalation exposure. QP, alveolar ventilation rate; C_{inh}, concentration in inhaled air; C_{alv}, concentration in exhaled air; QC, total cardiac output; QF, QM, QR, QL are the blood perfusion-rates to fat, muscle, richly perfused tissue, and liver, respectively; CF, CM, CR, CL are the venous blood concentrations leaving fat, muscle, richly perfused tissue, and liver, respectively; V_{max}, maximum reaction rate; K_M, Michaelis-Menten constant. From [3]; reproduced with permission from Elsevier.

ing tissue group (i.e., liver). Anatomical and physiological parameters, used to characterize each compartment, were obtained from published sources. The assumptions made in the construction of the Ramsey-Andersen model are: (1) intercompartmental transport is *via* blood flow alone, (2) no permeability barriers to distribution, (3) small tissue volumes contribute negligibly to the overall kinetic profile and can be lumped, and (4) elimination is *via* liver and lung. These features are described in further detail below.

Inputs to the Ramsey-Andersen model occur *via* the lungs. Pulmonary uptake is equal to the pulmonary ventilation rate (QP) times the concentration of VOC inhaled (C_{inh}):

$$Uptake = QP \times C_{inh}$$

(eq. 1)

In equation 1, rapid equilibration of VOCs takes place across the alveolar membrane such that the concentration of VOCs in alveolar blood and alveolar air maintains a constant ratio specified by the blood-to-air partition coefficient. It is assumed that neither storage nor metabolism in the lungs appreciably affect the uptake of VOCs into the systemic circulation.

The absorbed VOC is distributed to other tissue compartments at a rate equal to the blood flow rate to that tissue. The following MBDE describes the rate of input and output into a non-metabolizing tissue compartment.

$$VT \frac{dCT}{dt} = QT \times (CA - CT_L) \qquad \text{(eq. 2)}$$

where VT is the tissue volume, QT is the blood perfusion rate to the tissue, CA is the VOC concentration in arterial blood, and CT_L is the VOC concentration in venous blood leaving the tissue. CT_L is related to the tissue-to-blood partition coefficient (PT) by the following expression:

$$CT_L = \frac{CT}{PT} \qquad \text{(eq. 3)}$$

where CT is the VOC concentration in the tissue. VOCs are stored in various tissue compartments based on the physiological parameters of that compartment (i.e., tissue-to-blood partition coefficient, tissue volume, and blood perfusion rate). In most instances it is assumed that all tissues are perfusion limited and well mixed.

Elimination of VOCs proceeds by two significant mechanisms: exhalation and metabolism. The concentration of chemical in end-exhaled air is equal to the blood concentration divided by the blood-to-air partition coefficient (PB). Pulmonary clearance of VOCs is QP divided by PB. Metabolism of VOCs is assumed to occur primarily in the liver; thus, the kinetics of VOCs in liver are treated separately from storage compartments. The rate of metabolism of VOCs is described by a Michaelis-Menten type of equation:

$$\frac{dA_m}{dt} = \frac{V_{max} \times CL}{K_M + CL} \qquad \text{(eq. 4)}$$

where A_m is the amount metabolized, V_{max} is the maximum rate of metabolism, K_M is the Michaelis-Menten constant, and CL is the concentration in the liver.

The blood leaving each tissue compartment is mixed to yield the VOC concentration in mixed venous blood. The VOC concentration in mixed venous blood is described by the sum of the contributions from each of the tissue groups. The mixed venous blood returns to the lungs at a flow rate equal to the cardiac output. The following equation describes the kinetics of VOCs in mixed venous blood.

$$VB\frac{dCV}{dt} = \sum_{i=1}^{n} QT_i \times CT_{Li} - QC \times CV \qquad \text{(eq. 5)}$$

where VB is the volume of mixed venous blood, CV is the VOC concentration in mixed venous blood, i is the ith tissue compartment, n is the number of tissue compartments in the PBTK model, and QC is the total cardiac output.

Dermal exposure

The composition of the sc and ve, both of which a chemical must penetrate to become systemically available, is very different. The sc is comprised of dead cells and is very hydrophobic, whereas the ve is comprised of viable cells and is less hydrophobic. The sc and ve layers of the skin are shown schematically in Figure 2. Two types of models are used to quantify dermal absorption: (1) compartmental models, and (2) membrane models. Both types of models are based on Fick's laws of diffusion across a homogeneous membrane.

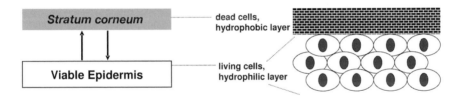

Figure 2. Schematic of the *stratum corneum* (*sc*) and viable epidermis (*ve*) layers of the epidermis layer. The *sc* is comprised of dead cells and is very hydrophobic, whereas the *ve* is comprised of viable cells and is less hydrophobic.

Chemical exposures on the skin set up a concentration gradient between the surface and the richly perfused dermis. This gradient results in a transfer of mass that is dependent on the properties of the chemical and the layer(s) through which the chemical must traverse. The mass transfer, or flux, of chemical across a thin homogeneous membrane was described by Adolf Fick in his first law, which states that:

$$J = -D\frac{\partial C}{\partial x} \qquad \text{(eq. 6)}$$

where J is the flux of material across the membrane (mass/surface area × time), D is the diffusion coefficient (surface area/time), ∂C is the concentration gradient across the membrane (mass/volume), and ∂x is the thickness of the membrane (length). The permeability coefficient (K_p) can be calculated by dividing the diffusion coefficient by the path length of absorption.

One- and two-compartment models have been used successfully to predict dermal absorption of a variety of compounds. The major difference between one- and two-compartment skin models is in the treatment of the skin. One-compartment models treat the skin as a single membrane (i.e., combine *sc* and *ve*). Two-compartment models separate the *sc* from the *ve*. Chemicals enter the *ve via* arterial blood; the concentration of chemicals leaving the *ve* is determined by the *ve*-to-blood partition coefficient. The MBDEs for one- and two-compartment models are:

1. one-compartment skin model

$$VS\frac{dCS}{dt} = QS\left(CA - \frac{CS}{P_{skin:blood}}\right) + K_pA_{exp}\left(C_{dermal} - \frac{CS}{P_{skin:surface}}\right) \quad \text{(eq. 7)}$$

2. two-compartment skin model

$$VS_{sc}\frac{dCS_{sc}}{dt} = K_{ps}A_{exp}\left(C_{dermal} - \frac{CS_{sc}}{P_{sc:surface}}\right) + K_{pv}A_{exp}\left(\frac{CS_{ve}}{P_{sc:ve}} - CS_{sc}\right) \quad \text{(eq. 8)}$$

$$VS_{ve}\frac{dCS_{ve}}{dt} = QS\left(CA - \frac{CS_{ve}}{P_{ve:blood}}\right) + K_{pv}A_{exp}\left(CS_{sc} - \frac{CS_{ve}}{P_{sc:ve}}\right) \quad \text{(eq. 9)}$$

where, for the one-compartment model, CS is the skin concentration, QS is the skin perfusion rate, $P_{skin:blood}$ is the skin-to-blood partition coefficient, $P_{skin:surface}$ is the skin-to-surface partition coefficient, and A_{exp} is the surface area of exposure. In the two-compartment model, CS_{sc} and CS_{ve} are the concentrations in the *sc* and *ve*, respectively; K_{ps} and K_{pv} are the permeability coefficients for the *sc* and *ve*, respectively; $P_{sc:surface}$, $P_{sc:ve}$, and $P_{ve:blood}$ are the skin-to-surface, *sc*-to-*ve*, and *ve*-to-blood partition coefficients, respectively. One- and two-compartment models are discussed in more detail by McCarley and Bunge [11].

Ingestion exposure

Both one- and two-compartment models of the GI tract have been implemented in oral exposure studies. Schematics of one- and two-compartment models are shown in Figure 3. These models consist of compartments representing the GI lumen and tissue in the one-compartment model, and the lumen of the stomach and small intestine, and the GI tissue in the two-compartment model. The MBDEs describing the rate of change of the amount of chemical in gut tissue are:

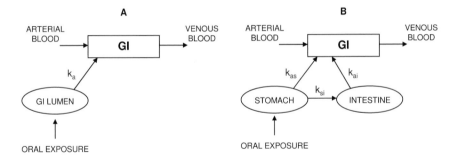

Figure 3. Schematics of one- and two-compartment models use for simulating oral exposure modified from Staats et al. [6].

1. one-compartment GI model

$$VG\frac{dCG}{dt} = QG\left(CA - \frac{CG}{P_{GI:blood}}\right) + k_a ASt \qquad \text{(eq. 10)}$$

2. two-compartment GI model

$$VG\frac{dCG}{dt} = QG\left(CA - \frac{CG}{P_{GI:blood}}\right) + k_{as} ASt + k_{ai} AIn \qquad \text{(eq. 11)}$$

where, for the one-compartment model, VG is the volume of GI tissue, CG is the GI tissue concentration, QG is the GI perfusion rate, $P_{GI:blood}$ is the GI-to-blood partition coefficient, k_a is the absorption constant, and ASt is the amount in the stomach. In the two-compartment model, ASt and AIn are the amounts of chemical in the stomach and small intestine, respectively. There are two sources of input to the GI tissue: (1) absorption from the stomach lumen (k_{as}), and (2) absorption from the intestinal lumen (k_{ai}). Each of these constants k_a, k_{as}, and k_{ai} quantify absorption into the GI tissue.

Emerging problems

Several emerging problems in toxicokinetics are reviewed. The emphasis of this section is to identify several issues (i.e., absorption, distribution, and excretion) that have been recently discussed in the peer-reviewed literature.

Membrane transporters

For many contaminants that are lipophilic and have a small molecular weight, passive diffusion is the primary mechanism of uptake; however, some chemicals can cross membranes by carrier-mediated transport mechanisms. The GI system contains specialized transport systems that have broad substrate specificity; examples are the peptide (PepT) transporter, and the organic anion (OAT) and organic cation (OCT) transporters [12]. These transporters are expressed in the brush-border membranes of intestinal epithelial cells. Activation of these transport systems can increase uptake of compounds into the systemic circulation. Transporters are expressed also at higher concentrations in the kidneys, liver, and brain, and they play a significant role in cellular distribution of chemicals. One of the more widely studied transporters, P-glycoprotein (P-gp), is responsible for the distribution profile of a broad class of compounds [13, 14]. P-gp can restrict access of many compounds to the central nervous systems, decrease the bioavailability of drugs, and limit the transport of teratogens to the fetus [15]. Single nucleotide polymorphisms (SNPs) in the multidrug resistance (MDR) gene, which encodes for P-gp, are associated with reduced intestinal expression of P-gp and increased oral bioavailability of substrates specific to P-gp [16, 17]. Further insight into the interaction of chemicals with transporters will be informative for predicting individual risks from environmental exposures.

Transporter systems can be divided into influx and efflux transporters where the influx transporters facilitate the entry of chemicals into cells and the efflux transporters limit entry of chemicals or enhance the removal of chemicals from cells. Based on location along the cell membrane, transport systems can function as either influx or efflux transporters. An important feature of transporters is that they are saturable at high concentrations. For example, the effective absorption of chemicals across the intestinal membrane can be depicted as follows:

$$J_e = J_p + J_c = k_a CIn + \frac{T_{\max} CIn}{K_T + CIn} \qquad \text{(eq. 12)}$$

where J_e, J_p, and J_c are the effective flux, diffusive flux, and carrier-mediated flux, respectively; k_a is the permeability across the intestinal membrane; T_{max} is the capacity of transport by the transporter; K_T is the transporter-substrate affinity constant; and CIn is the concentration of chemical in the intestinal lumen. In the presence of inhibitors, the expression for J_c can be derived similarly to enzymatic reactions [18]:

1. competitive inhibition

$$J_c = \frac{T_{\max} CIn}{K_T \left(1 + \dfrac{I}{K_i}\right) + CIn} \qquad \text{(eq. 13)}$$

2. non-competitive inhibition

$$J_c = \frac{T_{max} CIn}{K_T + \left(1 + \dfrac{I}{K_i}\right) CIn} \qquad \text{(eq. 14)}$$

where I is the inhibitor concentration and K_i is the inhibition-transporter affinity constant, respectively.

The primary efflux transporter expressed in human intestinal membrane is P-gp. The involvement of P-gp in intestinal efflux limits the entry of chemical or enhances the removal of chemicals. The net flux of chemicals across the intestinal membrane can be expressed as follows

$$J_e = J_p + J_c - J_{P\text{-}gp} \qquad \text{(eq. 15)}$$

where $J_{P\text{-}gp}$ represents the efflux of material. Equation 15 assumes a high concentration in the lumen and low concentration in the intestinal membrane/tissue. Various *in vitro* methods, including isolated tissues and established cell lines (e.g., Caco-2 cells derived from human colon carcinoma) have been used to predict the GI uptake of chemicals [19].

Carrier-mediated clearance

Kidney transporters may be involved in tubular secretion, and resorption from the urinary filtrate back into the systemic circulation. The major transporters expressed in kidneys are OATs (i.e., OAT1, OAT3, and OAT4) and OCTs (i.e., OCT2). An example of a PBTK model that incorporates a description of resorption by an apical transporter is shown in Figure 4 [20]. In their model, Andersen et al. [20] modeled resorption of perfluorooctanoate (PFOA) with T_{max} and K_T values of 4.73 mg/(kg·h) and 0.01 mg/L for intravenous input, and 5 mg/(kg·h) and 0.055 mg/L for oral dosing studies done in Cynomolgus monkeys. They postulated that the transporter responsible for resorption of anions from the renal filtrate is OAT1.

In another example, Lohitnavy et al. [21] incorporated a description for active biliary excretion of 3,3',4,4',5'-pentachlorobiphenyl (PCB126). The specific transporter involved in this process was postulated to be the MDR-associated protein 2 (Mrp2), which they determined based on molecular characteristics using a 3-D quantitative structure-activity relationship model. With their model, Lohitnavy et al. estimated a K_T value of 7760.0 nmol/L, and used the PBTK model to optimize T_{max} (64.6 nmol/h); experimental data were previously collected by the U.S. National Toxicology Program. With their description of active excretion of PCB126, Lohitnavy et al. were able to accurately simulate the time course of PCB126 in plasma and other tissues under multiple-dosing scenarios.

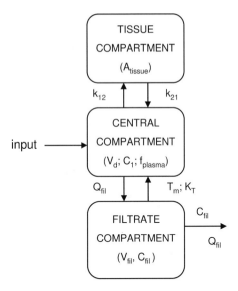

Figure 4. Schematic of a PBTK model that incorporates a description of renal resorption by an apical transporter [20]. V_d, volume of distribution; k_{12}, compartment transfer rate constant from central to tissue; k_{21}, compartment transfer rate constant from tissues back to central; f_{plasma}, proportion of compound free in blood; T_m, transport maximum; K_T, transport affinity constant; C_{fil}, concentration in filtrate compartment; Q_{fil}, renal plasma filtration rate; C_1, concentration in central compartment; A_{tissue}, amount in tissue compartment. From [20]; reproduced with permission from Elsevier.

Fractal kinetics

The MBDEs for describing carrier-mediated processes have relied on classical reaction kinetics where the constants (e.g., T_{max}) are not dependent on time. In this approach, ADME processes are considered to occur in homogeneous spaces. However, it is known that materials are distributed in tissues and organs that do not fulfill the condition of homogeneity. It has been suggested that compounds are distributed throughout the body by fractal networks of branching tubes [22]. Fractal analysis has been applied to examine irregular skin morphology following skin damage from chemical exposures [23]. Thus, Fick's laws of diffusion may not be valid in heterogeneous compartments like the skin and intestinal membrane, and fractal kinetics may be more appropriate for describing these uptake processes. In these heterogeneous conditions, the rate constants have been observed to be time dependent, following a power-law of time [24]:

$$\text{rate} = k_1 t^{-h} \quad (t \neq 0) \tag{eq. 16}$$

where k_1 is a constant and h is a dimensionless exponent. Time-dependent absorption models have been used to explain GI absorption of cyclosporine A [25] and propranolol [26].

The interaction of chemicals with transporters can also be considered heterogeneous and spatially constrained because of the restricted movement of the transporter in the membrane. Building on the work of Kopelman [27, 28] and Lopez-Quintela and Casado [29], who introduced a modified Michaelis-Menten equation for the kinetic study of enzyme catalysis, Macheras [24] derived the following expression for carrier-mediated transport:

$$J_c = \frac{T^{eff}_{max} \, CIn^{2-D}}{K^{eff}_T + CIn} \qquad (eq. 17)$$

where T^{eff}_{max} and K^{eff}_T are defined in terms of effective capacity and affinity, and D is the fractal dimension (see [22] for definition of D). When $D = 1$ a homogeneous condition is assumed. Larger values of D are a measure of deviation from homogeneity. The relationship between J_c and C for different values of D is shown in Figure 5 [24].

Metabolism is classically analyzed using the Michaelis-Menten formalism, which assumes that the substrate-enzyme reaction occurs in a homogeneous tissue compartment. However, enzymes are known to be localized in specific regions of the liver, and this zonal heterogeneity may have significant impact on chemical dosimetry. Several approaches have been proposed to study metabolism under heterogeneous conditions, one of which is the fractal structure to the liver compartment in a PBTK model developed for the drug mibefradil [30]. The MBDE for the homogeneous liver compartment is as follows:

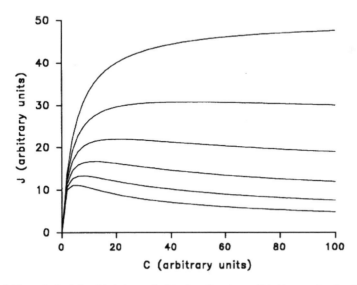

Figure 5. Theoretical relationship between the kinetics of carrier mediated transport and variations in the fractal dimension D in the range $1 \le D < 2$ (i.e., for effective $T_{max} = 50$ and effective $K_T = 5$ in equation 17). The values of D from top to bottom are: 1.0, 1.1, 1.2, 1.3, 1.4, and 1.5. Figure taken from Macheras [24]; reproduced with permission from the American Association of Pharmaceutical Scientists.

$$VL\frac{dCL}{dt} = QL \times \left(CA - \frac{CL}{PL}\right) - Cl_h CL \qquad \text{(eq. 18)}$$

where VL is the volume of liver, CL is the concentration of chemical in liver, QL is blood flow-rate to the liver, and PL is the liver-to-blood partition coefficient, and Cl_h is metabolic clearance. Invoking fractal kinetics, clearance is assumed to be time dependent. Thus, the MBDE is:

$$VL\frac{dCL}{dt} = QL \times \left(CA - \frac{CL}{PL}\right) - k\left(t^{-h}\right)CL \qquad \text{(eq. 19)}$$

where k is constant in time and $h \cdot \varepsilon$ [0,1]. Comparison of model simulations with data supported the hypothesis that mibefradil may experience a fractal-like environment within the liver.

Future needs

Human populations may exhibit large interindividual variation in the toxicokinetic behavior of chemicals. Continued investigations into toxicogenetics can help delineate toxicokinetic differences between individuals. This section reviews some of the current research needs for better understanding population toxicokinetics and heterogeneity.

Understanding toxicogenetics and individual differences in response to toxicant exposure and toxicity

Understanding individual genetic and epigenetic differences in relation to individual variable levels of acute and chronic responses to toxicants, based upon ADME and toxicity to target organ, will require identifying the role of the structure of the human genome as it has evolved and has become admixed over time within an exposed population. These genetic differences include inherited SNPs [31] and/or copy number variants (CNV), which include sequence deletions, insertions, gene duplications, etc. [32], and dynamic changes in the methylation pattern of imprinted genes (epigenome) [33, 34]. These are highly influenced by our environment as risk factors and are the basis for individual toxicogenetic differences observed within an exposed population.

Significant individual differences in genetic and structural variation exist in the human genome [35] and in DNA damage repair and detoxification enzymes (SNPs, microsatellite sequences, or mutations that alter both gene regulation and function) [36–39]. These individual differences may contribute greatly to the level of a toxic (parent) compound or its metabolite(s) (i.e., biomarkers of exposure) within the body of an exposed individual. The source of variation may be dependent upon SNPs in the highly conserved coding region of genes or in their regulatory control sequences, which might include locus-control regions and promoter (transcription factor response elements) as well as less conserved intron sequences, or splice variant sequences that affect gene expression, and function critical to xenobiotic metabolism and toxic response.

SNP variation occurs in both synonymous and non-synonymous codons in gene regulatory sequences (promoter, introns, and 3'-untranslated regions) as well as coding and non-coding exons. The haplotype (genes physically linked within a specific genetic sequence pattern) structure is a result of an inherited haplotype structure formed during meiotic recombination and/or maintained by evolutionary pressure [40–42]. The allelic frequency of SNPs of genes in linkage disequilibrium (not physically linked and frequently separated by meiotic recombination) make it possible to search and identify haplotypes highly associated (causal) with biological phenotypes [43–46]. The basis for functional differences in gene expression and protein function may be due to SNPs in regulatory (locus control regions, promoters, introns) and/or structural (within exons that cause amino acid substitutions that alter protein structure and function) sequences of the inherited gene or haplotype structure [36–39]. Variations exist particularly, in detoxification enzyme networks [mixed-function oxygenases, N-acetyltransferases (NAT), GSTs, as well as other conjugation enzymes] and in DNA damage repair gene networks. These individual differences (heritable traits) may result in either increased or decreased function of critical enzymes that alter the output of a toxic compound or its metabolite(s) (i.e., biomarkers) and may significantly affect the biomarker levels and, thus, confound exposure assessment.

Linking exposure to susceptibility and disease in risk assessment

Sources of individual variation in response to exposure to xenobiotics are rarely accounted for in current toxicokinetic and exposure biology assessment models. In part, this may be due to the difficulty in distinguishing between signal and noise [47] (even though significant individual genetic variation is known to exist in mixed-function oxygenases, GSTs, and other conjugation enzymes, as well as in DNA damage repair genes), and/or selection of appropriate statistical tests. For many of the genes known or suspected to be involved in xenobiotic metabolism, the haplotypes (associated SNP genotypes) and structural variants are unknown because insufficient numbers of individuals and their environmentally responsive genes have been sequenced and adequately genotyped for haplotype allelic frequency. Many studies of association between SNP genetic markers and disease phenotypes are negative or weak due, possibly, to both insufficient number of individuals sampled, and inadequate SNP genotyping to infer haplotype [43, 46, 48] and identify the quantitative trait nucleotide variant. Individual genetic differences should be investigated as a component of exposure variance (not only for association with a disease or toxicity phenotype) along with other critical determinants of exposure and toxicity. Thus, unexplained variance in exposure assessment and biomarkers of exposure (exposure biology) may be decreased and sufficient power gained relative to population sample size. Inclusion of the character and magnitude of the individual differences in exposure to xenobiotics and in xenobiotic detoxification will allow us to link individual genetic differences, as a variance component, to other predictors of exposure levels and, thus, to develop more predictive models for exposure and risk assessment.

Legal considerations (GINA signed into law by President Bush 2008)

Individual ADME and gene expression data may be used to investigate the role that genetic factors play in modifying the relationship between exposure to xenobiotics and subsequent health effects. Inclusion of ADME and gene expression data as well as macromolecule adduction and/or damage data in PBTK models to estimate the contribution of xenobiotic exposure through all routes of exposure is critical for overall risk assessment and scientifically based risk management and prevention to protect the public's health, including individuals that may be genetically susceptible.

Legislation that prohibits discrimination (e.g., insurance, occupational) against people having genetic and/or susceptibility traits (even behavior may be a genetic predisposition) has been enacted in the United States. H.R. 493; the Genetic Information Nondiscrimination Act of 2008 was signed into law on May 21, 2008. This legal protection may provide incentives for both the employee and employer to accept and allow the use of genetic-based markers for determination of individual differences in biomarkers of exposure and effect in risk assessment and identification of a subpopulation that may differ in susceptibility to exposure and risk.

Conclusions

The main contribution of toxicokinetic models to exposure assessment is to advance the understanding of absorption, distribution, and elimination of xenobiotics in the human body. This information is critical for the establishment of the relationship between exposure levels and health effects. This information can be used to establish the relationship between the degree of exposure and the biological levels of the exposure indicator, to determine allowable levels for xenobiotic exposure (e.g., Biological Exposure Indices [BEI] set by American Conference of Governmental Industrial Hygienists [49] or Biological Tolerance Values [BAT] set by Deutsche Forschungsgemeinschaft [50]), to select proper sampling time for biological markers, and to identify confounding factors such as interindividual and intraindividual differences in the body response to exposure.

The consequence of potential individual differences in ADME from exposure to multiple sites of contact is unknown. Furthermore, individual differences in DNA damage and repair, adduction (DNA and protein), and variable expression of detoxification enzymes on metabolites (biomarkers), may affect the detection and use of exposure biomarkers based upon genetic differences. Determination of genes responsive to xenobiotic exposure (detoxification as well as innate and/or acquired immunity) will provide focus for translational research on disease prediction and the development of preventive or intervention strategies and medicines for environmentally related diseases. Furthermore, demonstration and recognition of individual differences in detoxification and gene-expression patterns in response to xenobiotic exposure, and their contribution to toxicity and risk, will impact both prevention policy and translation studies that are crucial to the protection of health and treatment of disease of the exposed individuals, and making valid associations between exposure and disease. A multidisciplinary strategy, which includes exposure assessment with repeated measures of exposure through multiple routes (i.e., inhalation, dermal, and ingestion) and biomarker levels, individual genetic differences in detoxification of toxic compounds, and sophisticated exposure and toxicokinetic modeling, will be valuable for developing strategies to measure and prevent exposure and for understanding the significance of toxicity in risk classification.

References

1 Reddy MB, Yang RSH, Clewell HJ III, Andersen ME (2005) *Physiologically Based Pharmacokinetic Modeling: Science and Applications.* John Wiley & Sons, Hoboken, 420
2 Rowland M, McLachlan A (1996) Pharmacokinetic considerations of regional administration and drug targeting: influence of site of input in target tissue and flux of binding protein. *J Pharmacokinet Biopharm* 24: 369–387
3 Ramsey JC, Andersen ME (1984) A physiologically based description of the inhalation pharmacokinetics of styrene in rats and humans. *Toxicol Appl Pharmacol* 73: 159–175
4 Potts RO, Bommannan DB, Guy RH (1992) Percutaneous absorption. In: H Mukhtar (ed.): *Pharmacology of the Skin.* CRC Press, Boca Raton, 13–28
5 Johnson LR (2006) *Physiology of the Gastrointestinal Tract.* Elsevier Academic Press, Burlington
6 Staats DA, Fisher JW, Connolly RB (1991) Gastrointestinal absorption of xenobiotics in physiologically based pharmacokinetic models. A two-compartment description. *Drug Metab Dispos* 19: 144–148
7 Withey JR, Collins BT, Collins PG (1983) Effect of vehicle on the pharmacokinetics and uptake of four halogenated hydrocarbons from the gastrointestinal tract of the rat. *J Appl Toxicol* 3: 249–253
8 Vandam LD, Fox JA (1998) Adolf Fick (1829–1901), physiologist: A heritage for anesthesiology and critical care medicine. *Anesthesiology* 88: 514–518
9 Pfitzner J (1976) Poiseuille and his law. *Anaesthesia* 31: 273–275
10 Teorell T (1935) Studies on the "Diffusion Effect" upon ionic distribution. Some theoretical considerations. *Proc Natl Acad Sci USA* 21: 152–161
11 McCarley KD, Bunge AL (2001) Pharmacokinetic models of dermal absorption. *J Pharm Sci* 90: 1699–1719
12 Kim RB (2002) Transporters and xenobiotic disposition. *Toxicology* 181–182: 291–297
13 Raub TJ (2006) P-glycoprotein recognition of substrates and circumvention through rational drug design. *Mol Pharmacol* 3: 3–25

14 Sharom FJ (2006) Shedding light on drug transport: Structure and function of the P-glycoprotein
    multidrug transporter (ABCB1). *Biochem Cell Biol* 84: 979–992
15 Abu-Qare AW, Elmasry E, Abou-Donia MB (2003) A role for P-glycoprotein in environmental
    toxicology. *J Toxicol Environ Health B Crit Rev* 6: 279–288
16 Hoffmeyer S, Burk O, von Richter O, Arnold HP, Brockmöller J, Johne A, Cascorbi I, Gerloff T,
    Roots I, Eichelbaum M, Brinkmann U (2000) Functional polymorphisms of the human multidrug-
    resistance gene: Multiple sequence variations and correlation of one allele with P-glycoprotein
    expression and activity *in vivo*. *Proc Natl Acad Sci USA* 97: 3473–3478
17 Marzolini C, Paus E, Buclin T, Kim RB (2004) Polymorphisms in human MDR1 (P-glycoprotein):
    recent advances and clinical relevance. *Clin Pharmacol Ther* 75: 13–33
18 You G, Morris ME (2007) *Drug Transporters: Molecular Characterization and Role in Drug
    Disposition*. Wiley-Interscience, Hoboken
19 Meunier V, Bourrie M, Berger Y, Fabre G (1995) The human intestinal epithelial cell line Caco-2;
    pharmacological and pharmacokinetic applications. *Cell Biol Toxicol* 11: 187–194
20 Andersen ME, Clewell HJ 3rd, Tan YM, Butenhoff JL, Olsen GW (2006) Pharmacokinetic mod-
    eling of saturable, renal resorption of perfluoroalkylacids in monkeys – Probing the determinants
    of long plasma half-lives. *Toxicology* 227: 156–164
21 Lohitnavy M, Lu Y, Lohitnavy O, Chubb LS, Hirono S, Yang RS (2008) A possible role of mul-
    tidrug resistance-associated protein 2 (Mrp2) in hepatic excretion of PCB126, and environmental
    contaminant: PBPK/PD modeling. *Toxicol Sci* 104: 27–39
22 Bassingthwaighte JB, Liebovitch LS, West BJ (1994) *Fractal Physiology*. Oxford University
    Press, New York
23 Obata Y, Sesumi T, Takayama K, Isowa K, Grosh S, Wick S, Sitz R, Nagai T (2000) Evaluation of
    skin damage caused by percutaneous absorption enhancers using fractal analysis. *J Pharm Sci* 89:
    556–561
24 Macheras P (1995) Carrier-mediated transport can obey fractal kinetics. *Pharm Res* 12: 541–548
25 Caroli-Bosc FX, Iliadis A, Salmon L, Macheras P, Montet AM, Bourgeon A, Garraffo R, Delmont
    JP, Montet JC (2000) Ursodeoxycholic acid modulates cyclosporin A oral absorption in liver trans-
    plant recipients. *Fundam Clin Pharmacol* 14: 601–609
26 Higaki K, Yamashita S, Amidon GL (2001) Time-dependent oral absorption models. *J Pharmaco-
    kinet Pharmacodyn* 28: 109–128
27 Kopelman R (1986) Rate processes on fractals: Theory, simulations, and experiments. *J Stat Phys*
    41: 185–200
28 Kopelmam R (1988) Fractal reaction kinetics. *Science* 241: 1620–1626
29 Lopez-Quintela MA, Casado J (1989) Revision of the methodology in enzyme kinetics: A fractal
    approach. *J Theor Biol* 139: 129–130
30 Marsh RE, Tuszynski JA (2006) Fractal Michaelis-Menten kinetics under steady state conditions:
    Application to mibefradil. *Pharm Res* 23: 2760–2767
31 Paschou P, Ziv E, Burchard EG, Choudhry S, Rodriguez-Cintron W, Mahoney MW, Drineas P
    (2007) PCA-correlated SNPs for structure identification in worldwide human populations. *PLoS
    Genet* 3: 1672–1686
32 Estivill X, Armengol L (2007) Copy number variants and common disorders: Filling the gaps and
    exploring complexity in genome-wide association studies. *PLoS Genet* 3: 1787–1799
33 Bird A (2002) DNA methylation patterns and epigenetic memory. *Genes Dev* 16: 6–21
34 Laird P (2003) The power and the promise of DNA methylation markers. *Nat Rev Cancer* 3:
    253–266
35 Livingston RJ, von Niederhausern A, Jegga AG, Crawford DC, Carlson CS, Rieder MJ,
    Gowrisankar S, Aronow BJ, Weiss RB, Nickerson DA (2004) Pattern of sequence variation across
    213 environmental response genes. *Genome Res* 14: 1821–1831
36 Kim YD, Lee CH, Nan HM, Kang JW, Kim H (2003) Effects of genetic polymorphisms in meta-
    bolic enzymes on the relationships between 8-hydroxydeoxyguanosine levels in human leukocytes
    and urinary 1-hydroxypyrene and 2-naphthol concentrations. *J Occup Health* 45: 160–167
37 Lee CY, Lee JY, Kang JW, Kim H (2001) Effects of genetic polymorphisms of CYP1A1,
    CYP2E1, GSTM1, and GSTT1 on the urinary levels of 1-hydroxypyrene and 2-naphthol in air-
    craft maintenance workers. *Toxicol Lett* 123: 115–124
38 Nan HM, Kim H, Lim HS, Choi JK, Kawamoto T, Kang JW, Lee CH, Kim YD, Kwon EH (2001)
    Effects of occupation, lifestyle and genetic polymorphisms of CYP1A1, CYP2E1, GSTM1 and
    GSTT1 on urinary 1-hydroxypyrene and 2-naphthol concentrations. *Carcinogenesis* 22: 787–793

39 Yoshimura K, Hanaoka T, Ohnami S, Ohnami S, Kohno T, Liu Y, Yoshida T, Sakamoto H, Tsugane S (2003) Allele frequencies of single nucleotide polymorphisms (SNPs) in 40 candidate genes for gene-environment studies on cancer: Data from population-based Japanese random samples. *J Hum Genet* 48: 654–658

40 Eberle MA, Rieder MJ, Kruglyak L, Nickerson DA (2006) Allele frequency matching between SNPs reveals an excess of linkage disequilibrium in genic regions of the human genome. *PLoS Genet* 2: e142

41 Hinds DA, Stuve LL, Nilsen GB, Halperin E, Eskin E, Ballinger DG, Frazer KA, Cox DR (2005) Whole-genome patterns of common DNA variation in three human populations. *Science* 307: 1072–1079

42 Hinds DA, Kloek AP, Jen M, Chen X, Frazer KA (2006) Common deletions and SNPs are in linkage disequilibrium in the human genome. *Nat Genet* 38: 82–85

43 Elston RC (1995) Linkage and association to genetic markers. *Exp Clin Immunogen* 12: 129–140

44 Lawrence RW, Evans DM, Cardon LR (2005) Prospects and pitfalls in whole genome association studies. *Phil Trans Roy Soc London* 360: 1589–1595

45 Romero R, Kuivaniemi H, Tromp G, Olson J (2002) The design, execution, and interpretation of genetic association studies to decipher complex diseases. *Am J Obstet Gynecol* 187: 1299–1312

46 Zaykin DV, Zhivotovsky LA (2005) Ranks of genuine associations in whole-genome scans. *Genetics* 171: 813–823

47 Kimmel G, Sharan R, Shamir R (2004) Computational problems in noisy SNP and haplotype analysis: Block scores, block identification, and population stratification. *INFORMS J Comput* 16: 360–370

48 Devlin B, Roeder K, Bacanu SA (2001) Unbiased methods for population-based association studies. *Genet Epidemiol* 21: 273–284

49 ACGIH (2007) *Documentation of the TLVs® and BEIs®*. American Conference of Governmental Industrial Hygienist, Cincinnati

50 Deutsche Forschungsgemeinschaft, Commission for Investigation of Health Hazards of Chemical Compounds in the Work Area (2007) *List of MAK and BAT Values 2006: Maximum Concentrations and Biological Tolerance Values at the Workplace*. Wiley-VCH, Weinheim

Molecular, Clinical and Environmental Toxicology. Volume 1: Molecular Toxicology
Edited by A. Luch

The role of biotransformation and bioactivation in toxicity

Wolfgang Dekant

Department of Toxicology, University of Würzburg, Würzburg, Germany

Abstract. Biotransformation is essential to convert lipophilic chemicals to water-soluble and readily excretable metabolites. Formally, biotransformation reactions are classified into phase I and phase II reactions. Phase I reactions represent the introduction of functional groups, whereas phase II reactions are conjugations of such functional groups with endogenous, polar products. Biotransformation also plays an essential role in the toxicity of many chemicals due to the metabolic formation of toxic metabolites. These may be classified as stable but toxic products, reactive electrophiles, radicals, and reactive oxygen metabolites. The interaction of toxic products formed by biotransformation reactions with cellular macromolecules initiates the sequences resulting in cellular damage, cell death and toxicity.

Introduction

The biological effects initiated by a xenobiotic are not related simply to the inherent toxic properties of the xenobiotic; the initiation, intensity, and duration of a toxic response are a function of numerous factors intrinsic to the biological system as well as the administered dose. Each factor influences the ultimate interaction of the xenobiotic and the active site. Only when the toxic chemical has reached the specific site and interacted with that site, can the inherent toxicity be realized. The route a xenobiotic follows from the point of administration or absorption to the site of action usually involves many steps and is termed toxicokinetics. However, the same dose of a chemical administered by different routes may cause different toxic effects. Moreover, the same dose of two different chemicals may result in vastly different concentrations of the chemical or its biotransformation products in a particular target organ. This differential pattern is due to differences in the disposition of a xenobiotic [1–5].

The disposition of a xenobiotic consists of absorption, distribution, biotransformation and excretion. The complicated interactions between the different processes of distribution are major determinants for organ-specific toxicity. Most xenobiotics entering the body are lipophilic. This property enables them to penetrate lipid membranes, to be transported by lipoproteins with blood, and to be rapidly absorbed by the target organ. However, the efficient excretory mechanisms of the organism require a certain degree of hydrophilicity for efficient excretion. In the absence of efficient means for excretion constant exposure to a lipophilic chemical could result in accumulation of the xenobiotic in the organism. Therefore, animal organisms have developed a number of biochemical processes that convert lipophilic chemicals to hydrophilic chemicals and thus assist in their excretion. These enzymatic processes are termed biotransformation, the enzymes catalyzing biotransformation reactions are referred to as enzymes of biotransformation. The enzymes of biotransformation differ from most other enzymes active in organisms by having a broad substrate specificity and by catalyzing reactions at comparably low rates. The low rates of biotransformation reactions are often compensated by high concentrations of biotransformation enzymes. Biotransformation enzymes have evolved to facilitate the excretion of lipophilic chemicals present in the diet of animals. The broad substrate specificity helped to adjust to new dietary constituents and thus caused evolutionary advantages. Biotransformation is most often the sum of several processes by which the structure of a chemical is changed during passage through the organism [6].

Phase I and phase II reactions

Xenobiotic metabolism is catalyzed by a number of different enzymes. For solely operational purposes, the enzymes of biotransformation are separated into two phases. In phase I reactions, which involve oxidation, reduction and hydrolysis, a polar group is added to the xenobiotic or is exposed by the enzymes of biotransformation. Phase II reactions are biosynthetic and link the metabolite formed by phase I reactions to a polar endogenous molecule producing a conjugate. Various endogenous molecules with high polarity are utilized for conjugation, the formed conjugates are often ionized at physiological pH and thus highly water soluble. Moreover, the moieties used for conjugation are often recognized by specific active transport processes, which assist in their translocation across plasma membranes and thus further enhance the rate of excretion [7].

The enzymes of biotransformation are localized mainly in the liver. A significant fraction of the blood from the splanchnic area, which also contains xenobiotics absorbed from the intestine, enters the liver. Therefore, the liver has developed the capacity to enzymatically modify most of these chemicals. However, most other tissues also have the capacity to catalyze biotransformation reactions. The contribution of extrahepatic organs to the biotransformation of a chemical depends on many factors including chemical structure, dose and route of administration. However, biotransformation of a chemical within an extrahepatic tissue may cause toxic effects on this specific tissue and may thus have important toxicological consequences.

Inside cells, phase I enzymes are mainly present in the endoplasmic reticulum, lipoprotein membranes extending from the mitochondria and the nucleus to the plasma membranes of the cell. The presence of phase I enzymes within membranes has important implications since lipophilic chemicals will preferentially distribute into lipid membranes; thus high concentrations of lipophilic xenobiotics are present at this site of biotransformation.

The general purpose of biotransformation reactions is detoxication, since xenobiotics should be transformed to metabolites, which are more readily excreted. However, depending on the structure of the chemical and the enzyme catalyzing the biotransformation reaction, metabolites with a higher potential for toxicity than the parent compound are often formed. This process is termed bioactivation and is the basis for the toxicity and carcinogenicity of many xenobiotics with a low chemical reactivity. The interaction of the toxic metabolite initiates events that ultimately may result in cell death, cancer, organ failure and other manifestations of toxicity. Formation of reactive and more toxic metabolites is more frequently associated with phase I reactions; however, phase II reactions may also be involved in toxication as well as combinations of phase I and phase II reactions [8–10].

Phase I enzymes and their reactions

Enzymes that catalyze phase I reactions include microsomal monooxygenases and peroxidases, cytosolic and mitochondrial oxidases, reductases and hydrolytic enzymes. Microsomal monooxygenases are the cytochrome P450 enzymes and the mixed function amine oxidase or flavin-dependent monooxygenase. Both enzyme systems add a hydroxyl moiety to the xenobiotic. Cytochrome P450 is the most important enzyme system involved in phase I reactions. Actually, cytochrome P450 enzymes represent a coupled enzyme system composed of the heme-containing cytochrome P450 and the NADPH-containing cytochrome P450 reductase [11, 12]. This flavoprotein has a preference for NADPH as its cofactor and transfers either one or two electrons from NADPH to cytochrome P450. Cytochrome P450 and the reductase are embedded into the phospholipid matrix of the endoplasmic reticulum. The phospholipid matrix is crucial for enzymatic activity since it facilitates the interaction between both enzymes [13]. In vertebrates, the highest concentrations of cytochrome P450 are found in the liver; however, cytochrome P450 enzymes are also present in lung, kidney, testes, skin and gastrointestinal tract [14]. The presence of several forms of cytochrome P450 with different substrate specificity and different rates of biotransformation for certain xenobiotics was indicated by studies in the 1970s. All these cytochrome P450 enzymes share the heme, but they differ in both the composition and thus the structure of the polypeptide chain and in the reactions they catalyze [14]. Moreover, the individual enzymes are regulated in their expression by a variety of factors such as treatment with xenobiotics, species, organ, sex and diet. Because of the multitude of enzymes present, the term "superfamily" of cytochromes P450 is frequently used. A complex nomenclature, based on amino acid sequence similarity, has been developed to designate individual cytochrome P450 enzymes.

Reactions catalyzed by cytochrome P450

Cytochrome P450 enzymes are monooxygenases; thus, these enzymes utilize one of the oxygen atoms of molecular oxygen and incorporate it into the xenobiotic in the following stoichiometry (Scheme 1, RH is the substrate). The other oxygen atom is reduced to water under consumption of NADPH as reducing cofactor. Examples for oxidation reactions catalyzed by cytochromes P450 are shown in Table 1 [15–18].

Scheme 1. General mechanisms of cytochrome P450 catalyzed reactions.

$$RH + O_2 + NADPH + H^+ \longrightarrow ROH + H_2O + NADP^+$$

Table 1. Examples of oxidations catalyzed by cytochrome P450 enzymes (P450).

Type of reaction	Examples
aliphatic hydroxylation	$R\text{-}CH_2\text{-}CH_2\text{-}CH_3 \xrightarrow{P450} R\text{-}CH_2\text{-}CHOH\text{-}CH_3$
N-dealkylation	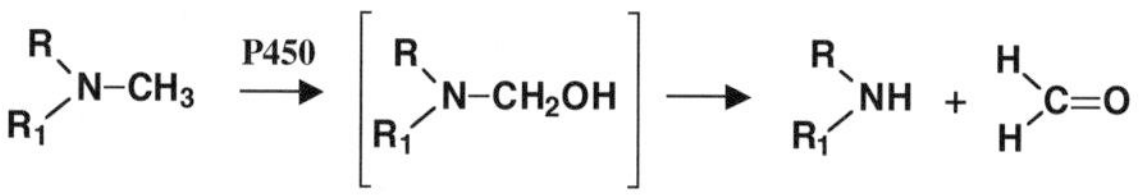
O-dealkylation	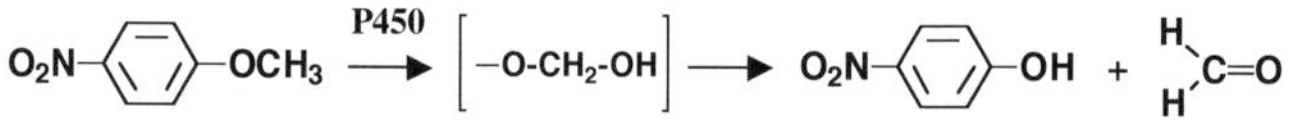
epoxidation	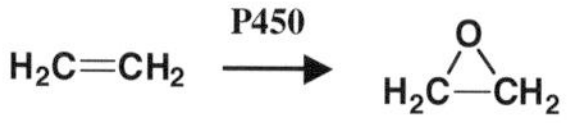

Cytochromes P450 may catalyze the hydroxylation of carbon-hydrogen bonds to transform hydrocarbons to the corresponding alcohols. The oxidative *N*-, *O*-, or *S*-dealkylation and the oxidative dehalogenation are similar in mechanism to the aliphatic hydroxylation but give different end products due to further reactions of the formed intermediate product. Olefins are also oxidized by cytochrome P450 and with some substrates, epoxides are formed as products. The reaction involves discrete ionic intermediates; these may also rearrange to products other than epoxides as shown with chloroolefins (Fig. 1). Oxidation of sulfur or nitrogen containing compounds occurs by the addition of oxygen at the unshared electron pair on the sulfur or nitrogen atom. The products formed (sulfoxides or hydroxylamines) may be stable, may be further oxidized by other enzymes in the organism (e.g., hydroxylamines) or may decompose to sulfur and the corresponding oxo-compound.

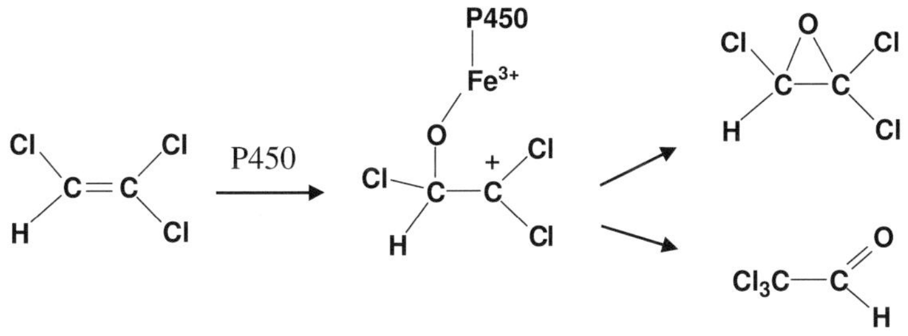

Figure 1. Mechanism of oxidation and rearrangement of trichloroethene to chloral and trichlorooxirane by cytochrome P450.

The above-mentioned reactions may be catalyzed by most cytochromes P450 involved in xenobiotic biotransformation; the type of reaction catalyzed seems to be more influenced by steric factors regarding the substrate binding site of individual cytochromes P450 than by electronic factors (Tab. 2).

In addition to promoting oxidative metabolism, cytochrome P450 enzymes may also catalyze reductive biotransformation reactions. These reactions are favored under reduced oxygen tension, or occur with xenobiotics lacking oxidizable C–H bounds or olefinic moieties. In these cases, the xenobiotic, instead of oxygen, accepts one or two electrons from the NADPH-dependent cytochrome P450 reductase or from cytochrome P450. Reductive biotransformation catalyzed by cytochrome P450 has been demonstrated with some azo dyes and several aromatic nitro compounds (Fig. 2). The double bond in azo compounds may be progressively reduced to give amine metabolites, aromatic nitro groups may also be reduced through the nitrone and the hydroxylamine to the corresponding amine.

The reductive biotransformation of polyhalogenated alkanes is exemplified by the one-electron reduction of carbon tetrachloride to the trichloromethyl radical and chloride; reductive biotransformation of carbon tetrachloride by a two-electron reduction results in formation of chloroform (Fig. 2).

Table 2. Human cytochromes P450 identified as major catalysts in the biotransformation of specific xenobiotics. The denoted enzymes seem to play major roles in the oxidation of the substrates listed.

Cytochrome P450			
1A1	1A2	2E1	3A4
benzo[a]pyrene	phenacetin	vinylchloride	aflatoxin B$_1$
other polycyclic hydrocarbons	1-aminofluorene	trichloroethene	17β-estradiol
	2-naphtylamine	halothane	6-aminochrysene
	2-amino-3-methyl-imidazo[4,5-f]quinoline	benzene	sterigmatocystin
		dimethylnitrosamine	nifedipin
		acetaminophen	ethinylestradiol

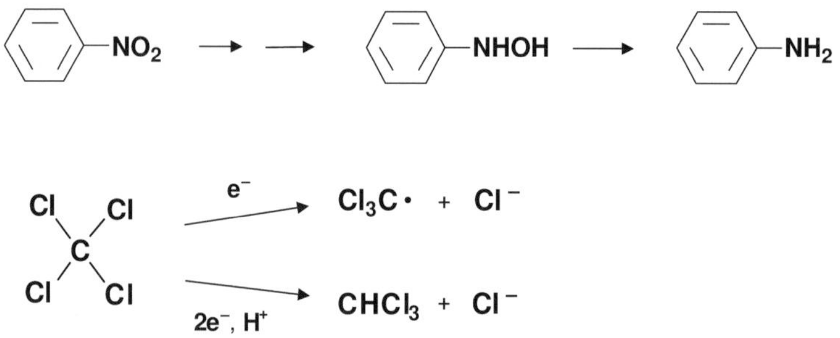

$$\text{—NO}_2 \longrightarrow \longrightarrow \text{—NHOH} \longrightarrow \text{—NH}_2$$

$$\xrightarrow{\text{e}^-} \text{Cl}_3\text{C} \cdot + \text{Cl}^-$$

$$\xrightarrow[\text{2e}^-, \text{H}^+]{} \text{CHCl}_3 + \text{Cl}^-$$

Figure 2. Reductive biotransformation reactions catalyzed by cytochrome P450.

Microsomal monooxygenases: Flavin-dependent monooxygenases

Tertiary amines and sulfur-containing drugs have been known to be metabolized to *N*-oxides or sulfoxides by a microsomal monooxygenase, which is not dependent on cytochrome P450. This enzyme is a flavoprotein localized in the endoplasmic reticulum. It is capable of oxidizing nucleophilic nitrogen and sulfur atoms, but shows a catalytic mechanism different from other heme- or flavin-containing enzymes. Flavin-dependent monooxygenases require molecular oxygen and NADPH as cofactors for oxygenation, but do not contain heme or iron, and the binding of the substrate is not required for the generation of the enzyme-bound oxygenating intermediate (Fig. 3).

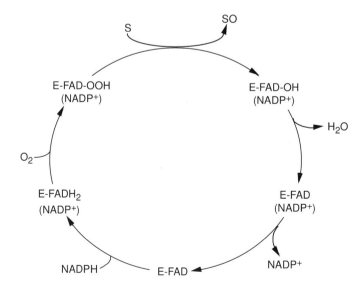

Figure 3. Mechanism of xenobiotic oxidation catalyzed by flavin-dependent monooxygenase.

The active, oxygenating form of the enzyme is present in the cell and any soft oxidizable nucleophile that can gain access to the enzyme-bound oxygenating intermediate will be oxidized. Precise fit of substrate to the enzyme is not necessary. This property seems to be largely responsible for the broad substrate specificity of flavin-dependent monooxygenases. These enzymes catalyze the oxidation of xenobiotics with few, if any common structural features, at maximum velocity, and catalyze the oxidation of a wide variety of xenobiotics (Tab. 3).

Table 3. Examples of oxidations catalyzed by flavin-dependent monooxygenases.

Type of reaction	Examples
amine oxidation	
hydroxylamine oxidation	
thioamide oxidation	
thiol oxidation	$2\ R\text{-}SH \longrightarrow R\text{-}S\text{-}S\text{-}R$
disulfide oxidation	$R\text{-}S\text{-}S\text{-}R \longrightarrow 2\ R\text{-}SO_2$

As with cytochromes P450, species- and tissue-specific forms of flavin-dependent monooxygenases have been described. Species differences in hepatic flavin-dependent monooxygenases seem to be quantitative rather than qualitative, whereas tissue-specific forms in same species are clearly distinct enzymes. For example, hepatic and pulmonary flavin-dependent monooxygenases in rabbits exhibit distinct, but overlapping substrate specificities and are different gene products. Several forms exist, and all forms show differences in their distribution in species and organs within species [19–22].

The regulation of expression of flavin-dependent monooxygenases seems to be complex. Xenobiotics, which were shown to increase the concentration of cytochromes P450 in mammals, did not influence flavin-dependent monooxygenase concentrations. Recent evidence suggest that soft nitrogen and sulfur nucleophiles in the diet may act as inducers of flavin-dependent monooxygenases. Since the dietary inducers are continuously taken up, flavin-dependent monooxygenases are present at maximal concentrations in animals on commercial rat chow.

Peroxidative biotransformation: Prostaglandin synthase

During biosynthesis of prostaglandins the polyunsaturated fatty acid arachidonic acid is oxidized to yield prostaglandin G_2, a hydroperoxy endoperoxide. This is further transformed to prostaglandin H_2. Both the formation of prostaglandin G_2 and its further transformation to prostaglandin H_2 are catalyzed by the same enzyme, prostaglandin synthase. This enzyme is a glycoprotein with a molecular mass of approximately 70 kDa and contains one heme per subunit. The enzyme is found in high concentrations in germinal vesicles and renal medulla, but also in several other tissues such as skin and adrenals [23]. The cyclooxygenase and peroxidase activity of prostaglandin synthase generate enzyme- and substrate-derived free radical intermediates (Fig. 4).

Biotransformation of xenobiotics may be associated both with the cyclooxygenase and the peroxidase activity of prostaglandin synthase (Fig. 4). During the cyclooxygenase-catalyzed conversion of arachidonic acid to prostaglandin G_2, peroxy radicals are formed as intermediates. These lipid per-

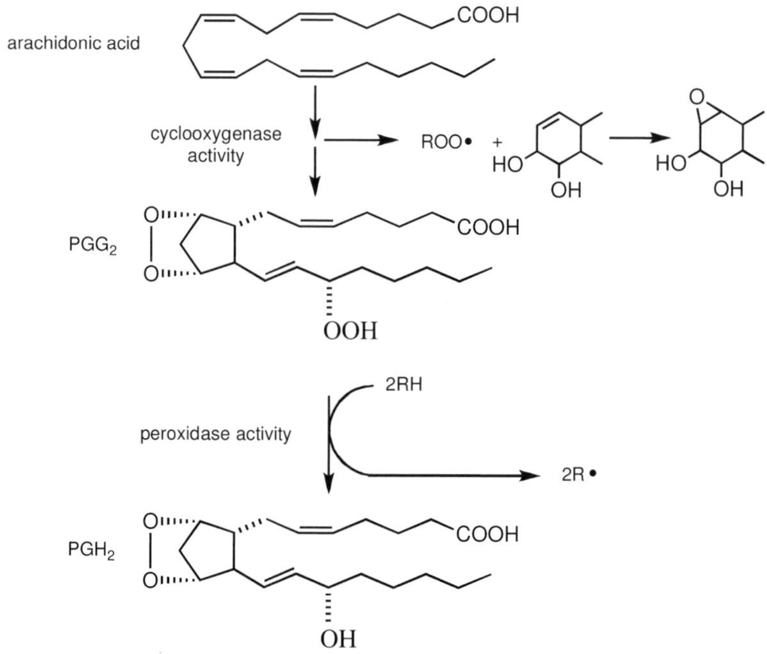

Figure 4. Possible mechanisms of cooxidation of xenobiotics during the biosynthesis of prostaglandins.

oxy radicals represent a source of reactive oxygen metabolites and can in turn biotransform xenobiotics. Oxidation by the cyclooxygenase activity of prostaglandin synthase is important in the oxidation of diols derived from carcinogenic polycyclic aromatic hydrocarbons and transforms these diols to diol-epoxides (Fig. 4, top).

During reduction of the peroxide prostaglandin G_2 to prostaglandin H_2, the peroxidase undergoes a two-electron oxidation. To return to the ground state, the enzyme requires two consecutive one-electron reductions, which are performed by abstracting electrons from available donors. In addition to endogenous substrates, xenobiotics may act as electron donors and may thus be oxidized to radicals (Fig. 4, bottom). This process is termed cooxidation of xenobiotics. Classes of xenobiotics undergoing cooxidation during prostaglandin syntheses are aromatic amines, phenols, hydroquinones and aminophenols. The role of prostaglandin synthase in the biotransformation of xenobiotics is somewhat unclear since many of the end products of xenobiotic oxidation catalyzed by prostaglandin synthase-mediated cooxidation are identical to those formed by cytochrome P450. Therefore, it is assumed that prostaglandin synthase may contribute to the oxidative biotransformation of xenobiotics in tissues low in monooxygenase activity [24].

In addition to prostaglandin synthase, other peroxidases may also participate in the oxidation of xenobiotics. For example, mammary gland epithelium contains lactoperoxidase and leukocytes contain myeloperoxidase. The general reaction catalyzed by these enzymes involves the reduction of hydroperoxide coupled to the oxidation of the substrate. The availability of peroxides in tissues likely controls the extent of peroxidative biotransformation; however, the availability of hydrogen peroxide is usually low due to efficient scavenging by catalase and glutathione peroxidase.

Non-microsomal oxidations

Several enzymes located in mitochondria or the cytoplasm of the cell may also catalyze oxidation of xenobiotics. In contrast to cytochromes P450 with their broad substrate specificity, most of the non-microsomal oxidases have a more narrow substrate specificity and accept only xenobiotics bearing specific functional groups as substrates. Alcohol dehydrogenases catalyze the oxidation of alcohols to aldehydes or ketones (Scheme 2). The enzyme is mainly found in the soluble fraction of liver, but also in other organs such as kidney and lung, and is responsible for the oxidation of ethanol. The expression of alcohol dehydrogenase is under genetic control, giving rise to a number of variants with differing activities. Usually, the oxidation of alcohols to aldehydes is

Scheme 2. Alcohol dehydrogenases catalyze the oxidation of alcohols to aldehydes or ketones.

$$RCH_2OH + NAD^+ \longrightarrow RCHO + NADH + H^+$$

reversible, since the reduction of aldehydes is also efficiently catalyzed by aldehyde reductases. However, *in vivo* the reaction proceeds in the direction of alcohol consumption since aldehydes are further oxidized by aldehyde dehydrogenases. These enzymes catalyze the formation of acids from aliphatic and aromatic aldehydes (Scheme 3). The reaction may be catalyzed by aldehyde dehydrogenase, which has a broad substrate specificity. Other enzymes in the soluble fraction of liver that may oxidize aldehydes are the flavoproteins aldehyde oxidase and xanthine oxidase.

Scheme 3. Oxidative formation of acids from aliphatic and aromatic aldehydes.

$$RCHO + NAD^+ \longrightarrow RCOOH + NADH + H^+$$

Epoxide hydrolase is an important enzyme cleaving aliphatic and aromatic epoxides. The enzyme hydrates arene oxides and aliphatic epoxides to the corresponding *trans*-1,2-diols [25]. Water is required as cofactor. The catalytic mechanism of epoxide hydrolases involves ester formation of the oxirane with a carboxylic acid function at the active site of the enzyme and hydrolysis of this ester by water; no metals or other cofactors are required (Scheme 4). Microsomal epoxide hydrolases are thought to be present in close proximity to microsomal cytochromes P450; in most cases, the reaction of the epoxide to the less reactive *trans*-diol is considered to represent an important detoxication reaction for metabolically formed oxiranes. Epoxide hydrolases are found in many tissues such as liver, kidney, testes, and intestine. Their distribution is heterogeneous between different cell types in a specific organ; in addition, several forms of microsomal epoxide hydrolase with broad substrate specificities have been found in different animal species. Moreover, in animals, in addition to the membrane-bound epoxide hydrolase, a soluble epoxide hydrolase is present in the cytoplasm of cells from several tissues [26, 27].

Scheme 4. Mechanism of epoxide hydrolase catalyzed hydrolysis of expoxides: ester formation from an oxirane with a carboxylate function at the active site of the enzyme and hydrolysis of this ester by water.

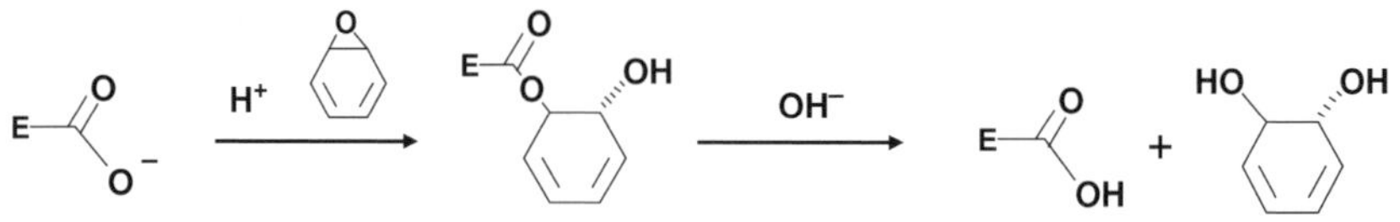

Phase II enzymes of biotransformation and their reactions

Products of phase I biotransformation reactions carrying functional groups such as hydroxyl, amino, or carboxyl groups often undergo a conjugation reaction with an endogenous substrate. In contrast to phase I biotransformation reactions, phase II reactions are biosynthetic and require energy to drive the reaction. Energy is usually consumed to generate a cofactor or an activated intermediate then utilized as cosubstrate. Thus, depleting of the cofactor or general interference with cellular energy status may interfere with the ability of cells to conduct phase II biotransformation reactions.

UDP-glucuronosyltransferases

Glucuronidation represents one or the main phase II biotransformation reaction in the conversion of both endogenous and exogenous compounds to water-soluble products. The formed glucuronides are excreted *via* bile or urine. The formation of an activated glucuronide (uridine diphosphate glucuronic acid, UDPGA) is required for glucuronide formation. UDPGA is formed in a sequential reaction from uridine and glucose-1-phosphate. The enzymes that carry out the coupling of the xenobiotic with UDPGA are termed UDP-glucuronosyltransferases. They couple D-glucuronic acid with a wide variety of xenobiotics carrying functional groups to give β-D-glucuronides [28]. These glucuronides are highly polar and ionized at physiological pH, thus they are rapidly excreted. The membrane-bound UDP-glucuronosyltransferases are found in highest concentration in the liver, but are also present in most other tissues studied. The reaction catalyzed involves a nucleophilic displacement (S_N2) of the functional group of the substrate with Walden inversion (Fig. 5). UDP-glucuronosyltransferases, like cytochromes P450, represent a large family of enzymes. The various forms respond differently to inducers and have preferences for certain classes of chemicals. UDP-glucuronosyltransferases catalyze the conjugation of numerous functional groups of xenobiotics with glucuronic acids. Some typical examples are shown in Table 4 [29].

Glucuronides formed in the liver are excreted with urine or bile. Aglycones with molecular weights higher than 300 will be transformed to glucuronides, surpassing the molecular weight threshold for biliary excretion and will thus be excreted with bile into the intestine. Subsequently, glucuronides may be cleaved in the gut by β-glucuronidase present in intestinal microflora to the respective aglycone, which then may be reabsorbed and translocated back to

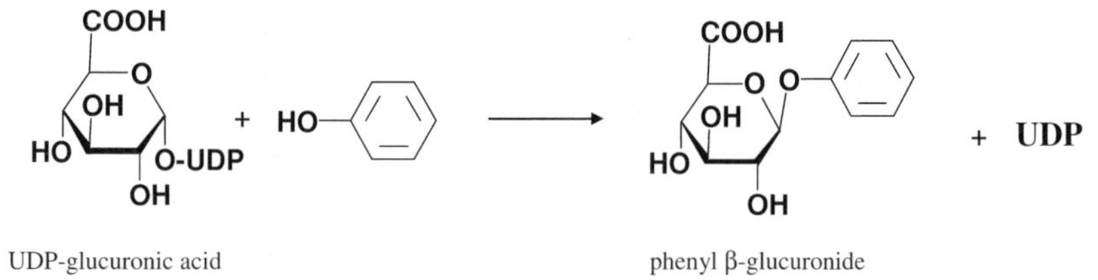

UDP-glucuronic acid   phenyl β-glucuronide

Figure 5. Conjugation of phenol to phenyl glucuronide catalyzed by UDP-glucuronosyltransferase (UDP, uridine diphosphate).

Table 4. Typical conjugation reactions catalyzed by UDP-glucuronosyltransferases (UGT).

Type of reaction	Examples

the liver with blood. The resulting cycle is called enterohepatic circulation. Compounds undergoing enterohepatic circulation are only slowly excreted.

Sulfate conjugation

The formation of water-soluble sulfate esters is observed with many xenobiotics carrying functional groups. These reactions are catalyzed by sulfotransferases, a large group of soluble enzymes found in many tissues [30–33]. Sulfotransferases catalyze the transfer of a sulfate group from the "active sulfate" 3'-phosphoadenosine-5'-phosphosulfate (PAPS) to hydroxyl groups and amines. The resulting products are referred to as sulfate esters of sulfamates (Fig. 6).

The products of this reaction are ionized at physiological pH; thus sulfate conjugation is an effective mechanism to enhance the rate of excretion for many xenobiotics. Sulfotransferases comprise a family of enzymes and at least four different classes of sulfotransferases are involved in biotransformation reactions. Sulfate conjugation requires the sulfate donor PAPS. PAPS is likely synthesized in the cytosol of most mammalian cells by a two-step reaction consuming ATP and utilizing inorganic sulfate coming from the catabolism of cysteine or from diet. In the first step of this sequence the sulfation of ATP to adenosine-5'-phosphosulfate is catalyzed by ATP sulfurylase. Adenosine-5'-

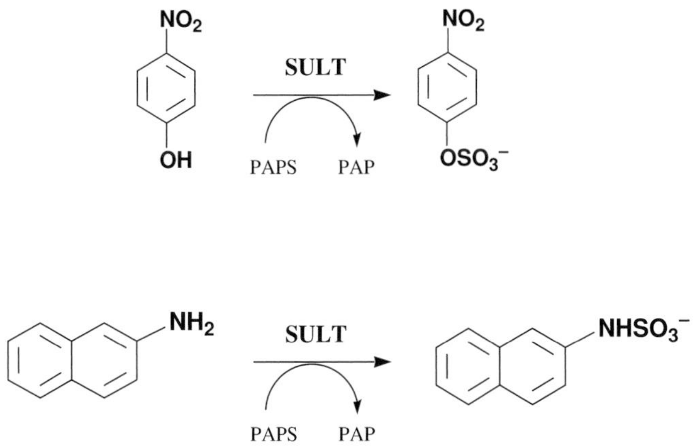

Figure 6. Representative conjugation reactions catalyzed by mammalian sulfotransferases (SULT).

phosphosulfate is further transformed to PAPS by an adenosine-5'-phospho-sulfate kinase. The equilibrium concentration of PAPS in mammalian cells may be low; however, due to the tight coupling of the two enzymes PAPS biosynthesis may proceed rapidly. Despite the rapid synthesis, the sulfation of xenobiotics may be limited by reduced availabily of PAPS. Further, the avail-abily of the cofactor of the synthesis reaction, sulfate, may be limited by con-sumption due to sulfation or the limited availability of free cysteine for trans-formation to sulfate. Therefore, the capacity of sulfation for certain xenobi-otics is dependent on dose. Following administration of low doses, the com-pound may be excreted as sulfate; after high doses, the capacity of sulfate con-jugation may be saturated and other biotransformation reactions such as glu-curonide formation may become more important for biotransformation.

N-Acetyltransferases

Aromatic amines, hydrazines, sulfonamides and certain aliphatic amines are biotransformed to amides in a reaction catalyzed by *N*-acetyltransferases. Enzymes that catalyze the acetylation of amines are designated as acetyl CoA:amine *N*-acetyltransferases. These enzymes utilize acetyl coenzyme A as cofactor [29]. The acetylation reaction of arylamines occurs in discrete steps. In the first step, the acetyl group from acetyl coenzyme A is transferred to the *N*-acetyltransferase, which then acetylates the arylamine thus regenerating the enzyme and forming the amide (Fig. 7).

N-Acetyltransferases are found in a number of different forms in the cytosol of cells of many tissues. In many species, the expression of *N*-acetyl-transferases is under genetic controls; polymorphisms for the expression of *N*-acetyltransferase have been found in several animal species and in humans.

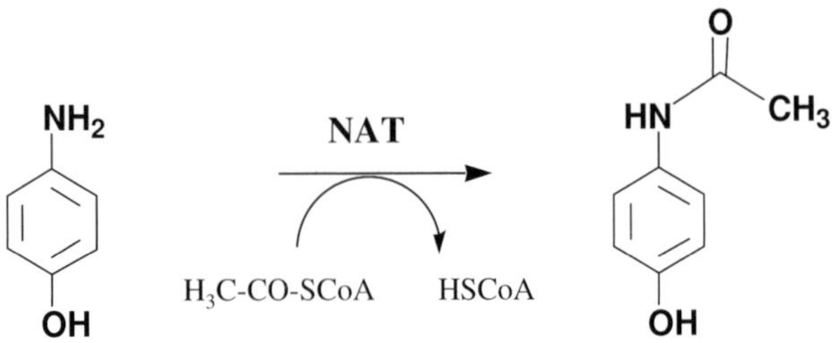

Figure 7. Mechanism of *N*-acetylation of amines catalyzed by *N*-acetyltransferase (NAT); -CoA, Coenzyme A.

The transfer of an acetyl group to amines is reversible; deacetylation of amides occurs in many species. There are large differences between strains, species and individuals in the extent of expression of amidases [34].

Glutathione conjugation of xenobiotics and mercapturic acid excretion

The conjugation of xenobiotics or their metabolites with the tripeptide glutathione is an important conjugation reaction. Glutathione is composed of the amino acids cysteine, glutamic acid and glycine (γ-glutamylcysteinyl-glycine) and is present in many cells in high concentrations (up to 10 mM in liver cells) [35]. Since glutathione conjugation captures reactive electrophiles and transforms them to stable, often non-toxic thioethers, the formation of glutathione conjugates serves as a major detoxication reaction. Glutathione conjugation is catalyzed by a family of enzymes termed glutathione-*S*-transferases, which are present in highest concentration in the liver, but they are also found in kidney, testes and lung in high activity [36]. Glutathione-*S*-transferases are both membrane-bound and soluble enzymes; with most substrates, the activity of the soluble glutathione-*S*-transferase is higher than that of the microsomal glutathione-*S*-transferase [37, 38].

Cytosolic glutathione-*S*-transferases exist in numerous different isoforms, each species being a dimer differing in subunit composition [32, 36]. The glutathione-*S*-transferase gene family exists of at least six different families. In contrast to the multiple forms of soluble glutathione-*S*-transferases, only one form of the membrane-bound enzyme is known. The glutathione-*S*-transferases catalyze the reaction of the sulfhydryl group of glutathione with chemicals containing electrophilic carbon atoms (Tab. 5).

Thioethers are formed by the reaction of the thiolate anion of glutathione with the electrophile; a spontaneous reaction, albeit at low rates, of the electrophile with glutathione without assistance by glutathione-*S*-transferases is required for enzymatic catalysis. Glutathione thioethers formed in the organism are not excreted, but further processed to excretable mercapturic acids. Mercapturic acids are thioethers derived from *N*-acetyl-L-cysteine.

Table 5. Substrates for mammalian glutathione-*S*-transferases (GST).

Type of reaction	Examples

aryl transferase

arylalkyl transferase

alkene transferase

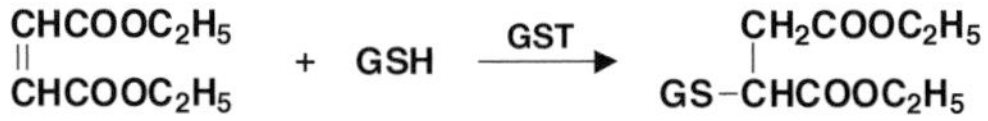

epoxide transferase

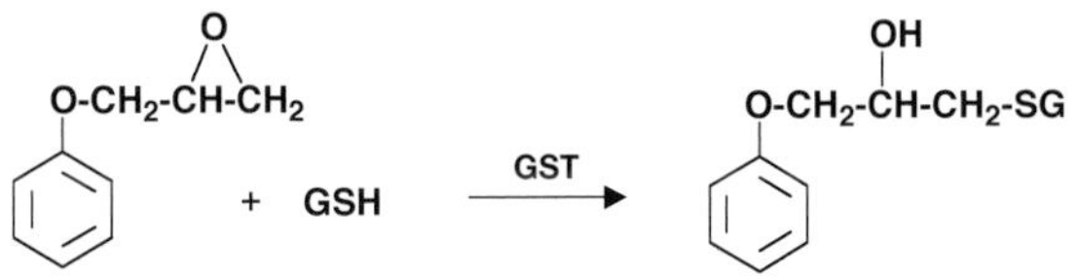

Mercapturic acid formation is initiated by conjugation of the xenobiotic or an electrophilic metabolite with glutathione (Fig. 8). This is followed by transfer of the glutamate by γ-glutamyl transpeptidase, an enzyme specifically recognizing γ-glutamyl peptides and found in high concentrations in the kidney and other excretory organs. Dipeptidases catalyze the loss of glycine from the intermediary cysteinylglycine *S*-conjugate to give the cysteine *S*-conjugate which, in the final step of mercapturic acid formation, is *N*-acetylated by a cysteine conjugate-specific *N*-acetyltransferase using acetyl coenzyme A as cofactor. The mercapturic acids formed are readily excreted into urine by active transport mechanisms in the kidney [39].

Bioactivation of xenobiotics

Many xenobiotics with low chemical reactivity (for example the solvent carbon tetrachloride, the environmental contaminant hexachlorobutadiene and the heat exchanger fluid tri-*o*-cresyl phosphate) cause toxic effects in cells. These toxic effects are initiated by the covalent binding to macromolecules of metabolites formed in the organisms by enzymatic biotransformation. This process is termed bioactivation. With many chemicals, reactive metabolites

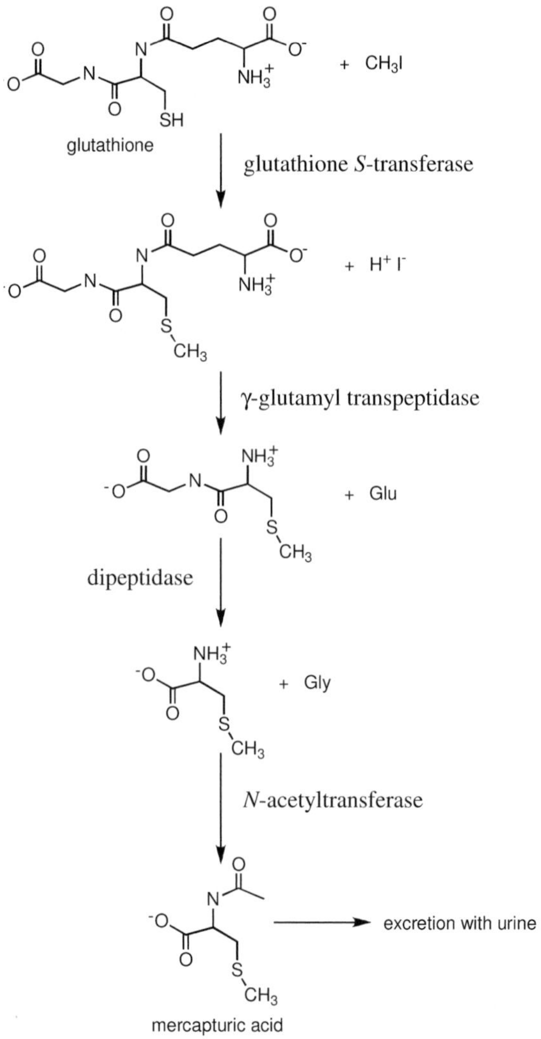

Figure 8. Formation of mercapturic acids by processing of glutathione S-conjugates as exemplified by the metabolism of methyl iodide.

formed during bioactivation may be efficiently detoxified; thus, toxic effects only occur when the balance between the production of reactive metabolites and their detoxication is disrupted. For example, toxic effects may be observed with a certain chemical only when the formation of reactive intermediates is enhanced or when the capacity for detoxication is diminished.

The mechanisms of bioactivation of xenobiotics may be classified into four categories describing the basic types of reactive intermediates formed and their potential reactivity (Tab. 6) [40, 41].

Table 6. Basic mechanisms involved in the bioactivation of xenobiotics based on the chemical reactivity of the intermediates formed [9].

Mechanism	Structure and reactivity of the intermediate	Examples
biotransformation to stable but toxic metabolites	different structures, selective interaction of formed metabolite with specific acceptors or disruption of specific biochemical pathways	dichloromethane acetonitrile parathion
biotransformation to electrophiles	reactive electrophiles	dimethylnitrosamine acetaminophen bromobenzene
biotransformation to free radicals	radicals	carbon tetrachloride
formation of reactive oxygen metabolites	oxygen radicals	paraquat aromatic nitro compounds

Formation of stable but toxic metabolites

This mechanism is limited to a few selected chemicals because there are only few xenobiotic metabolites that are both stable and toxic. The bioactivation mechanisms of the solvents n-hexane and dichloromethane may serve as examples to illustrate this mechanism. n-hexane produces a characteristic neuropathy and peripheral nerve injury after chronic exposure. The same typical manifestations of toxicity are also observed when the n-hexane metabolites 2-hexanone and 2,5-hexane-dione are administered to animals. The mechanism of n-hexane-dependent neuropathy thus involves oxidation of n-hexane by cytochromes P450 at both ends of the carbon chain (ω-1 hydroxylation) and further oxidation of the generated alcoholic function. The 2,5-hexane-dione formed reacts with critical lysine residues in axonal proteins by Schiff-base formation followed by cyclization to give pyrroles [42]. Oxidation of the pyrrole residues then causes cross-linking between two n-hexane modified proteins; the resulting changes in the three-dimensional structures of proteins perturb axonal transport and function and cause neuronal cell damage (Fig. 9).

Carboxyhemoglobin formation is seen after human exposure to dichloromethane. Dihalomethanes are oxidized by cytochrome P450, likely by the form P450 2E1, to carbon monoxide, which, due to its high affinity for iron(II)-containing porphyrins, binds to hemoglobin and interferes with oxygen transport in blood.

Biotransformation to reactive electrophiles

This is the most common pathway of bioactivation. The cytotoxicity and carcinogenicity of many chemicals is associated with the formation of elec-

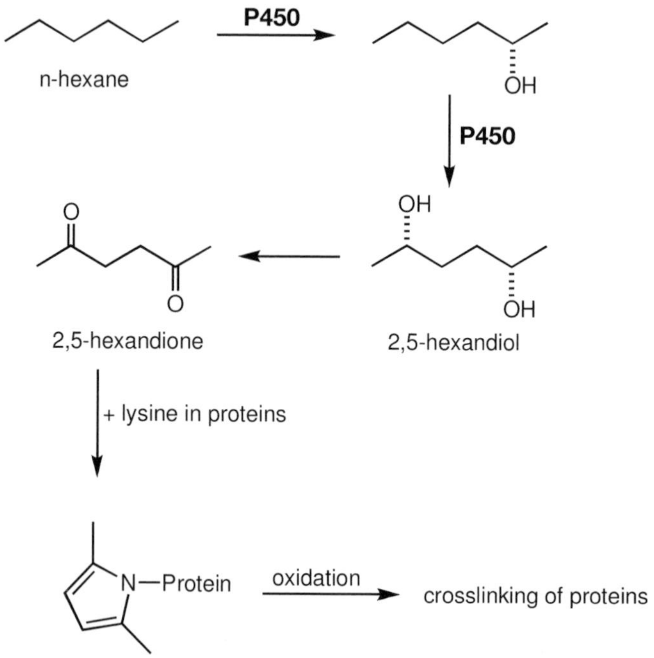

Figure 9. Bioactivation of n-hexane by cytochrome P450 to 2,5-hexanedione. Hexanedione reacts with lysine groups in proteins to pyrroles, oxidation of two neighboring pyrrole residues causes the cross-linking of proteins.

trophiles and the ensuing alkylation or acylation of tissue constituents such as protein, lipid or DNA. Reactive intermediates include chemically diverse functionalities such as epoxides, quinones, acyl halides, carbocations and nitrenium ions. The metabolic formation of electrophiles may be catalyzed by many different enzymes, although the majority of examples elucidated up to date involve cytochrome P450-mediated oxidations [43].

Cytochromes P450 catalyze the transformation of olefins to reactive and electrophilic oxiranes. For example, the carcinogenicity of the industrial intermediate vinyl chloride (Fig. 10) depends on its transformation to electrophilic oxiranes [44, 45]. Carbocations are formed during cytochrome P450-mediated oxidation of dialkyl nitrosamines. For example, the mutagen and potent carcinogen dimethylnitrosamine is hydroxylated by cytochrome P450 enzymes followed by loss of formaldehyde. The monomethylnitrosamine formed is unstable and rearranges to release an electrophilic carbocation (Fig. 11).

Besides cytochromes P450, other monooxygenases such as the flavin-dependent monooxygenases, or enzymes of phase II biotransformation such as UDP-glucuronosyltransferases, sulfotransferases or glutathione-S-transferases may catalyze the bioactivation of xenobiotics [46]. For example, N-acetyl-amidofluorene is oxidized to N-hydroxyacetylamidofluorene by cytochrome

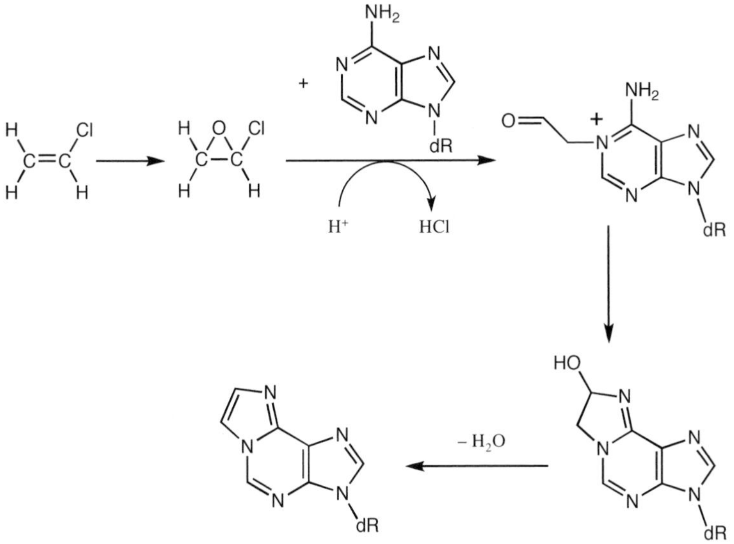

Figure 10. Bioactivation of vinyl chloride to chlorooxirane and reaction of the oxirane with deoxy-adenosine in the cell.

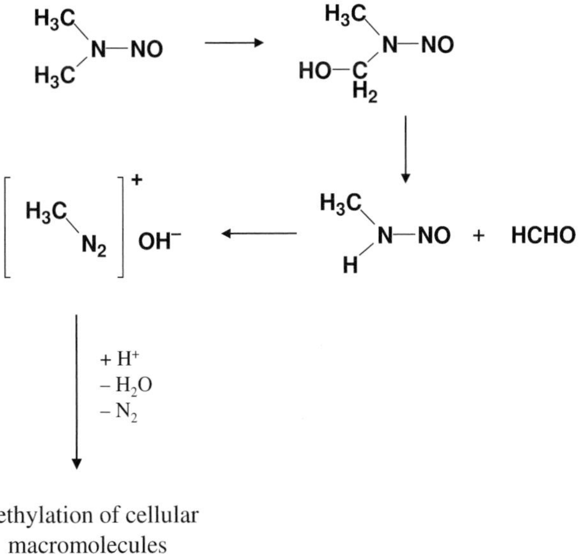

1,N⁶-ethenodeoxyadenosine

methylation of cellular
macromolecules

Figure 11. Bioactivation of dimethylnitrosamine by cytochrome P450 to a methylating agent.

P450. However, this metabolite is not electrophilic and requires further bio-transformation *via* sulfate conjugation to the highly reactive *O*-sulfate ester.

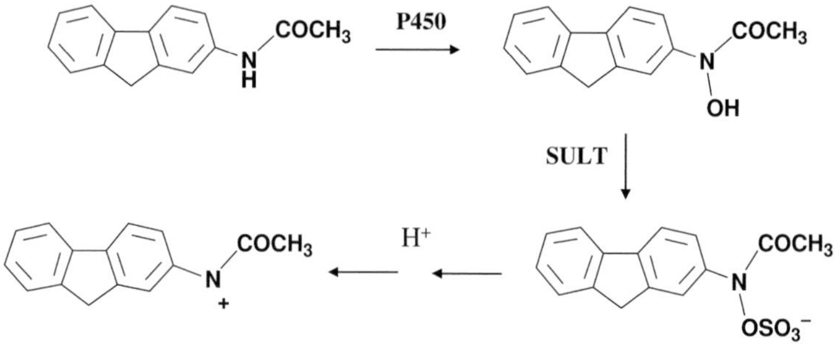

Figure 12. Bioactivation of *N*-acetylamidofluorene by cytochrome P450 (P450) and by sulfotransferases (SULT). The sulfate formed is acid-labile and decomposes to a nitrenium ion.

This sulfate ester fragments to a reactive intermediate (a nitrenium ion), which covalently binds to tissue constituents such as DNA (Fig. 12).

Some glutathione *S*-conjugates, which are biosynthesized to detoxify electrophiles, were found to be toxic and mutagenic [8, 9, 47]. 1,2-Dibromoethane is metabolized by glutathione conjugation to *S*-(2-bromoethyl)glutathione. Intramolecular displacement of the bromine on the adjacent carbon atom gives a highly strained, electrophilic episulfonium ion (Fig. 13). Other toxic glutathione *S*-conjugates require processing by the enzymes of mercapturic acid formation to give electrophiles. A minor pathway in perchloroethene biotransformation results in *S*-(1,2,2-trichlorovinyl)glutathione [48]. This glutathione *S*-conjugate is cleaved by γ-glutamyl transpeptidase and dipeptidases to *S*-(1,2,2-trichlorovinyl)-L-cysteine, which is a substrate for renal cysteine conjugate β-lyase and transformed to pyruvate, ammonia and a reactive thioketene

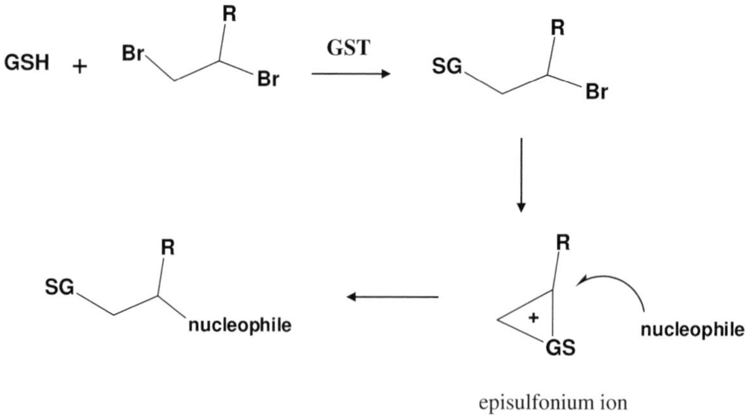

Figure 13. Bioactivation of 1,2-dibromoethane by glutathione conjugation to a reactive and electrophilic episulfonium ion.

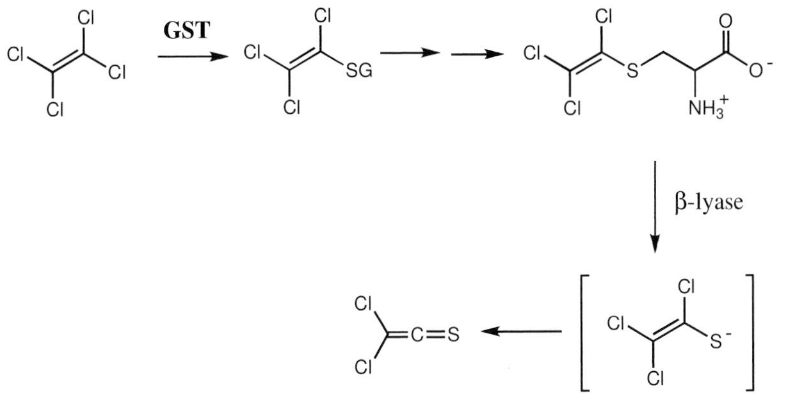

Figure 14. Bioactivation of perchloroethene by glutathione conjugation. The conjugate S-(1,2,2-trichlorovinyl)glutathione is biosynthesized in the liver, translocated to the kidney to be processed by γ-glutamyl transpeptidases and dipeptidases and finally cleaved by cysteine-conjugate β-lyase to give dichlorothioketene.

whose binding to cellular macromolecules is likely responsible for the renal toxicity of perchloroethene (Fig. 14). Due to high concentrations of S-(1,2,2-trichlorovinyl)-L-cysteine obtained by active transport into the kidney, covalent binding of the dichlorothioketene formed *via* this pathway occurs only in kidney; despite the presence of cysteine conjugate β-lyase in many other organs.

Biotransformation of xenobiotics to radicals

Free radicals are chemical species that may be formed by a one-electron oxidation to yield a cation radical. Free radicals are highly reactive and, when formed in biological systems, are expected to react with a variety of tissue molecules. Radicals may abstract hydrogen atoms, undergo oxidation-reduction reactions, dimerization and disproportionation reactions. Radicals may also participate in a chain mechanism, which is initiated by a reaction causing a free radical and propagated by a subsequence of reactions causing further radicals as products.

The toxic and tumorigenic solvent carbon tetrachloride is the outstanding example of a bioactivation reaction to a free radical. Carbon tetrachloride is biotransformed by one-electron reduction to yield the trichloromethyl radical and chloride (Scheme 5). The trichloromethyl radical may abstract hydrogen atoms from tissue macromolecules to give chloroform, a proven metabolite of

Scheme 5. Biotransformation of carbon tetrachloride by a one-electron reduction to give the trichloromethyl radical.

$$Cl_4C \;+\; e^- \longrightarrow Cl_3C\bullet \;+\; Cl^-$$

carbon tetrachloride, or may dimerize to give hexachloroethane, which is also a metabolite of carbon tetrachloride. Toxic effects of the formation of radicals during biotransformation reactions are lipid peroxidation and oxidative modification of proteins. Formation of radicals has been implicated in the bioactivation of many xenobiotics, but formation of free radicals from tissue constituents also plays an important role in the toxic effects of ionizing radiation [15, 49–51].

Formation of reactive oxygen metabolites by xenobiotics

Xenobiotic-induced formation of reduced oxygen metabolites such as the superoxide radical anion, hydrogen peroxide and the hydroxy radical have been implicated as a mechanism of producing cell damage, the so-called oxidative stress [52–55]. Biotransformation of certain xenobiotics that are able to undergo redox cycling or enzyme catalyzed oxidation-reduction reactions may be associated with the production of reduced oxygen metabolites. 2-Methylnaphtoquinone (menadione) has been intensively used to study the formation and cellular reactions of reduced oxygen metabolites (Fig. 15). Menadione and other quinones undergo enzymatic redox cycling; these one-electron oxidation reactions are associated with the formation of the superoxide radical anion by a one-electron reduction of triplet oxygen. In aqueous solution, superoxide is not impressively reactive; the dismutation or further reduction of superoxide may, however, give rise to hydrogen peroxide. Hydrogen peroxide is also a poor oxidant in biological systems, but suffi-

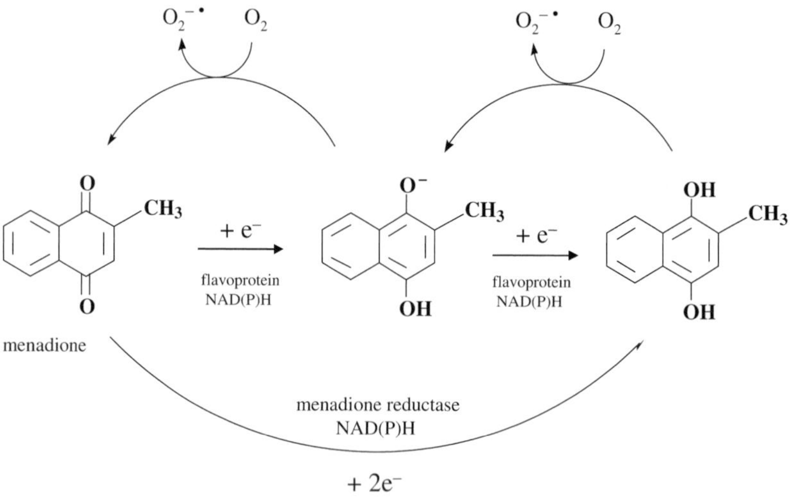

Figure 15. Biotransformation of menadione and the induction of oxidative stress by reduction of triplet oxygen to the superoxide radical anion $(O_2^{-\cdot})$.

ciently stable to cross biological membranes. The toxicity of hydrogen peroxide is attributed to the formation of the hydroxyl radical by the Fenton-reaction, catalyzed by metal ions such as Fe^{2+} (Scheme 6, M = transition metal).

Scheme 6. Formation of the hydroxyl radical from hydrogen peroxide *via* Fenton-reaction
(M = metal).

$$M^n + H_2O_2 \longrightarrow M^{n+1} + HO\bullet + HO^-$$

The highly reactive hydroxyl radical may then initiate cellular damage by
radical-based mechanisms. Besides menadione, oxidative stress may also be
initiated by other xenobiotics such as the bis-pyridinium herbicide paraquat
and nitroheterocycles. Moreover, the formation of reduced oxygen metabolites
plays an important role in host defense against infectious agents and in the initiation and propagation of certain diseases such as arteriosclerosis and poly-
arthritis. Since oxygen radicals are also formed in low concentrations during
cellular respiration, efficient mechanisms for their detoxication exist.
Oxidative stress is thus only observed when the equilibrium between oxidants
and reductants is disturbed and detoxication mechanisms are overwhelmed.

Detoxication and interactions of reactive metabolites with cellular
macromolecules

Reactive intermediates formed inside cells may react with low and high molecular weight cellular constituents. These interactions may result in formation
of less reactive chemicals and thus in detoxication, or may perturb important
cellular functions resulting in acute and/or chronic toxic effects such as necrosis or cancer. Usually, the interaction with low molecular weight constituents
in the cell results in detoxication, whereas the irreversible interaction with cellular macromolecules results in adverse effects [39, 56–58].
    Detoxication of reactive intermediates may be due to hydrolysis, glu-
tathione conjugation or interactions with cellular antioxidants. The reaction of
electrophilic xenobiotics with the nucleophile water, present in high concentrations in all cells, is the simplest detoxication. Many of the products thus
formed are of low reactivity and may be rapidly excreted. For example, acyl
halides formed by the oxidation of olefins such as perchloroethene are
hydrolyzed rapidly to halogenated carboxylic acids; only minor amounts of the
intermediary acyl halide react with proteins and lipids (Fig. 16).
    Glutathione-dependent reactions represent an important detoxication mech-
anism for metabolically formed electrophiles, free radicals and reduced oxygen metabolites [59–62]. Spontaneous reactions are only observed in appre-
ciable rates with soft electrophiles (glutathione is a soft nucleophile). The con-

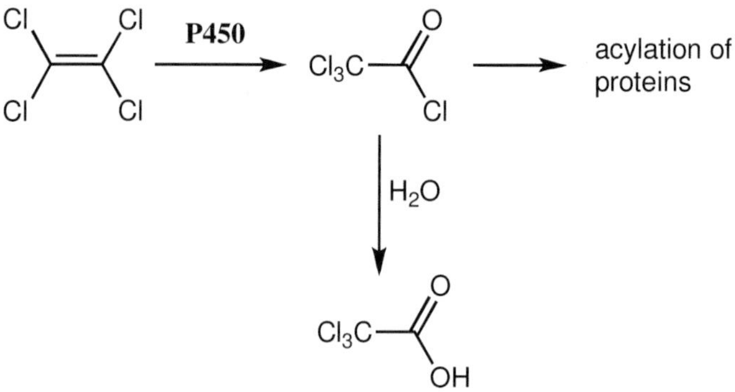

Figure 16. Biotransformation of perchloroethene to trichloroacetyl chloride followed by hydrolysis to trichloroacetic acid, the major urinary metabolite formed from perchloroethene. Only minor amounts of the intermediate acyl halide react with proteins.

jugation of hard electrophiles with glutathione requires enzymatic catalysis; usually, the glutathione-*S*-transferase-catalyzed rates of conjugation differ between hard and soft electrophiles, soft electrophiles being conjugated more efficiently. For example, the hard electrophile aflatoxin B_1-8,9-oxide does not spontaneously react with glutathione; the glutathione *S*-conjugate of aflatoxin B_1-8,9-oxide is formed only in the presence of a certain glutathione-*S*-transferase enzyme. Species differences in the tumorigenesis of aflatoxin B_1 may serve to illustrate the important role of glutathione-*S*-transferases in the expression of toxicity and carcinogenicity. Aflatoxin B_1 is a potent liver carcinogen in rats; in mice, aflatoxin B_1 is only weakly carcinogenic. The liver of mice contains a certain glutathione-*S*-transferase form that efficiently detoxifies aflatoxin B_1-8,9-oxide. This glutathione-*S*-transferase enzyme is not present in rat liver; thus, the binding of aflatoxin B_1-8,9-oxide to rat liver DNA and liver carcinogenicity of aflatoxin B_1 is much higher in rats as compared to mice.

Glutathione also plays a major role in the detoxication of reactive oxygen metabolites and radicals. Selenium-dependent glutathione peroxidases are important enzymes catalyzing the detoxication of hydrogen peroxide. In the glutathione peroxidase catalyzed reaction, two molecules of glutathione are oxidized to glutathione disulfide (Scheme 7). Glutathione may then be recycled through the reduction of glutathione disulfide by glutathione reductase.

Scheme 7. Glutathione peroxidase catalyzed decomposition of hydrogen peroxide.

$$H_2O_2 + 2\,GSH \xrightarrow{\text{GSH peroxidase}} GSSG + H_2O$$

GSH reductase

On the other hand, copper- and zinc-dependent cytosolic and the manganese-dependent mitochondrial superoxide dismutases detoxify superoxide radical anions. Hydrogen peroxide formed by dismutation of superoxide is transformed to water and oxygen and thus detoxified by catalase (Scheme 8).

Scheme 8. Hydrogen peroxide formed by dismutation of superoxide is transformed by catalase to water and oxygen.

$$2\ H_2O_2 \longrightarrow 2\ H_2O\ +\ {}^1\!/_2\ O_2$$

Several cellular antioxidants also play a role in the detoxication of radicals. α-Tocopherol is an important lipophilic antioxidant, whose presence in lipid membranes prevents radical-mediated damage to lipid constituents (e.g., unsaturated fatty acids). The hydroxyl radical, superoxide radical anion, and peroxy-radicals react with α-tocopherol to yield water, hydrogen peroxide and hydroperoxides, which may be detoxified further by catalase and glutathione peroxidase. α-Tocopherol is transformed during these reactions to give a stable radical of comparatively low reactivity. Ascorbic acid is an important antioxidant present in the cytoplasm of the cell and may also participate in the detoxication of radicals.

Interaction of reactive intermediates with cellular macromolecules

Although a substantial body of information is available on the biotransformation of xenobiotics to reactive metabolites and the chemical nature of those metabolites, considerable less is known about how reactive intermediates interact with cellular constituents and how those interactions cause cell injury and cell death. The reaction of toxic metabolites may result in the formation of covalent bonds between the molecule and a cellular target molecule, or they may alter the target molecule without formation of a covalent bond, usually by oxidation or reduction [63].

Electrophilic metabolites may react with different nucleophilic sites in cells. Nucleophilic sites in cellular macromolecules are thiols and amino groups present in proteins, amino groups present in lipids, and the oxygen and nitrogen atoms present in the purine and pyrimidine bases in DNA. The formation of a covalent bond may permanently alter the structure and/or activity of the modified macromolecule and thus result in a toxic response. The complexity of the reaction of electrophilic metabolites with the various nucleophilic sites in cells may be interpreted based upon the concept of hard and soft electrophiles/nucleophiles (hard and soft acids/bases). The donor atom of a soft nucleophile is of high polarizability and low electronegativity, and is easily oxidized; the donor atom of a hard nucleophile is of low polarizability and high electronegativity. Hard electrophiles carry a highly positive charge and have a small size; soft electrophiles are of low positive charge and larger size. Soft electrophiles react predominantly with soft nucleophiles and hard electrophiles with hard nucleophiles [64]. Thus, hard electrophiles formed during biotransformation reactions (e.g., carbocations formed from dialkylnitrosamines) predominantly react with hard nucleophiles such as oxygen and nitrogen atoms in DNA. In contrast, soft electrophiles such as α,β-unsaturated carbonyl compounds (e.g., acrolein or benzoquinone) react predominantly with soft tissue nucleophiles such as the sulfhydryl groups of cysteine in proteins (Tab. 7).

Table 7. Metabolically formed electrophiles and their primary targets for covalent binding in cells.

Soft	⟶			Hard
nucleophile:				
SH of cysteine or glutathione	sulfur of methionine	primary or secondary nitrogen atoms in peptides (lysine, arginine oder histidine)	amino groups of purine and pyrimidines in RNA and DNA	oxygen in purines and pyrimidines in DNA and RNA
electrophile:				
α,β-unsaturated carbonyls, quinones	epoxides, alkyl sulfates, alkyl halides	nitrenium ions	benzylic carbocations	aliphatic and aromatic carbocations

Covalent interactions of xenobiotics with proteins occur with several nucleophilic nitrogen atoms. Both alkylation and acylation reactions of amino acids have been reported as the consequences of formation of reactive intermediates in cells. Nitrogen atoms in the amino acids lysine, histidine and valine are, besides the sulfur atom of cysteine, frequent targets for electrophilic metabolites. Consequences of the modifications may be inactivation of enzymes important for cellular function, changes in the tertiary structure of proteins, or changes in gene expression. Alkylation of the sulfhydryl-dependent enzymes of mitochondrial respiration is thought to play an important role in the initiation of mitochondrial dysfunction and cell damage.

Some modified proteins may also serve as immunogens, and hypersensitivity reactions, formation of immune complexes and delayed hypersensitivities may be the consequences of protein adduct formation. Indeed, many drug- and chemical-related hypersensitivity reactions observed in clinical medicine are based on the formation of covalent protein adducts and their recognition as "foreign" by the immune system [65].

Oxidative stress that produces mixed disulfides of proteins with low molecular weight thiols such as glutathione also alters protein structure and function. In addition, oxidants and radicals promote the oxidation of amino acids in proteins, which may increase the susceptibility of these proteins to proteolysis [66, 67]. Increased protein oxidation has been implicated in cellular aging and in the mechanisms of toxicity of several redox active transition metals.

Radicals formed during the biotransformation of xenobiotics may abstract hydrogen atoms from cellular components [49]. The abstraction of hydrogen atoms from polyunsaturated fatty acids of lipids results in a process termed lipid peroxidation. The fatty acid radicals thus formed may react with molecular oxygen to give peroxy radicals and hydroperoxides. The initiated radical chain reactions cause the cleavage of carbon-carbon bonds in the fatty acids to short fragments such as α,β-unsaturated carbonyl compounds [68–70]. The disruption of membranes and the formation of toxic hydroperoxides and α,β-unsaturated carbonyl compounds may cause disruptions in cellular homeostasis and thus cause biochemical changes that ultimately lead to cell death [71, 72]. The reaction of electrophilic metabolites with DNA constituents results in the formation of altered purine and pyrimidine bases or other DNA damage such as DNA strand breaks or the loss of single bases from the double helix. Many of these modifications are "pre-mutagenic lesions". After gene expression, these lesions may be translated into mutations [73]. Mutations in certain genes are considered as the basis for the evolution of neoplastic cells and cancer and thus play a major role in chemical carcinogenesis. Other types of DNA damage may result in the activation of genes important for cellular differentiation or other regulatory functions. Radicals formed as reactive intermediates may also cause DNA damage. Besides DNA strand breaks, which have been frequently observed, the reaction of oxygen-derived radicals may also result in the oxidation of purine and pyrimidine nucleotides.

References

1 Medinsky M, Valentine JL (2001) Toxicokinetics. In: CD Klaassen (ed.): *Casarett and Doull's Toxicology. The Basic Science of Poisons*. McGraw-Hill, New York, 225–237
2 May DG (1994) Genetic differences in drug disposition. *J Clin Pharmacol* 34: 881–897
3 Pang KS, Xu X, St-Pierre MV (1992) Determinants of metabolite disposition. *Annu Rev Pharmacol Toxicol* 32: 623–669
4 Filser JG (2008) Toxicokinetics. In: H Greim, R Snyder (eds): *Toxicology and Risk Assessment*. John Wiley & Sons, Hoboken, 19–49
5 Gibson GG, Skett P (2001) *Introduction to Drug Metabolism*. Nelson Thornes, Cheltenham
6 Guengerich FP (2006) Cytochrome P450s and other enzymes in drug metabolism and toxicity. *AAPS J* 8: E101–111
7 Buters JTM (2008) Phase I metabolism. In: H Greim, R Snyder (eds): *Toxicology and Risk Assessment*. John Wiley & Sons, Hoboken, 49–74
8 Anders MW (1985) *Bioactivation of Foreign Compounds*. Academic Press, New York
9 Anders MW (1988) Bioactivation mechanisms and hepatocellular damage. In: IM Arias, WB Jakoby, H Popper, D Schachter, DA Shafritz (eds): *The Liver: Biology and Pathology*, 2nd edition. Raven Press, New York, 389–400
10 DeBethizy JD, Hayes JR (1994) Metabolism, a determinant of toxicity. In: AW Hayes (ed.): *Principles and Methods of Toxicology*. Raven Press, New York, 59–100
11 Guengerich FP (1991) Reactions and significance of cytochrome P450 enzymes. *J Biol Chem* 266: 10019–10022
12 Guengerich FP (2008) Cytochrome P450 and chemical toxicology. *Chem Res Toxicol* 21: 70–83
13 Guengerich FP (2007) Mechanisms of cytochrome P450 substrate oxidation: MiniReview. *J*

Biochem Mol Toxicol 21: 163–168

14 Ding X, Kaminsky LS (2003) Human extrahepatic cytochromes P450: Function in xenobiotic metabolism and tissue-selective chemical toxicity in the respiratory and gastrointestinal tracts. *Annu Rev Pharmacol Toxicol* 43: 149–173

15 Guengerich FP, Liebler DC (1985) Enzymatic activation of chemicals to toxic metabolites. *CRC Crit Rev Toxicol* 14: 259–307

16 Guengerich FP (2001) Common and uncommon cytochrome P450 reactions related to metabolism and chemical toxicity. *Chem Res Toxicol* 14: 611–650

17 Rendic S, Di Carlo FJ (1997) Human cytochrome P450 enzymes: A status report summarizing their reactions, substrates, inducers, and inhibitors. *Drug Metab Rev* 29: 413–580

18 Brown CM, Reisfeld B, Mayeno AN (2008) Cytochromes P450: A structure-based summary of biotransformations using representative substrates. *Drug Metab Rev* 40: 1–100

19 Cashman JR (2003) The role of flavin-containing monooxygenases in drug metabolism and development. *Curr Opin Drug Discov Dev* 6: 486–493

20 Cashman JR, Zhang J (2006) Human flavin-containing monooxygenases. *Annu Rev Pharmacol Toxicol* 46: 65–100

21 Cashman JR (2005) Some distinctions between flavin-containing and cytochrome P450 monooxygenases. *Biochem Biophys Res Commun* 338: 599–604

22 Krueger SK, Williams DE (2005) Mammalian flavin-containing monooxygenases: Structure/function, genetic polymorphisms and role in drug metabolism. *Pharmacol Ther* 106: 357–387

23 Vogel C (2000) Prostaglandin H synthases and their importance in chemical toxicity. *Curr Drug Metab* 1: 391–404

24 Smith BJ, Curtis JF, Eling TE (1991) Bioactivation of xenobiotics by prostaglandin H synthase. *Chem Biol Interact* 79: 245–264

25 Kettle AJ, Winterbourn CC (1992) Oxidation of hydrochinone by myeloperoxidase. *J Biol Chem* 267: 8319–8324

26 Arand M, Cronin A, Oesch F, Mowbray SL, Jones TA (2003) The telltale structures of epoxide hydrolases. *Drug Metab Rev* 35: 365–383

27 Arand M, Cronin A, Adamska M, Oesch F (2005) Epoxide hydrolases: Structure, function, mechanism, and assay. *Methods Enzymol* 400: 569–588

28 Sheweita SA, Tilmisany AK (2003) Cancer and phase II drug-metabolizing enzymes. *Curr Drug Metab* 4: 45–58

29 Tukey RH, Strassburg CP (2000) Human UDP-glucuronosyltransferases: Metabolism, expression, and disease. *Annu Rev Pharmacol Toxicol* 40: 581–616

30 Burchell B, Coughtrie WH (1989) UPD-glucuronosyltransferases. *Pharmacol Ther* 43: 261–289

31 Tephly TR, Burchell B (1990) UDP-glucuronosyltransferases: A family of detoxifying enzymes. *Trends Pharmacol Sci* 11: 276–279

32 Duffel MW, Marshal AD, McPhie P, Sharma V, Jakoby WB (2001) Enzymatic aspects of the phenol (aryl) sulfotransferases. *Drug Metab Rev* 33: 369–395

33 Gamage N, Barnett A, Hempel N, Duggleby RG, Windmill KF, Martin JL, McManus ME (2006) Human sulfotransferases and their role in chemical metabolism. *Toxicol Sci* 90: 5–22

34 Knights KM, Sykes MJ, Miners JO (2007) Amino acid conjugation: Contribution to the metabolism and toxicity of xenobiotic carboxylic acids. *Exp Opin Drug Metab Toxicol* 3: 159–168

35 Ritter JK (2000) Roles of glucuronidation and UDP-glucuronosyltransferases in xenobiotic bioactivation reactions. *Chem Biol Interact* 129: 171–193

36 Bock KW, Bock-Hennig BS, Fischer G, Ullrich D (1984) Role of glucuronidation and sulfation in the control of reactive metabolites. In: H Greim, R Jung, M Kramer, H Marquardt, F Oesch (eds): *Biochemical Basis of Chemical Carcinogenesis*. Raven Press, New York, 13–22

37 Mulder GJ (1981) *Sulfation of Drugs and Related Compounds*. CRC Press, Boca Raton

38 Hayes JD, Flanagan JU, Jowsey IR (2005) Glutathione transferases. *Annu Rev Pharmacol Toxicol* 45: 51–88

39 Krynetski EY, Evans WE (1999) Pharmacogenetics as a molecular basis for individualized drug therapy: The thiopurine *S*-methyltransferase paradigm. *Pharm Res* 16: 342–349

40 Evans DA (1989) *N*-acetyltransferase. *Pharmacol Ther* 42: 157–234

41 Meisel P (2002) Arylamine *N*-acetyltransferases and drug response. *Pharmacogenomics* 3: 349–366

42 Arias IM, Jakoby WB (1976) *Glutathione: Metabolism and Function*. Raven Press, New York

43 Guengerich FP (2003) Cytochrome P450 oxidations in the generation of reactive electrophiles:

Epoxidation and related reactions. *Arch Biochem Biophys* 409: 59–71

44 Ketterer B, Meyer DJ, Clark AG (1988) Soluble glutathione transferase isoenzymes. In: H Sies, B Ketterer (eds): *Glutathione Conjugation: Mechanisms and Biological Significance*. Academic Press, New York, 73–135

45 Morgenstern R, Lundqvist G, Andersson G, Balk L, DePierre JW (1984) The distribution of microsomal glutathione transferase among different organelles, different organs, and different organisms. *Biochem Pharmacol* 33: 3609–3614

46 Caldwell J, Jakoby WB (1983) *Biological Basis of Detoxification*. Academic Press, New York

47 Anders MW (2004) Glutathione-dependent bioactivation of haloalkanes and haloalkenes. *Drug Metab Rev* 36: 583–594

48 Bessems JG, Vermeulen NP (2001) Paracetamol (acetaminophen)-induced toxicity: Molecular and biochemical mechanisms, analogues and protective approaches. *Crit Rev Toxicol* 31: 55–138

49 Genter StClair MB, Amarnath V, Moody MA, Anthony DC, Anderson CW, Graham DG (1988) Pyrrole oxidation and protein cross-linking as necessary steps in the development of γ-diketone neuropathy. *Chem Res Toxicol* 1: 179–185

50 Liebler DC, Guengerich FP (1983) Olefin oxidation by cytochrome P450: Evidence for group migration in catalytic intermediates formed with vinylidene chloride and *trans*-1-phenyl-1-butene. *Biochemistry* 22: 5482–5489

51 Pohl LR, Branchflower RV, Highet RJ, Martin JL, Nunn DS, Monks TJ, George JW, Hinson JA (1981) The formation of diglutathionyl dithiocarbonate as a metabolite of chloroform, bromotrichloromethane, and carbon tetrachloride. *Drug Metab Dispos* 9: 334–339

52 Anders MW, Dekant W (1994) *Conjugation-Dependent Carcinogenicity and Toxicity of Foreign Compounds*. Academic Press, New York

53 Anders MW, Dekant W, Vamvakas S (1992) Glutathione-dependent toxicity. *Xenobiotica* 22: 1135–1145

54 Dekant W, Vamvakas S, Anders MW (1992) The kidney as a target organ for xenobiotics bioactivated by glutathione conjugation. In: W Dekant, HG Neumann (eds): *Tissue Specific Toxicity: Biochemical Mechanisms*. Academic Press, New York, 163–194

55 Monks TJ, Anders MW, Dekant W, Stevens JL, Lau SS, van Bladeren PJ (1990) Glutathione conjugate mediated toxicities. *Toxicol Appl Pharmacol* 106: 1–19

56 Aust SD, Chignell CF, Bray TM, Kalyanaraman B, Mason RP (1993) Free radicals in toxicology. *Toxicol Appl Pharmacol* 120: 168–178

57 Goldstein BD, Czerniecki B, Witz G (1989) The role of free radicals in tumor promotion. *Environ Health Perspect* 81: 55–57

58 Goldstein BD, Witz G. (1990) Free radicals and carcinogenesis. *Free Radic Res Commun* 11: 3–10

59 Kehrer JP, Mossman BT, Sevanian A, Trush MA, Smith MT (1988) Free radical mechanisms in chemical pathogenesis. *Toxicol Appl Pharmacol* 95: 349–362

60 Cerutti PA, Trump BF (1991) Inflammation and oxidative stress in carcinogenesis. *Cancer Cell* 3: 1–7

61 Chacon E, Acosta D (1991) Mitochondrial regulation of superoxide by Ca^{2+}: An alternate mechanism for the cardiotoxicity of doxorubicin. *Toxicol Appl Pharmacol* 107: 117–128

62 Burdon RH, Gill V, Rice-Evans C (1990) Oxidative stress and tumour cell proliferation. *Free Radic Res Commun* 11: 65–76

63 Pero RW, Roush GC, Markowitz MM, Miller DG (1990) Oxidative stress, DNA repair, and cancer susceptibility. *Cancer Detect Prev* 14: 551–561

64 Jakoby WB, Ziegler DM (1990) The enzymes of detoxication. *J Biol Chem* 265: 20715–20719

65 Sies H (1989) Zur Biochemie der Thiolgruppe: Bedeutung des Glutathions. *Naturwissenschaften* 76: 57–64

66 Ziegler DM (1980) Microsomal flavin-containing monooxygenase: Oxygenation of nucleophilic nitrogen and sulfur compounds. In: WB Jacoby (ed.): *Enzymic Basis of Detoxification*, vol 1. Academic Press, New York, 201–227

67 Boyland E, Chasseaud LF (1969) Role of glutathione and glutathione *S*-transferases in mercapturic acid biosynthesis. *Adv Enzymol* 32: 173–177

68 James RC, Harbison RD. (1982) Hepatic glutathione and hepatotoxicity. *Biochem. Pharmacol.* 31: 1829–1835

69 Chung FL, Pan J, Choudhury S, Roy R, Hu W, Tang MS (2003) Formation of *trans*-4-hydroxy-2-nonenal- and other enal-derived cyclic DNA adducts from omega-3 and omega-6 polyunsaturated

fatty acids and their roles in DNA repair and human p53 gene mutation. *Mutat Res* 531: 25–36
70 West JD, Marnett LJ (2006) Endogenous reactive intermediates as modulators of cell signaling and cell death. *Chem Res Toxicol* 19: 173–194
71 McLellan LI, Wolf CR, Hayes JD (1989) Human microsomal glutathione *S*-transferase. Its involvement in the conjugation of hexachlorobuta-1,3-diene with glutathione. *Biochem J* 258: 87–93
72 Hinson JA, Roberts DW (1992) Role of covalent and noncovalent interactions in cell toxicity: Effects on proteins. *Annu Rev Pharmacol Toxicol* 32: 471–510
73 Nelson SD, Pearson PG (1990) Covalent and noncovalent interactions in acute lethal cell injury caused by chemicals. *Annu Rev Pharmacol Toxicol* 30: 169–195

Genotoxicity: damage to DNA and its consequences

David H. Phillips and Volker M. Arlt

Institute of Cancer Research, Sutton, Surrey, United Kingdom

Abstract. A genotoxin is a chemical or agent that can cause DNA or chromosomal damage. Such damage in a germ cell has the potential to cause a heritable altered trait (germline mutation). DNA damage in a somatic cell may result in a somatic mutation, which may lead to malignant transformation (cancer). Many *in vitro* and *in vivo* tests for genotoxicity have been developed that, with a range of endpoints, detect DNA damage or its biological consequences in prokaryotic (e.g. bacterial) or eukaryotic (e.g. mammalian, avian or yeast) cells. These assays are used to evaluate the safety of environmental chemicals and consumer products and to explore the mechanism of action of known or suspected carcinogens. Many chemical carcinogens/mutagens undergo metabolic activation to reactive species that bind covalently to DNA, and the DNA adducts thus formed can be detected in cells and in human tissues by a variety of sensitive techniques. The detection and characterisation of DNA adducts in human tissues provides clues to the aetiology of human cancer. Characterisation of gene mutations in human tumours, in common with the known mutagenic profiles of genotoxins in experimental systems, may provide further insight into the role of environmental mutagens in human cancer.

Introduction

Cancer is a genetic disease arising from a series of somatic mutations. Mutations in DNA may arise spontaneously, or as a result of chemical action by agents of either endogenous or exogenous origin. Genetic toxicology is the study of agents that can damage the DNA and chromosomes of cells. In eukaryotic organisms, genetic damage in somatic cells may lead to malignancy. In germ cells it may adversely affect reproduction or provoke heritable mutations. Consequently, investigating the genotoxicity of a compound is often carried out in the context of seeking to understand its mechanism of carcinogenicity. This has become an essential component of the process of risk assessment for human exposure to a known animal carcinogen. Investigating genotoxicity is also important in assessing whether or not a new compound is a carcinogen and/or mutagen, and this process contributes to the more fundamental process of hazard identification.

Understanding mechanisms of carcinogenesis often relies on analysis of the molecular and cellular effects of carcinogens in laboratory experiments. This is a necessary simplification of a complex process and while such approaches often provide critical evidence for mechanisms, it must be recognised that laboratory models rarely cover all the possible facets of the process, and there are many instances in which the classification of a carcinogen is not a straight-forward matter.

A genotoxic carcinogen typically induces tumours in multiple organs of rodents, may be carcinogenic to more than one species and to both males and females. In addition, there is often evidence of a dose-response relationship for tumour induction, of the type that does not suggest evidence of a threshold. In contrast, non-genotoxic carcinogens are more likely to be characterised by tumour induction in a single species and/or in a single tissue and, commonly, in one sex only, often at low incidence and only at high dose with associated evidence of toxicity. Multi-species, multi-organ carcinogens are more likely to be human carcinogens and, indeed, most agents classified as human carcinogens by the International Agency for Research on Cancer (IARC) are genotoxic. It is thus a general principle that chemicals that are carcinogenic in animals by a genotoxic mechanism pose a greater potential risk to humans than non-genotoxic carcinogens, and the default assessment of genotoxins is that human exposure at any level poses a risk. For such agents exposure should be 'as low as reasonably achievable' (ALARA). For non-genotoxic carcinogens it may be possible to define a threshold, i.e. a level of exposure below which the agent does not present a carcinogenic risk to humans.

In some animal models carcinogenicity can be divided into an initiation phase, involving a single treatment with a genotoxic agent, followed by promotion, involving repeated treatments with a non-genotoxic agent. A common feature is that initiation is considered irreversible, such that the promotion phase can be delayed significantly yet still ultimately result in tumour formation. Furthermore, the tumour response may be absent or greatly reduced if the initiator is applied after the promoter, or if treatment is with either initiator or promoter alone. Although this model of initiation and promotion has served well as an experimental system for defining genotoxicity and for some studies of mechanisms of tumour formation, it appears to present an over-simplification of the process, particularly when considering the mechanism(s) of carcinogenesis in humans. Epidemiological evidence on the age distribution of many common cancers suggested that cancer induction is a multi-stage process, involving as many as five to seven distinct events [1]. Although this conclusion was reached more than 50 years ago, before anything was known about the changes associated with malignancy were identified or understood, genetic analysis of tumours over the last 10 or more years has borne out this. Current understanding is that the accumulation of a number of mutations (five to seven is a reasonable estimate) in critical genes in progenitor cells leads to the manifestation of the malignant phenotype.

Phenotypically, malignancy is characterised by six essential alterations in cell physiology: (i) self sufficiency in growth signals, (ii) insensitivity to growth-inhibitory signals, (iii) evasion of programmed cell death (apoptosis), (iv) limitless replicative potential, (v) angiogenesis and (vi) tissue invasion and metastasis [2]. Although there is not yet an exact match between the genotypic and phenotypic characteristics of tumours, it is logical to conclude that carcinogenesis is driven by the accumulation of critical mutations in cells that converts them and their progeny from normal cells to fully malignant ones. The identification of mutations in specific genes in human cancers has demonstrated that they are associated with both early and late stages of tumour progression [3]. Thus, while it is thought that DNA damage (often involving binding to DNA by carcinogens) occurs in the early, initiating stages of carcinogenesis, now that it is apparent that gene mutation is associated with several stages of carcinogenesis, it is probable that genotoxic events are also a feature of later stages of the multistage process.

Short-term tests for genotoxicity

In the context of short-term tests for mutagenicity and genotoxicity, a genotoxic agent is one that induces point mutations, deletions, insertions, gene amplifications, chromosomal rearrangements or numerical chromosomal changes (aneuploidy). The tests are therefore designed to detect one or more type of genetic alteration. Since such biological properties result directly or indirectly from DNA damage, other assays have been developed to identify this damage directly. No single assay, no matter how extensive the protocol, can detect all genotoxic chemicals [4]. Therefore, it is generally accepted that a number of tests must be conducted to evaluate whether a chemical is genotoxic or not, and often a weight-of-evidence approach must be taken to evaluate the results.

A number of organisations and advisory bodies have produced guidelines in the last 10 years. These include the International Programme on Chemical Safety (IPCS), the International Conference on Harmonisation of Technical Requirements for Registration of Pharmaceuticals for Human Use (ICH), and the International Workshop on Genotoxicity Testing (IWGT). These and other guidelines have been reviewed and compared by Cimino [5].

For a test to be useful it should be both sensitive and specific (Tab. 1). Several guidelines recommend that a test battery for genotoxicity should include: (1) a test for gene mutation in bacteria; (2) an *in vitro* test that includes cytogenetic evaluation of chromosomal damage in mammalian cells or an *in vitro* mammalian cell mutagenicity test such as the mouse lymphoma thymidine kinase (*Tk*) assay; and (3) an *in vivo* test for chromosomal damage using rodent haematopoietic cells. Some guidelines advise that negative results in both of the first two assays may in some cases (e.g. a low-volume chemical where the potential for human exposure is minimal) remove the necessity of conducting the third, *in vivo*, test. More recently, however, the Seventh Amendment to the European Union Cosmetics Directive that will ban the marketing of cosmetics and personal care products containing ingredients tested in animals has highlighted the need for better *in vitro* tests for toxicity and genotoxicity [6].

Table 1. Performance terms for short-term tests for genotoxicity.

Test outcome	Carcinogen		
	Yes	No	Total
Positive	a	b	a + b
Negative	c	d	c + d
Total	a + c	c + d	N = a + b + c + d

Term	Definition	Description
Sensitivity	a/(a + c)	number of carcinogens found positive / number of carcinogens tested
Specificity	d/(b + d)	number of non-carcinogens found negative / number of non-carcinogens tested
Positive predictivity	a/(a + b)	number of carcinogens found positive / number of positive results obtained
Negative predictivity	d/(c + d)	number of non-carcinogens found negative / number of negative results obtained
Accuracy	(a + d)/N	number of correct test results / number of chemicals tested

Adapted from Anon [102] and Shelby and Purchase [103].

Bacterial mutagenicity testing

The most widely used bacterial assay to detect chemically induced gene mutations is the Ames Salmonella assay developed by Bruce Ames [7, 8]. *Salmonella typhimurium* strains that contain defined mutations in the histidine locus form the basis of this 'reverse' mutation assay. In the assay, bacteria are incubated with a range of concentrations of the test compound to induce a second mutation that directly reverses or suppresses the original mutations and, thus, restores the biological function to the non-functional histidine gene. Strains of *S. typhimurium* used in the Ames assay are auxotrophic for histidine and revertants are selected by their ability to grow in the absence of this amino acid. Two of the most commonly used *S. typhimurium* strains are TA98 and TA100. *S. typhimurium* TA98 has a *hisD3052* mutation detecting frame-shift reversion events, whereas *S. typhimurium* TA100 has a *hisG46* mutation detecting base-pair substitution events [9]. The great strength of the assay is the ability to identify and score a small number of mutants from a relatively large population of unmutated cells. However, as bacteria lack many endogenous metabolic pathways that are required for the bioactivation of the test chemicals, extracts of mammalian liver (usually rat) are incorporated as an exogenous activation system [9]. Fractionated tissue homogenate such as the 9000 *g* supernatant (S9 fraction), prepared from the livers of rats pretreated with Aroclor 1254, provides a rich source of mixed-function monooxygenases required for bioactivation.

The Ames assay can either be used to assess the mutagenic potency of a chemical as part of the toxicological screening, or it can form part of a detailed mechanistic examination of the chemical's mutagenic potential. In the current ICH and OECD Guidelines, the use of five tester strains is recommended: TA98, TA100 and TA1535; TA1537 or TA97 or TA97a; TA100 (or alternatively one of several *Escherichia coli* WP2 strains). To make the assay more sensitive, these strains contain an *rfa* mutation resulting in defective lipopolysaccharide and increased permeability to large test molecules, or a deletion in the *uvrB* gene making the strains deficient in nucleotide excision repair. Some strains (e.g. *S. typhimurium* TA98 and TA100) include additionally a plasmid (pKM101) containing *umuDC* genes encoding for a translesion-synthesis DNA polymerase that elicits error-prone repair [10]. Using genetically engineered *S. typhimurium* strains that either overexpress or lack enzymes required for the bioactivation of different carcinogens can provide useful information on their metabolism [11, 12]. Moreover, 'humanised' *S. typhimurium* strains with defined human enzymes have been developed to identify which human enzymes are involved in bioactivation and to improve the relevance of Ames Salmonella assay for detecting agents hazardous to humans [13].

Mammalian mutagenicity testing

The mouse lymphoma assay (MLA) is the most widely used mammalian gene mutation assay [14]. It detects various mutation events involving the *Tk* gene in L5178Y/*Tk*$^{+/-}$ 3.6.2C mouse lymphoma cells [15, 16]. The gene coding for thymidine kinase is on mouse chromosome 11 and allows the cell to salvage nucleotides from the culture medium for reuse in metabolism but is not essential for cell survival. Since eukaryotic cells are diploid, heterozygous cells are used where two copies of the *Tk* gene are present but one copy has been inactivated. Otherwise, many mutations arising in mammalian cells cannot be selected directly, since the second copy of the gene would complement the first. Mutants in the MLA are detected by plating cells into medium containing trifluorothymidine (TFT), a thymidine analogue [17]. Thus, toxic TFT placed in the medium will be transported into normal *Tk*$^{+/-}$ (non-mutated) cells that consequently die, while *Tk*$^{-/-}$ mutants will be resistant to the toxic TFT and survive, and subsequently form clones that can be counted. The L5178Y system is the recommended *in vitro* mammalian cell mutation assay because it detects a wide range of genetic alterations, including both mutations and chromosomal damage [18].

Transgenic rodent mutation assays

Transgenic rodent mutation assays were first developed in the 1990s [19, 20], Muta^{TM} Mouse and Big Blue® Mouse and Rat being the assays most widely used [21, 22]. The Muta[TM] Mouse carries a recombinant λ-bacteriophage vector containing the entire *Escherichia coli lacZ* (β-galactosidase) [21]. Mutations occurring in the *lacZ* gene are measured by positive selection of *lacZ⁻* mutants on phenylgalactosidase (P-gal)-containing medium using an indicator bacteria strain (*E. coli lacZ⁻ galE⁻*). In the presence of P-gal, only *lacZ⁻* bacteria (mutants) will grow and produce plaques, whereas *lacZ⁺* (i.e. non-mutants) produce the enzyme β-galactosidase converting P-gal into galactose and subsequently into the toxic intermediate uridine diphosphate (EDP)-galactose, which accumulates in *E. coli galE⁻* and kills the cells. In the Big Blue® system, the reporter gene is *lacI* contained in a λ-bacteriophage vector [22]. Mutants occurring in the *lacI* gene are selected on 5-bromo-4-chloro-3-indolyl-β-D-galactopyranoside (X-gal)-containing medium using an *E. coli lacI⁻* indicator strain. Wild-type *lacI* (non-mutants) will repress the *lac* operon encoding for β-galactosidase forming clear plaques, whereas *lacI⁻* mutants will produce β-galactosidase that uses X-gal as substrate producing blue-coloured plaques that can be counted. Since the reporter genes that serve as targets for detecting mutations are incorporated into the chromosomes of the transgenic mice or rats, somatic mutations can be measured within any tissue of the exposed animal, and more importantly enables mutation induction and measurement in the actual target tissue for tumour development. General guidance on recommended protocols has been published [23, 24].

The *lacI/lacZ* models are well suited to detect point mutations but unsuited for the detection of large deletion mutations induced by clastogens [19]. However, the coding size of, for instance, the *lacZ* gene is about 3 kilo base (kb) pairs, which is not compact enough to routinely identify mutations by DNA sequencing. Thus, a new reporter gene, the *cII* gene of the λ phage, has been used and is applicable to both Muta[TM] Mouse and Big Blue® Mouse systems [25]. The *cII* gene is susceptible to mutagenesis, just as is *lacI* or *lacZ*, but has the advantage over them that the coding region is only 300 base pairs and can easily be sequenced in a single run. An alternative transgenic mouse model, *gpt* delta, is reported to be suitable for the detection of large deletions [26].

Test for chromosome damage

Structural chromosome changes that can be detected by conventional *in vitro* cytogenetics are chromosome aberrations (CA), micronuclei (MN) and sister chromatid exchanges (SCE).

Structural CA are changes in normal chromosome structure or number that can occur in cells after chemical exposure or radiation. They result from direct DNA breakage, replication on a damaged DNA template, inhibition of DNA synthesis, and other mechanisms (e.g. inhibition of topoisomerase II) [27]. Cells commonly used to measure structural CA are human peripheral blood lymphocytes or established lymphoblastoid cell lines [28]. Peripheral lymphocytes are popular cells for *in vitro* studies because they are human primary cells, have a low spontaneous rate of chromosomal damage, and can be easily cultured with a stable karyotype ($2n = 46$). Structural CA are generally scored in metaphase-arrested cells after Giemsa staining [29]. For over 30 years structural CA in human peripheral blood lymphocytes have been used in occupational and environmental settings as a biomarker of exposure and a marker of cancer risk [30, 31].

The *in vitro* micronucleus test allows the detection of both structural (clastogenic) and numerical (aneugenic) chromosome changes using interphase cells [32, 33]. Thus, MN represents a measure of both chromosome breakage and chromosome loss, and an increased frequency of micronucleated cells is a biomarker of genotoxic effects. MN formed by clastogenicity induction can be distinguished from those produced by aneugenic activity by the absence of centrometric DNA or kinetochore proteins in the MN using centrometric probes or kinetochore antibodies [34]. The standard *in vitro* micronucleus test is usually performed in lymphocytes [35], the cytokinesis-block micronucleus assay being the most widely used method. This assay is specifically restricted to once-divided cells and these cells are recognised by their binucleated appearance after inhibition of cytokinesis by cytochalasin-B [36]. Restricting the scoring of MN to binucleate cells prevents confounding effects that can be a major variable in the assay. The use of MN as a measure of chromosomal damage has become a standard assay both in genotoxicity testing (although an OECD guideline protocol has yet to be adopted) and human biomonitoring studies [37, 38].

The rodent micronucleus test is a widely used and extensively validated assay to assess chromosome damage *in vivo* and has been incorporated into standard rodent toxicology screening assays [39, 40]. For the analysis, immature erythrocytes (i.e. polychromatic erythrocytes; reticulocytes) in either bone marrow or peripheral blood have been found equally acceptable when the rodents have been exposed to the test compound by an appropriate route. A detailed description of the study design and experimental procedure has been published [40, 41].

Comet assay

The comet assay (or single-cell gel electrophoresis assay) is a simple and sensitive method for measuring alkali labile sites and strand breaks in the DNA of mammalian cells [42]. In the assay, a small number of cells suspended in agar are lysed under alkaline conditions, subjected to electrophoresis, neutralised and stained with a fluorescent DNA dye, such as propidium iodide [43, 44]. Cells with increased DNA damage display increased migration of the chromosomal DNA from the nucleus resembling the shape of a 'comet'. Using image analysis software parameters such as percentage of DNA in the tail (percent migrated DNA), tail length, and tail moment (fraction of migrated DNA multiplied by some measure of tail length) can be determined as a measure of DNA damage. As an *in vivo* genotoxicity assay, it has the advantage that it can be applied to a single-cell suspension of material from any animal tissue, allowing consideration of potential target tissues and also taking account of possible inaccessibility to exposure of tissues (e.g. bone marrow) required for other *in vivo* assays.

Under alkaline conditions, the assay detects overt strand breaks, which can include single and double strand breaks, as well as transient repair-induced breaks. It also detects lesions that are alkali-labile, which includes AP (apurinic/apyrimidinic) sites (see below). More precise information on the nature of the lesions detected can be obtained by the inclusion of lesion-specific endonucleases in the assay protocol, which convert some types of DNA damage to strand breaks [45]. Formamidopyrimidine-DNA-glycosylase (FPG) has been used to detect oxidised DNA damage, principally 7,8-dihydro-8-oxo-2'-deoxyguanosine (8-oxo-dGuo), with high sensitivity [46]. More recently, it has been reported that FPG also detects some types of alkylation damage, and that the human homologue of FPG, hOGG1, is a more specific endonuclease for oxidation products [47]. Another enzyme, endonuclease III, converts oxidised pyrimidines to strand breaks [45].

The comet assay is used in many studies to assess DNA damage and repair and has widespread application in genotoxicity testing *in vitro* and *in vivo* [48]. Since virtually any cell population or single-cell suspension from any tissue type can be used for analysis, the assay is widely used in environmental biomonitoring and human population monitoring [45, 49, 50].

Correlations of mutagenicity and carcinogenicity

Clearly a major purpose of conducting the foregoing assays is to be able to predict whether or not a chemical is a carcinogen without conducting a costly and time-consuming animal bioassay. The reliability of such tests, both in terms of the specificity and selectivity, is a matter of ongoing debate, subject to the continual accumulation of new data. Where a compound known to be carcinogenic is not detected as a mutagen or genotoxin in such assays, or where a mutagen (or genotoxin) has been found to be non-carcinogenic, some sort of explanation needs to be sought. A carcinogen may be non-mutagenic or non-genotoxic because its mechanism of action does not involve DNA damage (i.e. it is a non-genotoxic carcinogen that may, for example, act as a tumour promoter or by inhibiting DNA methylation). On the other hand it may be that mutagenic activity of the genotoxic carcinogen is limited to the chromosomal level, or that unusual metabolic activation for activity is required and that this is not achieved in the *in vitro* test. The addition of external enzymatic activation systems (e.g. rat liver S9) may not be adequate for some compounds, particularly where phase II enzymes are required or where the half-life of the reactive species may be short or its cell permeability limited. Strategies to overcome these shortcomings include the use of human liver S9 [51] and the engineering of bacteria or mammalian cells to express human xenobiotic metabolising genes

[13, 52], but these approaches are, as yet, research tools and not part of the regulatory armoury.

Non-carcinogens may test positive as mutagens if the activity in the test system is due to a metabolic pathway not exhibited *in vivo* or due to the absence of a competing detoxification pathway or lack of DNA repair; or it may be that mutagenicity is limited to a particular type of genetic damage (e.g. aneuploidy). Alternatively it may be that mutagenicity is insufficient for carcinogenicity, which may require accompanying cell proliferation in the target tissue; or it may simply be that the *in vivo* rodent models used for carcinogenicity testing may not be sufficiently sensitive for some weak mutagens.

As stated earlier, it is widely recognised that no single assay can detect all genotoxic carcinogens, hence the evolution of the standard battery of tests. But even then, there are some genotoxins that go undetected by all these assays [53].

DNA adducts formed by chemical carcinogens

Genotoxic chemical carcinogens are either directly or indirectly DNA reactive. Most chemical carcinogens are not chemically reactive as such, but undergo metabolic activation in mammalian cells to reactive intermediates that react with DNA (Tab. 2), hence the requirement for inclusion of metabolising enzymes in many *in vitro* genotoxicity assay systems. Carcinogen-induced DNA damage can take several forms. It can result in breaks in the sugar-phosphate backbone of the molecule, either in one of the two strands of the double helix (forming single-strand breaks), or in both (causing double-strand breaks). Covalent binding of the carcinogen results in the formation of a chemically altered base (or, occasionally, phosphate group) in DNA that is termed an 'adduct'. Formation of adducts at some positions of the DNA bases (for example at the N7 position of guanine) can render the base-sugar bond unstable and lead to loss of the adducted base (depurination or depyrimidination). The resulting modification to DNA is the formation of an AP site. Some carcinogens are bifunctional and can give rise to both monoadducts and crosslinks in DNA, the latter being either intrastrand or interstrand crosslinks. Many cancer chemotherapeutic agents have this property, and it is widely held that interstrand crosslinks are cytotoxic (accounting for the therapeutic properties of the drugs), while the monoadducts and intrastrand crosslinks are potentially mutagenic and carcinogenic.

DNA adducts can also originate from endogenous processes, including normal metabolism, oxidative stress and chronic inflammation [54]. The most abundant oxidation lesion in DNA is 8-oxo-dGuo, which can be formed by free radical attack on DNA or through normal aerobic metabolism. It is suspected that some genotoxic carcinogens that do not appear to directly modify DNA instead damage it through inducing oxidative stress leading to increased oxidative damage to DNA.

Table 2. Some representative carcinogens, their environmental sources, their active metabolites, sites of modification of DNA, and major type of induced mutation.

Carcinogen	Environmental source	Major active metabolite	Sites of modification	Major type of mutation
Benzo[a]pyrene (B[a]P)	Tobacco smoking Combustion processes	B[a]P-7,8-dihydrodiol 9,10-epoxide (BPDE)	N^2-Guanine N^6-Adenine	GC → TA
Aflatoxin B$_1$ (AFB$_1$)	Mycotoxin	AFB$_1$ 8,9-epoxide	N7-Guanine	GC → TA
2-Amino-1-methyl-6-phenylimidazo-[4,5-b]pyridine (PhIP)	Food processing	N-Acetoxy-PhIP	C8-Guanine	GC → TA
Tamoxifen	Anticancer drug	α-Hydroxytamoxifen sulphate	N^2-Guanine N^6-Adenine	GC → TA
3-Nitrobenzanthrone	Diesel exhaust Urban air pollution	N-Hydroxy-3-aminobenzanthrone	N^2-Guanine C8-Guanine N^6-Adenine	GC → TA
Aristolochic acid I	Aristolochia species	N-Hydroxyaristolactam I	N^2-Guanine N^6-Adenine	AT → TA

In experimental studies where multiple doses of carcinogens have been administered to animals and both tumour outcome and DNA adduct levels have been determined, in general a linear relationship has been found between dose and both these parameters at low dose, although deviations from linearity may be observed at higher doses [55]. Nevertheless the low-dose effects are more relevant to human exposure scenarios than the high-dose effects, which may be explained in part by the influence of toxicity.

Methods for adduct detection

A number of sensitive methods have been developed for the detection and characterisation of DNA adducts (Tab. 3) [55, 56]. For an assay to be applicable to human exposure, it must (i) be sensitive enough to detect low levels of adducts; (ii) require only microgram quantities of DNA; (iii) provide results quantitatively related to the exposure; (iv) be applicable to unknown adducts that may be formed from complex mixtures; and (v) be able to resolve, quantitate and identify adducts.

Most of the early work on adducts required the use of radiolabelled compounds (labelled either with ^{3}H or ^{14}C) at a position of the molecule where the isotope is not lost during metabolic activation and binding to DNA [56]. The DNA binding is then measured by the detection of radioactivity in DNA isolated from exposed animals or cells in culture achieving sensitivities of detection of 1 adduct in 10^8 nucleotides with ^{3}H labelling, although ^{14}C labelling is less sensitive due to the lower specific activity of ^{14}C-labelled compounds

Table 3. DNA adduct detection methods applicable to human biomonitoring and their limits of detection.

Method	Variations	Amount of DNA required	Approximate detection limits
^{32}P-postlabelling	Nuclease P$_1$ digestion, butanol extraction, HPLC	1–10 µg	1 adduct per 10^9–10^{10} nucleotides
Immunoassay	ELISA, DELFIA, CIA, IHC	20 µg	1.5 adducts per 10^9 nucleotides
Fluorescence	HPLC fluorescence, SFS	100–1000 µg	1 adduct per 10^9 nucleotides
Mass spectrometry		up to 100 µg	1 adduct per 10^8 nucleotides
AMS [a]		up to 100 µg	1 adduct per 10^{11}–10^{12} nucleotides

HPLC: high-performance liquid chromatography; ELISA: enzyme-linked immunosorbent assay; DELFIA: dissociation-enhanced lanthanide fluoroimmunoassay; CIA: chemoluminescence immunoassay; IHC: immunohistochemistry; SFS: synchronous fluorescence spectroscopy.
[a]Accelerator mass spectrometry; requires use of radiolabelled compounds; reproduced from [104].

compared with ^{3}H-labelled ones (a consequence of the much longer half-life of ^{14}C compared with that of ^{3}H) [56]. However, due to the highly radioactive test compounds it was not possible to use this approach in studies involving humans.

In 1981, the ^{32}P-postlabelling technique was developed [57–59]. The method comprises a four-step process that involves (i) DNA digestion, (ii) a procedure that isolates or selects the adducts for preferential labelling, (iii) the introduction of a radiolabel into the DNA adducts using enzymatic [^{32}P]phosphorylation of the nucleotide adduct and (iv) separation of the ^{32}P-labelled adducts using thin layer (TLC) or high-performance liquid chromatography (HPLC) [60, 61]. The assay requires only small (1–10 μg) quantities of DNA and is capable of detecting adducts at frequencies as low as 1 adduct in 10^{10} nucleotides, making it widely applicable in human biomonitoring [62]. It can be used for a wide variety of classes of compounds [including polycyclic aromatic hydrocarbons (PAHs), aromatic and heterocyclic amines, unsaturated aldehydes, simple alkylating agents, reactive oxygen species], for the effects of ultraviolet (UV) light and for the detection of adducts formed by complex mixtures [55, 63]. A limitation of the method is that it does not provide structural information; identification of adducts is reliant on co-chromatography using characterised synthetic standards [60]. A different approach using a similar experimental protocol is the chemical linkage of a fluorescent dye (e.g. BODIPY) to the DNA adducts, which can subsequently be separated by capillary electrophoresis and detected by laser-induced fluorescence (CE-LIF) [64]. Although this methodology is not yet sensitive enough to be applied to human samples (detection limit 1 adduct per 10^7 nucleotides) [64, 65], it has proved to be a suitable technique to determine global DNA methylation levels [66].

Mass spectrometry (MS) coupled with liquid chromatography-electrospray ionisation spectrometry (ESI-LC-MS) is becoming increasingly used for the detection of DNA adducts and provides unequivocal identification of the nature of an adduct [56, 67, 68]. The sensitivities achieved are normally lower than with ^{32}P-postlabelling but, with the detection of 1 adduct per 10^8–10^9 nucleotides using 50–100 μg DNA, they are sufficient to give useful data on human environmental or dietary exposures [68]. Accurate quantitation of DNA adduct levels is achieved by the use of a stable isotope internal standard (e.g. labelled either with ^{15}N or ^{13}C). Although mostly applied to the detection of specific well-characterised lesions, more recent techniques allow for the simultaneous detection of multiple adducts, and this 'adductome' approach has potential for the detection and characterisation of DNA adducts in human tissues [69]. Accelerator mass spectrometry (AMS), which measures isotope ratios, represents the most sensitive analytical method so far for detecting DNA adducts, with limits of adduct detection as low as 1 adduct in 10^{11} nucleotides [67, 70]. The main limitation of the technique is that it depends on the presence of an isotope such as ^{14}C or ^{3}H in the compound of interest. However, because of the high sensitivity of AMS, it has been possible to obtain ethical approval to give minute amounts of a radioactive carcinogen, for example, 2-amino-1-methyl-6-phenylimidazo[4,5-*b*]pyridine (PhIP) or tamoxifen, to human individuals prior to surgery and to detect DNA adducts in the excised tissue (Tab. 2) [71, 72].

Other physicochemical methods for the detection of DNA adducts are based on the fact that some adducts are highly fluorescent, enabling their detection by fluorescence spectroscopy [56, 73]. Combining the fluorescent characteristics (specific excitation and emission wavelengths) with HPLC separation techniques make it even possible to detect stereoisomers. Adducts with fluorescent properties include those formed by PAHs and aflatoxins, cyclic (etheno) adducts, and some methylated adducts [73]. Other adducts, notably 8-oxo-dGuo, are readily detected by HPLC coupled with electrochemical detection (ECD) [74]. All these methods can provide a sensitivity of detection of around 1 adduct per 10^8 nucleotides, while requiring relative large quantities (100–1000 µg) of sample DNA. Overall, the major limitations of these methods are the required spectral (e.g. intrinsic fluorescence) and physicochemical properties of the adducts.

Immunoassays have also been used for the detection of DNA adducts in human and experimental samples [75, 76]. Antibodies have been raised against a variety of carcinogen-modified DNAs, including those containing adducts of PAHs, aromatic amines, methylating agents, tamoxifen, or modified by UV radiation, and oxidative damage. Immunoassays are highly sensitive, but have generally been less sensitive than ^{32}P-postlabelling and usually require more DNA for analysis, although some recent developments have both increased sensitivity and reduced the amount of DNA required, improving the sensitivity to a level closer to that of ^{32}P-postlabelling [77]. When combined with histochemistry, cell-specific localisation of adducts in paraffin-embedded tissue is possible [78, 79]. However, antibodies can show cross-reactivity with adducts formed by the same class of compounds, which can obscure both the nature of the adducts and the levels at which they are present.

Biological significance of DNA adducts

While it is evident that DNA damage and binding by carcinogens occurs in the early, initiating stages of carcinogenesis, it has become increasingly clear that damage to DNA is also a feature of later stages of the multistage process, now that it is known that mutations in some genes are associated with later stages of progression of some types of tumour. It is also evident that the formation of DNA adducts is by no means a sufficient event for carcinogenesis, as DNA adducts are frequently detectable in both target and non-target tissues. Nevertheless, inhibition of DNA adduct formation will decrease the incidence of tumours formed subsequently, and increasing the adduct levels generally leads to a higher tumour yield. Other evidence that strongly links DNA adduct formation to tumour initiation is the demonstration that XPA knockout mice, which are deficient in nucleotide excision repair, are highly sensitive to tumour induction by carcinogens that form stable adducts that would be removed from DNA in normal mice by this repair mechanism [80].

That chemical modification of DNA can result in the same alterations as observed in mutated genes in tumours was observed with the H-*ras*-1 proto-oncogene transfected into NIH3T3 cells [81]. Prior modification (by reaction with benzo[*a*]pyrene diol-epoxide, the reactive metabolite of benzo[*a*]pyrene; Fig. 1, Tab. 2) of the plasmid containing the gene resulted in mutations occurring in the DNA after transfection and replication of the host cells, manifested as the appearance of transformed foci. Mutations that activate proto-oncogenes such as *ras* genes occur in a few codons in the gene, so correlations between the sites of mutations in such experiments may not be very informative. In contrast, for tumour suppressor genes there may be many possible sites of DNA damage and mutation that can lead to altered function of the gene product that contributes to malignant transformation. Such a gene is *TP53*, which has been found to be mutated in ~50% of human tumours. Correlations can be usefully sought between the mutation spectra observed in different human tumours to provide clues to the nature of the initiating agent(s) [82]. This approach has led to evidence for the involvement of the mycotoxin aflatoxin B_1 in the initiation of liver cancer in regions of high incidence in China, where a G→T transversion in codon 249 of *TP53* is a common mutation in the disease (Tab. 2) [83]. In lung cancer, codons 157, 248 and 273 of the gene are frequently mutated; G→T transversions are much more common in cases of lung cancer among smokers than among non-smokers, and these types of mutation are characteristic of bulky carcinogens, such as the PAHs, which are present in tobacco smoke. When the sites of DNA adduct formation by benzo[*a*]pyrene diol-

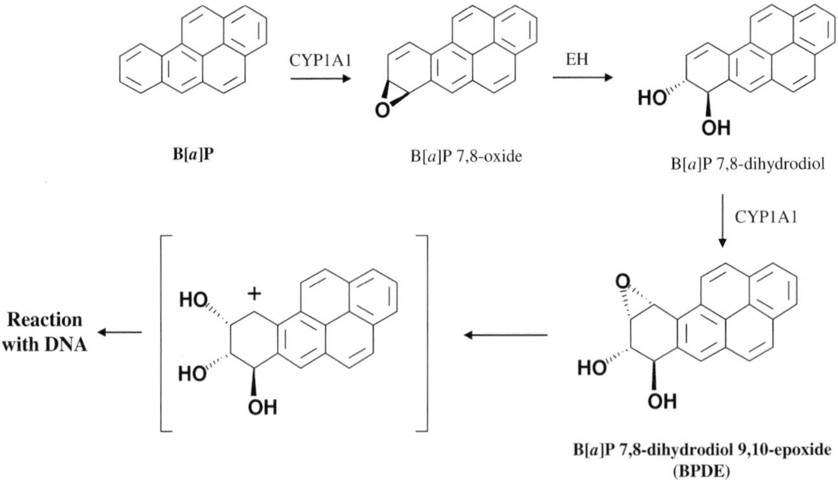

Figure 1. Major pathway of metabolic activation and DNA adduct formation of benzo[*a*]pyrene (B[*a*]P) (see text for details). CYP1A1, cytochrome P450 1A1; EH, epoxide hydrolase.

epoxide in the *TP53* gene in HeLa cells and bronchial epithelial cells were determined, it was found that codons 157, 248 and 273 were preferentially modified, correlating with the frequently mutated sites in lung tumours of smokers [84].

UV causes DNA damage chiefly by dimerisation of adjacent pyrimidines in the same DNA strand. The biological importance of these lesions is illustrated by the fact that sufferers of Xeroderma pigmentosum (XP), who have a deficiency in nucleotide excision repair mechanisms that remove pyrimidine dimers and other bulky adducts from DNA (also deficient in XPA knockout mice mentioned above), are prone to sunlight-induced skin cancer. Moreover, the type of *TP53* mutation found commonly in such tumours, but rarely in tumours of internal organs, is a tandem mutation occurring at pyrimidine pairs (CC→TT transitions), highly suggestive that it arose from UV-induced pyrimidine dimers [85].

Thus, there are examples of genetic changes in tumours that closely match the genetic changes that can be induced experimentally in cellular DNA by specific genotoxic agents. These tumour-specific mutations in *TP53* and the demonstration that chemically modified DNA transforms cells show that the mutations observed in human tumours could have arisen from the formation of carcinogen-DNA adducts *in vivo*. Clonal expansion of the mutated cells and the acquisition of further genetic alterations eventually leads to malignancy [3].

Adducts as biomarkers of occupational and environmental exposure to carcinogens

Sensitive DNA adduct detection methods, not requiring the use of radiolabelled carcinogens, make it possible to monitor DNA isolated from human tissues for evidence of prior exposure to carcinogens. Many different tissues have provided DNA for such studies [86], including blood, sputum, buccal mucosa, cervical mucosa, sperm, bladder (exfoliated urothelial cells in urine), placenta and hair roots. DNA from these accessible sources of human cells have been used in many studies, but by far the most commonly used tissue source has been blood cells (either lymphocytes or the whole fraction of nucleated white blood cells). While these are not target cells for malignancy, they are useful surrogates and are known to display evidence of genotoxic exposure using other, less sensitive, endpoints, such as micronucleus formation, chromosomal aberrations and mutation in reporter genes such as *HPRT*.

Heavy industries where an increased risk of lung and other cancers has been observed include iron and steel production, aluminium production, coke ovens and graphite electrode manufacture. The principal genotoxic exposure in these industries is to PAHs. Many studies have investigated DNA adduct formation in workers in these industries, using white blood cells as the monitored tissue. In general, the results of such studies have been the demonstration of statistically significant increases in the level of DNA adducts in the exposed workers, compared with controls [87]. Other industrial workforces studied, with similar results, include roofers, chimney sweeps, incinerator workers, petrol refinery workers, traffic police and bus maintenance workers [87].

DNA adduct detection can also be used to investigate environmental exposure to genotoxic carcinogens [87]. For example, chronic environmental exposure to industrial sources of carcinogens has occurred in Upper Silesia and the Krakow region of Poland, and in Northern Bohemia in the Czech Republic. Studies on these populations have revealed significantly elevated levels of DNA adducts in blood cells compared with control populations from rural areas of the same countries. In Xuan Wei province of China, the practice of using smoky coal for cooking and heating in unventilated houses leads to a high level of smoke indoors and high incidences of lung cancer, particularly in the women (very few of whom smoke tobacco). Placental, blood and lung (from bronchoalveolar lavage) cells have all been used as sources of DNA to compare exposed female residents of Xuan Wei with a control group from Beijing, and in each case evidence for elevated levels of adducts was obtained [88]. In Henan province of China, there is an exceptionally high prevalence of oesophageal cancer and, among several suspected environmental factors, the high content of PAHs in the diet has recently become of interest; evidence that this may play a role in the aetiology of the disease is supported by the observation of high levels of PAH-DNA adducts, detected by immunohistochemistry, in archived surgical specimens of oesophagi from the region [89].

A recent example of the identification of a human carcinogen is aristolochic acid (AA), a constituent of plants of the genus *Aristolochia* (Tab. 2). AA is genotoxic, being positive in many short-term tests, and forms covalent DNA adducts in tissues of rodents, and in human cells in culture [90]. An outbreak of renal failure, followed by urothelial cancer in some of the patients, occurred among individuals in Belgium who took a slimming regimen containing Chinese herbs, one of which turned out to be an *Aristolochia* species. ^{32}P-Postlabelling analysis of DNA from the tissues of these patients revealed the presence of AA-DNA adducts, implicating the compound as the genotoxic agent involved in the carcinogenic process leading to urothelial tumours [91]. The renal disease, now known as Aristolochic Acid Nephropathy (AAN), is pathologically similar to Balkan endemic nephropathy (BEN), in which AA is also implicated [92]. The source is thought to be *Aristolochia clematitis,* which grows wild in the Balkans and whose seeds may contaminate wheat flour in the region. The detection of AA-DNA adducts in renal tissues from BEN sufferers provides strong evidence for the involvement of AA in the aetiology of the disease [93]. Furthermore, analysis of *TP53* mutations in BEN tumours, and in one AAN tumour, shows a preponderance of AT→TA transversion mutations, which is the predominant mutation that AA causes in experimental studies [93, 94].

The relationship between DNA adduct formation and tobacco smoking has been widely studied and used to validate the biomarker (see below). Tobacco smoke contains at least 50 compounds that are known to be carcinogenic, including representatives of several distinct classes of compounds (PAHs, aromatic amines, *N*-nitrosamines, aza-arenes, aldehydes, other organic compounds and inorganic compounds). Most of these compounds are genotoxic carcinogens that form DNA adducts. In many studies that have compared DNA from smokers, ex-smokers and non-smokers, higher levels of adducts have been found in many target tissues of smokers: lung, bronchus, larynx, bladder, cervix and oral mucosa [95]. In some of these studies a linear correlation between estimated tobacco smoke exposure and adduct levels has been observed. In tissues of the respiratory tract, adduct levels in ex-smokers tend to be intermediate between smokers and non-smokers, indicating that adducts are removed through DNA repair and/or cell turnover. The half-life of adduct persistence appears to be between 1 and 2 years.

For some of these studies specific adducts have been detected, but in others a more general measure of DNA damage has been made, namely aromatic/hydrophobic adducts detected by ^{32}P-postlabelling, or PAH-DNA adducts detected by immunoassay. Recent studies have found that when adduct levels are adjusted to take account of the level of tobacco smoke exposure, lung DNA from women smokers is more highly adducted than that of male smokers. This finding is interesting in view of epidemiological evidence suggesting that women are at a 1.5–2-fold greater risk of lung cancer from smoking. It would appear that the adduct analysis provides biochemical and mechanistic evidence to support the morbidity data [96].

Some, but not all, studies have shown elevated levels of lung adducts in cancer cases compared with controls. The relationship between adduct levels in target tissues (e.g. lung) and other tissues (e.g. blood) has been investigated to see whether the latter can serve as a valid accessible surrogate source of DNA for the former. Results for smoking-related adducts have been inconsistent [55, 97], perhaps because other sources of exposure to some classes of carcinogens, such as PAHs, which are also ingested as dietary contaminants, may contribute to the overall level of adducts in the blood but not to the same extent in the lung.

DNA adducts in prospective studies

When measuring adducts in smokers at the time of cancer diagnosis (e.g. in case-control studies), investigators are not looking at the biochemical events causal in the initiation of those tumours, as these would have occurred decades earlier. However, because smoking is addictive and habitual for the vast majority of tobacco users, DNA adducts in tumour-adjacent tissue at the time of tumour manifestation can still serve as a useful biomarker that gives an indication of an individual's probable steady-state level of DNA damage maintained over a long period of time. To determine whether DNA adducts have predictive value in cancer risk, it is necessary to conduct prospective studies in which DNA samples are collected and stored from a large cohort of individuals who are then followed up to determine who does and who does not develop cancer in the future. It is then possible to perform a nested case-control study within the cohort to determine whether DNA adduct analysis of the stored samples reveals whether differences between the two groups were evident prior to the onset of disease.

The first example of this approach was a study conducted in Shanghai, China, where a high incidence of liver cancer is associated with dietary exposure to aflatoxin B_1 (Tab. 2); 18 244 men provided a single urine sample and detailed dietary questionnaire data, in addition to which food analyses were carried out [98]. When 55 cases of liver cancer subsequently arose in the cohort, these were matched to 267 disease-free controls and their urine samples analysed by HPLC-fluorescence to detect the presence of aflatoxin derivatives. A significant association was found between the presence of aflatoxin metabolites, including the aflatoxin-N7-guanine adduct, and liver cancer. Interestingly, when data obtained from questionnaires and food analyses were considered without the biomarker data, no association between exposure and liver cancer was evident. Thus, in this case the power of biomarkers of exposure showed a clear advantage over more traditional means of exposure assessment to show a causal association.

The ability of DNA adducts to predict lung cancer risk was investigated in tobacco smokers [99]. From a follow-up of a cohort of 15 700 males who had provided blood samples at the outset of the study, 93 cases of lung cancer were identified and matched to 173 controls. Analysis of white blood cell DNA by ^{32}P-postlabelling revealed that smokers who got lung cancer had 2-fold higher levels of bulky/hydrophobic DNA adducts than smokers who did not. The smokers who had elevated levels of adducts were approximately three times more likely to be diagnosed with lung cancer 1–13 years later than the smokers with lower adduct concentrations.

The predictive power of DNA adducts to distinguish groups of individuals who developed cancer from those who did not was also investigated in two recent studies that measured bulky DNA adducts in leukocytes by ^{32}P-postlabelling analysis. In the first, 115 cases of lung cancer were matched with twice the number of controls from European cohorts totalling more than 500 000 people [100]. Detectable DNA adducts were significantly more common in non-smokers and long-term ex-smokers who developed lung cancer than in those who did not. The second study investigated 245 individuals with lung cancer and 255 without, from a population-based cohort of 53 689 men and women [101]. The median level of DNA adducts was significantly higher for smokers who developed lung cancer than for those that did not. Although adduct levels were statistically significantly higher in the cases in both these studies, the numerical differences from the controls were somewhat small. Thus, the ability to predict cancer risk from DNA adduct measurements on an individual basis will be very limited, despite the collective differences between the cases and the controls. Nevertheless, DNA adduct analysis should have applications in investigating the efficacy of chemoprevention strategies by, for example, documenting a reduction in adduct levels concomitant with a reduction in cancer risk in interventions in an occupationally or environmentally exposed population.

Summary

DNA adduct formation, or the causation of DNA damage by less direct means, is an important property of genotoxic agents. The strategies that have been developed for determining the carcinogenic potential of chemicals, using short-term tests, are based on detecting evidence of either DNA damage or its biological consequences. Although it is well recognised that the carcinogenic activity of some chemicals is the result of non-genotoxic mechanisms, the majority of known human carcinogens are genotoxic. Early studies on DNA adducts required use of radioactively labelled compounds, but alternative methods with a high degree of sensitivity and selectivity have since been developed, enabling their wider application, including the monitoring of human exposure to environmental carcinogens and in providing clues to the aetiology of some cancers. Experimental interventions that reduce DNA adduct formation also reduce carcinogenicity, while enhancing DNA adduct formation has the opposite effect. In prospective studies, elevated DNA adducts have been found in individuals who subsequently developed cancer relative to those who did not. Continuing research into the detection and characterisation of DNA adducts in human tissues will shed further light on the causative agents of human cancers.

Acknowledgements
We thank Dr. Stan Venitt for critical reading of the manuscript. The work of the authors is supported by Cancer Research UK.

References

1 Armitage P, Doll R (1954) The age distribution of cancer and a multi-stage theory of carcinogenesis. *Br J Cancer* 8: 1–12
2 Hanahan D, Weinberg RA (2000) The hallmarks of cancer. *Cell* 100: 57–70
3 Fearon ER, Vogelstein B (1990) A genetic model for colorectal tumorigenesis. *Cell* 61: 759–767
4 Kirkland DJ, Hayashi M, MacGregor JT, Muller L, Schechtman LM, Sofuni T (2003) Summary of major conclusions. *Mutat Res* 540: 123–125
5 Cimino MC (2006) Comparative overview of current international strategies and guidelines for genetic toxicology testing for regulatory purposes. *Environ Mol Mutagen* 47: 362–390
6 Tweats DJ, Scott AD, Westmoreland C, Carmichael PL (2007) Determination of genetic toxicity and potential carcinogenicity *in vitro* – Challenges post the Seventh Amendment to the European Cosmetics Directive. *Mutagenesis* 22: 5–13
7 Ames BN, Durston WE, Yamasaki E, Lee FD (1973) Carcinogens are mutagens: A simple test system combining liver homogenates for activation and bacteria for detection. *Proc Natl Acad Sci USA* 70: 2281–2285
8 Maron DM, Ames BN (1983) Revised methods for the Salmonella mutagenicity test. *Mutat Res* 113: 173–215
9 Mortelmans K, Zeiger E (2000) The Ames Salmonella/microsome mutagenicity assay. *Mutat Res* 455: 29–60

10 Mortelmans K (2006) Isolation of plasmid pKM101 in the Stocker laboratory. *Mutat Res* 612: 151–164

11 Einisto P, Watanabe M, Ishidate M Jr, Nohmi T (1991) Mutagenicity of 30 chemicals in *Salmonella typhimurium* strains possessing different nitroreductase or *O*-acetyltransferase activities. *Mutat Res* 259: 95–102

12 Yamada M, Espinosa-Aguirre JJ, Watanabe M, Matsui K, Sofuni T, Nohmi T (1997) Targeted disruption of the gene encoding the classical nitroreductase enzyme in *Salmonella typhimurium* Ames test strains TA1535 and TA1538. *Mutat Res* 375: 9–17

13 Glatt H, Meinl W (2005) Sulfotransferases and acetyltransferases in mutagenicity testing: Technical aspects. *Methods Enzymol* 400: 230–249

14 Clements J (2000) The mouse lymphoma assay. *Mutat Res* 455: 97–110

15 Clive D, Flamm WG, Machesko MR, Bernheim NJ (1972) A mutational assay system using the thymidine kinase locus in mouse lymphoma cells. *Mutat Res* 16: 77–87

16 Clive D, Johnson KO, Spector JF, Batson AG, Brown MM (1979) Validation and characterization of the L5178Y/TK$^{+/-}$ mouse lymphoma mutagen assay system. *Mutat Res* 59: 61–108

17 Moore MM, Clive D, Hozier JC, Howard BE, Batson AG, Turner NT, Sawyer J (1985) Analysis of trifluorothymidine-resistant (TFTr) mutants of L5178Y/TK$^{+/-}$ mouse lymphoma cells. *Mutat Res* 151: 161–174

18 Moore MM, Honma M, Clements J, Bolcsfoldi G, Burlinson B, Cifone M, Clarke J, Clay P, Doppalapudi R, Fellows M et al (2007) Mouse lymphoma thymidine kinase gene mutation assay: Meeting of the International Workshop on Genotoxicity Testing, San Francisco, 2005, recommendations for 24-h treatment. *Mutat Res* 627: 36–40

19 Nohmi T, Suzuki T, Masumura K (2000) Recent advances in the protocols of transgenic mouse mutation assays. *Mutat Res* 455: 191–215

20 Lambert IB, Singer TM, Boucher SE, Douglas GR (2005) Detailed review of transgenic rodent mutation assays. *Mutat Res* 590: 1–280

21 Gossen JA, de Leeuw WJ, Tan CH, Zwarthoff EC, Berends F, Lohman PH, Knook DL, Vijg J (1989) Efficient rescue of integrated shuttle vectors from transgenic mice: A model for studying mutations *in vivo*. *Proc Natl Acad Sci USA* 86: 7971–7975

22 Kohler SW, Provost GS, Fieck A, Kretz PL, Bullock WO, Sorge JA, Putman DL, Short JM (1991) Spectra of spontaneous and mutagen-induced mutations in the lacI gene in transgenic mice. *Proc Natl Acad Sci USA* 88: 7958–7962

23 Heddle JA, Dean S, Nohmi T, Boerrigter M, Casciano D, Douglas GR, Glickman BW, Gorelick NJ, Mirsalis JC, Martus HJ et al (2000) *In vivo* transgenic mutation assays. *Environ Mol Mutagen* 35: 253–259

24 Thybaud V, Dean S, Nohmi T, de Boer J, Douglas GR, Glickman BW, Gorelick NJ, Heddle JA, Heflich RH, Lambert I et al (2003) *In vivo* transgenic mutation assays. *Mutat Res* 540: 141–151

25 Jakubczak JL, Merlino G, French JE, Muller WJ, Paul B, Adhya S, Garges S (1996) Analysis of genetic instability during mammary tumor progression using a novel selection-based assay for *in vivo* mutations in a bacteriophage λ transgene target. *Proc Natl Acad Sci USA* 93: 9073–9078

26 Nohmi T, Katoh M, Suzuki H, Matsui M, Yamada M, Watanabe M, Suzuki M, Horiya N, Ueda O, Shibuya T et al (1996) A new transgenic mouse mutagenesis test system using Spi- and 6-thioguanine selections. *Environ Mol Mutagen* 28: 465–470

27 Albertini RJ, Anderson D, Douglas GR, Hagmar L, Hemminki K, Merlo F, Natarajan AT, Norppa H, Shuker DE, Tice R et al (2000) IPCS guidelines for the monitoring of genotoxic effects of carcinogens in humans. International Programme on Chemical Safety. *Mutat Res* 463: 111–172

28 Parry EM, Parry JM (1995) *In vitro* cytogenetics and aneuploidy. In: DH Phillips, S Venitt (eds): *Environmental Mutagenesis*. BIOS Scientific Publishers, Oxford, 121–139

29 Dean BJ, Danford N (1984) Assays for the detection of chemically-induced chromosome damage in cultured mammalian cells. In: S Venitt, JM Parry (eds): *Mutagenicity Testing – A Practical Approach*. IRL Press, Oxford, 187–232

30 Bonassi S, Ugolini D, Kirsch-Volders M, Stromberg U, Vermeulen R, Tucker JD (2005) Human population studies with cytogenetic biomarkers: Review of the literature and future prospectives. *Environ Mol Mutagen* 45: 258–270

31 Norppa H, Bonassi S, Hansteen IL, Hagmar L, Stromberg U, Rossner P, Boffetta P, Lindholm C, Gundy S, Lazutka J et al (2006) Chromosomal aberrations and SCEs as biomarkers of cancer risk. *Mutat Res* 600: 37–45

32 Fenech M, Morley AA (1985) Measurement of micronuclei in lymphocytes. *Mutat Res* 147:

29–36

33 Kirsch-Volders M, Vanhauwaert A, De Boeck M, Decordier I (2002) Importance of detecting numerical *versus* structural chromosome aberrations. *Mutat Res* 504: 137–148

34 Eastmond DA, Pinkel D (1990) Detection of aneuploidy and aneuploidy-inducing agents in human lymphocytes using fluorescence *in situ* hybridization with chromosome-specific DNA probes. *Mutat Res* 234: 303–318

35 Van Hummelen P, Kirsch-Volders M (1990) An improved method for the '*in vitro*' micronucleus test using human lymphocytes. *Mutagenesis* 5: 203–204

36 Fenech M (2006) Cytokinesis-block micronucleus assay evolves into a "cytome" assay of chromosomal instability, mitotic dysfunction and cell death. *Mutat Res* 600: 58–66

37 Decordier I, Kirsch-Volders M (2006) The *in vitro* micronucleus test: From past to future. *Mutat Res* 607: 2–4

38 Corvi R, Albertini S, Hartung T, Hoffmann S, Maurici D, Pfuhler S, van Benthem J, Vanparys P (2008) ECVAM retrospective validation of *in vitro* micronucleus test (MNT). *Mutagenesis* 23: 271–283

39 Hayashi M, Tice RR, MacGregor JT, Anderson D, Blakey DH, Kirsh-Volders M, Oleson FB Jr, Pacchierotti F, Romagna F, Shimada H et al (1994) *In vivo* rodent erythrocyte micronucleus assay. *Mutat Res* 312: 293–304

40 Hayashi M, MacGregor JT, Gatehouse DG, Adler ID, Blakey DH, Dertinger SD, Krishna G, Morita T, Russo A, Sutou S (2000) *In vivo* rodent erythrocyte micronucleus assay. II. Some aspects of protocol design including repeated treatments, integration with toxicity testing, and automated scoring. *Environ Mol Mutagen* 35: 234–252

41 Hayashi M, Tice RR, MacGregor JT, Anderson D, Blakey DH, Kirsh-Volders M, Oleson FB Jr, Pacchierotti F, Romagna F, Shimada H et al (1994) *In vivo* rodent erythrocyte micronucleus assay. *Mutat Res* 312: 293–304

42 Singh NP, McCoy MT, Tice RR, Schneider EL (1988) A simple technique for quantitation of low levels of DNA damage in individual cells. *Exp Cell Res* 175: 184–191

43 Speit G, Hartmann A (2005) The comet assay: A sensitive genotoxicity test for the detection of DNA damage. *Methods Mol Biol* 291: 85–95

44 Olive PL, Banath JP (2006) The comet assay: A method to measure DNA damage in individual cells. *Nat Protoc* 1: 23–29

45 Collins AR, Oscoz AA, Brunborg G, Gaivao I, Giovannelli L, Kruszewski M, Smith CC, Stetina R (2008) The comet assay: Topical issues. *Mutagenesis* 23: 143–151

46 Collins AR (2004) The comet assay for DNA damage and repair: Principles, applications, and limitations. *Mol Biotechnol* 26: 249–261

47 Smith CC, O'Donovan MR, Martin EA (2006) hOGG1 recognizes oxidative damage using the comet assay with greater specificity than FPG or ENDOIII. *Mutagenesis* 21: 185–190

48 Tice RR, Agurell E, Anderson D, Burlinson B, Hartmann A, Kobayashi H, Miyamae Y, Rojas E, Ryu JC, Sasaki YF (2000) Single cell gel/comet assay: Guidelines for *in vitro* and *in vivo* genetic toxicology testing. *Environ Mol Mutagen* 35: 206–221

49 Moller P, Knudsen LE, Loft S, Wallin H (2000) The comet assay as a rapid test in biomonitoring occupational exposure to DNA-damaging agents and effect of confounding factors. *Cancer Epidemiol Biomarkers Prev* 9: 1005–1015

50 Jha AN (2008) Ecotoxicological applications and significance of the comet assay. *Mutagenesis* 23: 207–221

51 Hakura A, Shimada H, Nakajima M, Sui H, Kitamoto S, Suzuki S, Satoh T (2005) Salmonella/human S9 mutagenicity test: A collaborative study with 58 compounds. *Mutagenesis* 20: 217–228

52 Doehmer J (2006) Predicting drug metabolism-dependent toxicity for humans with a genetically engineered cell battery. *Altern Lab Anim* 34: 561–575

53 Brambilla G, Martelli A (2004) Failure of the standard battery of short-term tests in detecting some rodent and human genotoxic carcinogens. *Toxicology* 196: 1–19

54 Marnett LJ (2000) Oxyradicals and DNA damage. *Carcinogenesis* 21: 361–370

55 Poirier MC, Santella RM, Weston A (2000) Carcinogen macromolecular adducts and their measurement. *Carcinogenesis* 21: 353–359

56 Phillips DH, Farmer PB, Beland FA, Nath RG, Poirier MC, Reddy MV, Turteltaub KW (2000) Methods of DNA adduct determination and their application to testing compounds for genotoxicity. *Environ Mol Mutagen* 35: 222–233

57 Randerath K, Reddy MV, Gupta RC (1981) ^{32}P-labeling test for DNA damage. *Proc Natl Acad Sci USA* 78: 6126–6129

58 Gupta RC, Reddy MV, Randerath K (1982) ^{32}P-postlabeling analysis of non-radioactive aromatic carcinogen-DNA adducts. *Carcinogenesis* 3: 1081–1092

59 Gupta RC (1985) Enhanced sensitivity of ^{32}P-postlabeling analysis of aromatic carcinogen:DNA adducts. *Cancer Res.* 45: 5656–5662

60 Phillips DH, Arlt VM (2007) The ^{32}P-postlabeling assay for DNA adducts. *Nat Protoc* 2: 2772–2781

61 Phillips DH, Castegnaro M (1999) Standardization and validation of DNA adduct postlabelling methods: Report of interlaboratory trials and production of recommended protocols. *Mutagenesis* 14: 301–315

62 Phillips DH (1997) Detection of DNA modifications by the ^{32}P-postlabelling assay. *Mutat Res* 378: 1–12

63 Beach AC, Gupta RC (1992) Human biomonitoring and the ^{32}P-postlabeling assay. *Carcinogenesis* 13: 1053–1074

64 Schmitz OJ, Worth CC, Stach D, Wiessler M (2002) Capillary electrophoresis analysis of DNA adducts as biomarkers for carcinogenesis. *Angew Chem Int Ed* 41: 445–448

65 Wirtz M, Schumann CA, Schellentrager M, Gab S, Vom Brocke J, Podeschwa MA, Altenbach HJ, Oscier D, Schmitz OJ (2005) Capillary electrophoresis-laser induced fluorescence analysis of endogenous damage in mitochondrial and genomic DNA. *Electrophoresis* 26: 2599–2607

66 Stach D, Schmitz OJ, Stilgenbauer S, Benner A, Dohner H, Wiessler M, Lyko F (2003) Capillary electrophoretic analysis of genomic DNA methylation levels. *Nucleic Acids Res* 31: E2

67 Farmer PB, Brown K, Tompkins E, Emms VL, Jones DJ, Singh R, Phillips DH (2005) DNA adducts: Mass spectrometry methods and future prospects. *Toxicol Appl Pharmacol* 207: 293–301

68 Singh R, Farmer PB (2006) Liquid chromatography-electrospray ionization-mass spectrometry: The future of DNA adduct detection. *Carcinogenesis* 27: 178–196

69 Kanaly RA, Hanaoka T, Sugimura H, Toda H, Matsui S, Matsuda T (2006) Development of the adductome approach to detect DNA damage in humans. *Antioxid Redox Signal* 8: 993–1001

70 White IN, Brown K (2004) Techniques: The application of accelerator mass spectrometry to pharmacology and toxicology. *Trend Pharmacol Sci* 25: 442–447

71 Dingley KH, Curtis KD, Nowell S, Felton JS, Lang NP, Turteltaub KW (1999) DNA and protein adduct formation in the colon and blood of humans after exposure to a dietary-relevant dose of 2-amino-1-methyl-6-phenylimidazo[4,5-*b*]pyridine. *Cancer Epidemiol Biomarkers Prev* 8: 507–512

72 Brown K, Tompkins EM, Boocock DJ, Martin EA, Farmer PB, Turteltaub KW, Ubick E, Hemingway D, Horner-Glister E, White IN (2007) Tamoxifen forms DNA adducts in human colon after administration of a single [^{14}C]-labeled therapeutic dose. *Cancer Res* 67: 6995–7002

73 Weston A (1993) Physical methods for the detection of carcinogen-DNA adducts in humans. *Mutat Res* 288: 19–29

74 ESCODD (2003) Measurement of DNA oxidation in human cells by chromatographic and enzymic methods. *Free Radic Biol Med* 34: 1089–1099

75 Poirier MC (1994) Human exposure monitoring, dosimetry, and cancer risk assessment: The use of antisera specific for carcinogen-DNA adducts and carcinogen-modified DNA. *Drug Metab Rev* 26: 87–109

76 Santella RM (1999) Immunological methods for detection of carcinogen-DNA damage in humans. *Cancer Epidemiol Biomarkers Prev* 8: 733–739

77 Divi RL, Beland FA, Fu PP, Von Tungeln LS, Schoket B, Camara JE, Ghei M, Rothman N, Sinha R, Poirier MC (2002) Highly sensitive chemiluminescence immunoassay for benzo[*a*]pyrene-DNA adducts: Validation by comparison with other methods, and use in human biomonitoring. *Carcinogenesis* 23: 2043–2049

78 van Gijssel HE, Divi RL, Olivero OA, Roth MJ, Wang GQ, Dawsey SM, Albert PS, Qiao YL, Taylor PR, Dong ZW et al (2002) Semiquantitation of polycyclic aromatic hydrocarbon-DNA adducts in human esophagus by immunohistochemistry and the automated cellular imaging system. *Cancer Epidemiol Biomarkers Prev* 11: 1622–1629

79 Pratt MM, Sirajuddin P, Poirier MC, Schiffman M, Glass AG, Scott DR, Rush BB, Olivero OA, Castle PE (2007) Polycyclic aromatic hydrocarbon-DNA adducts in cervix of women infected with carcinogenic human papillomavirus types: An immunohistochemistry study. *Mutat Res* 624: 114–123

80 van Steeg H, Klein H, Beems RB, van Kreijl CF (1998) Use of DNA repair-deficient XPA transgenic mice in short-term carcinogenicity testing. *Toxicol Pathol* 26: 742–749
81 Marshall CJ, Vousden KH, Phillips DH (1984) Activation of c-Ha-*ras*-1 proto-oncogene by *in
   vitro* modification with a chemical carcinogen, benzo[*a*]pyrene diol-epoxide. *Nature* 310:
   586–589
82 Petitjean A, Mathe E, Kato S, Ishioka C, Tavtigian SV, Hainaut P, Olivier M (2007) Impact of
   mutant p53 functional properties on *TP53* mutation patterns and tumor phenotype: Lessons from
   recent developments in the IARC *TP53* database. *Hum Mutat* 28: 622–629
83 Besaratinia A, Pfeifer GP (2006) Investigating human cancer etiology by DNA lesion footprint-
   ing and mutagenicity analysis. *Carcinogenesis* 27: 1526–1537
84 Denissenko MF, Pao A, Tang M, Pfeifer GP (1996) Preferential formation of benzo[*a*]pyrene
   adducts at lung cancer mutational hotspots in p53. *Science* 274: 430–432
85 Dumaz N, Drougard C, Sarasin A, Daya-Grosjean L (1993) Specific UV-induced mutation spectrum in the p53 gene of skin tumors from DNA-repair-deficient *Xeroderma pigmentosum*
   patients. *Proc Natl Acad Sci USA* 90: 10529–10533
86 Phillips DH (1996) DNA adducts in human tissues: Biomarkers of exposure to carcinogens in
   tobacco smoke. *Environ Health Perspect* 104 Suppl. 3: 453–458
87 Phillips DH (2005) Macromolecular adducts as biomarkers of human exposure to polycyclic aromatic hydrocarbons. In: A Luch (ed.): *The Carcinogenic Effects of Polycyclic Aromatic
   Hydrocarbons*. Imperial College Press, London, 137–169
88 Mumford JL, Lee X, Lewtas J, Young TL, Santella RM (1993) DNA adducts as biomarkers for
   assessing exposure to polycyclic aromatic hydrocarbons in tissues from Xuan Wei women with
   high exposure to coal combustion emissions and high lung cancer mortality. *Environ Health
   Perspect* 99: 83–87
89 van Gijssel HE, Schild LJ, Watt DL, Roth MJ, Wang GQ, Dawsey SM, Albert PS, Qiao YL,
   Taylor PR, Dong ZW, Poirier MC (2004) Polycyclic aromatic hydrocarbon-DNA adducts determined by semiquantitative immunohistochemistry in human esophageal biopsies taken in 1985.
   Mutat Res 547: 55–62
90 Arlt VM, Stiborova M, Schmeiser HH (2002) Aristolochic acid as a probable human cancer hazard in herbal remedies: A review. *Mutagenesis* 17: 265–277
91 Nortier JL, Martinez MC, Schmeiser HH, Arlt VM, Bieler CA, Petein M, Depierreux MF, De
   Pauw L, Abramowicz D, Vereerstraeten P, Vanherweghem JL (2000) Urothelial carcinoma associated with the use of a Chinese herb (*Aristolochia fangchi*). *N Engl J Med* 342: 1686–1692
92 Arlt VM, Ferluga D, Stiborova M, Pfohl-Leszkowicz A, Vukelic M, Ceovic S, Schmeiser HH,
   Cosyns JP (2002) Is aristolochic acid a risk factor for Balkan endemic nephropathy-associated
   urothelial cancer? *Int J Cancer* 101: 500–502
93 Grollman AP, Shibutani S, Moriya M, Miller F, Wu L, Moll U, Suzuki N, Fernandes A,
   Rosenquist T, Medverec Z et al (2007) Aristolochic acid and the etiology of endemic (Balkan)
   nephropathy. *Proc Natl Acad Sci USA* 104: 12129–12134
94 Arlt VM, Stiborova M, vom Brocke J, Simoes ML, Lord GM, Nortier JL, Hollstein M, Phillips
   DH, Schmeiser HH (2007) Aristolochic acid mutagenesis: Molecular clues to the aetiology of
   Balkan endemic nephropathy-associated urothelial cancer. *Carcinogenesis* 28: 2253–2261
95 Phillips DH (2002) Smoking-related DNA and protein adducts in human tissues. *Carcinogenesis*
   23: 1979–2004
96 Mollerup S, Ryberg D, Hewer A, Phillips DH, Haugen A (1999) Sex differences in lung CYP1A1
   expression and DNA adduct levels among lung cancer patients. *Cancer Res* 59: 3317–3320
97 Perera FP (2000) Molecular epidemiology: On the path to prevention? *J Natl Cancer Inst* 92:
   602–612
98 Qian GS, Ross RK, Yu MC, Yuan JM, Gao YT, Henderson BE, Wogan GN, Groopman JD (1994)
   A follow-up study of urinary markers of aflatoxin exposure and liver cancer risk in Shanghai,
   People's Republic of China. *Cancer Epidemiol Biomarkers Prev* 3: 3–10
99 Tang D, Phillips DH, Stampfer M, Mooney LA, Hsu Y, Cho S, Tsai WY, Ma J, Cole KJ, She MN,
   Perera FP (2001) Association between carcinogen-DNA adducts in white blood cells and lung
   cancer risk in the physicians health study. *Cancer Res* 61: 6708–6712
100 Peluso M, Munnia A, Hoek G, Krzyzanowski M, Veglia F, Airoldi L, Autrup H, Dunning A, Garte
   S, Hainaut P et al (2005) DNA adducts and lung cancer risk: A prospective study. *Cancer Res* 65:
   8042–8048
101 Bak H, Autrup H, Thomsen BL, Tjonneland A, Overvad K, Vogel U, Raaschou-Nielsen O, Loft

S (2006) Bulky DNA adducts as risk indicator of lung cancer in a Danish case-cohort study. *Int J Cancer* 118: 1618–1622

102 Anon (1999) Consensus report. In: DB McGregor, JM Rice, S Venitt (eds): *The Use of Short- and Medium-Term Tests for Carcinogens and Data on Genetic Effects in Carcinogenic Hazard Evaluation.* IARC, Lyon, 1–18

103 Shelby MD, Purchase IFH (1981) Assay systems and criteria for their comparisons. In: FJ de Serres, J Ashby (eds): *Evaluation of Short-term Tests for Carcinogens.* Elsevier/North Holland, New York, 16–20

104 Phillips DH (2008) Biomarkers of exposure: Adducts. In: CP Wild, P Vineis, S Garte (eds): *Molecular Epidemiology of Chronic Diseases.* Wiley, New York, 111–125

Molecular, Clinical and Environmental Toxicology. Volume 1: Molecular Toxicology
Edited by A. Luch

Role of DNA repair in the protection against genotoxic stress

Ulrike Camenisch and Hanspeter Naegeli

Institute of Pharmacology and Toxicology, University of Zürich-Vetsuisse, Zürich, Switzerland

Abstract. The genome of all organisms is constantly attacked by a variety of environmental and endogenous mutagens that cause cell death, apoptosis, senescence, genetic diseases and cancer. To mitigate these deleterious endpoints of genotoxic reactions, living organisms have evolved one or more mechanisms for repairing every type of naturally occurring DNA lesion. For example, double-strand breaks are rapidly religated by non-homologous end-joining. Homologous recombination is used for the high-fidelity repair of interstrand cross-links, double-strand breaks and other DNA injuries that disrupt the replication fork. Some genotoxic lesions inflicted by alkylating agents can be repaired by direct reversal of DNA damage. The base excision repair pathway takes advantage of multiple DNA glycosylases to remove modified or incorrect bases. Finally, the nucleotide excision repair machinery provides a versatile strategy to monitor DNA quality and eliminate all forms of helix-distorting DNA lesions, including a wide diversity of carcinogen adducts. The efficiency of DNA repair responses is enhanced by their coupling to transcription and coordination with the cell cycle circuit.

Introduction

The integrity of DNA, the molecule of heredity, is essential for normal development, viability, longevity and the health of organisms. It was once thought that the faithful maintenance of genetic information over generations is attributable to an inherent stability of DNA, but later it became clear that this molecule is subject to spontaneous decay due to the intrinsic reactivity of many functional groups [1]. In addition, the DNA structure is highly susceptible to damage caused by endogenous metabolic byproducts as well as a variety of environmental insults including radiation and genotoxic chemicals [2]. Thus, in a typical mammalian cell, chemical reactions modify or destroy thousands of DNA nucleotides every day. Some DNA lesions, for example double strand breaks (DSBs), can lead to chromosomal aberrations (deletions, translocations), whereas structural alterations of the bases interfere with transcription and DNA replication [2, 3]. In particular, when DNA polymerases encounter lesion sites in their template, the replication fork may collapse giving rise to intermediates that, if not processed, evolve to the formation of further chromosomal breaks [4]. Alternatively, disruption of the normal replication process leads to the recruitment of a subset of DNA polymerases that are tolerant to DNA damage and, hence, capable of elongating DNA strands across base lesions in the template sequence. Such specialized mechanisms of translesion DNA synthesis are frequently associated with replication errors that alter the coding sequence [5]. The most frequent biological endpoint of all these genetic insults is cell death, but the resulting chromosomal aberrations or gene mutations may also trigger the development of cancer through inactivation of tumor suppressors and activation of oncogenes.

To counteract the manifold genotoxic reactions that compromise genome stability, living organisms are equipped with a network of DNA damage processing pathways. In this context, DNA repair is defined as a biochemical mechanism during which alterations in the chemistry of DNA are removed to restore the native double-helical molecule. Mammalian cells employ the following major DNA repair strategies to cope with hazardous lesions of the bases or the DNA backbone [2, 3]: (i) Non-homologous end-joining (NHEJ) to repair DSBs, (ii) homologous recombination (HR) to repair DSBs, interstrand cross-links (ICLs) as well as collapsed replication forks, (iii) DNA damage reversal to remove alkyl groups, (iv) base excision repair (BER) to eliminate incorrect or modified bases, and (v) nucleotide excision repair (NER) to remove helix-distorting (bulky) DNA adducts.

Repair of double strand breaks

Of all kinds of DNA damage, the simultaneous breakage of both strands constitutes the most deleterious lesion because no intact complementary sequence is available that could be used as a template to repair the genetic defect [6]. DSBs can be induced by mechanical stress, ionizing radiation (X-rays, γ-rays), chemicals with radiomimetic properties (bleomycin, neocarcinostatin) or by topoisomerase II inhibitors such as etoposide [7, 8]. In addition, DSBs are constantly induced by free radical byproducts arising from normal metabolic reactions. Also, DSBs are commonly formed during physiological processes such as meiosis in germ cells or V(D)J recombination in lymphoid lineages, an essential aspect of the developing immune system. Further DSBs are generated during DNA synthesis when the replication fork collides with a single strand break or other preexisting lesions [9]. The immediate detection of DSBs is crucial because these molecular defects cause chromosome breakage and, during the subsequent cell division, the resulting chromosomal fragments may be distributed unequally over the daughter cells or they can be rearranged to aberrant sites in the genome. Thus, it is not surprising to find that living organisms have evolved different repair strategies to handle DSBs [6, 10].

The two main DSB repair pathways, NHEJ and HR, differ in the requirement for a homologous template and in the fidelity of their reactions. It is not yet clear which factors determine the pathway choice, but the cell cycle stage must play an important role in this decision. During G1 and early S phase, DSBs are religated by the error-prone NHEJ process. Instead, in late S phase and G2 phase, the presence of sister chromatids (identical copies of the same chromosome generated upon DNA replication) favors the error-free repair of DSBs through the HR pathway [10, 11].

DNA repair by non-homologous end-joining

This DSB repair mechanism rejoins free DNA ends within minutes of their occurrence. However, it is an inherently error-prone process that frequently involves the loss or alteration of nucleotide sequences [12]. At least five core factors are required for NHEJ: DNA-dependent protein kinase (DNA-PK), a DNA polymerase, the DNA ligase IV complex, Artemis and the more recently identified Cernunnos protein [13, 14]. DNA-PK consists of a large catalytic subunit (DNA-PK$_{CS}$) and two smaller regulatory components (Ku70 and Ku80) with high affinity for break sites [15].

The first step of NHEJ involves recognition of DSBs by the Ku heterodimer (Fig. 1). Subsequently, the DNA-bound Ku proteins recruit DNA-PK$_{CS}$ and thereby translocate into the duplex by one helical turn, leaving DNA-PK$_{CS}$ near the nucleic acid termini [16]. The role of DNA-PK$_{CS}$ is not completely understood but it has become clear that the protein kinase activity of this subunit, and in particular its autophosphorylation reaction, is necessary to modulate the reversible association with DNA ends and ensure efficient NHEJ activity [17]. Several functions have been attributed to DNA-PK$_{CS}$. Electron microscopy analyses revealed conglomerates of juxtaposed DNA ends connected by two DNA-PK$_{CS}$ molecules, indicating that the enzyme forms an intermediate molecular bridge, or synaptic complex, that keeps the broken DNA fragments in close proximity before religation can take place [18, 19]. In addition, DNA-PK$_{CS}$ is thought to mediate the alignment of DNA strands in search for local sequence micro-homologies, i.e., short sequence similarities involving just a few nucleotides. Also, DNA-PK$_{CS}$ serves as a landing platform for DNA polymerases or ligation factors and, finally, may be required to suppress improper recombination events such as for example intrachromosomal strand exchanges between repetitive sequences [6, 11, 12].

When genotoxic agents induce strand breaks, the deoxyribose or base rings near the DNA ends are often fragmented, or these reactions leave terminal 5' hydroxyl or 3' phosphate residues, none of which are compatible with direct enzymatic religation. Thus, further processing of such "dirty" DNA ends is necessary to convert intractable termini to regular DNA ends that can be easily rejoined. It is this step of the NHEJ pathway that is responsible for the occasional loss of nucleotides during DSB repair (Fig. 1). Multiple enzymes are involved in the processing of DNA ends. In addition to Artemis, the generation of suitable DNA ends is dependent on apurinic/apyrimidinic (AP) endonucleases, the tyrosyl-DNA phosphodiesterase TDP1, aprataxin as well as polynucleotide kinase [10]. On its own, Artemis displays a 5' to 3' exonuclease activity. However, in association with DNA-PK$_{CS}$, the enzymatic properties of Artemis shift from exonucleolytic to endonucleolytic, thus gaining a suitable

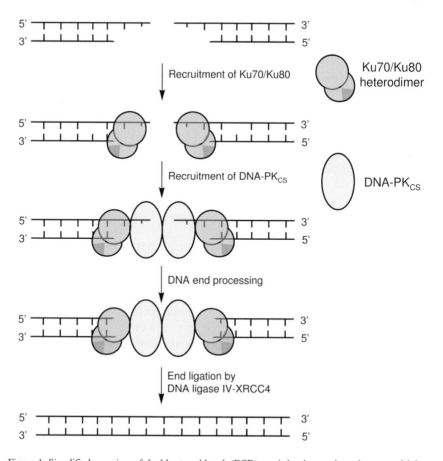

Figure 1. Simplified overview of double strand break (DSB) repair by the non-homologous end-joining (NHEJ) pathway. After DSB formation, the Ku proteins recognize free DNA ends. These regulatory subunits mediate the recruitment of DNA-PK$_{CS}$ (catalytic subunit of DNA-dependent protein kinase), which is probably required for the juxtaposition of the DNA termini. "Dirty" DNA ends might be processed by nucleolytic activity, resulting in the removal or addition of base pairs. Finally, end-to-end ligation takes place by the DNA ligase IV-X-ray repair cross-complementing-4 (XRCC4) complex.

activity for the resection of larger DNA overhangs [20]. Presumably, short gaps are filled in by error-prone DNA synthesis mediated by a mixture of enzymes including terminal deoxynucleotidyltransferase (TdT), DNA polymerase λ or DNA polymerase μ [10, 21]. The final rejoining of DNA ends (Fig. 1) is catalyzed by a dimeric factor that consists of DNA ligase IV and X-ray repair cross-complementing-4 (XRCC4) [22]. The conclusive ligation step of NHEJ is enhanced by the Cernunnos protein, also known as XLF (for XRCC4-like factor). This newly discovered subunit appears to stimulate the NHEJ pathway by promoting the ligation of mismatched or non-cohesive overhanging strands, thus minimizing the need for extensive DNA resections [13, 14, 23].

DNA repair by homologous recombination

In principle, HR is a cellular process that ensures genetic diversity by mediating the exchange of homologous chromosomal sequences during meiosis. However, the same mechanism promotes genomic stability in mitotic cells by repairing severe DNA discontinuities such as gaps, DSBs and ICLs [8]. In addition, the mitotic HR pathway is critically involved in the recovery of stalled or broken DNA replication forks. Unlike NHEJ, it is a very accurate repair mechanism because a homologous sequence of DNA is transiently recruited to guide repair of the lesion. HR is most efficient in the S and G2 phases of the cell cycle when newly copied sister chromatids are readily available as suitable repair templates. In fact, during normal mitotic replication in somatic cells, sister chromatids participate in HR events at least 100-fold more efficiently then the homologous chromosome. Nevertheless, the less frequent recombination between homologous chromosomes (non-identical copies of each chromosome in diploid cells) is significant as this mechanism results in the somatic loss of heterozygosity, a genetic process that accelerates the development of cancer by eliminating functional tumor suppressor genes [24, 25].

First, DSBs have to be processed by nucleolytic activity to generate protruding single-stranded overhangs (Fig. 2). This initial step is likely mediated by the MRN (MRE11/RAD50/NBS1) nuclease complex [26]. Subsequently, RAD51 protein molecules associate with single-stranded overhangs and catalyze the central HR reaction, i.e., the search for DNA sequences that share homology and subsequent strand invasion [27]. RAD51 molecules polymerize on single-stranded DNA in a cooperative and ATP-dependent manner to form a right-handed helical filament with a protein to nucleotide stoichiometry of 1:3 [28, 29]. This reaction is stimulated by several cofactors, or mediator proteins, which are required to form effective RAD51 nucleoprotein filaments in the physiological cellular environment. In humans, the mediator proteins include replication protein A (RPA), five RAD51 paralogs (RAD51B, RAD51C, RAD51D, XRCC2 and XRCC3), RAD52, RAD54 as well as the breast cancer gene products BRCA1 and BRCA2 [8]. When a homologous DNA sequence is available, the process of homology search and strand invasion results in the formation of a joint DNA molecule. In this synaptic complex, the homologous sequence is used for template-directed DNA synthesis, culminating in the high-fidelity repair of DSBs and restoration of any lost sequences at the break site (Fig. 2). DNA ligation following successful repair synthesis results in the formation of branched structures called Holliday junctions in which the two DNA duplexes are linked together (Fig. 2). The Holliday junctions can migrate (a process called branch migration) and, finally, they are dissolved by a "resolvase" complex that reconstitutes intact double-stranded DNA molecules [30].

An important manifestation of defective DSB repair is chromosomal instability characterized by gross chromosomal rearrangements. Several human hereditary disorders associated with DSB repair have been identified. For

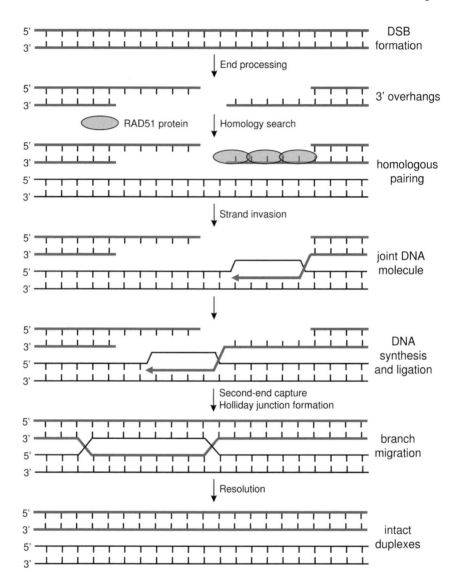

Figure 2. Schematic overview illustrating the main structural intermediates of DSB repair occurring by the homologous recombination (HR) process. The two double-stranded DNA molecules represent homologous sequences (sister chromatids or homologous chromosomes). The pathway is initiated by nucleolytic processing of DNA ends, generating a 3' overhang. RAD51 protein, with the aid of mediator proteins, polymerizes on this single-stranded DNA overhang to form a nucleoprotein filament. This RAD51 nucleofilament searches for homologous sequences and mediates strand invasion, thus giving rise to a joint molecule between the damaged and undamaged duplexes. This strand exchange mechanism provides the template for error-free DNA repair synthesis and results in the formation of Holliday junctions, the universal intermediate of all HR subpathways. Finally, resolution of the Holliday junctions restores intact chromosomes.

example, the significance of the HR process in maintaining genome stability is demonstrated by the Nijmegen breakage syndrome (NBS), which is caused by germline mutations that inactivate NBS1, a subunit of the initial nuclease complex. NBS is characterized by radiosensitivity, growth retardation, immunodeficiency and cancer predisposition. Similarly, mutations in the tumor suppressor genes *BRCA1* and *BRCA2*, involved in the regulation of RAD51 filaments and strand exchanges, predispose humans to malignancies of breast, ovary and prostate tissues [2].

DSBs have also been identified as intermediates in the repair of toxic ICLs caused for example by antitumor drugs (*cis*-diamminedichloroplatinum, mitomycin C), photoactivated natural psoralens or endogenous cross-linking agents such as malondialdehyde, a byproduct of lipid peroxidation and prostaglandin biosynthesis [31]. Most current models suggest that recognition and repair of ICLs take place during S phase, where these lesions pose an insurmountable block to the replication fork. Besides the core HR complex, a number of DNA endonucleases are necessary to separate the two covalently linked DNA strands. One of several structure-specific endonucleases involved in ICL repair is MUS81-EME1, which is required for proper DSB formation after treatment with cross-linking agents [32]. Efficient processing of ICLs is also dependent on the Fanconi anemia (FA) system, a signaling pathway that mediates protein ubiquitylation and RAD51 recruitment after the successful conversion of ICLs to repairable DSB intermediates. The significance of this "signalosome" complex is demonstrated by the rare recessive condition known as FA syndrome. Mutations in one of several FA genes cause developmental abnormalities and chromosomal instability with an exquisite sensitivity to cross-linking agents [33].

Direct reversal of base damage

Base lesions are eliminated from DNA by three basic mechanisms: direct reversal, BER and NER (Fig. 3). In the repair by damage reversal, the chemical bonds that form the lesion are broken in a one-step reaction catalyzed by a single enzyme that restores native DNA. This process differs from excision repair, where DNA damage is eliminated by a cascade of enzymatic activities that replace one or more nucleotides in the defective sequence. BER is initiated by the removal of abnormal or incorrect bases through the activity of a DNA glycosylase (Fig. 3). In contrast, DNA endonucleases make the first step in the NER pathway by incising damaged strands on either side of the lesion, thus releasing the damaged base as the component of an oligonucleotide segment [2, 34, 35]. Before outlining in detail these key excision repair processes, the significance of DNA damage reversal is demonstrated by reviewing the action of two different enzymes targeting methylated bases.

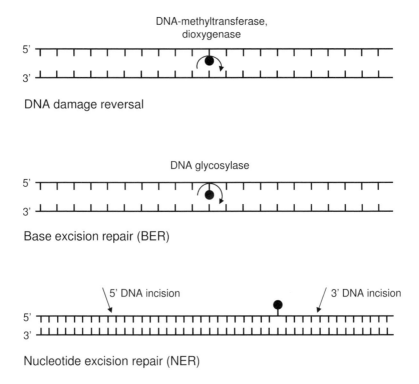

Figure 3. Basic strategies used by mammalian cells for the repair of base lesions. In DNA damage reversal, the offending modification is removed from the base itself, thus restoring the intact double helix in a one-step enzymatic reaction. For example, DNA-methyltransferases or dioxygenases remove methyl groups directly from alkylated bases. In the base excision repair (BER) pathway, abnormal bases are removed by DNA glycosylases that cleave the base-sugar N-glycosidic bond, thereby leaving a potentially mutagenic abasic (apurinic or apyrimidinic) intermediate that needs to be further processed. In nucleotide excision repair (NER), a multiprotein complex leads to dual incision of damaged strands, thereby releasing the damaged residue as the component of an oligonucleotide fragment.

DNA damage reversal by methyltransferases

Alkylating agents are not only widespread environmental carcinogens but are also formed endogenously by metabolic transactions. For example, the coenzyme S-adenosylmethionine is used for the methyl transfer to numerous cellular constituents and is thought to represent a physiological source of DNA methylation [1, 2]. Typical exogenous methylating agents (for example N-methyl-N'-nitro-N-nitrosoguanidine, N-methyl-N-nitrosurea or methyl methanesulfonate) react with DNA to generate multiple O-alkylated and N-alkylated products, but only two of these manifold base lesions are able to induce mutations directly, namely guanine methylated at the O^6 position (O^6-methylguanine) and thymine methylated at the O^4 position (O^4-methylthymine). These two highly mutagenic base lesions are shown in Figure 4.

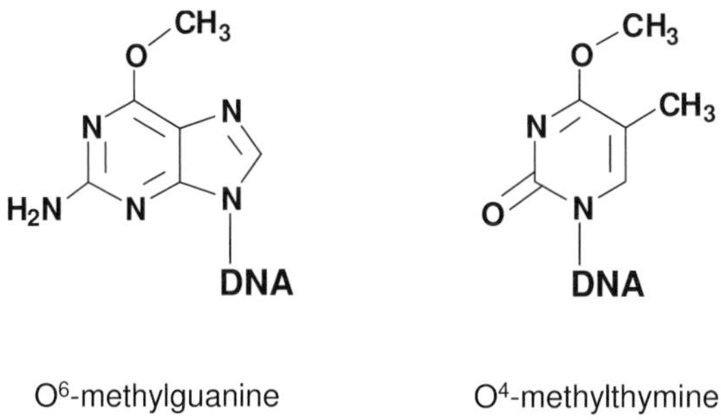

O⁶-methylguanine   O⁴-methylthymine

Figure 4. Methylated substrates repaired by a one-step damage reversal process catalyzed by O^6-methylguanine-DNA methyltransferase (MGMT).

During DNA replication, O^6-methylguanine can pair with thymine and O^4-methylthymine can pair with guanine, generating G:C→A:T or A:T→G:C transition mutations [36–38]. As a consequence, DNA methyltransferases that reconvert O^6-methylguanines and O^4-methylthymines to native guanine or thymine bases, exert a prominent protective function against the mutagenic effects of methylating agents.

The human O^6-methylguanine-DNA methyltransferase (MGMT, also known as AGT for alkylguanine transferase) is a small enzyme of 21.7 kDa that catalyzes the transfer of methyl groups from the O^6-guanine, and less efficiently from the O^4-thymine, to a cysteine moiety residing in the active site of the protein itself. MGMT also repairs larger alkylated residues such as O^6-ethylguanine or O^6-butylguanine, but at considerably lower rates [39, 40]. Apparently, no mechanism exists to remove the resulting S-alkylcysteine moiety and regenerate the catalytic center. Therefore, DNA repair methyltransferases have also been termed suicide enzymes because the transfer of alkyl groups from DNA to the cysteine acceptor leads to their irreversible inactivation and degradation after a single reaction turnover [41].

Targeted deletion of the *MGMT* gene in mice conferred a strong hypersensitivity to treatment with chemotherapeutic alkylating agents or nitrosamines. The histological examination of exposed animals revealed suppression of the lymphoreticular system as well as dysplastic changes in the intestinal mucosa, liver and lung [42–44]. On the other hand, overexpression of MGMT in transgenic mice protected from thymic lymphomas induced by methylnitrosourea [45], from liver cancer induced by dimethyl- or diethylnitrosamine [46], from skin cancer in two-stage carcinogenesis assays [47], and from lung tumors induced by 4-(methylnitrosamino)-1-butanone [48]. These findings support the notion that MGMT plays a significant role in cancer prevention by preventing mutations caused by alkylation of guanine. Interestingly, many human tumor cell lines defined phenotypically as Mer(−) (methylation repair minus), do not possess detectable MGMT activity and, accordingly, are unable to remove O^6-methylguanine from DNA [49]. Their conversion to a Mer(+) phenotype, by *MGMT* gene induction, confers tumor resistance against classical alkylating drugs such as carmustine (a chloroethylating agent), procarbazine, streptozotocin or temozolomide [50, 51].

DNA damage reversal by AlkB homologs

The AlkB protein of *Escherichia coli,* directed against ring-methylated bases (Fig. 5) provides another example of direct DNA damage reversal. In the presence of 2-oxoglutarate, this protein displays a dioxygenase activity that, in conjunction with Fe(II) as a cofactor, converts N1-methyladenine and N3-methylcytosine residues into normal adenine and cytosine bases [52]. In the course of this reaction, the oxidized methyl group is released as form-aldehyde.

Eight possible human homologs of AlkB, designated ABH1 to ABH8, have been identified [53]. In two cases (ABH2, ABH3), it has been demonstrated that these homologs indeed possess an oxidative demethylation activity that, like AlkB, removes methyl groups from the ring nitrogens of adenine and cytosine [54, 55]. Mice deficient in ABH2, ABH3, or both, develop normally and show no obvious phenotype. ABH2$^{(-/-)}$ mice accumulate ring-methylated bases in their genome, indicating that ABH2 protein plays a pivotal role in eliminating endogenously generated N1-methyled purines and N3-methylated pyrimidines [56]. Interestingly, both the *E. coli alkB* and human ABH3 proteins catalyze the removal of N1-methyladenine and N3-methylcytosine not only from DNA but also from RNA [54]. Subcellular localization studies demonstrated that many human AlkB homologs are found both in the nucleus and in the cytoplasm, whereas ABH8 is found only in the cytoplasm, lending support to the notion that methylated RNA molecules may represent an additional repair substrate for this class of enzymes [53, 54]. However, the biological significance of RNA repair processes remains to be fully established.

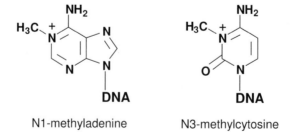

N1-methyladenine   N3-methylcytosine

Figure 5. Ring-methylated substrates of a one-step base damage reversal process catalyzed by mammalian AlkB homologs.

DNA repair by enzymatic photoreactivation

Photoreactivation is the light-induced reversal of helix-distorting DNA lesions [cis-syn cyclobutane pyrimidine dimers and pyrimidine-pyrimidone (6-4) photoproducts] resulting from exposure to ultraviolet (UV) radiation. By generating such pyrimidine cross-links, generally known as pyrimidine dimers, the UV component of sunlight constitutes the major trigger of skin malignancies in humans [57]. Photoreactivation is a rapid and highly efficient DNA repair process that was described in 1949 by Kelner [58], who observed that visible light protects microorganisms from the lethal effects of UV radiation. At the same time, Dulbecco [59] noted that bacteriophages recovered from the effects of UV radiation when exposed to visible light. Nearly 10 years later, Rupert and coworkers [60] showed that photoreactivation is an enzymatic process, and the enzymes that catalyze this repair reaction are now referred to as DNA photolyases. These DNA repair enzymes bind to UV-irradiated DNA and, using a light-harvesting moiety as cofactor, convert photons in the near-UV or blue-wavelength range into chemical energy, which is finally used to split pyrimidine dimers into native monomeric bases [61, 62]. DNA photolyases directed against cyclobutane pyrimidine dimers or (6-4) photoproducts occur in almost all living organisms exposed to sunlight, including non-placental mammals [63]. However, placental mammals and humans lack functional photolyases [64] and, hence, the phenomenon of photoreactivation is not discussed further.

DNA excision repair

As a basic principle, DNA excision repair relies on the redundant information of the DNA duplex to replace damaged residues. If the nucleotides of one strand are damaged, they are excised and the intact complementary strand serves as the template for DNA repair synthesis. The two major DNA excision repair pathways differ mainly in their mechanism of initiation. In BER, the lesion is removed by one of several DNA glycosylases, leaving an apurinic or apyrimidinic (AP) site that is further trimmed by a downstream sequence of enzymatic activities (Fig. 6). In NER, a multi-enzyme machine promotes the release of short oligonucleotide segments carrying the damaged residue. In both excision pathways, the double-helical integrity of DNA is subsequently restored by the synthesis of a repair patch, which is finally ligated to complete the repair cycle [2, 34, 35, 65, 66]. Although there is substantial overlap in the substrate range of these two excision repair systems, the BER pathway is particularly active on subtle base lesions that fail to distort the DNA double helix, whereas NER is more specialized on the removal of helix-distorting adducts resulting from exposure to UV irradiation or bulky electrophilic chemicals. Accordingly, BER and NER employ completely different strategies for DNA damage recognition. In the BER pathway, each target base has to be accommodated into the active site pocket of a specific DNA glycosylase. In contrast,

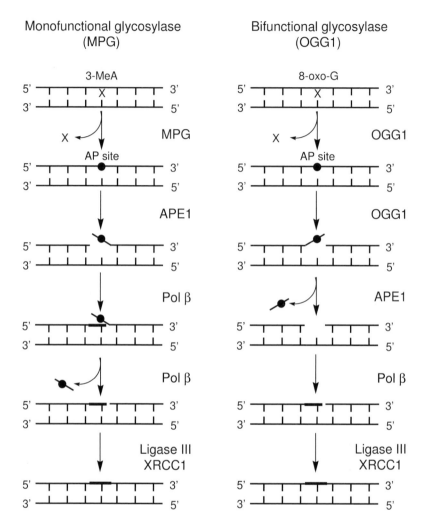

Figure 6. Main base excision repair (BER) pathways initiated by methylpurine-DNA glycosylase (MPG) and 8-oxo-guanine-DNA glycosylase-1 (OGG1). MPG is a monofunctional DNA glycosylase that eliminates for example N3-methyladenine (3-MeA), generating an apurinic/apyrimidinic (AP) site that is incised by AP endonuclease-1 (APE1). DNA polymerase β (Pol β) removes the residual deoxyribose-phosphate moiety before replacing the degraded nucleotide. Instead, OGG1 is a bifunctional enzyme with an associated AP lyase activity that converts the remaining AP site to a 3'-phospho-α,β-unsaturated aldehyde residue. Subsequently, APE1 processes the modified single-strand break by removing the aldehyde-phosphate moiety to form a normal 3'-hydroxyl end. Pol β inserts the missing nucleotide and the nick is ligated by a heterodimer consisting of DNA ligase III and X-ray repair cross-complementing-1 (XRCC1).

damage recognition during the NER pathway is achieved by a more versatile mechanism based on sensor subunits that do not need to make such intimate contacts with the offending adduct.

Base excision repair

Damage to DNA resulting from base deamination, oxidation or alkylation is mainly processed through the BER pathway [1, 2, 65]. As mentioned before, BER activity is initiated by the removal of abnormal or inappropriate bases through the action of DNA glycosylases (Fig. 6). This base excision reaction is followed by strand cleavage in the vicinity of the resulting abasic site by an AP endonuclease, which hydrolyzes the phosphodiester bond immediately 5' to the AP site. The missing nucleotide is replaced by DNA polymerase β, which also removes the remaining deoxyribose-phosphate residue by cleavage of the phosphodiester bond 3' to the AP site. Alternatively, some DNA glyco-sylases (Tab. 1) also have an intrinsic AP lyase activity that is able to cleave the 3' phosphodiester bond [2, 34, 66, 67]. In this case, the deoxyribo-phos-phodiesterase activity of DNA polymerase β is no longer necessary (Fig. 6). Finally, the BER pathway is concluded by DNA ligase III, which ligates the single-nucleotide repair patch in conjunction with X-ray repair cross-comple-menting-1 (XRCC1) protein.

There are multiple well-characterized human DNA glycosylases (Tab. 1) and, in general, these enzymes display a limited substrate range because they have to undergo close interactions with the modified base targeted for excision [68]. After binding to damaged sites, each DNA glycosylase catalyzes the hydrolysis of the N-glycosyl bond (also called N-glycosidic bond) between the base and the sugar-phosphate backbone to generate an abasic site. The known DNA glycosylases differ in the type of nucleophilic moiety used to attack the N-glycosyl bond. Glycosylases with associated AP lyase activity employ an own nucleophilic residue in their active site, while the enzymes lacking AP lyase activity catalyze this reaction by the addition of a hydroxyl group derived from the aqueous solvent [67]. Here, the importance of these BER enzymes is exemplified by two relevant representatives, i.e., methylpurine-DNA glycosylase (MPG) and 8-oxo-guanine-DNA glycosylase-1 (OGG1), which have a key function in protecting the genome from alkylating agents and oxidative stress.

Methylpurine-DNA glycosylase

The bacterial AlkA protein and human MPG excise N3-methyladenine, N7-methylguanine (Fig. 7), other minor forms of alkylated bases as well as hypox-anthine, which occurs spontaneously by deamination of adenine or exposure to nitric oxide. MPG is also active on the lipid peroxidation product $1,N^6$-ethenoadenine, an exocyclic DNA adduct that is further generated by industri-al pollutants including vinyl chloride [69]. It is an example of DNA glycosy-lase that lacks a concomitant AP lyase activity and, hence, depends on an AP endonuclease enzyme for subsequent incision of the abasic site. Transgenic mice with disruption of the *MPG* gene develop normally and are viable. These

Table 1. Human DNA glycosylases.

Glycosylase	Abbreviations	Major substrates	AP lyase activity
Endonuclease III	NTH1, NTHL1	oxidized pyrimidines (thymine and uracil glycol, thymine and cytosine hydrates, urea, etc.)	Yes
Endonuclease VIII-like DNA glycosylase-1	NEIL1	oxidized pyrimidines, 8-oxo-guanine	Yes
Endonuclease VIII-like DNA glycosylase-2	NEIL2	oxidized pyrimidines, 8-oxo-guanine	Yes
Endonuclease VIII-like DNA glycosylase-3	NEIL3	FaPy-G	Yes
Methyl-binding domain glycosylase-4	MBD4, MED1	uracil or T opposite G, T opposite O^6-methylguanine, 5-fluorouracil, $3,N^4$-ethenocytosine	Yes
Methylpurine-DNA glycosylase	MPG, AAG, MDG, ANPG	3-methylpurines, 7-methylguanine, hypoxanthine, $1,N^6$-ethenoadenine	No
MutY homolog DNA glycosylase	MYH	A opposite 8-oxo-guanine	No
8-oxo-guanine-DNA glycosylase-1	OGG1	8-oxo-guanine, FaPy-G, 8-oxo-adenine, urea	Yes
SMUG DNA glycosylase-1	SMUG1	uracil, 5-hydroxyuracil, 5-hydroxymethyluracil, 5-formyluracil	No
Thymine-DNA glycosylase	TDG	uracil or T opposite G, $3,N^4$-ethenocytosine	No
Uracil-DNA glycosylase	UNG	uracil	No

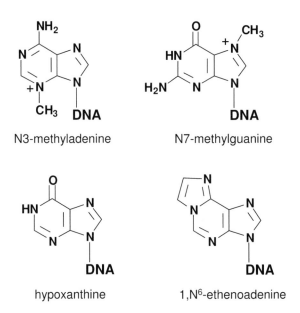

N3-methyladenine   N7-methylguanine

hypoxanthine   1,N^6-ethenoadenine

Figure 7. Major substrates of the promiscuous DNA glycosylase activity of human methylpurine-DNA glycosylase (MPG).

MPG-deficient mice display only a slight increase in spontaneous mutagenesis but no increase in tumorigenesis, although their cells are moderately sensitive to killing by alkylating agents. In rodents, MPG turned out to be the major DNA glycosylase not only for the removal of cytotoxic N3-methyladenine lesions, but also for the excision of mutagenic 1,N^6-ethenoadenine and hypoxanthine lesions [70, 71].

A structural feature shared by many DNA glycosylases (including AlkA and OGG1) is a central DNA-binding domain known as helix-hairpin-helix motif. It consists of two α-helices, connected by a hairpin turn, which mediate interactions with the negatively charged backbone of the nucleic acid substrate [72–74]. Interestingly, all DNA glycosylases make contacts with the backbone of only the lesion-containing strand [75, 76]. Another common motif of most DNA glycosylases is that the target base is flipped-out of the double helix to be inserted into a deep recognition pocket of the enzyme [68]. In many cases, the rigid substrate-binding pocket is specifically tailored for the dimension, shape and chemical properties of a narrow selection of modified targets. However, the crystal structure of AlkA protein revealed a more flexible substrate recognition cleft lined with the aromatic (electron-rich) side chains of the amino acids phenylalanine, tryptophan and tyrosine. Another exceptional feature of AlkA is its ability to adapt to different substrates by a hinge-like motion, thereby conferring flexibility to the aromatic recognition pocket. Widening or narrowing of this promiscuous active site helps to accommodate diverse forms of electron-deficient alkylated bases as well as other modified purines and pyrimidines [74].

Human OGG1

Probably the most abundant lesion induced in DNA by oxidative stress is 8-oxo-guanine (Fig. 8) [77]. A glycosylase that removes this mutagenic injury from DNA was originally been identified in *E. coli* as an activity that releases 2,3-diamino-4-hydroxy-5-formamido-pyrimidine (FaPy-G), a ring-opened purine derivative (Fig. 8) that arises from exposure to ionizing radiation and as a secondary product of alkylating agents [78, 79]. This enzyme responsible for the release of FaPy-G was named FaPy-DNA glycosylase and the *fpg* gene was cloned. Subsequent studies showed that the FaPy-DNA glycosylase is identical to an enzyme that removes 8-oxo-guanine residues [63, 80]. A few years later, a eukaryotic *OGG1* gene was detected in the yeast *Saccharomyces cerevisiae* by its ability to complement bacterial *fpg* mutant strains [81].

To prevent mutations, OGG1 eliminates 8-oxo-guanine before the DNA is replicated, as DNA polymerases would misread the damaged template and insert an adenine instead of cytosine opposite to 8-oxo-guanine [82]. Human OGG1 is a bifunctional DNA-glycosylase/AP lyase that hydrolyses the *N*-glycosyl bond and subsequently cleaves the sugar-phosphate backbone directly 3' to the remaining abasic site (Fig. 6). The structural characterization of OGG1 revealed a helix-hairpin-helix fold, used for interactions with the DNA backbone [73], and a substrate-binding pocket where the flipped-out 8-oxo-guanine is sandwiched between phenylalanine and cysteine side chains [83–85]. Unlike other DNA glycosylases, OGG1 displays a second recognition site where several specific hydrogen bonds are formed with the opposing cytosine base [84]. This additional interaction with the complementary base accounts for the strong discrimination of OGG1 against 8-oxo-guanine residues that are mispaired with adenine, guanine or thymine [83]. The preference of OGG1 for 8-oxo modifications in G:C substrates, and its aversion to process mispaired

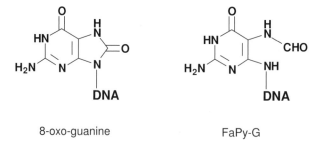

8-oxo-guanine   FaPy-G

Figure 8. Substrates of BER initiated by 8-oxo-guanine-DNA glycosylase-1 (OGG1). FaPy-G, 2,3-diamino-4-hydroxy-5-formamido-pyrimidine.

8-oxo-guanines, represents an important principle against mutagenesis caused by oxidative stress. This notion is supported by the finding that a single point mutation in the cytosine-binding site confers a more relaxed substrate specificity for the base opposite 8-oxo-guanine and results in cancer cells with a mutator phenotype [86, 87].

The structures of guanine and 8-oxo-guanine differ only in the presence of one additional oxygen atom. This minimal variation prompted a series of molecular studies focused on the mechanism by which OGG1 can distinguish between the native base and its oxidized guanine derivative. It is generally thought that DNA glycosylases locate their target bases by sliding along the DNA until a damaged site is detected [85, 88]. However, it was not clear whether normal bases are also flipped-out of the double helix and probed by the enzyme during this scanning process. In the crystal structure, binding of OGG1 to damaged DNA induces a kink of about 70° in the nucleic acid backbone [84]. The finding that the OGG1 enzyme also bends undamaged DNA has been taken as evidence that the search for lesions involves extrusion of normal bases [89]. Indeed, guanine is flipped-out from its helical conformation and presented to the enzyme, but this normal base does not gain access to the substrate recognition pocket. Instead, the native guanine is trapped within an alternative binding surface located outside the lesion recognition cleft [90]. Through a series of conformational changes in the dynamic nucleoprotein complex, damaged bases are forwarded into the active site pocket, whereas the normal guanine is excluded from the catalytic center [90]. The offending 8-oxo-guanine is then excised by attacking the N-glycosylic bond with a conserved lysine residue. Subsequently, OGG1 carries out an AP lyase reaction by β-elimination, generating a 3'-phospho-α,β-unsaturated aldehyde. AP endonuclease-1 recognizes the OGG1-DNA complex, actively displaces OGG1 from the lesion and converts the modified single-strand break to a normal 3'-hydroxyl end by removing the residual aldehyde-phosphate moiety [91, 92].

Transgenic mice lacking OGG1 show an increased accumulation of 8-oxo-guanine in their genome [93] and, in one study [94], suffer from spontaneous lung tumors. Together with the observation that inactivating *OGG1* mutations are found in sporadic cases of human lung, kidney and gastric cancer [86, 87, 95], the phenotype of these knockout mice indicates that OGG1 plays a major role in removing endogenously generated oxidative base lesions. Surprisingly, OGG1$^{(-/-)}$ cells still are able to excise 8-oxo-guanine to a considerable extent, pointing to the existence of redundant repair pathways (Tab. 1). A backup activity for the excision of oxidized bases is for example provided by the DNA glycosylases NEIL1 and 2, which are also able to process 8-oxo-guanine residues [96–98]. Alzheimer's disease is associated with an accumulation of oxidative damage, including 8-oxo-guanine lesions, in the brain [99, 100] and a fraction of Alzheimer's disease patients exhibit mutations in *OGG1* that inactivate or greatly reduce its glycosylase activity [101]. These findings indicate that DNA repair enzymes directed against 8-oxo-guanine protect not only from cancer but also from premature aging and neurodegenerative diseases.

Nucleotide excision repair

Living cells must be capable of dealing with a wide diversity of chemical modifications in DNA with only a limited number of repair factors. Thus, to compensate for the rather narrow substrate range of the previously discussed DNA repair enzymes such as methyltransferases or DNA glycosylases, an alternative repair process has evolved that accommodates a broad spectrum of damaged DNA substrates, primarily helix-distorting (bulky) base lesions induced by short-wavelength UV light and a wide array of chemical carcinogens [34, 35, 102]. Other known NER substrates include a subset of oxidative lesions, protein-DNA cross-links and intrastrand DNA cross-links induced by chemotherapeutic drugs [103–106].

General outline of the NER pathway

The NER reaction consists of a cut-and-patch mechanism that involves excision of single-stranded DNA followed by enzymatic restoration of the intact double helix (Fig. 9). This process is initiated by the recognition of bulky DNA lesions through a multi-protein complex, and culminates in the incision of damaged strands on either side of the lesion, and at some distance from it, through the action of two distinct DNA endonucleases. The injured site is released as the component of an oligonucleotide segment and the pathway is concluded by DNA repair synthesis and DNA ligation [107].

Xeroderma pigmentosum and transgenic mouse models

The importance of preventing genetic mutations caused by UV photoproducts and other helix-distorting DNA adducts is illustrated by a direct link between defects in the NER pathway and a devastating cancer-prone disorder in humans. Indeed, many NER proteins are encoded by genes that, when mutated, give rise to Xeroderma pigmentosum (XP), a recessively inherited disease characterized by extreme photosensitivity and a 2000-fold increased incidence of sunlight-induced skin cancer [2, 108]. Besides tumors of the skin (basal cell carcinomas, squamous cell carcinomas and melanomas), appearing at an average age of 10 years, XP patients develop malignancies of the eyes and tongue. They also have a higher risk of contracting internal tumors and, in some cases, neurological complications, probably reflecting the essential role of the NER pathway in the removal of 5',8-purine cyclodeoxynucleosides or other bulky oxidative DNA lesions [103, 109]. By cell fusion experiments, individuals suffering from XP have been assigned to different complementation groups reflecting mutations in distinct genes. The respective NER proteins (XPA–XPG) were named after the corresponding genetic complementation group. An additional variant form of the disease is due to mutations in a gene

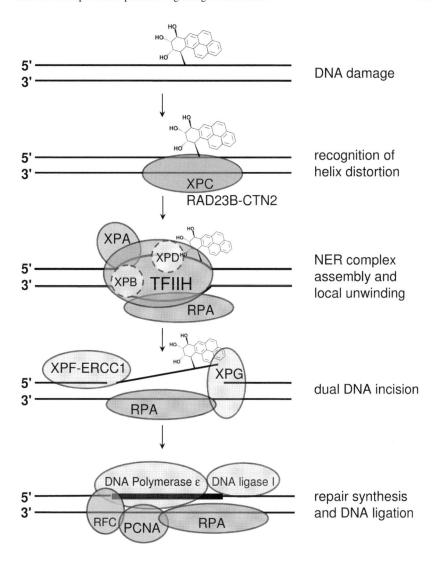

Figure 9. Simplified scheme illustrating the removal of a polycyclic aromatic hydrocarbon adduct by the nucleotide excision repair (NER) pathway. DNA damage recognition by the action of the XPC complex is followed by local DNA unwinding and double DNA incision through the endonucleases XPF-ERCC1 and XPG. The synthesis of repair patches is dependent on RFC, a matchmaker that binds to the remaining excision gap and loads PCNA, which in turn acts as a sliding clamp for DNA polymerases. Finally, the newly synthesized repair patches are sealed by DNA ligase I. XPA-XPF, Xeroderma pigmentosum complementation group A-F proteins; CTN2, centrin-2; TFIIH, transcription factor IIH; RPA, replication protein A; ERCC1, excision repair cross complementing-1 protein; RFC, replication factor C; PCNA, proliferating cell nuclear antigen.

(*XPV*) coding for DNA polymerase η, which is involved in translesion synthesis over UV-damaged DNA templates [2, 3].

Rodents have been genetically engineered to further analyze the pathophysiological consequences of an NER defect. For example, transgenic mice lacking functional XPA protein developed normally, were fertile and showed no obvious clinical abnormalities as long as they were not challenged with DNA-damaging agents. Unexposed XPA-deficient mice only showed a low background incidence of hepatocellular adenomas that were not life threatening. However, UV irradiation of XPA$^{(-/-)}$ mice resulted in acute dermal effects such as erythema and hyperkeratosis, followed by a high incidence of skin and eye malignancies. These XPA$^{(-/-)}$ animals were not only hypersensitive to UV light but also susceptible to tumors of the lymphoid system and liver after oral administration of benzo[a]pyrene. The XPA-deficient mice were also prone to skin cancer following painting with dimethylbenz[a]anthracene [110–112]. Similarly, UV-irradiated XPC$^{(-/-)}$ mice developed skin tumors and pathological eye changes, including keratitis and corneal ulceration, that are reminiscent of the ocular abnormalities found in XP patients [113]. After oral administration of N-acetoxy-acetylaminofluorene or its metabolite N-hydroxy-acetylaminofluorene, XPC$^{(-/-)}$ mice were more susceptible than normal or heterozygous littermates to develop liver and lung tumors [114]. To summarize, transgenic rodent models recapitulate the skin malignancies of XP patients and, in addition, highlight the important protective function of the NER pathway in preventing tumors of internal organs by eliminating DNA lesions caused by chemical carcinogens.

Human core NER reaction

The *XPA–XPG* gene products are part of the human NER complex [115–117]. The order of arrival of individual NER factors on damaged DNA substrates is still debated, but a favored model illustrated in Figure 9 proposes the assembly of a multi-subunit complex triggered by XPC protein together with RAD23B (one of two mammalian homologs of yeast RAD23) and the calcium-binding protein centrin-2 (CTN2) [118, 119]. After initial recognition of damaged sites by the XPC subunit, the human NER pathway proceeds with the sequential recruitment of transcription factor IIH (TFIIH, containing the two DNA helicases XPB and XPD together with eight other subunits), XPA, replication protein A (RPA, consisting of three subunits), XPG and XPF-ERCC1 (a dimer composed of XPF and excision repair cross complementing-1 protein). XPC protein behaves like a "molecular matchmaker" as it initiates the assembly of a repair complex but leaves the DNA substrate before completion of the incision reaction.

In conjunction, TFIIH, XPA and RPA participate in the formation of a stable recognition intermediate [120, 121] characterized by partial unwinding of the duplex substrate (Fig. 9). The XPD subunit of TFIIH, an ATP-dependent DNA helicase with 5' to 3' polarity [122], is mainly responsible for local unwinding, thus generating an open DNA structure flanked by "Y-shaped" double to single strand junctions. The structure-specific endonucleases XPF-ERCC1 and XPG act as "scissors" to cut out DNA damage by cleaving the damaged strand at each of the "Y-shaped" transitions of this open intermediate. XPF-ERCC1 makes the 5' cleavage, whereas XPG introduces the 3' incision [123]. The DNA scissions occur 15–25 nucleotides away from the damaged base on the 5' side but only 3–9 nucleotides away on the 3' side. By this double DNA incision reaction, injured residues are released as part of an oligomeric segment of 24–32 nucleotides in length [124]. Additional factors (XPA and RPA) are required at this stage to coordinate the action of the two endonucleases and to protect the undamaged strand from spurious degradation [125]. The excised 24–32 nucleotides are replaced by a DNA repair synthesis machinery initiated by replication factor C (RFC), which loads proliferating cell nuclear antigen (PCNA) onto the DNA substrate. The trimeric PCNA forms a molecular clamp that supports the processive chain elongation by one of several possible DNA polymerases (δ, ε or κ). Finally, the newly synthesized repair patch is rejoined to the preexisting strand by DNA ligase I [115, 117].

A versatile recognition assay

The question of how the NER complex is able to detect a wide range of carcinogen-DNA adducts, using only a limited number of recognition subunits, has been the subject of intense scrutiny. Work in our laboratory demonstrated that this substrate versatility is achieved by a unique bipartite discrimination strategy that exploits different characteristic features of damaged DNA.

Using an *in vitro* system based on precisely defined DNA substrates, we observed that the efficiency of bulky lesion recognition by the human NER complex correlates with the degree of helical destabilization resulting from the loss of base pairing at DNA adducts [126]. This finding led to the expectation that the NER factors responsible for damage recognition would show an affinity for helical distortions caused by DNA lesions. Base mismatches or nucleotide bulges are, however, not processed by the NER machinery, indicating that the simple thermodynamic destabilization of duplex DNA is not sufficient to qualify as an excision substrate [126, 127].

Further *in vitro* studies demonstrated that the human NER complex remains inactive on DNA duplexes containing a "non-distorting" DNA adduct that preserves normal hydrogen bonds between complementary bases. As indicated in Figure 10, however, such a "non-distorting" DNA adduct in conjunction with local disruption of base pairs, caused by mismatches (substrate 2) or a DNA bulge (substrate 3), gains the ability to induce strong NER reactions. In summary, these *in vitro* experiments revealed that the molecular determinants leading to NER activity consists of two distinct elements, i.e., altered chemistry of the damaged deoxyribonucleotide residue and disruption of Watson-Crick base pairing [126, 127]. Neither defective chemistry alone, in the absence of helical distortions, nor defective base pairing in the absence of bulky adducts, is

Substrate:

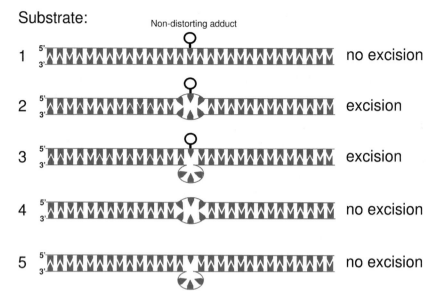

Figure 10. Highly defined DNA substrates constructed to demonstrate the mechanism of bipartite discrimination in the NER pathway. An artificial "non-distorting" adduct is recognized by the NER system only in combination with a concomitant helix distortion induced by base mismatches (substrate 2) or a short DNA loop (substrate 3). However, undamaged homoduplexes or heteroduplexes containing mismatches or loops (substrates 4 and 5, respectively) are not excised and, hence, do not constitute a substrate for the NER process.

sufficient to induce excision responses, but the combination of these two substrate alterations results in the assembly of active repair complexes. Thus, the term of "bipartite recognition" has been introduced to indicate that NER factors use at least two principal levels of discrimination to recognize damaged substrates.

Additional support for a bipartite mechanism of substrate discrimination was sought by studying the excision of DNA adducts caused by bay region benzo[a]pyrene diol-epoxides (B[a]PDE) and the related fjord region benzo[c]phenanthrene diol-epoxides (B[c]PhDE). A fjord region (+)-*trans-anti*-B[c]PhDE-N^6-dA adduct, which retains normal Watson-Crick base pairing, is not excised during incubation in human cell extracts when placed in a fully complementary DNA duplex. However, the same (+)-*trans-anti*-B[c]PhDE-N^6-dA adduct in combination with three mismatched bases stimulates NER activity in human cell extracts [128]. Also in agreement with the hypothesis of bipartite recognition are the diverging excision rates of B[a]PDE adducts that differ only in their stereochemical configuration. In fact, *cis-anti*-B[a]PDE-N^2-dG adducts, which disrupt the G:C base pairing properties, are removed from DNA more rapidly than the isomeric *trans-anti*-B[a]PDE-N^2-dG adducts, which maintain Watson-Crick geometry throughout the modified duplexes [129].

A common conformational feature of NER substrates

Do diverse DNA lesions induce a common structural deformation that may explain how they are all recognized by the same NER system? The native double-stranded DNA is not a static molecule because the DNA strands are constantly in motion due to thermal oscillations, such that the distance between two complementary strands exhibits fast and small variations [130]. In the absence of DNA damage, the extremely short timescale of these small vibrations is probably too short to be recognized by DNA repair factors. However, molecular dynamics simulations predict that the introduction of a single DNA lesion provokes longer-lived and larger openings of the double helix relative to undamaged DNA [131]. In the case of a UV-induced photoproduct, for example, the covalently bonded residues move together in phase, forcing the undamaged bases in the opposite strand to synchronize and give rise to more prominent oscillations compared to native sites. These large fluctuations between complementary strands appear 25 times more frequently at the position of a DNA lesion than in undamaged sequences. Also, the amplitude of these oscillations is drastically increased because the strength of interactions between the two complementary strands is weakened. Interestingly, these dynamic changes triggered by base damage generate mainly oscillations of the intact complementary sequence across lesion sites, as the strand containing base adducts is less flexible than native DNA. Thus, the paradigm of a UV photoproduct shows that DNA lesions have the ability to induce dynamic fluctuations of the DNA duplex, involving transient openings between complementary sequences [131]. The appearance of oscillations mainly in the undamaged strand is also consistent with the conformational heterogeneity detected across *cis-anti*-B[*a*]PDE-N^2-dG adducts [132]. In addition, the analysis of DNA duplexes containing an acetylaminofluorene adduct revealed that only the complementary sequence is susceptible to incision by endonuclease VII, an enzyme that cleaves distorted DNA, thus supporting the notion that bulky DNA lesions generate dynamic deformations particularly on the undamaged side of the duplex [133].

XPC protein is a sensor of abnormal strand oscillations

The key initiator of the NER pathway, XPC protein, is found in complexes with RAD23B, a human homolog of the yeast NER protein RAD23 [134], and CTN2, a calcium-binding protein [119]. XPC protein possesses DNA-binding activity, whereas the RAD23B and CTN2 partners exert accessory functions in stabilizing the complex and stimulating its action in DNA repair. XPC protein alone or in conjunction with RAD23B binds preferentially to damaged DNA substrates containing, for example, (6-4) photoproducts, acetylaminofluorene adducts or *cis*-diamminedichloroplatinum (cisplatin) cross-links [135–137]. Scanning force microscopy studies showed that the binding of XPC protein to damaged double-stranded DNA induces a kink in the nucleic acid backbone [138].

Analysis of the XPC sequence revealed a similarity to the transglutaminase fold of peptide-N-glycanases, which remove glycan modifications from glycoproteins during their degradation. However, the amino acid sequence of XPC protein lacks the predicted catalytic center (Cys-His-Asp) characteristic of this family of enzymes, suggesting that the XPC subunit emerged during evolution through duplication of an ancient peptide-N-glycanase, followed by the loss of enzymatic activity [139]. In addition, we recently discovered an amino acid sequence similarity between one of the single-stranded DNA-binding domains of human RPA (RPA-B), and an XPC region of approximately 100 amino acids. These sequences share 76% similarity and include most of the conserved secondary structure elements characteristic of the oligonucleotide/oligosaccharide-binding fold (OB-fold) responsible for a tight interaction of RPA with single-stranded DNA. A sequence similarity of 64% and 66%, respectively, has also been observed between the same XPC region and two distinct OB-folds of BRCA2 [107].

In agreement with the presence of a single-stranded DNA-binding motif, XPC protein displays a strong bias for single-stranded over double-stranded oligonucleotides, indicating that it senses DNA damage by recognizing the local single-stranded character of DNA duplexes containing bulky lesions. Surprisingly, XPC exhibits an unfavorable binding to damaged single-stranded oligonucleotides compared to the more efficient interaction with undamaged counterparts [107]. To summarize, the exquisite affinity of XPC protein for single-stranded oligonucleotides, in combination with its aversion to interact with damaged DNA strands, points to an "inverted" lesion recognition strategy guided by an association with the intact native strand of the duplex substrate. Further support for this unprecedented model of DNA damage recognition came from the structure of RAD4 protein, a yeast XPC homolog, which has been crystallized in complex with damaged double-stranded DNA [140]. Analysis of the co-crystal revealed that RAD4 binds to damaged DNA in two parts (Fig. 11). The first module, consisting of a large transglutaminase domain (TGD) and a small β-hairpin domain (BHD1), interacts with portions of the native duplex flanking the lesion on the 3' side. The second module, containing two consecutive β-hairpin domains (BHD2 and BHD3), interacts with the undamaged complementary strand opposite to the modified bases. As expected from the biochemical characterization of XPC protein [141], the RAD4 homolog makes no contacts with the modified bases, which are fully expelled from the double helix [140].

Future analyses will show whether the structural basis of damage recognition determined for the yeast RAD4 protein can be fully extrapolated to its XPC homolog. In any case, the indirect mechanism of action established for XPC/RAD4 protein fits with the appearance of large and long-lived oscillation in the native DNA strand across lesion sites, thus predicting that this early sensor operates at sites of bulky lesions by capturing the local and transient for-

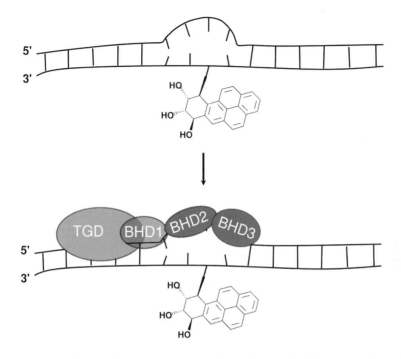

Figure 11. Recognition of DNA sites containing a bulky base lesions. Simplified scheme illustrating how yeast RAD4 protein (a yeast homolog of XPC) may be able to detect DNA oscillations in the undamaged strand opposite to a polycyclic aromatic hydrocarbon adduct. RAD4 protein binds to the native double helix flanking the lesion site through two subdomains designated TGD (transglutami-nase-like domain) and BHD1 (β-hairpin domain-1). Two additional β-hairpin domains (BHD2 and BHD3) bind to the complementary strand opposite to the lesion, thus detecting the single-stranded character of DNA across damaged sites.

mation of a single-stranded configuration in the undamaged complementary sequence (Fig. 11). One advantage of this inverted mode of substrate recognition is that the initial protein sensor operates independently of the variable chemistry of lesion sites, thus explaining the unique substrate versatility of the NER pathway in recognizing a wide array of carcinogen-DNA adducts.

An accessory subunit for the detection of cyclobutane pyrimidine dimers

Biochemical studies suggested that XPC protein on its own is unable to recognize the poorly distorting cyclobutane pyrimidine dimer lesion, which is the most frequent UV radiation product induced by sunlight [142]. However, an accessory subunit has been isolated by virtue of its ability to interact preferentially with DNA fragments containing cyclobutane pyrimidine dimers [143]. This factor, termed UV-DDB for UV-damaged DNA binding, accelerates the excision of cyclobutane dimers in mammalian cells [144]. It is a heterodimer of p127 (DDB1) and p48 (DDB2), with the smaller subunit being encoded by the *XPE* gene [145].

The association of UV-DDB with UV-irradiated substrates leads to bending of the DNA duplex by an angle of 55°, indicating that UV-DDB may recognize cyclobutane pyrimidine dimers and distort the DNA around the lesion to mediate the subsequent recruitment of XPC protein [146]. However, this attractive model fails to explain the efficient excision of cyclobutane pyrimidine dimers in a number of *in vitro* systems lacking UV-DDB [142, 147]. Contradictory results have been reported as to how UV-DDB participates in the NER reaction and it remains possible that UV-DDB represents a chromatin modification factor or that it plays an independent signaling role in the mammalian response to DNA damage [147]. UV-DDB also recognizes other lesions including abasic sites, (6-4) photoproducts and some chemical DNA adducts such as those generated by cisplatin, nitrogen mustard or *N*-methyl-*N'*-nitro-*N*-nitrosoguanidine [148]. DDB1, the large subunit of UV-DDB, forms a molecular adaptor for the Cul4A-Roc1 ubiquitin ligase complex [149] and mediates the modification of target proteins comprising DDB2 itself [150] and XPC protein [151]. Ubiquitylation regulates the turnover of these factors or their recruitment to lesion sites. Following UV radiation, for example, ubiquitylated DDB2 loses its damaged DNA-binding activity [151] and is rapidly degraded [152], in apparent contradiction with its presumed role in the recognition of cyclobutane pyrimidine dimers because the majority of these lesions are still not repaired when most DDB2 is already destroyed. Thus, further studies are necessary to clarify the role of UV-DDB2 in the NER reaction.

TFIIH is a sensor of defective deoxyribonucleotide chemistry

The multifunctional TFIIH complex shuttles between transcription and DNA repair. It has two main components: the core complex consisting of six polypeptides (XPB, TTDA, p62, p52, p44, p34), which assemble in a ring-like structure with a central hole [153] and a protruding CAK (Cdk-activating kinases) complex containing cdk7, cyclin H and MAT1 [154]. TFIIH is presumably recruited to NER sites by XPC protein through interactions with the XPB and p62 subunits [155]. TFIIH then separates the two strands around the lesion, until an approximately 30-nucleotide "bubble" is formed. This unwinding activity generates double-stranded to single-stranded transitions on either side of the lesion, thus providing the substrates for double DNA incision by structure-specific endonucleases XPF-ERCC1 and XPG [120, 121].

The local unwinding process is mediated by the two DNA helicases, XPD and XPB. The XPD subunit functions primarily in DNA repair because its helicase activity, which has 5' to 3' polarity, is required for NER but is dispensable for transcription [156]. In transcription, XPD is thought to have a more structural role by acting as a bridge between the core TFIIH ring and the CAK protrusion [153]. XPB unwinds double-stranded DNA with opposite 3' to 5' polarity and this activity is required for both transcription and the NER process. By analogy with other known DNA helicase complexes [157], a bipolar pair of DNA helicases may serve to unwind the double helix by a joint mechanism whereby the two enzymes translocate with opposite polarity, but in the same direction, on each strand of the antiparallel DNA duplex (Fig. 12A). When tested in bidirectional DNA unwinding assays, the TFIIH complex containing both XPD and XPB only displays 5' to 3' polarity and a side-by-side comparison of purified recombinant proteins showed that XPD has a higher specific activity than XPB [122]. These results indicate that XPD is the main molecular engine that drives translocation of the whole TFIIH complex along DNA.

By interacting predominantly with the undamaged strand, XPC protein may facilitate the subsequent loading of XPD onto the damaged strand, such that this DNA helicase constitutes the first subunit that comes in direct contact with the offending residue. Biochemical experiments with a yeast homolog of XPD, the RAD3 helicase of *S. cerevisiae*, revealed that this molecular engine of the TFIIH complex is arrested by DNA lesions located on the strand along which the enzyme translocates. Like XPD, purified RAD3 protein requires single-stranded regions to initiate unwinding and this loading strand must be free of

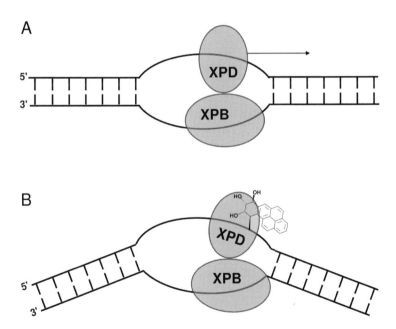

Figure 12. Model illustrating how, in conjunction, the two DNA helicases of TFIIH may cooperate to unwind double-stranded DNA and locate a polycyclic aromatic hydrocarbon adduct. (A) Action of DNA helicases with opposite polarity on undamaged duplex substrates. (B) Obstruction of DNA helicase activity on damaged substrate containing a bulky base adduct. Inhibition of the XPD helicase generates an open and kinked intermediate with double-stranded to single-stranded junctions flanking the lesion.

damage to promote full ATPase and DNA helicase activity. If, however, the loading strand contains DNA lesions, both enzymatic activities are stopped when RAD3 protein encounters the damaged residue. In addition, the presence of base damage induces the formation of stable RAD3 protein-DNA complexes, indicating that RAD3 protein becomes sequestered on DNA at lesion sites [158, 159]. In contrast, lesions in the opposite complementary strand have no effect on this tracking mechanism [158]. More recent studies demonstrated that XPD possesses an iron-sulfur cluster that is required for strand separation [160, 161]. This structural element, where the iron atoms are held in the protein by cysteine residues, may be employed as a molecular "ploughshare" that moves in the 5' to 3' direction to search for DNA lesions. Thus, we propose that inhibition of the 5' to 3' helicase activity serves to detect damaged deoxyribonucleotides during the NER process. Figure 12B illustrates how the strand-specific block of one of the two DNA helicases (XPD) may generate a recognition complex in which the DNA is denatured around the lesion site. This one-dimensional scanning activity provides a mechanism to confirm the presence of DNA bulky lesions before recruiting the endonucleases responsible of double DNA incision.

Summary: Bipartite DNA damage recognition

To initiate repair, DNA lesions must first be recognized and located. Several lines of experimentation led to the hypothesis that the NER system achieves its molecular versatility in damage recognition by a bipartite discrimination mechanism, in which XPC protein constitutes the sensor of abnormal DNA oscillations, whereas TFIIH functions as a tracking enzyme that locates the chemically damaged residues. A salient feature of this bipartite model of substrate discrimination is that the early sensor of DNA damage avoids direct contacts with bulky lesions and, instead, recognizes dynamic deformations of the double helix. XPC protein then attracts TFIIH and loads the ring-like subcomplex of TFIIH, including XPD, onto the damaged strand. Driven by the 5' to 3' helicase activity of XPD protein, TFIIH moves up to a distance of about 15 nucleotides along the damaged strand [127] in search of the lesion. The activity of TFIIH promotes partial unwinding by 20–25 base pairs, thereby separating the duplex. Using this tracking activity, the enzyme probes the chemical composition of the target strand and determines the exact location of the modified residue. Damage recognition is completed when XPD encounters an obstructing adduct and becomes sequestered on the damaged strand.

Transcription-coupled repair

In the NER process, the critical step of DNA damage recognition is accomplished through two different routes. In one subpathway, referred to as global genome repair (GGR), the lesion site is recognized by the XPC complex as outlined above. GGR takes place throughout the genome, irrespective of whether the target sequence is silent or actively transcribed, and this XPC-dependent process constitutes an important mechanism for the removal of potentially mutagenic DNA lesions that ultimately lead to cancer [34, 35, 162]. However, unrepaired DNA lesions not only cause mutations but also interfere with the transcriptional process, thus perturbing normal gene functions and triggering cell senescence or apoptosis [162, 163]. Therefore, living organisms have developed the fast transcription-coupled repair (TCR) subpathway to remove DNA lesions from the transcribed strand of active genes [164]. In this second subpathway, lesion recognition is initiated by RNA polymerase II complexes stalling at sites of damage along the template strand [163]. This XPC-independent mechanism is important for the rapid recovery of transcriptional activity after DNA damage. Due to the TCR process, DNA lesions are cleared more rapidly from transcribed genes than from transcriptionally silent regions [165] and, in proliferating cells, this difference is attributable to the more efficient excision of DNA lesions from the template strand of RNA polymerase II-transcribed genes compared to the non-transcribed coding strand [166].

Recognition of DNA lesions by RNA polymerase II

The progression of RNA polymerases along transcribed strands is obstructed by DNA damage resulting from exposure to UV light [163, 164]. In addition, several genotoxic chemicals, including vinyl chloride, polycyclic aromatic hydrocarbons, aminofluorene, acetylaminofluorene and cisplatin have been shown to induce DNA lesions that pose a physical block to the progression of RNA polymerase II [167–169]. Cisplatin is a widely used and effective anti-cancer drug that forms cross-links between purine bases, with 1,2-d(GpG) intrastrand cross-links being the most frequent adduct [170]. Recent structure-function analyses with RNA polymerase II blocked at a cisplatin 1,2-d(GpG) lesion revealed that the stalling itself results primarily from an insurmountable translocation barrier, which prevents delivery of the cisplatin cross-link into the active site of the polymerase enzyme. The damage-induced inhibition of template translocation is accompanied by misincorporation of AMP opposite the first guanine of the cisplatin cross-link [171]. This mechanism is distinctly different from that of RNA polymerase II stalling at a UV-induced cyclobutane pyrimidine dimer. In this case, after the correct incorporation of AMP opposite the first thymine, the template strand slowly overcomes the translocation barrier and enters the active catalytic center. However, the second thymine of the cyclobutane dimer induces the misincorporation of UTP and the resulting T:U mismatch impedes further elongation of the nascent RNA transcript. Thus, when RNA polymerase II encounters a cyclobutane dimer, it is the damaged-induced misincorporation opposite to the lesion, rather than a physical barrier, that accounts for the stalling of transcription [172]. These findings indicate that the mechanism of transcript elongation by RNA polymerases provides several different opportunities to control the molecular quality of DNA templates.

Coupling of RNA polymerase II to DNA repair

In TCR, the exquisite sensitivity of RNA polymerases to the presence of DNA lesions provides a facilitated mechanism of damage recognition. After immobilization at damaged sites, the stalled RNA polymerase II serves as a molecular "bait" for the recruitment of TFIIH, XPA, RPA, as well as the endonucleases XPG and XPF [163, 164], whereas the XPC and UV-DDB damage recognition subunits are no longer necessary for the formation of an incision intermediate [173]. The damaged site is physically shielded by the RNA polymerase, which prevents access of downstream repair factors. Thus, DNA excision repair can only occur if the RNA polymerase complex is displaced either by complete removal from damaged DNA or through transient regression from the lesion.

The poorly understood TCR process involves the Cockayne syndrome (CS) complementation group A and B proteins [174, 175]. The characteristic molecular hallmark underlying the hereditary condition known as CS is a defect in the recovery of mRNA synthesis after UV irradiation [176]. The delayed resumption of transcriptional activity following genotoxic stress causes a clinical phenotype of postnatal growth failure, progressive neurodegeneration and other symptoms that are reminiscent of segmental premature aging [177]. CSA protein is a member of the WD40-repeat family that regulates protein ubiquitylation reactions by acting as a cofactor for Cul4A-containing E3 ubiquitin ligase enzymes [178]. CSB, on the other hand, is a member of the SWI2/SNF2 family of DNA-dependent ATPases that display chromatin remodeling activity [177, 179].

Several scenarios have been proposed to explain the molecular mechanism of the TCR process. For example, one hypothesis suggests that the arrested RNA polymerase II induces a prolonged interaction with CSB, which then recruits the CSA ubiquitylation complex [180, 181]. The sequestered RNA polymerase enzyme is then phosphorylated and polyubiquitylated [182]. These reports prompted two main interpretations of the role of protein ubiquitylation in the TCR reaction. On the one hand, its has been suggested that ubiquitin residues mark the RNA polymerase II molecule for degradation, thus leaving the damaged strand accessible for repair factors [183]. On the other hand, another hypothesis suggests that a blocked RNA polymerase II molecule does not need to be degraded and that ubiquitylation is a signal for activation of DNA repair or other cellular responses [184]. Yet another study indicates that CSB, TFIIH and XPG cooperate to remodel the structure of the RNA polymerase II complex in an ATP-dependent manner. The resulting conformational shift would allow access to the DNA lesion without removal or disruption of the entire transcription machinery [185].

Domain-associated repair

The vast majority of spontaneous or damage-induced mutations arise during DNA replication. Therefore, it can be expected that terminally differentiated cells, which no longer replicate, may accumulate DNA lesions in the genome without severe consequences, at least as long as they maintain the integrity of those genes needed for their specialized function. In fact, another level of DNA repair complexity has been discovered in post-mitotic cells, for example neurons or macrophages, which neither replicate nor change their gene expression pattern. These terminally differentiated cells are characterized by a general reduction in the overall efficiency of several DNA repair processes, including NER activity. However, the repair of both strands of actively transcribed genes remains proficient in such terminally post-mitotic cells [186, 187]. At the time of its discovery, this preferential repair of active genes was termed "differentiation-associated repair" (DAR). Later, the term has been changed to "domain-associated repair" to indicate more precisely the observation that DNA repair is confined to specific chromatin domains [188]. DAR is not identical to TCR because it depends on the transcription machinery but is not simply triggered by RNA polymerase II encountering a lesion. In fact, DAR also includes the non-transcribed strand and a possible explanation for this phenomenon is the existence of subnuclear compartments or chromatin domains containing active cellular genes. It appears that the DNA substrate is more accessible to repair processes when located in such active subnuclear compartments where transcription usually takes place. Down-regulation of various NER factors with small interfering RNA confirmed that DAR relies on the same subunits, including for example XPC protein, as the GGR pathway [187, 188]. A possible advantage of repairing not only the transcribed strand is that DAR maintains lesion-free templates for the synthesis of repair patches during the TCR process in differentiated cells.

Coordination of DNA repair with cell cycle

The cell division cycle is under the control of cell cycle checkpoints in the G1, S, G2 and M phases and the activation of these checkpoints by DNA damage leads to arrest of the normal division process [2, 3, 6, 189]. It is generally thought that transient cell cycle arrest in response to DNA damage is necessary to provide the cell with enough time to repair DNA lesions, or to resolve replication problems, before entering the next phase of the division cycle [189]. Thus, these checkpoint mechanisms prevent cells from starting DNA replication (the G1/S checkpoint), from progressing with replication (the intra S checkpoint) or from going into mitotic division (the G2/M checkpoint), if they contain DNA damage, collapsed replication forks or broken chromosomes.

In principle, all cellular DNA damage response pathways operate through the action of DNA damage sensors, signal transducers and downstream effectors. One of the key initiators of checkpoint responses is the MRN (MRE11/RAD50/NBS1) complex (see section on HR), which acts as a sensor of DSBs [190]. If, on the other hand, the DNA replication fork becomes arrested, RPA polymerizes on abnormally long stretches of single-stranded DNA and recruits the PCNA-related RAD9/RAD1/HUS1 complex, also known as 911, which acts as sensor of replication problems [189]. The recognition of DSBs or replication blocks (by MRN and 911 complexes, respectively) activates a signal transduction cascade that is primarily triggered by members of the phosphatidylinositol 3-kinase-like kinase (PIKK) family [191]. For example, activation of the Ataxia telangiectasia-mutated (ATM) and Ataxia telangiectasia-related (ATR) kinases, two different PIKK family members, leads to phosphorylation of downstream transducers such as the Chk1 and Chk2 kinases, which in turn activate p53 protein and other effectors that induce cell cycle arrest [2, 3, 189]. Once repair is completed, the cells are able to exit the checkpoint and resume cell cycle progression. Alternatively, unsuccessful DNA repair leads to p53-dependent apoptosis or cellular senescence in order to avoid the propagation of genetically defective cells [3, 177, 190].

Acknowledgements
H.N. is supported by the Swiss National Science Foundation (grant 3100A0-113694) and Oncosuisse (grant KLS-01827-02-2006). U.C. is supported by a research grant from the University of Zürich.

References

1 Lindahl T (1993) Instability and decay of the primary structure of DNA. *Nature* 362: 709–715
2 Friedberg EC, Walker GC, Siede W, Wood RD, Schultz RA, Ellenberger T (2006) *DNA Repair and Mutagenesis.* ASM Press, Washington D.C.
3 Hanawalt PC (2007) Paradigms for the three Rs: DNA replication, recombination, and repair. *Mol Cell* 28: 702–707
4 Lopes M, Foiani M, Sogo JM (2006) Multiple mechanisms control chromosome integrity after replication fork uncoupling and restart at irreparable UV lesions. *Mol Cell* 21: 15–27
5 Naegeli H (1994) Roadblocks and detours during DNA replication: Mechanisms of mutagenesis in mammalian cells. *BioEssays* 16: 557–564
6 van Gent DC, Hoeijmakers JH, Kanaar R (2001) Chromosomal stability and the DNA double-stranded break connection. *Nat Rev Genet* 2: 196–206
7 Povirk LF (1996) DNA damage and mutagenesis by radiomimetic DNA-cleaving agents: Bleomycin, neocarzinostatin and other enediynes. *Mutat Res* 355: 71–89
8 Li X, Heyer W-D (2008) Homologous recombination in DNA repair and DNA damage tolerance. *Cell Res* 18: 99–113
9 Cox MM (2002) The nonmutagenic repair of broken replication forks *via* recombination. *Mutat Res* 510: 107–120
10 Weterings E, Chen DJ (2008) The endless tale of non-homologous end-joining. *Cell Res* 18: 114–124
11 Dip R, Naegeli H (2005) More than just strand breaks: The recognition of structural DNA discontinuities by DNA-dependent protein kinase catalytic subunit. *FASEB J* 19: 704–715
12 Lieber MR, Ma Y, Pannicke U, Schwarz K (2003) Mechanisms and regulation of human non-homologous DNA end-joining. *Nat Rev Mol Cell Biol* 4: 712–720
13 Buck D, Malivert L, de Chasseval R, Barraud A, Fondanèche MC, Sanal O, Plebani A, Stéphan JL, Hufnagel M, le Deist F et al (2006) Cernunnos, a novel nonhomologous end-joining factor, is mutated in human immunodeficiency with microcephaly. *Cell* 124: 287–299

14 Ahnesorg P, Smith P, Jackson SP (2006) XLF interacts with the XRCC4-DNA ligase IV complex to promote DNA nonhomologous end-joining. *Cell* 124: 301–313

15 Downs JA, Jackson SP (2004) A means to a DNA end: The many roles of Ku. *Nat Rev Mol Cell Biol* 5: 367–378

16 Smith GC, Jackson SP (1999) The DNA-dependent protein kinase. *Genes Dev* 13: 916–934

17 Reddy YV, Ding Q, Lees-Miller SP, Meek K, Ramsden DA (2004) Non-homologous end joining requires that the DNA-PK complex undergo an autophosphorylation-dependent rearrangement at DNA ends. *J Biol Chem* 279: 39408–39413

18 DeFazio LG, Stansel RM, Griffith JD, Chu G (2002) Synapsis of DNA ends by DNA-dependent protein kinase. *EMBO J* 21: 3192–3200

19 Weterings E, Verkaik NS, Bruggenwirth HT, Hoeijmakers JH, van Gent DC (2003) The role of DNA-dependent protein kinase in synapsis of DNA ends. *Nucleic Acids Res* 31: 7238–7246

20 Ma Y, Pannicke U, Schwarz K, Lieber MR (2002) Hairpin opening and overhang processing by an Artemis/DNA-dependent protein kinase complex in nonhomologous end joining and V(D)J recombination. *Cell* 108: 781–794

21 Mahajan KN, McElhinny SA, Mitchell BS, Ramsden DA (2002) Association of DNA polymerase μ (pol μ) with Ku and ligase IV: Role of pol μ in end-joining double-strand break repair. *Mol Cell Biol* 22: 5194–5202

22 van Heemst D, Brugmans L, Verkaik NS, van Gent DC (2004) End-joining of blunt DNA double-strand breaks in mammalian fibroblasts is precise and requires DNA-PK and XRCC4. *DNA Rep* 3: 43–50

23 Tsai CJ, Kim SA, Chu G (2007) Cernunnos/XLF promotes the ligation of mismatched and non-cohesive DNA ends. *Proc Natl Acad Sci USA* 104: 7851–7856

24 Tischfield JA (1997) Loss of heterozygosity or: How I learned to stop worrying and love mitotic recombination. *Am J Hum Genet* 61: 995–999

25 Helleday T (2003) Pathways for mitotic homologous recombination. *Mutat Res* 532: 103–115

26 Trujillo KM, Yuan SS, Lee EY, Sung P (1998) Nuclease activities in a complex of human recombination and DNA repair factors Rad50, Mre11, and p95. *J Biol Chem* 273: 21447–21450

27 Sung P (1994) Catalysis of ATP-dependent homologous DNA pairing and strand exchange by yeast RAD51 protein. *Science* 265: 1241–1243

28 Ogawa T, Yu X, Shinohara A, Egelman EH (1993) Similarity of the yeast RAD51 filament to the bacterial RecA filament. *Science* 259: 1896–1899

29 Conway AB, Lynch TW, Zhang Y, Fortin GS, Fung CW, Symington LS, Rice PA (2004) Crystal structure of a Rad51 filament. *Nat Struct Mol Biol* 11: 791–796

30 Constantinou A, Chen XB, McGowan CH, West SC (2002) Holliday junction resolution in human cells: Two junction endonucleases with distinct substrate specificities. *EMBO J* 21: 5577–5585

31 Niedernhofer LJ, Daniels JS, Rouzer CA, Greene RE, Marnett LJ (2003) Malodialdehyde, a product of lipid peroxidation, is mutagenic in human cells. *J Biol Chem* 278: 31426–31433

32 Hanada K, Budzowska M, Modesti M, Maas A, Wyman C, Essers J, Kanaar R (2006) The structure-specific endonuclease Mus81-Eme1 promotes conversion of interstrand DNA crosslinks into double-strand breaks. *EMBO J* 25: 4921–4932

33 Kennedy RD, D'Andrea AD (2005) The Fanconi Anemia/BRCA pathway: New faces in the crowd. *Genes Dev* 19: 2925–2940

34 Sancar A (1996) DNA excision repair. *Annu Rev Biochem* 65: 43–81

35 Wood RD (1997) Nucleotide excision repair in mammalian cells. *J Biol Chem* 272: 23465–23468

36 Loechler EL, Green CL, Essigmann J (1984) *In vivo* mutagenesis by O^6-methylguanine built into a unique site in a viral genome. *Proc Natl Acad Sci USA* 81: 6271–6275

37 Preston BD, Singer B, Loeb LA (1986) Mutagenic potential of O^4-methylthymine *in vivo* determined by an enzymatic approach to site-specific mutagenesis. *Proc Natl Acad Sci USA* 83: 8501–8505

38 Mitra S, Kaina B (1993) Regulation of repair of alkylation damage in mammalian genomes. *Prog Nucleic Acids Res Mol Biol* 44: 109–142

39 Pegg AE, Byers TL (1992) Repair of DNA containing O^6-alkylguanine. *FASEB J* 6: 2302–2310

40 Samson L, Han S, Marquis JC, Rasmussen LJ (1997) Mammalian DNA repair methyltransferases shield O^4MeT from nucleotide excision repair. *Carcinogenesis* 18: 919–924

41 Moore MH, Gulbis JM, Dodson EJ, Demple B, Moody PC (1994) Crystal structure of a suicidal DNA repair protein: The Ada O^6-methylguanine-DNA methyltransferase from *E. coli*. *EMBO J* 13: 1495–1501

42 Tsuzuki T, Sakumi K, Shiraishi A (1996) Targeted disruption of the DNA repair methyltransferase gene renders mice hypersensitive to alkylating agents. *Carcinogenesis* 17: 1215–1220

43 Iwakuma T, Sakumi K, Nakatsuru Y, Kawate H, Igarashi H, Shiraishi A, Tsuzuki T, Ishikawa T, Sekiguchi M (1997) High incidence of nitrosamine-induced tumorigenesis in mice lacking DNA repair methyltransferase. *Carcinogenesis* 18: 1631–1635

44 Glassner BJ, Weeda G, Allan JM, Broekhof JL, Carla NH, Donker I, Engelward BP, Hampson RJ, Hersmus R, Hickman MJ et al (1999) DNA repair methyltransferase (Mgmt) knockout mice are sensitive to the lethal effects of chemotherapeutic alkylating agents. *Mutagenesis* 14: 339–347

45 Dumenco LL, Allay E, Norton K, Gerson SL (1993) The prevention of thymic lymphomas in transgenic mice by human O^6-akylguanine DNA alkyltransferase. *Science* 259: 219–222

46 Nakatsuru Y, Matsukuma S, Nemoto N, Sugano H, Sekiguchi M, Ishikawa T (1993) O^6-methyl-guanine-DNA methyltransferase protects against nitrosamine-induced hepatocarcinogenesis. *Proc Natl Acad Sci USA* 90: 6468–6472

47 Becker K, Dosch J, Gregel CM, Martin BA, Kaina B (1996) Targeted expression of human O(6)-methylguanine-DNA methyltransferase (MGMT) in transgenic mice protects against tumor initiation in two-stage skin carcinogenesis. *Cancer Res* 56: 3244–3249

48 Liu L, Qin X, Gerson SL (1999) Reduced lung tumorigenesis in human methylguanine DNA-methyltransferase transgenic mice achieved by expression of transgene within the target cell. *Carcinogenesis* 20: 279–284

49 Sklar RM, Strauss BS (1981) Removal of O^6-methylguanine from DNA of normal and *Xeroderma pigmentosum*-derived lymphoblastoid lines. *Nature* 289: 417–420

50 Esteller M, Garcia-Foncillas J, Andion E, Goodman SN, Hidalgo OF, Vanaclocha V, Baylin SB, Herman JG (2000) Inactivation of the DNA-repair gene MGMT and the clinical response of gliomas to alkylating agents. *N Engl J Med* 343: 1350–1354

51 Gerson SL (2002) Clinical relevance of MGMT in the treatment of cancer. *J Clin Oncol* 20: 2388–2399

52 Trewick SC, Henshaw TF, Hausinger RP, Lindahl T, Sedgwick B (2002) Oxidative demethylation by Escherichia coli AlkB directly reverts DNA base damage. *Nature* 419: 174–178

53 Tsujikawa K, Koike K, Kitae K, Shinkawa A, Arima H, Suzuki T, Tsuchiya M, Makino Y, Furukawa T, Konishi N, Yamamoto H (2007) Expression and sub-cellular localization of human ABH family molecules. *J Cell Mol Med* 11: 1105–1116

54 Aas PA, Otterlei M, Falnes PO, Vagbo CB, Skorpen F, Akbari M, Sundheim O, Bjoras M, Slupphaug G, Seeberg E, Krokan HE (2003) Human and bacterial oxidative demethylase repair alkylation damage in both RNA and DNA. *Nature* 421: 859–863

55 Duncan T, Trewick SC, Koivisto P, Bates PA, Lindahl T, Sedgwick B (2002) Reversal of DNA alkylation damage by two human dioxygenases. *Proc Natl Acad Sci USA* 99: 16660–16665

56 Ringvoll J, Nordstrand LM, Vagbo CB, Talstad V, Reite K, Aas PA, Lauritzen KH, Liabakk NB, Bjork A, Doughty RW et al (2006) Repair deficient mice reveal mABH2 as the primary oxidative demethylase for repairing 1meA and 3meC lesions in DNA. *EMBO J* 25: 2189–2198

57 Mitchell D (2006) Revisiting the photochemistry of solar UVA in human skin. *Proc Natl Acad Sci USA* 103: 13567–13568

58 Kelner A (1949) Effect of visible light on the recovery of *Streptomyces griseus* conidia from ultra-violet irradiation injury. *Proc Natl Acad Sci USA* 35: 73–79

59 Dulbecco R (1949) Reactivation of ultraviolet-inactivated bacteriophage with visible light. *Nature* 163: 949–950

60 Rupert CS, Goodgal SH, Herriott RM (1958) Photoreactivation *in vitro* of ultraviolet-inactivated *Hemophilus influenzae* transforming factor. *J Gen Physiol* 41: 451–471

61 Park HW, Kim ST, Sancar A, Deisenhofer J (1995) Crystal structure of DNA photolyase from *Escherichia coli*. *Science* 268: 1866–1872

62 Essen LO, Klar T (2006) Light-driven DNA repair by photolyases. *Cell Mol Life Sci* 63: 1266–1277

63 Kato T, Todo T, Ayaki H, Ishizaki K, Morita T, Mitra S, Ikenaga M (1994) Cloning of a marsupial DNA photolyase gene and the lack of related nucleotide sequences in placental mammals. *Nucleic Acids Res* 22: 4119–4124

64 Li YF, Kim ST, Sancar A (1993) Evidence for lack of photoreactivating enzyme in humans. *Proc Natl Acad Sci USA* 90: 4389–4393

65 Naegeli H (1997) *Mechanisms of DNA Damage Recognition in Mammalian Cells.* Springer, New York

66 Demple B, Harrison L (1994) Repair of oxidative damage to DNA: Enzymology and biology. *Annu Rev Biochem* 63: 915–948

67 Dodson ML, Michaels ML, Lloyd RS (1994) Unified catalytic mechanism for DNA glycosylases. *J Biol Chem* 269: 32709–32712

68 Slupphaug G, Mol CD, Kavli B, Arvai AS, Krokan HE, Tainer JA (1996) A nucleotide-flipping mechanism from the structure of human uracil-DNA glycosylase bound to DNA. *Nature* 386: 87–92

69 Saparbaev M, Kleibl K, Laval J (1995) *Escherichia coli, Saccharomyces cerevisiae*, rat, and human 3-methyladenine DNA glycosylase repair 1,N^6-ethenoadenine when present in DNA. *Nucleic Acids Res* 23: 3750–3755

70 Engelward BP, Weeda G, Wyatt MD, Broekhof JL, de Wit J, Donker I, Allan JM, Gold B, Hoeijmakers JH, Samson LD (1997) Base excision repair deficient mice lacking the Aag alkyladenine DNA glycosylase. *Proc Natl Acad Sci USA* 94: 13087–13092

71 Hang B, Singer B, Margison GP, Elder RH (1997) Targeted deletion of alkylpurine-DNA-*N*-glycosylase in mice eliminates repair of 1,N^6-ethenoadenine and hypoxanthine but not of 3,N^4-ethenocytosine or 8-oxoguanine. *Proc Natl Acad Sci USA* 94: 12869–12874

72 Thayer MM, Ahern H, Xing D, Cunningham RP, Tainer JA (1995) Novel DNA binding motifs in the DNA repair enzyme endonuclease III crystal structure. *EMBO J* 14: 4108–4120

73 Nash HM, Bruner SD, Schärer OD, Kawate T, Addona TA, Spooner E, Lane WS, Verdine GL (1996) Cloning of a yeast 8-oxoguanine DNA glycosylase reveals the existence of a base-excision DNA-repair protein superfamily. *Curr Biol* 6: 968–980

74 Labahn J, Schärer OD, Long A, Ezaz-Nikpay K, Verdine GL, Ellenberger TE (1996) Structural basis for the excision repair of alkylation-damaged DNA. *Cell* 86: 321–329

75 Parikh SS, Mol CD, Slupphaug G, Bharati S, Krokan HE, Tainer JA (1998) Base excision repair initiation revealed by crystal structures and binding kinetics of human uracil-DNA glycosylase with DNA. *EMBO J* 17: 5214–5226

76 Lau AY, Schärer OD, Samson L, Verdine GL, Ellenberger T (1998) Crystal structure of a human alkylbase-DNA repair enzyme complexed to DNA: Mechanisms for nucleotide flipping and base excision. *Cell* 95: 249–258

77 Cooke MS, Evans MD, Dizdaroglu M, Lunec J (2003) Oxidative DNA damage: Mechanisms, mutation, and disease. *FASEB J* 17: 1195–1214

78 Aruoma OI, Halliwell B, Dizdaroglu M (1989) Iron ion-dependent modification of bases in DNA by the superoxide radical-generating system hypoxanthine/xanthine oxidase. *J Biol Chem* 264: 13024–13028

79 Gajewski E, Rao G, Nackerdien Z, Dizdaroglu M (1990) Modification of DNA bases in mammalian chromatin by radiation-generated free radicals. *Biochemistry* 29: 7876–7882

80 Sung JS, Demple B (2006) Roles of base excision repair subpathways in correcting oxidized abasic sites in DNA. *FEBS J* 273: 1620–1629

81 van der Kemp PA, Thomas D, Barbey R, de Oliveira R, Boiteux S (1996) Cloning and expression in *Escherichia coli* of the OGG1 gene of *Saccharomyces cerevisiae*, which codes for a DNA glycosylase that excises 7,8-dihydro-8-oxoguanine and 2,6-diamino-4-hydroxy-5-*N*-methylformamidopyrimidine. *Proc Natl Acad Sci USA* 93: 5197–5202

82 Hsu GW, Ober M, Carell T, Beese LS (2004) Error-prone replication of oxidatively damaged DNA by a high-fidelity DNA polymerase. *Nature* 431: 217–221

83 Bjoras M, Seeberg E, Luna L, Pearl LH, Barrett TE (2002) Reciprocal "flipping" underlies substrate recognition and catalytic activation by the human 8-oxo-guanine DNA glycosylase. *J Mol Biol* 317: 171–177

84 Bruner SD, Norman DPG, Verdine GL (2000) Structural basis for recognition and repair of the endogenous mutagen 8-oxoguanine in DNA. *Nature* 403: 859–866

85 Stivers JT, Jiang YL (2003) A mechanistic perspective on the chemistry of DNA repair glycosylases. *Chem Rev* 103: 2729–2759

86 Shinmura K, Kohno T, Kasai H, Koda K, Sugimura H, Yokota J (1998) Infrequent mutations of the *hOGG1* gene, that is involved in the excision of 8-hydroxyguanine in damaged DNA, in human gastric cancer. *Jpn J Cancer Res* 89: 825–828

87 Audebert M, Radicella JP, Dizdaroglu M (2000) Effect of single mutations in the *OGG1* gene found in human tumors on the substrate specificity of the Ogg1 protein. *Nucleic Acids Res* 28: 2672–2678

88 Kuznetsov NA, Koval VV, Nevinsky GA, Douglas KT, Zharkov DO, Fedorova OS (2007) Kinetic

conformational analysis of human 8-oxoguanine-DNA glycosylase. *J Biol Chem* 282: 1029–1038

89 Chen L, Haushalter KA, Lieber CM, Verdine GL (2002) Direct visualization of a DNA glycosylase searching for damage. *Chem Biol* 9: 345–350

90 Banerjee A, Yang W, Karplus M, Verdine GL (2005) Structure of a repair enzyme interrogating undamaged DNA elucidates recognition of damaged DNA. *Nature* 434: 612–618

91 Sidorenko VS, Nevinsky GA, Zharkov DO (2007) Mechanism of interaction between human 8-oxoguanine-DNA glycosylase and AP endonuclease. *DNA Rep* 6: 317–328

92 Wilson DM, Bohr VA (2007) The mechanics of base excision repair, and its relationship to aging and disease. *DNA Rep* 6: 544–559

93 Klungland A, Rosewell I, Hollenbach S, Larsen E, Daly G, Epe B, Seeberg E, Lindahl T, Barnes DE (1999) Accumulation of premutagenic DNA lesions in mice defective in removal of oxidative base damage. *Proc Natl Acad Sci USA* 96: 13300–13305

94 Sakumi K, Tominaga Y, Furuichi M, Xu P, Tsuzuki T, Sekiguchi M, Nakabeppu Y (2003) *Ogg1* knockout-associated lung tumorigenesis and its suppression by *Mth1* gene disruption. *Cancer Res* 63: 902–905

95 Chevillard S, Radicella JP, Levalois C, Lebeau J, Poupon MF, Oudard S, Dutrillaux B, Boiteux S (1998) Mutations in OGG1, a gene involved in the repair of oxidative DNA damage, are found in human lung and kidney tumours. *Oncogene* 16: 3083–3086

96 Dou H, Mitra S, Hazra TK (2003) Repair of oxidized bases in DNA bubble structures by human DNA glycosylases NEIL1 and NEIL2. *J Biol Chem* 278: 49679–49684

97 Hu J, de Souza-Pinto NC, Haraguchi K, Hogue BA, Jaruga P, Greenberg MM, Dizdaroglu M, Bohr VA (2005) Repair of formamidopyrimidines in DNA involves different glycosylases: Role of the OGG1, NTH1, and NEIL1 enzymes. *J Biol Chem* 280: 40544–40551

98 Hazra TK, Das A, Das S, Choudhury S, Kow YW, Roy R (2006) Oxidative DNA damage repair in mammalian cells: A new perspective. *DNA Rep* 6: 470–480

99 Lovell MA, Markesbery WR (2001) Ratio of 8-hydroxyguanine in intact DNA to free 8-hydroxyguanine is increased in Alzheimer disease ventricular cerebrospinal fluid. *Arch Neurol* 58: 392–396

100 Wang J, Xiong S, Xie C, Markesbery WR, Lovell MA (2005) Increased oxidative damage in nuclear and mitochondrial DNA in Alzheimer's disease. *J Neurochem* 93: 953–962

101 Mao G, Pan X, Zhu BB, Zhang Y, Yuan F, Huang J, Lovell MA, Lee MP, Markesbery WR, Li GM, Gu L (2007) Identification and characterization of OGG1 mutations in patients with Alzheimer's disease. *Nucleic Acids Res* 35: 2759–2766

102 de Laat WL, Jaspers NG, Hoeijmakers JH (1999) Molecular mechanism of nucleotide excision repair. *Genes Dev* 13: 768–785

103 Satoh MS, Jones CJ, Wood RD, Lindahl T (1993) DNA excision-repair defect of *Xeroderma pigmentosum* prevents removal of a class of oxygen free radical-induced base lesions. *Proc Natl Acad Sci USA* 90: 6335–6339

104 Reardon JT, Bessho T, Kung HC, Bolton PH, Sancar A (1997) *In vitro* repair of oxidative DNA damage by human nucleotide excision repair system: Possible explanation for neurodegeneration in *Xeroderma pigmentosum* patients. *Proc Natl Acad Sci USA* 94: 9463–9468

105 Kuraoka I, Bender C, Romieu A, Cadet J, Wood RD, Lindahl T (2000) Removal of oxygen free-radical-induced 5',8-purine cyclodeoxynucleosides from DNA by the nucleotide excision-repair pathway in human cells. *Proc Natl Acad Sci USA* 97: 3832–3837

106 Reardon JT, Sancar A (2006) Repair of DNA-polypeptide crosslinks by human excision nuclease. *Proc Natl Acad Sci USA* 103: 4056–4061

107 Maillard O, Solyom S, Naegeli H (2007) An aromatic sensor with aversion to damaged strands confers versatility to DNA repair. *PLoS Biol* 5: e79

108 Kraemer KH, Lee MM, Scotto J (1984) DNA repair protects against cutaneous and internal neoplasia: Evidence from *Xeroderma pigmentosum*. *Carcinogenesis* 5: 511–514

109 D'Errico M, Parlanti E, Teson M, de Jesus BM, Degan P, Calcagnile A, Jaruga P, Bjoras M, Crescenzi M, Pedrini AM et al (2006) New functions of XPC in the protection of human skin cells from oxidative damage. *EMBO J* 25: 4305–4315

110 Nakane H, Takeuchi S, Yuba S, Saijo M, Nakatsu Y, Murai H, Nakatsuru Y, Ishikawa T, Hirota S, Kitamura Y et al (1995) High incidence of ultraviolet-B- or chemical-carcinogen-induced skin tumors in mice lacking the *Xeroderma pigmentosum* group A gene. *Nature* 377: 165–168

111 de Vries A, van Oostrom CT, Hofhuis FM, Dortant PM, Berg RJ, de Gruijl FR, Wester PW, van

Kreijl CF, Capel PJ, van Steeg H, Verbeck SJ (1995) Increased susceptibility to ultraviolet-B and carcinogens of mice lacking the DNA excision repair gene *XPA*. *Nature* 377: 169–173

112 de Vries A, van Oostrom CT, Dortant PM, Beems RB, van Kreijl CF, Capel PJ, van Steeg H (1997) Spontaneous liver tumors and benzopyrene-induced lymphomas in XPA-deficient mice. *Mol Carcinog* 19: 46–53

113 Sands AT, Abuin A, Sanchez A, Conti CJ, Bradley A (1995) High susceptibility to ultraviolet-induced carcinogenesis in mice lacking XPC. *Nature* 377: 162–165

114 Cheo DL, Burns DK, Meira LB, Houle JF, Friedberg EC (1999) Mutational inactivation of the *Xeroderma pigmentosum* group C gene confers predisposition to 2-acetylaminofluorene-induced liver and lung cancer and to spontaneous testicular cancer in Trp53$^{-/-}$ mice. *Cancer Res* 59: 771–775

115 Aboussekhra A, Biggerstaff M, Shivji MK, Vilpo JA, Moncollin V, Podust VN, Protic M, Hubscher U, Egly JM, Wood RD (1995) Mammalian DNA nucleotide excision repair reconstituted with purified protein components. *Cell* 80: 859–868

116 Mu D, Park CH, Matsunaga T, Hsu DS, Reardon JT, Sancar A (1995) Reconstitution of human DNA repair excision nuclease in a highly defined system. *J Biol Chem* 270: 2415–2418

117 Araujo SJ, Tirode F, Coin F, Pospiech H, Syvaoja JE, Stucki M, Hubscher U, Egly JM, Wood RD (2000) Nucleotide excision repair of DNA with recombinant human proteins: Definition of the minimal set of factors, active forms of TFIIH, and modulation by CAK. *Genes Dev* 14: 349–359

118 Volker M, Moné MJ, Karmakar P, van Hoffen A, Schul W, Vermeulen W, Hoeijmakers JH, van Driel R, van Zeeland AA, Mullenders LH (2001) Sequential assembly of the nucleotide excision repair factors *in vivo*. *Mol Cell* 8: 213–224

119 Nishi R, Okuda Y, Watanabe E, Mori T, Iwai S, Masutani C, Sugasawa K, Hanaoka F (2005) Centrin 2 stimulates nucleotide excision repair by interacting with *Xeroderma pigmentosum* group C protein. *Mol Cell Biol* 25: 5664–5674

120 Mu D, Wakasugi M, Hsu DS, Sancar A (1997) Characterization of reaction intermediates of human excision repair nuclease. *J Biol Chem* 272: 28971–28979

121 Evans E, Moggs JG, Hwang JR, Egly JM, Wood RD (1997) Mechanism of open complex and dual incision formation by human nucleotide excision repair factors. *EMBO J* 16: 6559–6573

122 Coin F, Marinoni JC, Rodolfo C, Fribourg S, Pedrini AM, Egly JM (1998) Mutations in the XPD helicase gene result in XP and TTD phenotypes, preventing interaction between XPD and the p44 subunit of TFIIH. *Nat Genet* 20: 184–188

123 O'Donovan A, Davies AA, Moggs JG, West SC, Wood RD (1994) XPG endonuclease makes the 3' incision in human DNA nucleotide excision repair. *Nature* 371: 432–435

124 Huang JC, Svoboda DL, Reardon JT, Sancar A (1992) Human nucleotide excision nuclease removes thymine dimers from DNA by incising the 22nd phosphodiester bond 5' and the 6th phosphodiester bond 3' to the photodimer. *Proc Natl Acad Sci USA* 89: 3664–3668

125 Missura M, Buterin T, Hindges R, Hubscher U, Kasparkova J, Brabec V, Naegeli H (2001) Double-check probing of DNA bending and unwinding by XPA-RPA: An architectural function in DNA repair. *EMBO J* 20: 3554–3564

126 Hess MT, Schwitter U, Petretta M, Giese B, Naegeli H (1997) Bipartite substrate discrimination by human nucleotide excision repair. *Proc Natl Acad Sci USA* 94: 6664–6669

127 Buschta-Hedayat N, Buterin T, Hess MT, Missura M, Naegeli H (1999) Recognition of nonhybridizing base pairs during nucleotide excision repair of DNA. *Proc Natl Acad Sci USA* 96: 6090–6095

128 Buterin T, Hess MT, Luneva N, Geacintov NE, Amin S, Kroth H, Seidel A, Naegeli H (2000) Unrepaired fjord region polycyclic aromatic hydrocarbon-DNA adducts in *ras* codon 61 mutational hot spots. *Cancer Res* 60: 1849–1856

129 Hess MT, Gunz D, Luneva N, Geacintov NE, Naegeli H (1997) Base pair conformation-dependent excision of benzo[*a*]pyrene diol epoxide-guanine adducts by human nucleotide excision repair enzymes. *Mol Cell Biol* 17: 7069–7076

130 Choi CH, Kalosakas G, Rasmussen KO, Hiromura M, Bishop AR, Usheva A (2004) DNA dynamically directs its own transcription initiation. *Nucleic Acids Res* 32: 1584–1590

131 Blagoev KB, Alexandrov BS, Goodwin EH, Bishop AR (2006) Ultra-violet light induced changes in DNA dynamics may enhance TT-dimer recognition. *DNA Rep* 5: 863–867

132 Cosman M, Hingerty BE, Luneva N, Amin S, Geacintov NE, Broyde S, Patel DJ (1996) Solution conformation of the (–)-*cis-anti*-benzo[*a*]pyrenyl-dG adduct opposite dC in a DNA duplex: Intercalation of the covalently attached BP ring into the helix with base displacement of the mod-

ified deoxyguanosine into the major groove. *Biochemistry* 35: 9850–9863

133 Bertrand-Burggraf E, Kemper B, Fuchs RP (1994) Endonuclease VII of phage T4 nicks *N*-2-acetylaminofluorene-induced DNA structures *in vitro*. *Mutat Res* 314: 287–295

134 Masutani C, Sugasawa K, Yanagisawa J, Sonoyama T, Ui M, Enomoto T, Takio K, Tanaka K, van der Spek PJ et al (1994) Purification and cloning of a nucleotide excision repair complex involving the *Xeroderma pigmentosum* group C protein and a human homologue of yeast RAD23. *EMBO J* 13: 1831–1843

135 Sugasawa K, Ng JM, Masutani C, Iwai S, van der Spek PJ, Eker AP, Hanaoka F, Bootsma D, Hoeijmakers JH (1998) *Xeroderma pigmentosum* group C protein complex is the initiator of global genome nucleotide excision repair. *Mol Cell* 2: 223–232

136 Batty D, Rapic'-Otrin V, Levine AS, Wood RD (2000) Stable binding of human XPC complex to irradiated DNA confers strong discrimination for damaged sites. *J Mol Biol* 300: 275–290

137 Kusumoto R, Masutani C, Sugasawa K, Iwai S, Araki M, Uchida A, Mizukoshi T, Hanaoka F (2001) Diversity of the damage recognition step in the global genomic nucleotide excision repair *in vitro*. *Mutat Res* 485: 219–227

138 Janicijevic A, Sugasawa K, Shimizu Y, Hanaoka F, Wijgers N, Djurica M, Hoeijmakers JH, Wyman C (2003) DNA bending by the human damage recognition complex XPC-HR23B. *DNA Rep* 2: 325–336

139 Anantharaman V, Koonin EV, Aravind L (2001) Peptide-*N*-glycanases and DNA repair proteins, Xp-C/Rad4, are, respectively, active and inactivated enzymes sharing a common transglutaminase fold. *Hum Mol Genet* 10: 1627–1630

140 Min JH, Pavletich NP (2007) Recognition of DNA damage by the Rad4 nucleotide excision repair protein. *Nature* 449: 570–575

141 Buterin T, Meyer C, Giese B, Naegeli H (2005) DNA quality control by conformational readout on the undamaged strand of the double helix. *Chem Biol* 12: 913–922

142 Reardon JT, Sancar A (2003) Recognition and repair of the cyclobutane thymine dimer, a major cause of skin cancers, by the human excision nuclease. *Genes Dev* 17: 2539–2551

143 Feldberg RS, Grossman L (1976) A DNA binding protein from human placenta specific for ultraviolet damaged DNA. *Biochemistry* 15: 2402–2408

144 Tang JY, Hwang BJ, Ford JM, Hanawalt PC, Chu G (2000) *Xeroderma pigmentosum* p48 gene enhances global genomic repair and suppresses UV-induced mutagenesis. *Mol Cell* 5: 737–744

145 Rapic-Otrin V, Navazza V, Nardo T, Botta E, McLenigan M, Bisi DC, Levine AS, Stefanini M (2003) True XP group E patients have a defective UV-damaged DNA binding protein complex and mutations in *DDB2* which reveal the functional domains of its p48 product. *Hum Mol Genet* 12: 1507–1522

146 Fujiwara Y, Masutani C, Mizukoshi T, Kondo J, Hanaoka F, Iwai S (1999) Characterization of DNA recognition by the human UV-damaged DNA-binding protein. *J Biol Chem* 274: 20027–20033

147 Kulaksiz G, Reardon JT, Sancar A (2005) *Xeroderma pigmentosum* complementation group E protein (XPE/DDB2): Purification of various complexes of XPE and analyses of their damaged DNA binding and putative DNA repair properties. *Mol Cell Biol* 25: 9784–9792

148 Payne A, Chu G (1994) *Xeroderma pigmentosum* group E binding factor recognizes a broad spectrum of DNA damage. *Mutat Res* 310: 89–102

149 Shiyanov P, Nag A, Raychaudhuri P (1999) Cullin 4A associates with the UV-damaged DNA-binding protein DDB. *J Biol Chem* 274: 35309–35312

150 Nag A, Bondar T, Shiv S, Raychaudhuri P (2001) The *Xeroderma pigmentosum* group E gene product DDB2 is a specific target of cullin 4A in mammalian cells. *Mol Cell Biol* 21: 6738–6747

151 Sugasawa K, Okuda Y, Saijo M, Nishi R, Matsuda N, Chu G, Mori T, Iwai S, Tanaka K, Hanaoka F (2005) UV-induced ubiquitylation of XPC protein mediated by UV-DDB-ubiquitin ligase complex. *Cell* 121: 387–400

152 Rapic-Otrin V, McLenigan MP, Bisi DC, Gonzalez M, Levine AS (2002) Sequential binding of UV DNA damage binding factor and degradation of the p48 subunit as early events after UV irradiation. *Nucleic Acids Res* 30: 2588–2598

153 Schultz P, Fribourg S, Poterszman A, Mallouh V, Moras D, Egly JM (2000) Molecular structure of human TFIIH. *Cell* 102: 599–607

154 Zurita M, Merino C (2003) The transcriptional complexity of the TFIIH complex. *Trends Genet* 19: 578–584

155 Yokoi M, Masutani C, Maekawa T, Sugasawa K, Ohkuma Y, Hanaoka F (2000) The *Xeroderma*

pigmentosum group C protein complex XPC-HR23B plays an important role in the recruitment of transcription factor IIH to damaged DNA. *J Biol Chem* 275: 9870–9875

156 Guzder SN, Qiu H, Sommers CH, Sung P, Prakash L, Prakash S (1994) DNA repair gene RAD3 of *S. cerevisiae* is essential for transcription by RNA polymerase II. *Nature* 367: 91–94

157 Dillingham MS, Spies M, Kowalczykowski SC (2003) RecBCD enzyme is a bipolar DNA helicase. *Nature* 423: 893–897

158 Naegeli H, Bardwell L, Friedberg EC (1992) The DNA helicase and adenosine triphosphatase activities of yeast Rad3 protein are inhibited by DNA damage. *J Biol Chem* 267: 392–398

159 Naegeli H, Bardwell L, Friedberg EC (1993) Inhibition of Rad3 DNA helicase activity by DNA adducts and abasic sites: Implications for the role of a DNA helicase in damage-specific incision of DNA. *Biochemistry* 32: 613–621

160 Rudolf J, Makrantoni V, Ingledew WJ, Stark MJ, White MF (2006) The DNA repair helicases XPD and FancJ have essential iron-sulfur domains. *Mol Cell* 23: 801–808

161 Pugh RA, Honda M, Leesley H, Thomas A, Lin Y, Nilges MJ, Cann IK, Spies M (2008) The iron-containing domain is essential in Rad3 helicases for coupling of ATP hydrolysis to DNA translocation and for targeting the helicase to the single-stranded DNA-double-stranded DNA junction. *J Biol Chem* 283: 1732–1743

162 Mitchell JR, Hoeijmakers JH, Niedernhofer LJ (2003) Divide and conquer: Nucleotide excision repair battles cancer and ageing. *Curr Opin Cell Biol* 15: 232–240

163 Kalogeraki VS, Tornaletti S, Hanawalt PC (2003) Transcription arrest at a lesion in the transcribed DNA strand *in vitro* is not affected by a nearby lesion in the opposite strand. *J Biol Chem* 278: 19558–19564

164 Hanawalt PC (1994) Transcription-coupled repair and human disease. *Science* 266: 1957–1958

165 Bohr VA, Smith CA, Okumoto DS, Hanawalt PC (1985) DNA repair in an active gene: Removal of pyrimidine dimers from the DHFR gene of CHO cells is much more efficient than in the genome overall. *Cell* 40: 359–369

166 Mellon I, Spivak G, Hanawalt PC (1987) Selective removal of transcription-blocking DNA damage from the transcribed strand of the mammalian DHFR gene. *Cell* 51: 241–249

167 Dimitri A, Goodenough AK, Guengerich FP, Broyde S, Scicchitano DA (2008) Transcription processing at $1,N^2$-ethenoguanine by human RNA polymerase II and bacteriophage T7 RNA polymerase. *J Mol Biol* 375: 353–366

168 Schinecker TM, Perlow RA, Broyde S, Geacintov NE, Scicchitano DA (2003) Human RNA polymerase II is partially blocked by DNA adducts derived from tumorigenic benzo[*c*]phenanthrene diol epoxides: Relating biological consequences to conformational preferences. *Nucleic Acids Res* 31: 6004–6015

169 Tornaletti S, Patrick SM, Turchi JJ, Hanawalt PC (2003) Behavior of T7 RNA polymerase and mammalian RNA polymerase II at site-specific cisplatin adducts in the template DNA. *J Biol Chem* 278: 35791–35797

170 Rabik CA, Dolan ME (2007) Molecular mechanisms of resistance and toxicity associated with platinating agents. *Cancer Treat Rev* 33: 9–23

171 Damsma GE, Alt A, Brueckner F, Carell T, Cramer P (2007) Mechanism of transcriptional stalling at cisplatin-damaged DNA. *Nat Struct Mol Biol* 14: 1127–1133

172 Brueckner F, Hennecke U, Carell T, Cramer P (2007) CPD damage recognition by transcribing RNA polymerase II. *Science* 315: 859–862

173 Venema J, van Hoffen A, Karcagi V, Natarajan AT, van Zeeland AA, Mullenders LH (1991) *Xeroderma pigmentosum* complementation group C cells remove pyrimidine dimers selectively from the transcribed strand of active genes. *Mol Cell Biol* 11: 4128–4134

174 Troelstra C, van Gool A, de Wit J, Vermeulen W, Bootsma D, Hoeijmakers JH (1992) ERCC6, a member of the subfamily of putative helicases, is involved in *Cockayne's syndrome* and preferential repair of active genes. *Cell* 71: 939–953

175 van Hoffen A, Natarajan AT, Mayne LV, van Zeeland AA, Mullenders LH, Venema J (1993) Deficient repair of the transcribed strand of active genes in *Cockayne's syndrome* cells. *Nucleic Acids Res* 21: 5890–5895

176 Mayne LV, Lehmann AR (1982) Failure of RNA synthesis to recover after UV irradiation: An early defect in cells from individuals with *Cockayne's syndrome* and *Xeroderma pigmentosum*. *Cancer Res* 42: 1473–1478

177 Andressoo JO, Hoeijmakers JH (2005) Transcription-coupled repair and premature ageing. *Mutat Res* 577: 179–194

178 Groisman R, Polanowska J, Kuraoka I, Sawada J, Saijo M, Drapkin R, Kisselev AF, Tanaka K, Nakatani Y (2003) The ubiquitin ligase activity in the DDB2 and CSA complexes is differentially regulated by the COP9 signalosome in response to DNA damage. *Cell* 13: 357–367

179 Citterio E, Van Den Boom V, Schnitzler G, Kanaar R, Bonte E, Kingston RE, Hoeijmakers JH, Vermeulen W (2000) ATP-dependent chromatin remodeling by the *Cockayne syndrome* B DNA repair-transcription-coupling factor. *Mol Cell Biol* 20: 7643–7653

180 van den Boom V, Citterio E, Hoogstraten D, Zotter A, Egly JM, van Cappellen WA, Hoeijmakers JH, Houtsmuller AB, Vermeulen W (2004) DNA damage stabilizes interaction of CSB with the transcription elongation machinery. *J Cell Biol* 166: 27–36

181 Fousteri M, Vermeulen W, van Zeeland AA, Mullenders LH (2006) *Cockayne syndrome* A and B proteins differentially regulate recruitment of chromatin remodeling and repair factors to stalled RNA polymerase II *in vivo*. *Mol Cell* 23: 471–482

182 Bregman DB, Halaban R, van Gool AJ, Henning KA, Friedberg EC, Warren SL (1996) UV-induced ubiquitination of RNA polymerase II: A novel modification deficient in *Cockayne syndrome* cells. *Proc Natl Acad Sci USA* 93: 11586–11590

183 Lee KB, Wang D, Lippard SJ, Sharp PA (2002) Transcription-coupled and DNA damage-dependent ubiquitination of RNA polymerase II *in vitro*. *Proc Natl Acad Sci* USA 99: 4239–4244

184 Anindya R, Aygün O, Svejstrup JQ (2007) Damage-induced ubiquitylation of human RNA polymerase II by the ubiquitin ligase Nedd4, but not *Cockayne syndrome* proteins or BRCA1. *Mol Cell* 28: 386–397

185 Sarker AH, Tsutakawa SE, Kostek S, Ng C, Shin DS, Peris M, Campeau E, Tainer JA, Nogales E, Cooper PK (2005) Recognition of RNA polymerase II and transcription bubbles by XPG, CSB, and TFIIH: Insights for transcription-coupled repair and *Cockayne Syndrome*. *Mol Cell* 20: 187–198

186 Nouspikel T, Hanawalt PC (2000) Terminally differentiated human neurons repair transcribed genes but display attenuated global DNA repair and modulation of repair gene expression. *Mol Cell Biol* 20: 1562–1570

187 Nouspikel T, Hanawalt PC (2002) DNA repair in terminally differentiated cells. *DNA Rep* 1: 59–75

188 Nouspikel T, Hyka-Nouspikel N, Hanawalt PC (2006) Transcription domain-associated repair in human cells. *Mol Cell Biol* 26: 8722–8730

189 Harper JW, Elledge SJ (2007) The DNA damage response: Ten years after. *Mol Cell* 28: 739–744

190 Lee JH, Paull TT (2005) ATM activation by DNA double-strand through the Mre11-Rad50-Nbs1 complex. *Science* 308: 551–554

191 Bakkenist CJ, Kastan MB (2003) DNA damage activates ATM through intermolecular autophosphorylation and dimer dissociation. *Nature* 421: 499–506

Molecular, Clinical and Environmental Toxicology. Volume 1: Molecular Toxicology
Edited by A. Luch

On the impact of the molecule structure in chemical carcinogenesis

Andreas Luch

German Federal Institute for Risk Assessment, Berlin, Germany

Abstract. Cancer is as a highly complex and multifactorial disease responsible for the death of hundreds of thousands of people in the western countries every year. Since cancer is clonal and due to changes at the level of the genetic material, viruses, chemical mutagens and other exogenous factors such as short-waved electromagnetic radiation that alter the structure of DNA are among the principal causes. The focus of this present review lies on the influence of the molecular structure of two well-investigated chemical carcinogens from the group of polycyclic aromatic hydrocarbons (PAHs), benzo[*a*]pyrene (BP) and dibenzo[*a,l*]pyrene (DBP). Although there is only one additional benzo ring present in the latter compound, DBP exerts much stronger genotoxic and carcinogenic effects in certain tumor models as compared to BP. Actually, DBP has been identified as the most potent tumorigen among all carcinogenic PAHs tested to date. The genotoxic effects of both compounds investigated in mammalian cells in culture or in animal models are described. Comparison of enzymatic activation, DNA binding levels of reactive diol-epoxide metabolites, efficiency of DNA adduct repair and mutagenicity provides some clues on why this compound is about 100-fold more potent in inducing tumors than BP. The data published during the past 20 years support and strengthen the idea that compound-inherent physicochemical parameters, along with inefficient repair of certain kinds of DNA lesions formed upon metabolic activation, can be considered as strong determinants for high carcinogenic potency of a chemical.

Introduction

Humans are inevitably exposed to foreign (environmental and industrial) chemicals in varying doses mainly through incorporation *via* intestinal tract (diet), lungs (air) and skin. In addition to xenobiotics, from which a large fraction of species or mixtures were proven to convey a significant cancer risk to humans, products of endogenous and, in part, age-related imbalances of metabolism and tissue inflammation may also contribute to the overall cancer burden in the human population. Epidemiological analyses based on geographic and temporal variations in cancer incidences and studies of migrant populations pointed to environmental exposures as having a substantial impact on the generation of human cancer [1–4]. Such environmental factors may even represent the overwhelmingly leading cause for the great majority of human cancers when human lifestyle is included into the picture. However, so-called lifestyle factors comprise extremely diverse factors such as tobacco consumption, lifelong patterns of diet, obesity, alcohol intake, use of pharmaceutical agents and exogenous hormones (e.g., replacement therapy), sun exposure, ionizing radiation, chronic infections (e.g., hepatitis), and so forth [5]. Among these factors, involuntary occupational and environmental exposures to chemical carcinogens contribute to cancer incidences to certain extents. In a review of available data compiled by Peto in 2001 [6], it was concluded that a huge number of all cancer deaths in the U.S., about 75% in smokers and 50% in non-smokers could actually be avoided simply by elimination of known risk factors (cancer prevention strategy).

Since 1971, the International Agency for Research on Cancer (IARC, www.iarc.fr) performed structured-data summaries mostly but not exclusively based on prospective cohort studies to associate certain environmental factors with cancer incidence. More than 900 factors encompassing chemicals, complex mixtures, occupational exposures, physical (e.g., radiation) and biological agents (e.g., viruses), and lifestyle factors, have been evaluated, of which approximately 400 were identified as carcinogenic or potentially carcinogenic to humans. The IARC has developed a classification scheme that divides carcinogenic factors into groups with varying degrees of evidence for carcinogenicity in humans based on data available from human epidemiology, animal studies, or short-term tests in cell culture and *in vitro*. Here, a factor is considered to be a definite human carcinogen when an association has been established between exposure and outcome, and chance, bias and confounding can be ruled out with reasonable confidence. To date, 105 agents (e.g., formaldehyde, 2-naphthylamine), mixtures (e.g., coal tar pitches, wood dusts) and exposure circumstances (e.g., rubber industries, tobacco smoking) have been recognized as being carcinogenic to humans (http://monographs.iarc.fr/ENG/Classification/index.php) (Group 1). In addition, factors 'probably' carcinogenic to humans (Group 2A), 'possibly' carcinogenic to humans (Group 2B), or 'as-yet-not-classifiable' as human carcinogens (Group 3) currently encompass 66, 248, and 515 different types, respectively, most of which are of chemical nature (Tab. 1). It is further important to note that most chemicals released to our environment or at workplaces never have been evaluated for their adverse effects in chronic toxicity bioassays. In the European Community, for instance, all chemicals that were reported as being on the market until 1981 (so-called "old" or "existing" chemicals) numbered more than 100 000 different compounds. While "new" chemicals that have been introduced to the market since then (about 3800) are sufficiently tested and characterized with regard to their acute or chronic toxicological (and ecotoxicological) effects, there were no such provisions for old chemicals produced prior to 1981 (http://ecb.jrc.it/reach/). Although there has become some limited information available in the meantime for these chemicals due to their long-term use on the market, sufficient information for assessment and control is usually absent, no matter whether compounds have been introduced and released in small amounts or as high-volume chemicals with several tons per year. Most of these chemicals surely can be further used without any concern, yet it seems likely that a number among them ultimately will be identified as human carcinogens in the future. Some selected chemical carcinogens are compiled in Table 1.

Table 1. IARC chemical carcinogen classification.[a]

Group 1 (carcinogenic to humans)	Group 2A (probably carcinogenic to humans)	Group 2B (possibly carcinogenic to humans)
aflatoxins	acrylamide	*p*-aminoazobenzene
asbestos	*bis*(chloroethyl)nitrosurea	carbon tetrachloride
benzene	cisplatin	DDT
benzidine	cyclopenta[*c,d*]pyrene	*p*-dichlorobenzene
benzo[*a*]pyrene (BP)	**dibenzo[*a,l*]pyrene (DBP)**	dimethylhydrazine
bis(chloromethyl)ether	dimethylhydrazine	benz[*a*]anthracene
1,3-butadiene	IQ	benzo[*c*]phenanthrene
DES	methyl methanesulfonate	hexachlorocyclohexanes
estrogen	phenacetin	naphthalene
ethanol	polychlorinated biphenyls	phenobarbital
ethylene oxide	styrene-7,8-oxide	titanium dioxide
formaldehyde	trichloroethylene	
NNK		
NNN		
TCDD		

DES, diethylstilbestrol; DDT, dichloro-diphenyl-trichloroethane; IQ, 2-amino-3-methylimidazo-[4,5-*f*]quinoline; NNK, 4-(*N*-nitroso-methylamino)-1-(3-pyridyl)-1-butanone, NNN, *N'*-nitroso-nor-nicotine, TCDD, 2,3,7,8-tetrachlorodibenzo-*p*-dioxin
[a] Only some examples of single agents considered to belong to group 1, group 2A, or group 2B are presented (for complete list see: http://monographs.iarc.fr/ENG/Classification/index.php).

What chemicals are human carcinogens?

Cancer as a chronic disease results from a long-term process in which a large number of endogenous and exogenous factors interact, simultaneously or in sequence, to alter normal cell growth and cell division [7, 8]. Tumor formation induced by chemicals is therefore not simply an irrevocable fate of those tissues exposed to molecules carrying inherent carcinogenic properties. Rather, tumorigenesis is to be considered as an outcome of the complex interaction between carcinogenic chemicals and the biological system with its individual genetic make-up [9].

From the IARC list of 105 human carcinogenic agents identified today, some were attributed to single chemicals (36) and complex mixtures of chemicals (16) based on epidemiological evidence only (http://monographs.iarc.fr/ENG/Classification/index.php). By contrast, mechanistic data led to the assumption that the following chemicals are carcinogenic in humans as well (Tab. 1): ethylene oxide, benzo[*a*]pyrene (BP), benzene, benzidine, 2,3,7,8-tetrachlorodibenzo-*p*-dioxin (TCDD), diethylstilbestrol (DES), methylene-*bis*(chloroaniline), *bis*(chloromethyl)ether, *N'*-nitroso-nornicotine (NNN) and 4-(*N*-nitroso-methylamino)-1-(3-pyridyl)-1-butanone (NNK) [10].

The inclusion of a chemical into the list of "known" human carcinogens usually relies on adequate epidemiological data and sufficient evidence for a causal relationship between exposure and biological outcome. Since cancer may derive from clonal expansion of a single cell, most chemical carcinogens induce tumorigenesis through interaction with cellular components involved in mediating a 'heritable' loss of growth control. In principle, such carcinogens that covalently interact with genomic DNA can be separated from non-genotoxic carcinogens that target signaling networks directly or through protein receptors, thereby altering cell proliferation, migration and communication [11]. Although the latter group of chemicals (e.g., TCDD, DES) do not bind to DNA and are not mutagenic, they are still being capable of inducing tumors in animals and most likely also in humans. On the other hand, genotoxic carcinogens induce genomic damage through either one or more of the following mechanisms: (i) direct binding to DNA (mutagens), (ii) breakage, fusion or loss of chromosome fragments (clastogens), or (iii) altering chromosome segregation during mitosis/induction of non-disjunction (aneugens) [7, 11]. Based on their ability to trigger the DNA damage checkpoint *via* stabilization of p53 protein and other damage-inducible signaling proteins, it has been proposed that genotoxic carcinogens can be distinguished from non-genotoxic carcinogens through expression profiling [12].

Genotoxic potency as function of chemical structure

To react with nucleophilic centers in macromolecules such as DNA, most chemical carcinogens require enzymatic derivatization to gain electrophilicity [11]. Chemical carcinogens that require initial enzymatic conversion into their ultimate DNA-damaging forms are designated 'procarcinogens'. On the other side, some known human chemical carcinogens such as ethylene oxide, *bis*(chloromethyl)ether and others, directly interact with DNA and are therefore termed 'direct carcinogens'. When a procarcinogen enters the biological system, the overall DNA damage that occurs depends on its structure-related toxicokinetic and toxicodynamic properties, including the effectiveness of its enzymatic activation prior to DNA binding, the effectiveness of enzymatic detoxification of reactive intermediates generated during biotransformation, as well as the efficiency of the repair of those DNA lesions that have been formed. Covalent carcinogen-DNA adducts can induce permanent damage to genomic DNA when the modifications formed are capable of escaping cellular repair mechanisms.

Sterical crowding in carcinogenic polycyclic aromatic hydrocarbons: Bay versus fjord region

In the remainder of this article, two potent chemical carcinogens from the class of polycyclic aromatic hydrocarbons (PAHs) are compared to each other in terms of their biological effects. This example provides some interesting insights on the toxicological consequences of rather small changes in chemical structure. The difference between both compounds, BP and dibenzo[*a,l*]pyrene (DBP), only lies in the absence or presence of one additional benzo ring (Fig. 1, Tab. 1). While BP belongs to the group of pentacyclic hydrocarbons ($C_{20}H_{12}$, m.w. 252), its high molecular weight relative, DBP ($C_{24}H_{14}$, m.w. 302), is made of six benzo rings fused together in a certain configuration. The latter compound is also known under the name dibenzo[*def,p*]chrysene. As indicated by their names, both PAHs can be considered as higher homologues of pyrene (Fig. 1). Condensation of one additional benzo ring at position *a* of pyrene creates BP that contains a so-called bay region, which is made of angular condensation of three rings resembling the structural configuration of phenanthrene (Fig. 2). Another addition of one ring at position *l* creates an even more crowded fjord region as part and biologically important structural feature of DBP (Figs 1 and 2).

In the early 1930s, James W. Cook and his co-workers at The Cancer Hospital in London were able to isolate some pure crystals of BP from 2 tons of coal tar pitch and to establish its role as an extremely potent chemical carcinogen [13, 14]. Since then this compound, which was formerly known under the name *3,4-benzpyrene*, became famous as worldwide reference compound in a countless number of studies on chemical-induced carcinogenesis [15]. By contrast, the first report on the carcinogenic power of DBP, initially designated *1,2:3,4-dibenzopyrene*, has been widely ignored by the scientific community [16]. It was not before the late 1950s that several reports on its presence in cigarette smoke [17, 18] and further evidence for its extremely potent biological activity [19–21] attracted some attention. However, all of these data published up to 1966 are dubious due to the kind of synthetic preparation of DBP [22, 23]. Since the synthesis route included a *peri*-condensation step of 12-phenylbenz[*a*]anthracene, actually large amounts of dibenzo[*a,e*]fluoranthene were generated in the mixture *via* rearrangement. The prerequisite for a reliable biological characterization of DBP was its pure

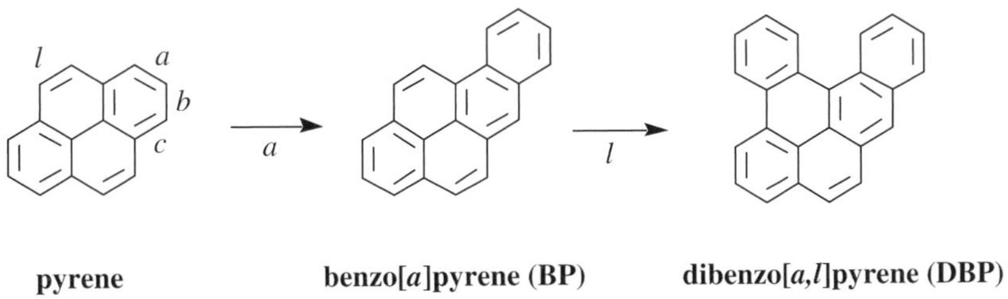

pyrene   benzo[*a*]pyrene (BP)   dibenzo[*a,l*]pyrene (DBP)

Figure 1. Chemical structures of benzo[*a*]pyrene (BP) and dibenzo[*a,l*]pyrene (DBP). Both compounds can be viewed as higher homologues of pyrene. Condensation of one additional benzo ring at position *a* of pyrene creates the pentacyclic BP and another addition of one ring at position *l* creates the hexacyclic DBP.

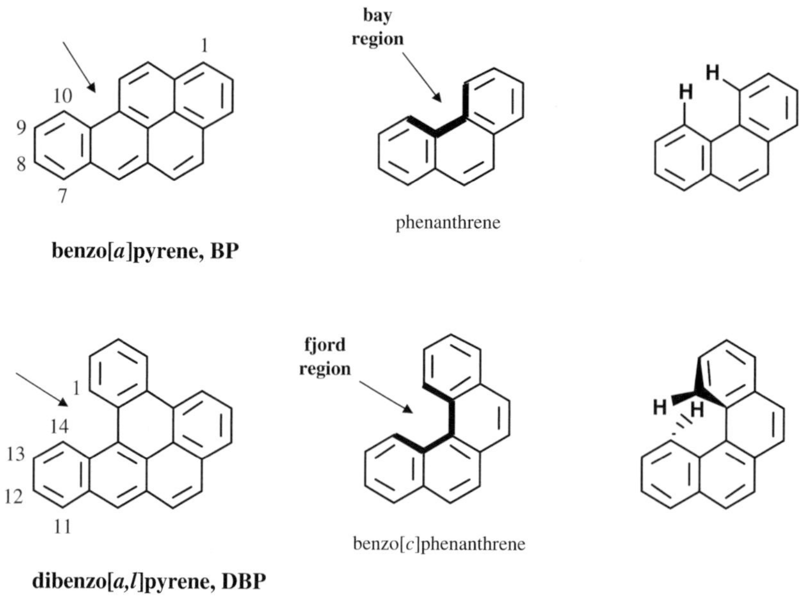

Figure 2. BP contains a so-called bay region, which is made of angular condensation of three rings resembling the structural configuration of phenanthrene. DBP contains a so-called fjord region, which is made of angular condensation of four rings resembling the structural configuration of benzo[c]phenanthrene. Due to repulsion forces between the hydrogens located in the fjord region, DBP is distorted out-of-plane and accommodates in a spatially demanding three-dimensional config- uration. Bay and fjord regions are marked by an arrow (see text for further explanations).

preparation without any contamination first accomplished in 1966 [22, 24]. Subsequent carcinogenicity studies in 1968 [25] and 1972 [26] provided unequivocal proof of the extraordinary strong sarcomagenic and carcinogenic activity of DBP in subcutaneous and intraepidermal tissue of mice, respec- tively. Still, after Masuda and Kagawa had noticed in 1972 that 'dibenzo[a,l]pyrene has much stronger carcinogenic potency on mouse skin than that of the well-known carcinogen, benzo[a]pyrene', it took almost two further decades until this hydrocarbon again shifted into the focus of chemi- cal cancer research. Several comparative tumorigenicity studies then revealed that DBP exerts the strongest carcinogenic activity in mouse skin and rat mammary gland among all individual PAHs tested to date [27–30]. Its excep- tionally high tumorigenicity can be roughly estimated in these bioassays as about 100-fold higher than that of BP (Tab. 2). Today, it is well established that this carcinogenic compound, similar to BP, is present in environmental samples such as cigarette smoke [31–33], smoky coal particulates [34, 35], and soil or sediment samples [36].

The central role of cytochrome P450-dependent monooxygenases in toxification

Metabolic activation of PAHs such as BP or DBP is mainly catalyzed by cytochrome P450-dependent monooxygenases (CYPs) [37]. CYPs work in close association with NADPH-dependent cytochrome P450 reductase as a multimeric complex located in the lipid bilayer of the endoplasmic reticulum membrane. In mammals, these enzymes are mainly expressed (or induced) in the liver, but can also be detected in the majority of extrahepatic tissues [38]. CYPs are considered the mainstay in biotransformation of PAHs, thereby possessing a broad and overlapping substrate specificity. From 57 different forms detected in the human genome, CYP1A1, CYP1A2, CYP1B1 and CYP3A4 are most important in PAH metabolism [39]. Whereas expression of CYP1A1 is virtually nil or only detectable at very low levels in mammalian tissues under physiological conditions, CYP1B1 is constitutively expressed at high basal levels in most extrahepatic tissues [40]. Conversely, CYP3A4 is the most abundant form in mammalian liver tissue, followed by CYP1A2. A broad range of carcinogenic PAHs has been shown to significantly increase the expression levels of CYP1A1 and CYP1B1 in liver, lung and most extrahepatic tissues through binding to the cytosolic aryl hydrocarbon receptor (AhR) [41]. This ligand-activated transcription factor mediates regulation of the CYP1 family (CYP1A1, CYP1A2, CYP1B1) and other xenobiotic-metabolizing enzymes such as certain glutathione-*S*-transferases (GSTs) and other transferases. To bind to the receptor with high affinity, the ligands have to meet certain structural requirements such as a large molecular area/depth ratio (planarity) and lipophilicity that is fulfilled by carcinogenic PAHs such as BP [41, 42]. For instance, the expression levels of CYP1A1 and CYP1B1 were both found elevated in human mammary carcinoma MCF-7 cells or in mouse skin upon exposure to BP, but not DBP [43, 44]. This difference can possibly be attributed to the out-of-plane distortion of DBP driven by hydrogen repulsion forces in the sterically crowded fjord region rendering the molecule three dimensional (Fig. 2) [45].

Bioactivation towards ultimate carcinogenic species: PAH bay or fjord region diol-epoxides

Studies, mainly from the 1970s and 1980s, have demonstrated that the standard carcinogenic PAH, BP, would be activated *via* its 7,8-diol towards bay region 7,8-diol-9,10-epoxides (BPDEs) [46]. Synthesis of the authentic BPDEs finally provided the proof that binding of BP to DNA in cells in culture and *in vivo* occurs mainly through these chemical species [47]. Subsequent studies on the effects of PAH metabolites both in cell culture and *in vivo* supported the assumption that a multistep enzymatic activation pathway with a sequence of epoxidation, subsequent hydrolysis of the primary

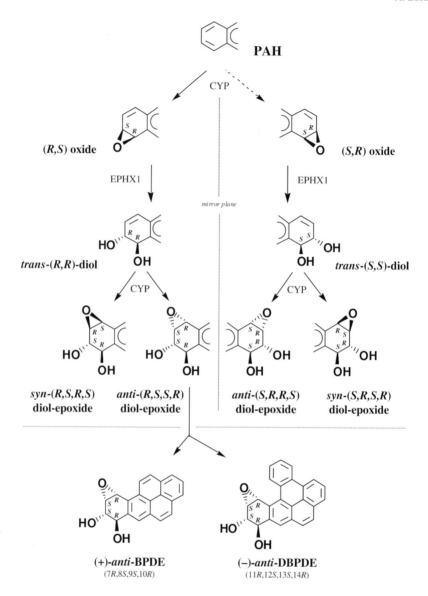

Figure 3. Cytochrome P450-dependent monooxygenases (CYP)-mediated bioactivation of polycyclic aromatic hydrocarbons (PAHs) towards bay or fjord region diol-epoxides. CYP and epoxide hydrolase (EPHX1) enzymes operate with high enantio- and diastereoselectivity. Initial CYP-mediated conversion principally can produce two enantiomeric epoxides (oxides). In the case of BP, oxidation at its 7,8-position mainly generates the (+)-(R,S)-oxide (left column). Subsequent EPHX1-mediated hydrolysis selectively affords the diol with *trans*-oriented hydroxy groups, i.e. (−)-(R,R)-diol. Conversely, the (−)-(S,R)-oxide and the corresponding (+)-(S,S)-diol are generated only in small amounts (right column). Secondary epoxidation of each *trans*-diol at its vicinal double can give rise to two diastereomeric diol-epoxides that would have the epoxide ring either at the same (*syn*) or at the opposite (*anti*) side of the aromatic plane (compared to the hydroxy group at the benzylic carbon atom 7 in BP, cf. Fig. 2). In fact, the (−)-(R,R)-diol is preferentially converted to the (+)-*anti*-7R,8S-diol

epoxide into a diol, and further epoxidation at the adjacent double bond produces vicinal diol-epoxides as the carcinogenic metabolites of a range of different PAHs (Fig. 3). After CYP-mediated initial monooxygenation, the arene oxides undergo stereoselective enzymatic hydrolysis to *trans*-diols. This reaction is catalyzed by the microsomal epoxide hydrolase (EPHX1) [48]. CYP enzymes are again responsible for subsequent epoxidation of *trans*-diols to vicinal bay or fjord region diol-epoxides (Fig. 3) [39, 40]. Besides CYP1A1 and 1B1, which are the predominant forms involved in both initial oxidation of PAHs to arene oxides and further oxidation of *trans*-diols to diol-epoxides, evidence from recombinant human CYP systems indicated that also CYP1A2 and, to small extents, CYP2C9 and CYP3A4 are capable of catalyzing oxidation at olefinic double bonds adjacent to diol moieties [49]. Similar to EPHX1, CYP enzymes operate with high enantio- and diastereoselectivity. Soon after the first identification of vicinal diol-epoxides as the crucial metabolites responsible for DNA binding of BP *in vitro* and *in vivo*, it was realized that not only the reactions of enzymes involved in this pathway were highly stereoselective [50, 51], but also the biological activity of individual stereoisomers differed markedly from each other [52, 53]. In principle, CYP-mediated integration of an oxygen atom can occur from both sides of the aromatic ring system. However, initial CYP-mediated epoxidation of BP at its 7,8-position generates the optically active (+)-7*R*,8*S*-oxide in high excess of up to 98% (Fig. 3) [54]. Subsequent EPHX1-mediated hydrolysis selectively affords the diol with *trans*-oriented hydroxy groups, i.e., (−)-7*R*,8*R*-diol. Conversely, the (−)-7*S*,8*R*-oxide and the corresponding (+)-7*S*,8*S*-diol are generated only in small amounts. Secondary epoxidation of the BP 7,8-diol at its vicinal double bond again occurs with high stereoselectivity. Each of the two *trans*-diol enantiomers theoretically could give rise to a set of two diastereomeric diol-epoxides that would have the epoxide ring either at the same (*syn*) or at the opposite (*anti*) side of the aromatic plane as compared to the hydroxy group in the benzylic position (carbon atom 7 in BP, Figs 2 and 3). In fact, the (−)-7*R*,8*R*-diol is preferentially converted to the (+)-*anti*-7*R*,8*S*-diol 9*S*,10*R*-epoxide [(+)-*anti*-BPDE = (*R*,*S*,*S*,*R*) diol-epoxide]. On the other hand, the small amounts of (+)-7*S*,8*S*-diol metabolically formed during primary metabolism are further oxidized mainly to the diastereomeric (+)-*syn*-7*S*,8*R*-diol 9*S*,10*R*-epoxide [(+)-*syn*-BPDE = (*S*,*R*,*S*,*R*) diol-epoxide] [54, 55]. In light of the high

---

9*S*,10*R*-epoxide [(+)-*anti*-BPDE = (*R*,*S*,*S*,*R*) diol-epoxide] (left column), while the small amounts of (+)-(*S*,*S*)-diol metabolically formed during primary metabolism are mainly converted to the diastereomeric (+)-*syn*-7*S*,8*R*-diol 9*S*,10*R*-epoxide [(+)-*syn*-BPDE = (*S*,*R*,*S*,*R*) diol-epoxide] (right column). In light of the high enantioselectivity during formation of the *trans*-diol and the diastereoselectivity during formation of the diol-epoxide, the main product in bay region metabolism of BP catalyzed by CYP enzymes appears to be the bay region (+)-*anti*-BPDE. By contrast, the three other diastereomeric diol-epoxides are formed only in minor amounts (cf. Fig. 4). Similarly, DBP is predominantly metabolized to fjord region (+)-*syn*- and (−)-*anti*-DBPDE enantiomers with (*S*,*R*,*S*,*R*) and (*R*,*S*,*S*,*R*) configuration. At low treatment doses of DBP in cells in culture or in animal target tissues, the (−)-*anti*-11*R*,12*S*-diol 13*S*,14*R*-epoxide [(−)-*anti*-DBPDE = (*R*,*S*,*S*,*R*) diol-epoxide] prevails (see text for further explanations).

enantioselectivity during formation of the *trans*-diol and the diastereoselectivity during formation of the diol-epoxide, the main product in bay region
metabolism of BP catalyzed by CYP enzymes appears to be the bay region
(+)-*anti*-BPDE (Fig. 3). By contrast, the three other diastereomeric diol-epoxides are formed only in minor amounts (Fig. 4).

Follow up studies with various other carcinogenic PAHs revealed that
CYP-mediated metabolism mainly produces the bay or fjord region (R,S,S,R)

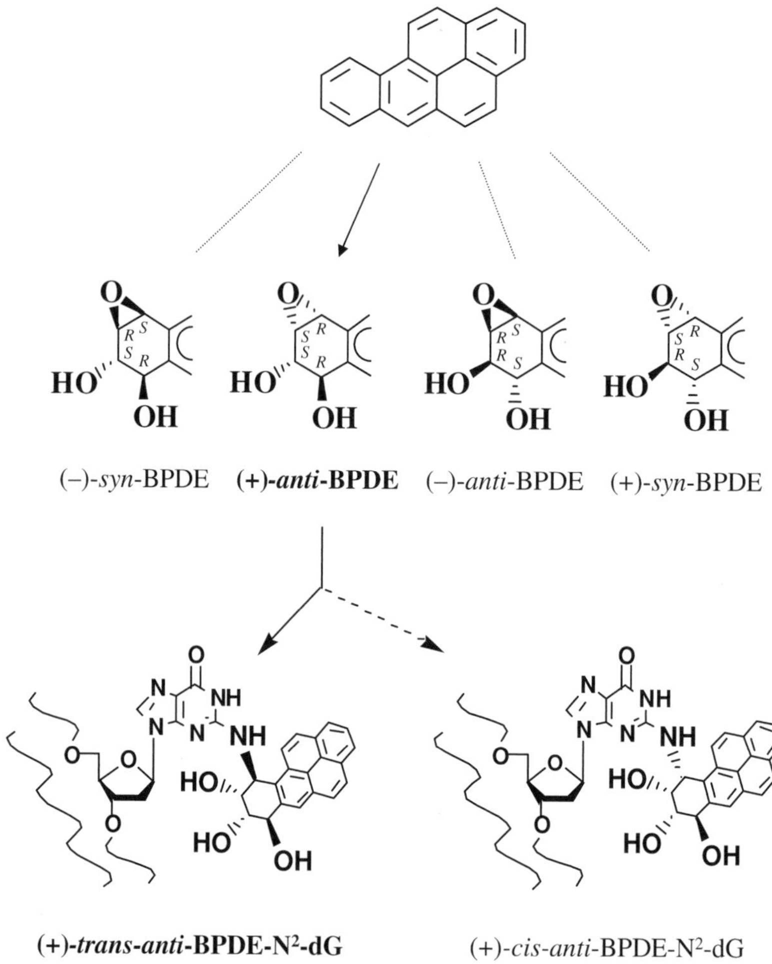

(−)-*syn*-BPDE   **(+)-*anti*-BPDE**   (−)-*anti*-BPDE   (+)-*syn*-BPDE

(+)-*trans-anti*-BPDE-N²-dG   (+)-*cis-anti*-BPDE-N²-dG

Figure 4. Bay region diol-epoxides formed by CYP-catalyzed biotransformation of BP mainly bind
to the exocyclic amino group of 2'-deoxyguanosine (dG) in DNA. Upon interaction with DNA, the
predominant stereoisomer (+)-*anti*-BPDE (cf. Fig. 3) gives rise to (+)-*anti*-BPDE-N²-dG adducts.
Only minor amounts of adducts are formed from (+)-*syn*-BPDE. BPDE-N²-dG DNA adducts are preferentially generated through *trans* opening of the epoxide ring [adducts referred to as (+)-*trans-anti*-
BPDE-N²-dG]. Conversely, adducts resulting from *cis* opening [referred to as (+)-*cis-anti*-BPDE-N²-
dG] are minor (see text for further explanations).

anti-diol-epoxides *via* the corresponding (*R,R*) *trans*-diol precursors, followed by some amounts of the (*S,R,S,R*) *syn*-diol-epoxides generated *via* (*S,S*) *trans*-diols [37]. It further became clear that vicinal PAH diol-epoxides are the main DNA-binding metabolites that mediate the biological effects of their parents (cf. below). With regard to DBP, metabolism and DNA binding studies performed with different cell lines including human mammary carcinoma MCF-7 cells [56, 57] or liver preparations from rats pretreated with different inducers of CYP1A1 and CYP1B1 such as 3-methylcholanthrene [58, 59] or Aroclor 1254 [60] revealed that this hydrocarbon is predominantly metabolized to genotoxic fjord region *syn*- and *anti*-11,12-diol 13,14-epoxides (DBPDEs, Fig. 3). According to the general rule, the activation route again resembles the stereoselectivity found for BP and various other PAHs: only (+)-*syn*- and (−)-*anti*-DBPDE enantiomers with (*S,R,S,R*) and (*R,S,S,R*) configuration are formed (Fig. 3). Using recombinant cell lines or subcellular fractions for analysis of the capacity of individual CYP enzymes to mediate bay and fjord region metabolism of BP and DBP, respectively, revealed highest catalytic turnover rates for both human CYP1A1 and CYP1B1 for either PAH [61–66].

By direct application of bay or fjord region diol-epoxides of various carcinogenic PAHs, it has been shown that their mutagenic potency correlates nicely to their level of DNA binding (see below) [67–69]. Further, the tumor-inducing potencies of a range of different PAHs including BP and DBP in epithelial organs of mice and rats, or in the lungs of strain A/J mice were found to correlate well with the total amounts of DNA adducts detected in the respective tissues [70–73]. Congruently, administration of racemic *anti*-DBPDE to target tissues led to strong tumor formation in mouse skin [74], rat mammary gland [75] and newborn mice [76], thereby confirming that the tumorigenicity of the parent compound can be rationalized through toxication *via* fjord region monooxygenation *in vivo*. The intimate relationship between the diol-epoxide DNA binding levels and PAH-mediated mutagenicity or carcinogenicity observed in mice *in vivo* (Tab. 2) support the notion that this parameter can serve as an important surrogate (biomarker) of the tumor threat that may result from certain exposures to carcinogenic PAHs such as BP or DBP [77].

PAH diol-epoxide DNA binding: Target sites and configurational aspects

Bay or fjord region diol-epoxides of various carcinogenic PAHs were identified as the ultimate DNA damaging metabolites that mediate the genotoxicity of their parents [78]. In the case of BP, the bay region (+)-*anti*-BPDE is the predominant species that covalently interacts with the exocyclic amino group of 2'-deoxyguanosine (dG) to form (+)-*anti*-BPDE-N^2-dG adducts (Fig. 4) [73, 79–81]. Only some minor adducts are formed from the diastereomeric (+)-*syn*-BPDE or from reactions at different positions in nucleobases. Similarly, the DNA binding of DBP is mainly mediated by (−)-*anti*-DBPDE and, to a small-

Table 2. Comparison of the biological activity of benzo[a]pyrene (BP) and dibenzo[a,l]pyrene (DBP).

Bioassay		BP	DBP
Carcinogenicity	1 nmol	0	4 (3/1)
in mouse skin	4 nmol	0	83 (49/19)
(TBA [%]) [a,b]	8 nmol	0	91 (86/20)
	33.3 nmol	43 (15/10)	96 (162/23)
Carcinogenicity	250 nmol	35 (8/7)	100 (131/20)
in rat mammary	1000 nmol	45 (11/9)	100 (206/19)
gland (TBA[%]) [c]			
Carcinogenicity	2 nmol	n.t.	100 (242/30)
in mouse skin	200 nmol	90 (241/27)	n.t.
(TBA [%]) [d]			
DNA binding	2 nmol	n.t.	10.0
in mouse skin	200 nmol	14.4	n.t.
(pmol/mg DNA) [e]			
DNA adducts	50 nmol	14 ± 2	21 ± 8
in mouse skin	400 nmol	22 ± 3	43 ± 6
(adducts/10^6 nt) [f]			
DNA binding in	1 μM	18 ± 8	255 ± 9
human MCF-7 cells	2 μM	34 ± 7	350 ± 18
(adducts/10^6 nt) [g]			
DNA binding	CYP1A1	46.8	10.0
in human CYP-	CYP1B1	86.1	85.7
expressing V79 cells			
(pmol/mg DNA) [h]			
Cytotoxicity	CYP1A1	300	12
in human CYP-	CYP1B1	>1000	45
expressing V79 cells			
(IC_{50} [nM]) [i]			
Mutagenicity	CYP1A1	720	150,000
in human CYP-	CYP1B1	3100	400,000
expressing V79 cells	[...rat S9	0.9	49...]
(mutants × 10^6/μM) [j]			

[a] Eight-week-old female Swiss mice were treated with BP or DBP at dose levels of 1, 4 or 8 nmol in 100 μl acetone twice weekly for 40 weeks. All surviving animals were killed at 48 weeks. All of the tumors were histologically verified and characterized as carcinomas, squamous papillomas or sebaceous gland adenomas (DBP-treated groups). No tumors were observed in the BP-treated groups. TBA, tumor-bearing animals; numbers in bracket: no. of tumors per no. of TBA [30].

[b] Eight-week-old female SENCAR mice were initiated with 33.3 nmol BP or DBP in 100 μl acetone. After 1 week, twice-weekly promotion with 12-O-tetradecanoylphorbol-13-acetate (TPA, 3.24 nmol/100 μl acetone) was started. After a break of 4 weeks (due to erythemas formed), promotion with TPA was continued for 11 weeks. Mice were killed after 16 weeks. The tumor latency was significantly shorter for DBP than that for BP. TBA, tumor-bearing animals; numbers in bracket: no. of tumors per no. of TBA [28].

[c] Eight-week-old female Sprague-Dawley rats were treated with intramammillary injection at indicated doses of BP or DBP in 50 μl trioctanoin per gland. After 24 weeks, surviving rats were killed. The tumor latency was significantly shorter at both doses for DBP versus BP. Tumors were histo-

logically confirmed as adenocarcinomas (DBP) and mesenchymal tumors (BP and DBP). TBA, tumor-bearing animals; numbers in bracket: no. of tumors per no. of TBA [28].

[d] Six to seven-week-old female SENCAR mice were treated with 200 nmol BP or 2 nmol DBP in 100 µl acetone. At 2 weeks after initiation, twice-weekly promotion was started with 1 µg TPA/200 µl acetone per mouse for 25 weeks. Tumors were histologically confirmed as skin papillomas. TBA, tumor-bearing animals; numbers in bracket: no. of tumors per no. of TBA; n.t., not treated [43].

[e] Six to seven-week-old female SENCAR mice were treated with 200 nmol BP or 2 nmol DBP in 100 µl acetone. After 24 h mice were killed and DNA adducts were quantified by ^{33}P-postlabeling and HPLC analysis as pmol adducts per mg DNA (mean values from two independent experiments) [43].

[f] Five to six-week-old female SENCAR mice were treated with 50 or 400 nmol BP or DBP in 100 µl acetone. After 24 h mice were killed and DNA adducts were quantified by ^{33}P-postlabeling and HPLC analysis as no. of adducts per 10^6 nucleotides (nt) [80].

[g] MCF-7 cells were treated with 1 or 2 µM BP or DBP for 24 h. DNA adducts were quantified by ^{33}P-postlabeling and HPLC analysis. Mean values ± SD from three independent experiments are given as no. of adducts per 10^6 nt [81].

[h] Human CYP1A1- or CYP1B1 expressing V79 cells were treated with 4 µM BP or 40 nmol DBP. DNA adducts were quantified by ^{33}P-postlabeling and HPLC analysis. Mean values (three independent experiments) from the maximum levels during a time course of 72 h are given as pmol adducts per mg DNA [119]. A 100-fold higher dose of BP was required to reach about the same adduct levels as for DBP (see text for further explanations).

[i] Human CYP1A1- or CYP1B1-expressing V79 cells were treated with various concentrations of BP or DBP. IC_{50} values (concentration of BP or DBP at 50% cell death) were calculated from the dose-response curves [64, 66].

[j] Human CYP1A1- or CYP1B1-expressing V79 cells were treated with various concentrations of BP or DBP. Specific mutagenicities, given as no. of mutants $×10^6$ per µM compound, were calculated from dose-response curves. About 100-fold higher doses of BP were required to achieve a similar mutagenic response as obtained for DBP. In a control experiment (bracket), normal V79 cells were supplemented in culture with S9 liver preparation from Aroclor 1254-induced rats and mutagenicity of BP or DBP was measured accordingly [134, 135].

er extent, by (+)-*syn*-DBPDE (Fig. 5) [56, 57]. This PAH, however, rather targets the exocyclic amino group of 2'-deoxyadenosine (dA) to form (–)-*anti*-DBPDE-N^6-dA adducts *in vitro* and *in vivo* in high excess (see below) [72, 80–82].

In-depth analyses of the DNA adducts generated *in vitro* revealed that N^2-dG and N^6-dA adducts from *anti*-(*R,S*)-diol (*S,R*)-epoxides are always preferentially generated through *trans* opening of the epoxide ring (Fig. 4) [83]. For instance, from all (+)-*anti*-BPDE-N^2-dG adducts formed about 99% result from *trans* opening of the epoxide ring [adducts referred to as (+)-*trans-anti*-BPDE-N^2-dG]. Conversely, only 1% of the DNA adducts result from *cis*-attack of the nucleophile [adducts referred to as (+)-*cis-anti*-BPDE-N^2-dG]. Similarly, covalent interaction of (–)-*anti*-DBPDE with dA produced (–)-*trans-anti*-DBPDE-N^6-dA adducts in high excess, whereas (–)-*cis-anti*-DBPDE-N^6-dA adducts were only minor [84, 85].

Mutagenicity of bay region BP diol-epoxides and fjord region DBP diol-epoxides

In the mammalian V79/*hprt* mutagenicity assay, bay and fjord region PAH diol-epoxides have been proven to exert extraordinarily strong mutagenic

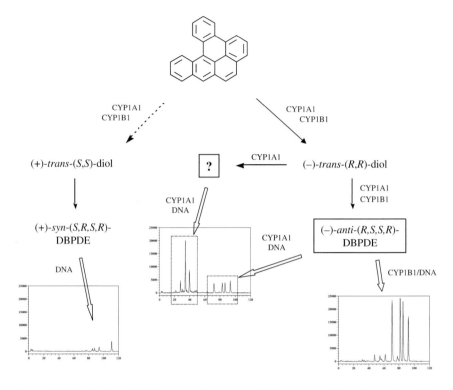

Figure 5. DNA binding of DBP investigated in human CYP-expressing V79 cells in culture. The majority of DNA adducts are formed by (–)-*anti*-DBPDE (right side), only small amounts by (+)-*syn*-DBPDE (left side). The HPLC profiles of ^{33}P-labeled DNA adducts are depicted. Using these recombinant V79 cells it became clear that DBP is activated towards (–)-*anti*- and (+)-*syn*-DBPDE by both human CYP1A1 and CYP1B1. In CYP1A1-expressing cells, however, some additional DNA adducts emerged that resulted from further activation of the (–)-*trans*-11,12-diol. The DNA-reactive intermediate responsible for the generation of highly hydrophilic DNA adducts (eluting from the HPLC column at early time points) is not identical to (–)-*anti*-DBPDE; yet, its chemical identity has not been uncovered until now (question mark) [64, 121].

effects as compared to other metabolites originating from the same parent compounds [68, 69, 86]. Further, carcinogenic PAH diol-epoxide derivatives vary considerably in their potency to induce mutations at the *hprt* locus: (i) (R,S,S,R) *anti*-diol-epoxides from bay region PAHs (e.g., BP) are less active than corresponding species originating from fjord region PAHs (e.g., DBP); and (ii) (S,R,S,R) *syn*-diol-epoxides are usually less potent in inducing DNA adducts and mutations when compared to their corresponding *anti*-diasteromeric counterparts (Fig. 6) [86–88]. For example, the mutant frequencies in V79 cells induced by (+)-*anti*-BPDE, (+)-*syn*-BPDE, (–)-*anti*-DBPDE and (+)-*syn*-DBPDE were as follows: 2000; 230; 124 000; and $51\,000 \times 10^6/\mu M$, respectively (Fig. 6) [87]. Thus, diol-epoxides from different PAHs and different stereoisomeric diol-epoxides from the same PAH exert (sometimes markedly) different efficiencies in mutation induction. On the other hand,

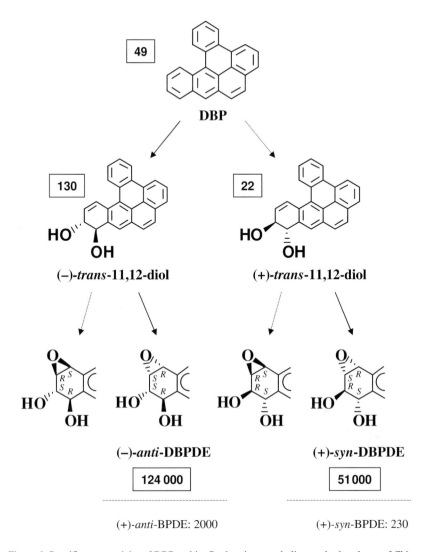

Figure 6. Specific mutagenicity of DBP and its fjord region metabolites at the *hprt* locus of Chinese hamster V79 cells. The numbers provided represent the count of mutants $\times 10^6$ per µM compound. In the case of DBP and its enantiomeric *trans*-11,12-diols, the medium has been supplemented with S9 liver preparation from Aroclor 1254-induced rats (cf. Tab. 2). The fjord region (–)-*anti*- and (+)-*syn*-DBPDE exert exceptionally high mutagenic activities in this assay system without the need for any further activation. The values given are significantly higher than all numbers measured for the same kinds of metabolites of any other carcinogenic PAH before. Compared to corresponding derivatives of BP, i.e. (+)-*anti*- and (+)-*syn*-BPDE, the specific mutagenicity of DBP metabolites is about 60- and 220-fold higher, respectively (see text for further explanations).

binding studies with isolated DNA in buffered solution did not mirror these results. While the overall binding of all four configurational isomers was found to be 10–15% in the case of BP, about 15–20% of the DBP diol-epoxides became covalently trapped by DNA [89–91]. For example, 14% of (+)-*anti*-BPDE and 20% of (–)-*anti*-DBPDE bind to DNA in buffered solution *in vitro*. Assuming a similar binding efficiency of diol-epoxides in cells in culture as compared to isolated DNA in buffered solution (an assumption that actually is inappropriate; see below), adducts formed by (–)-*anti*-DBPDE are therefore much more mutagenic than adducts formed by (+)-*anti*-BPDE.

If not repaired, bulky PAH-DNA adducts at N^2-dG or N^6-dA sites are likely to cause nucleotide misincorporation at the opposite DNA strand during the next round of DNA replication, thus conferring the potential of inducing mutations. Analysis of the mutational spectra induced by stereoisomeric bay region BPDEs *in vitro* or *in vivo* confirmed a propensity for inducing dG to dT transversions [92–95]. Conversely, fjord region PAHs such as DBP and their diol-epoxides predominantly induce dA to dT transversions [94–97]. At the *hprt* locus of V79 cells or the *cII* transgene in Big-Blue mouse cells, bay or fjord region *anti*-(*R,S*)-diol (*S,R*)-epoxides from BP and DBP cause a variety of different kinds of mutations with base substitutions prevailing (Fig. 7) [93, 96, 97]. Irrespective of the hydrocarbon, exposure of V79 cells to increasing concentrations of the *anti*-diol-epoxide resulted in a decrease in the fraction of base substitutions at dA sites and an increase in the fraction of base substitutions at dG sites [93]. Later work revealed that the observed shift of base substitutions from dA to dG at higher *anti*-(*R,S*)-diol (*S,R*)-epoxide concentrations relied on an intact DNA damage repair system [98]. In the absence of DNA repair the effect disappeared.

DNA damage analysis	Major adduct	Adduct mapping	Base Substitutions					
			V79/*hprt*			MEF/*cII*		
			total	dG	dA	total	dG	dA
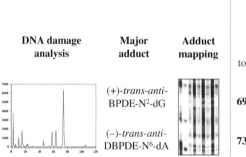	(+)-*trans-anti*-BPDE-N^2-dG		69%	63%		94%	91%	
	(–)-*trans-anti*-DBPDE-N^6-dA		73%		82%	89%		61%

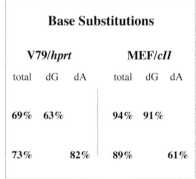

Figure 7. Analysis of the mutational spectra induced by stereoisomeric bay region (+)-*anti*-BPDE and fjord region (–)-*anti*-DBPDE at the *hprt* locus of V79 cells or the *cII* transgene in Big-Blue mouse embryonic fibroblasts (MEF). The diol-epoxides of BP and DBP caused a variety of different kinds of mutations, but base substitutions were prevailing in both systems (69–94%). From all base substitutions induced, dG → dT transversions prevailed in cells exposed to (+)-*anti*-BPDE (63–91%), whereas dA → dT transversions prevailed in cells exposed to (–)-*anti*-DBPDE (61–82%) (see text for further explanations) [96, 97].

Repair of PAH diol-epoxide-DNA adducts: The influence of structure and configuration

Bulky PAH-DNA adducts such as (+)-*trans-anti*-BPDE-N^2-dG or (–)-*trans-anti*-DBPDE-N^6-dA are subject to nucleotide excision repair (NER, Fig. 8) [99, 100]. Enzymatic repair of PAH-damaged DNA depends on a variety of different factors such as the DNA sequence context of the adduct and its geometrical structure and stereochemistry. In addition, the efficiency of PAH-DNA adducts in inducing mutations depends on the nature of the DNA strand affected (DNA strand bias). PAH adducts located in the transcriptionally active DNA strand are removed by a separate and less well characterized variant of the NER pathway referred to as 'transcription-coupled repair'. Since this mode of repair has been proven to be fast and efficient, PAH-DNA adducts located in the transcribed strand are unlikely to contribute considerably to the overall mutagenicity of the hydrocarbon. Rather, the slow and

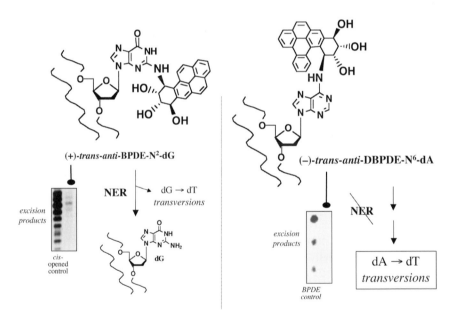

Figure 8. Investigation of the enzymatic repair of chemically defined PAH-DNA adducts *in vitro* employing nucleotide excision repair (NER) proteins present in human HeLa cell extracts. Repair of the lesion site is indicated by small excision products appearing as black bands in the thin-layer chromatograms shown. Site-specific introduction of a stereochemically defined diol-epoxide adduct in a certain DNA sequence context revealed that (+)-*trans-anti*-BPDE-N^2-dG adducts are much less efficiently repaired than the corresponding *cis*-opened adducts, i.e., (+)-*cis-anti*-BPDE-N^2-dG. If not efficiently repaired, dG → dT transversions can be induced during the next replication cycle (left column). In the sequence context of the *ras* oncogene, fjord region (–)-*trans-anti*-DBPDE-N^6-dA adducts are resistant to DNA repair, whereas stereochemically corresponding (+)-*trans-anti*-BPDE-N^6-dA adducts are removed from the same site quite effectively (BPDE control). If not repaired, dA → dT transversions will be created during the next replication cycle at the damaged site (right column). The picture shown is based on data from [100, 108]. See text for further explanations.

error-prone repair pathway operating at transcriptionally silent (non-transcribed) DNA strands (global NER pathway) is thought to be more decisive with regard to PAH-induced mutagenesis [100]. Induction of human NER activity requires both the disruption of normal base pairing and the presence of a chemical modification (so-called bipartite damage recognition) [101]. Its efficiency depends on the degree of local DNA distortion. PAH adducts that cause a displacement of the modified nucleobase and its counterpart from their normal intrahelical position fulfill the structural requirements for activating the NER machinery. For instance, it could be demonstrated that (+)-anti-BPDE induces a considerably different degree of NER activity depending on the way of epoxide ring opening during adduct formation (Figs 4 and 8) [102]. While the cis-opened adduct [(+)-cis-anti-BPDE-N^2-dG] adopts an intercalative, internal adduct conformation with the benzo[a]pyrenyl moiety inserted into the double helix and concomitant displacement of the modified base, the (+)-trans-anti-BPDE-N^2-dG displays an external conformation with the aromatic ring system accommodated in the minor DNA groove [103]. Hence, the local DNA distortion induced by the cis product is much more severe as compared to the trans product, thereby resulting in a 10-fold faster removal through NER. While this example demonstrates the importance of the kind of covalent linkage (cis versus trans) of a given stereoisomer, differences in the repairability of bay versus fjord region diol-epoxide-DNA adducts were also observed (cf. below, Fig. 8).

Previous studies have demonstrated the much higher tumorigenicity of racemic anti-DBPDE and other corresponding fjord region PAH diol-epoxide isomers (e.g., from benzo[c]phenanthrene) in lung and liver of newborn mice compared to anti-BPDE (>25-fold) [76]. This exceptional tumorigenicity exerted by the fjord region anti-DBPDE compared to configurational related bay region anti-BPDE is of critical importance for understanding the mechanisms of chemical tumor induction and somehow depends, at least in part, on the structural characteristics of the DNA adducts formed.

Members of the proto-oncogene family H-ras are commonly mutated in human cancers and animal models for chemical-induced carcinogenesis [104], and certain PAH-DNA adducts have been implicated in H-ras-mediated cell transformation. In the 1980s, exposure of plasmids carrying a H-ras proto-oncogene to BP diol-epoxides generated an activated oncogene that could transform fibroblasts in vitro [105]. Subsequent analysis of these cells revealed that the mutations were mainly confined to codon 61 of H-ras [106]. Since the same mutations were also found in PAH-mediated rodent tumors, it became clear that particularly codon 61 of H-ras is an important molecular target of carcinogenic PAHs (Fig. 9) [107]. About 15 years later, more specific in vitro studies with human HeLa cell extracts were conducted to assess the NER-mediated repair efficiency of a chemically synthesized and defined diol-epoxide-DNA adducts in the context of ras-specific sequence motifs (Fig. 8). At the second base (dA) of codon 61 of the human proto-oncogenes N-ras or H-ras, (–)-trans-anti-DBPDE-N^6-dA adducts were found to be virtually resistant to

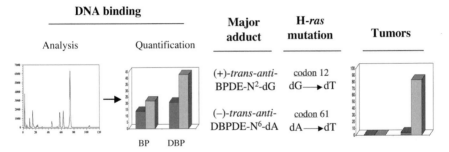

Figure 9. Genotoxicity and tumorigenicity of BP and DBP in mouse skin. The DNA binding levels of BP and DBP at doses of 50 and 400 nmol in mouse skin 24 h after treatment are depicted. Values are given in the number of adducts per 10^6 nucleotides (cf. Tab. 2) [80]. Nucleotide transversions within codons 12 or 61 of cellular H-*ras* were found in mouse skin upon exposure to both BP and DBP. Treatment with BP mainly results in the formation (+)-*trans-anti*-BPDE-N^2-dG adducts at early time points and subsequent induction of dG → dT transversions at codon 12 (and 13), whereas DBP predominantly forms (–)-*trans-anti*-DBPDE-N^6-dA adducts and induces dA → dT mutations at codon 61. The tumor yield shown for treatment doses of 1 and 4 nmol PAH is taken from [30] and also listed in Table 2 (see text for further explanations).

DNA repair, whereas stereochemically corresponding (+)-*trans-anti*-BPDE-N^6-dA adducts were readily removed from the same sites (Fig. 8) [108, 109]. It was demonstrated that (+)-*trans-anti*-BPDE-N^6-dA adducts were integrated into the double helix in a way that distorts and destabilizes the regular Watson-Crick base pairing at this site. Conversely, (–)-*trans-anti*-DBPDE-N^6-dA adducts were incorporated by an intercalative mode that retains normal base pairing [100, 101, 110]. Thus, bay region BP diol-epoxide adducts reduced the stability of the double helix, whereas fjord region DBP diol-epoxide adducts did not [111]. This notion was supported by differences in the thermal stabilities of short double-stranded DNA fragments containing single bay or fjord region diol-epoxide-dA adducts. Whereas the DNA duplex melting point was found to be significantly lowered in the presence of a bay region (+)-*trans-anti*-BPDE-N^6-dA lesion, it remained unchanged in the presence of the corresponding fjord-region (–)-*trans-anti*-DBPDE-N^6-dA adduct [100]. In the *ras* gene sequence context (–)-*trans-anti*-DBPDE-N^6-dA adducts from the non-planar fjord region DBP are therefore more resistant to repair. Located in the *ras* gene of a living cell, these adducts would be good candidate lesions for being converted into mutations during the next replication cycle (Fig. 9). If not repaired prior to replication, PAH-damaged templates induce nucleotide mis-incorporation opposite the lesion and have therefore the potential of inducing mutations in critical genes involved in tumorigenesis. In adenocarcinomas of the human lung and pancreas, for example, the incidences of point mutations in cellular *ras* may reach levels as high as 30% and 90%, respectively [112]. The principal activating (point) mutations in *ras* genes are almost exclusively located in codons 12, 13 and 61, rendering the GTPase activity of the corresponding p21 protein inactive. As consequence, p21 will be arrested in a GTP-bound activated mode and constantly trigger kinase signaling cascades leading to aberrant cell proliferation. In mouse models, it has been shown that the differential occurrence of mutations at codons 12, 13 or 61 depends on the particular PAH and the kind of the tissue targeted (Fig. 9). Nucleotide transversions within codons 12 or 61 of cellular H-*ras* were found in mouse skin upon exposure to both BP and DBP [95, 113, 114]. Treatment with BP mainly results in the formation of dG to dT transversions at codon 12 (and 13), whereas DBP predominantly induce dA to dT mutations at codon 61 (Fig. 9). On the other hand, K-*ras* is the main family member targeted by carcinogenic PAHs in lung tissue [72, 115]. Similar to skin, BP again displays a bias for inducing dG to dT transversions at codon 12, while DBP-induced K-*ras* mutations in lung tissue are more equally distributed between codons 12 and 61. In a compilation of the literature on PAH-induced *ras* oncogene mutations in animal models it has been noticed that there is a strong qualitative relationship between the kinds of DNA adducts formed in a certain tissue type (i.e., diol-epoxide-N^2-dG and/or diol-epoxide-N^6-dA adducts) and the kinds of mutations (at codons 12 and 61) induced in tumors that originate from this tissue (Fig. 9) [104].

Reprise

In rodent bioassays, DBP has been found to be the much more potent carcinogen compared to BP (Tab. 2). As a prerequisite to exert strong tumorigenic activity both compounds require metabolic activation by CYP1A1 or CYP1B1 enzymes into their ultimate DNA-damaging forms, the bay or fjord region diol-epoxides (Fig. 3). In mouse skin it is well established that CYP1B1-catalyzed activation plays a predominant role in DBP-induced DNA damage and carcinogenesis, while murine CYP1A1 is central in the bioactivation of BP and other planar carcinogenic PAHs [116, 117]. The almost exclusive enzymatic formation of (+)-*anti*-BPDE-N^2-dG adducts in skin cells of mice topically treated with BP [43, 80] has been confirmed in studies using cells in culture including human cell lines or cells expressing recombinant CYP enzymes (Tab. 2) [81, 118, 119]. On the other hand, metabolic activation of DBP in human cells, mouse embryo cells, and recombinant cell lines expressing human CYP1A1 or CYP1B1 resulted in the predominant formation of (–)-*anti*-DBPDE-DNA adducts [57, 81, 119, 120], thus confirming the result obtained in mouse skin studies [43, 80, 117]. Some small amounts of (+)-*syn*-DBPDE-related DNA adducts obtained upon high treatment doses of the parent hydrocarbon DBP were not detectable anymore at low doses (Fig. 5) [121, 122]. Comparison of the DNA adduct levels induced by BP or DBP revealed always higher values for the latter compound, an effect found much more pronounced in human MCF-7 cells or recombinant cell lines expressing CYP1A1 or CYP1B1 when compared to mouse skin *in vivo* (Tab. 2) [43, 80, 119]. For instance, while 4 μM BP induced DNA adduct levels in CYP1A1-expressing cells with a maximum value of about 47 pmol adducts/mg DNA at the end of the observation period (72 h post treatment), in the same cells DBP induced 10 pmol adducts/mg DNA at 100-fold lower concentrations (40 nM) at 12 h post treatment [119]. Much lower treatment doses were required for DBP due to its extremely high toxicity in cells and in mouse skin [43, 64, 66]. Using the same treatment doses as in CYP1A1-expressing cells, in CYP1B1-expressing cells the values were about the same for both PAHs: ~86 pmol adducts/mg DNA (4 μM BP/48 h time point and 40 nM DBP/12 h time point, respectively) [119].

DNA binding experiments with (+)-*anti*-BPDE and (–)-*anti*-DBPDE in buffered solutions *in vitro* revealed no significant differences between both chemical species under these artificial conditions (cf. above). However, in human cells in culture, (–)-*anti*-DBPDE was found to bind at least 10-fold more efficient to DNA than (+)-*anti*-BPDE [123, 124] and about 2–4-fold stronger than the isomeric (+)-*syn*-DBPDE [125]. At the same time it became obvious that intracellular formation of (–)-*anti*-DBPDE-DNA adducts occurs more slowly compared to the formation of (+)-*anti*-BPDE-DNA adducts (maximum binding level at about 1 h and 20 min post treatment, respectively). The much longer intracellular half-life of the fjord region diol-epoxide (about 180 min) compared to the bay region diol-epoxide (about 30 min), most likely due to the higher lipophilicity and lower solvolytic reactivity of the hexacyclic isomer [123], provides a rationale for this observation made in cell culture. At most, the 10-fold higher reactivity of (–)-*anti*-DBPDE towards DNA compared to (+)-*anti*-BPDE in cells therefore seems to result from its intrinsic physicochemical properties and its non-enzymatic fate in the intracellular compartment. By contrast, the efficiencies in cellular detoxification of (–)-*anti*-DBPDE and (+)-*anti*-BPDE *via* glutathione conjugation are comparable for both chemical species, although significant differences in the catalytic capacity of individual GST isoforms or different polymorphic variants of one and the same particular isoform have been described for purified enzymes *in vitro* [123, 126–128]. While certain isoforms may prefer (+)-*anti*-BPDE (P1-1) as substrate and others rather (–)-*anti*-DBPDE (A1-1), the overall amount of diol-epoxide-glutathione conjugates formed in mammalian cells in culture [123] and the reduction in mutagenicity of these species [87, 128, 129] were found in about the same range.

Given the ~10-fold stronger DNA binding and the ~60-fold higher levels of mutations induced in cells in culture by (–)-*anti*-DBPDE as compared to (+)-*anti*-BPDE (Fig. 6), it becomes obvious that DNA adduct recognition and subsequent removal catalyzed by the NER machinery most likely represent a significant factor that contributes to the differences in tumorigenicity observed for the parent compounds in rodent tumor models (Tab. 2). Studying the formation and removal of diol-epoxide-DNA adducts in human lung cancer A549 cells confirmed the repair resistance of (–)-*anti*-DBPDE-DNA adducts found earlier with NER proteins extracted from human HeLa cells. While treatment with 100 nM (–)-*anti*-DBPDE resulted in an increase of DNA adducts to a maximum level of about 144 pmol/mg DNA after 1 h of incubation followed by a slow removal of damaged sites, similar maximum DNA adduct levels (140 pmol/mg DNA) were already reached at 20 min upon exposure to 1 µM (+)-*anti*-BPDE [124]. This initial rapidness in (+)-*anti*-BPDE-DNA adduct formation in A549 cells was replaced by a similar fastness in the removal of these adducts. After 1 h of incubation only about 55% of adducts remained and 25% after 6 h, whereas 80% of (–)-*anti*-DBPDE-DNA adducts were still present at this later time point [124]. In a subsequent study on the repair and mutagenicity of diol epoxide-DNA adducts at the *hrpt* locus in V79 cells, it was demonstrated that the repair of racemic *anti*-DBPDE-DNA adducts is about 5 times less efficient as compared to (+)-*anti*-BPDE-DNA adducts in NER-proficient cells [130]. In NER-deficient cells this difference disappeared and the mutagenicity per target dose was even slightly higher in case of (+)-*anti*-BPDE-DNA adducts.

Conclusions

The fjord region PAH DBP is considerably more carcinogenic compared to the bay region PAH BP. The strong difference in the carcinogenic potencies of both compounds (~100-fold) can be ascribed to differences in DNA binding efficiency of their ultimate carcinogenic diol-epoxide metabolites and differences in structural features of the diol-epoxide-DNA adducts that result in differences in DNA adduct recognition, subsequent downstream signaling [131], and removal of the lesions by the NER machinery [130]. Given the preferential binding site and the kind of epoxide ring opening, (–)-*trans-anti*-DBPDE-N^6-dA adducts are much more efficient in escaping recognition and thus refractory to subsequent removal by the cellular DNA repair system compared to (+)-*trans-anti*-BPDE-N^2-dG adducts [100, 124]. In view of the 10-fold more effective DNA binding of (–)-*anti*-DBPDE and the at least 5-fold lower efficiency in removal of the corresponding adducts formed, the 60-fold difference in mutant frequency observed in V79 cells upon exposure to (+)-*anti*-BPDE and (–)-*anti*-DBPDE (Fig. 6) can be nicely substantiated. Altogether these results support the notion that the high carcinogenicity of the nonplanar fjord region PAH DBP arise from the ability of the preferentially formed non-distorting (–)-*trans-anti*-DBPDE-N^6-dA adducts to escape recognition by surveillance systems and NER-coupled repair. Based on the high reactivity of the potent carcinogen 7,12-dimethylbenz[*a*]anthracene at N^6-dA residues in DNA, already in 1983 it had been anticipated that N^6-dA adducts of nonplanar PAHs might have the greater potency for tumor induction in mouse skin than N^2-dG adducts of planar PAHs [132]. The exceptionally potent mouse skin tumorigen DBP and its mode of DNA binding *via* (–)-*anti*-DBPDE now adds some further support to this early proposal and strengthens the idea that inefficient excision repair is an important determinant of high tumorigenic potency [133].

References

1 Doll R, Peto R (1981) The causes of cancer: Quantitative estimates of avoidable risks of cancer in the United States today. *J Nat Cancer Inst* 66: 1191–1308
2 Lichtenstein P, Holm, NV, Verkasalo PK, Iliadou A, Kaprio J, Koskenvuo M, Pukkala E, Skytthe A, Hemminki K (2000) Environmental and heritable factors in the causation of cancer. *N Engl J Med* 343: 78–85
3 Czene K, Lichtenstein P, Hemminki K (2002) Environmental and heritable causes of cancer among 9.6 million individuals in the Swedish family-cancer database. *Int J Cancer* 99: 260–266
4 Kolonel LN, Altshuler D, Henderson BE (2004) The multiethic cohort study: Exploring genes, lifestyle and cancer risk. *Nat Rev Cancer* 4: 519–527
5 Colditz GA, Sellers TA, Trapido E (2006) Epidemiology – Identifying the causes and preventability of cancer? *Nat Rev Cancer* 6: 75–83
6 Peto F (2001) Cancer epidemiology in the last century and the next decade. *Nature* 411: 390–395
7 Luch A (2002) Cell cycle control and cell division: Implications for chemically induced carcinogenesis. *ChemBioChem* 3: 506–516
8 Chen F, Shi X (2002) Intracellular signal transduction of cells in response to carcinogenic metals. *Crit Rev Oncol Hematol* 42: 105–121
9 Wilson S, Jones L, Coussens C, Hanna K (2002) *Cancer and the Environment: Gene-Environment Interaction.* National Academy Press, Washington D.C.
10 Cogliano VJ, Baan RA, Straif K, Grosse Y, Secretan B, El Ghissassi F (2008) Use of mechanistic data in IARC evaluations. *Environ Mol Mutag* 49: 100–109
11 Luch A (2006) The mode of action of organic carcinogens on cellular structures. *EXS* 96: 65–95
12 van Delft JHM, van Agen E, van Breda SGJ, Herwijnen MH, Staal YCM, Kleinjans JCS (2004) Discrimination of genotoxic from non-genotoxic carcinogens by gene expression profiling. *Carcinogenesis* 25: 1265–1276
13 Cook JW, Hewett CL, Hieger I (1933) The isolation of a cancer-producing hydrocarbon from coal tar. *J Chem Soc* 395–405
14 Cook JW, Haslewood GAD, Hewett CL, Hieger I, Kennaway EL, Mayneord WV (1937) Chemical compounds as carcinogenic agents. *Am J Cancer* 29: 219–259
15 Phillips DH (1983) Fifty years of benzo[*a*]pyrene. *Nature* 303: 468–472
16 Bachmann WE, Cook JW, Dansi A, De Worms CGM, Haslewood GAD, Hewett CL, Robinson AM (1937) Production of cancer by pure hydrocarbons. IV. *Proc R Soc (Lond) B* 123: 343–368
17 Wynder EL, Wright G (1957) A study of tobacco carcinogenesis. *Cancer* 10: 255–271
18 Orris L, van Duuren BL, Kosak AI, Nelson N, Schmitt FL (1958) The carcinogenicity for mouse skin and the aromatic hydrocarbon content of cigarette-smoke condensates. *J Natl Cancer Inst* 21: 557–561
19 Wynder EL, Hoffmann D (1961) Carcinogenicity of dibenzo[*a,l*]pyrene. *Nature* 192: 1092–1093
20 Lacassagne A, Buu-Hoï NP, Zajadela F, Lavit-Lamy D (1963) Activité cancérogène élevée du 1.2: 3.4-dibenzopyrène et 1.2:4.5-dibenzopyrène. *Compt Rend Acad Sci Paris* 256: 2728–2730
21 Hoffmann D, Wynder EL (1966) Beitrag zur carcinogenen Wirkung von Dibenzpyrenen. *Z Krebsforsch* 68: 137–149
22 Vingiello FA, Yanez F, Greenwood EJ (1966) The synthesis of dibenzo[*a,l*]pyrene. *J Chem Soc Chem Commun* 863–864
23 Vingiello FA, Youssef AK (1967) The isomerization of 12-phenylbenz[*a*]anthracene. *J Chem Soc Chem Commun* 863–864
24 Carruthers W (1966) Synthesis of dibenzo[*a,l*]pyrene. *J Chem Soc Chem Commun* 548–549
25 Lacassagne A, Buu-Hoï NP, Zajadela F, Vingiello FA (1968) The true dibenzo[*a,l*]pyrene, a new, potent carcinogen. *Naturwissenschaften* 55: 43
26 Masuda Y, Kagawa R (1972) A novel synthesis and carcinogenicity of dibenzo[*a,l*]pyrene. *Chem Pharm Bull* 20: 2736–2737
27 Cavalieri EL, Rogan EG, Higginbotham S, Cremonesi P, Salmasi S (1989) Tumor-initiating activity in mouse skin and carcinogenicity in rat mammary gland of dibenzo[*a*]pyrenes: The very potent environmental carcinogen dibenzo[*a,l*]pyrene. *J Cancer Res Clin Oncol* 115: 67–72
28 Cavalieri EL, Higginbotham S, RamaKrishna NVS, Devanesan PD, Todorovic R, Rogan EG, Salmasi S (1991) Comparative dose-response tumorigenicity studies of dibenzo[*a,l*]pyrene *versus* 7,12-dimethylbenz[*a*]anthracene, benzo[*a*]pyrene and two dibenzo[*a,l*]pyrene dihydrodiols in mouse skin and rat mammary gland. *Carcinogenesis* 12: 1939–1944

29 LaVoie EJ, He ZM, Meegalla RL, Weyand EH (1993) Exceptional tumor-initiating activity of
    4-fluorobenzo[*j*]fluoranthene on mouse skin: Comparison with benzo[*j*]fluoranthene, 10-fluoro-
    benzo[*j*]fluoranthene, benzo[*a*]pyrene, dibenzo[*a,l*]pyrene and 7,12-dimethylbenz[*a*]anthracene.
    Cancer Lett 70: 7–14
30 Higginbotham S, RamaKrishna NVS, Johansson SL, Rogan RG, Cavalieri EL (1993) Tumor-ini-
    tiating activity and carcinogenicity of dibenzo[*a,l*]pyrene *versus* 7,12-dimethylbenz[*a*]anthracene
    and benzo[*a*]pyrene at low doses in mouse skin. *Carcinogenesis* 14: 875–878
31 Swauger JE, Steichen TJ, Murphy PA, Kinsler S (2002) An analysis of the mainstream smoke
    chemistry of samples of the U.S. cigarette market acquired between 1995 and 2000. *Regul Toxicol
    Pharmacol* 35: 142–156
32 Rodgman A (2003) The composition of cigarette smoke: Problems with lists of tumorigens. *Beitr
    Tabakforsch Int* 20: 402–437
33 Seidel A, Frank H, Behnke A, Schneider D, Jacob J (2004) Determination of dibenzo[*a,l*]pyrene
    and other fjord-region PAH isomers with MW 302 in environmental matrices. *Polycyclic Aromat
    Compd* 24: 759–771
34 Mumford JL, Harris DB, Williams K, Chuang JC, Cooke M (1987) Indoor air sampling and muta-
    genicity studies of emissions from unvented coal combustion. *Environ Sci Technol* 21: 308–311
35 Mumford JL, Xueming L, Fuding H, Xu BL, Chuang JC (1995) Human exposure and dosimetry
    of polycyclic aromatic hydrocarbons in urine from Xuan Wei, China with high lung cancer mor-
    tality associated with exposure to unvented coal smoke. *Carcinogenesis* 16: 3031–3036
36 Kozin IS, Gooijer C, Velthorst NH (1995) Direct determination of dibenzo[*a,l*]pyrene in crude
    extracts of environmental samples by laser excited Shpol'skii spectroscopy. *Anal Chem* 67:
    1623–1626
37 Luch A, Baird WM (2005) Metabolic activation of polycylic aromatic hydrocarbons. In: A Luch
    (ed.): *The Carcinogenic Effects of Polycyclic Aromatic Hydrocarbons.* Imperial College Press,
    London, 19–96
38 Brown CM, Reisfeld B, Mayeno AN (2008) Cytochromes P450: A structure-based summary of
    biotransformation using representative substrates. *Drug Metab Rev* 40: 1–100
39 Shimada T, Fujii-Kuriyama Y (2004) Metabolic activation of polycyclic aromatic hydrocarbons to
    carcinogens by cytochromes P450 1A1 and 1B1. *Cancer Sci* 95: 1–6
40 Shimada T (2006) Xenobiotic-metabolizing enzymes involved in activation and detoxification of
    carcinogenic polycyclic aromatic hydrocarbons. *Drug Metab Pharmacokinet* 21: 257–276
41 Nebert DW, Dalton TP, Okey AB, Gonzalez FJ (2004) Role of aryl hydrocarbon receptor-mediat-
    ed induction of the CYP1 enzymes in environmental toxicity and cancer. *J Biol Chem* 279:
    23847–23850
42 Nebert DW, Roe AL, Dieter MZ, Solis WA, Yang Y, Dalton TP (2000) Role of the aromatic hydro-
    carbon receptor and [*Ah*] gene battery in the oxidative stress response, cell cycle control and apop-
    tosis. *Biochem Pharmacol* 59: 65–85
43 Marston CP, Pereira C, Ferguson J, Fischer K, Hedstrom O, Dashwood WM, Baird WM (2001)
    Effect of a complex environmental mixture from coal tar containing polycyclic aromatic hydro-
    carbons (PAH) on the tumor initiation, PAH-DNA binding and metabolic activation of carcino-
    genic PAH in mouse epidermis. *Carcinogenesis* 22: 1077–1086
44 Mahadevan B, Marston CP, Dashwood WM, Li Y, Pereira C, Baird WM (2005) Effect of a stan-
    dardized complex mixture derived from coal tar on the metabolic activation of carcinogenic poly-
    cyclic aromatic hydrocarbons in human cells in culture. *Chem Res Toxicol* 18: 224–231
45 Katz AK, Carrell HL, Glusker JP (1998) Dibenzo[*a,l*]pyrene (dibenzo[*def,p*]chrysene): Fjord-
    region distortions. *Carcinogenesis* 19: 1641–1648
46 Sims P, Grover PL, Swaisland A, Pal K, Hewer A (1974) Metabolic activation of benzo[*a*]pyrene
    proceeds by a diol-epoxide. *Nature* 252: 326–328
47 Yagi H, Thakker DR, Hernandez O, Koreeda M, Jerina DM (1977) Synthesis and reactions of the
    highly mutagenic 7,8-diol 9,10-epoxides of the carcinogen benzo[*a*]pyrene. *J Am Chem Soc* 99:
    1604–1611
48 Morisseau C, Hammock BD (2005) Epoxide hydrolases: Mechanisms, inhibitor designs, and bio-
    logical roles. *Annu Rev Pharmacol Toxicol* 45: 311–333
49 Shimada T, Oda Y, Gillam EMJ, Guengerich FP, Inoue K (2001) Metabolic activation of poly-
    cyclic aromatic hydrocarbons and their dihydrodiol derivatives and other procarcinogens by
    cytochrome P450 1A1 and 1B1 allelic variants and other human cytochrome P450 enzymes in
    Salmonella typhimurium NM2009. *Drug Metab Dispos* 29: 1176–1182

50 Yang SK, McCourt DW, Roller PP, Gelboin HV (1976) Enzymatic conversion of benzo[*a*]pyrene leading predominantly to the diol-epoxide *r*-7,*t*-8-dihydroxy-*t*-9,10-oxy-7,8,9,10-tetrahydrobenzo[*a*]pyrene through a single enantiomer of *r*-7,*t*-8-dihydroxy-7,8-dihydrobenzo[*a*]pyrene. *Proc Natl Acad Sci USA* 73: 2594–2598

51 Yang SK, McCourt DW, Leutz JC, Gelboin HV (1977) Benzo[*a*]pyrene diol epoxides: Mechanisms of enzymatic formation and optically active intermediates. *Science* 196: 1199–1201

52 Wood AW, Chang RL, Levin W, Yagi H, Thakker DR, Jerina DM, Conney AH (1977) Differences in mutagenicity of the optical enantiomers of the diastereomeric benzo[*a*]pyrene 7,8-diol-9,10-epoxides. *Biochem Biophys Res Commun* 77: 1389–1396

53 Buening MK, Wislocki PG, Levin W, Yagi H, Thakker DR, Akagi H, Kooreda M, Jerina DM, Conney AH (1978) Tumorigenicity of the optical enantiomers of the diastereomeric benzo[*a*]pyrene 7,8-diol-9,10-epoxides in newborn mice: Exceptional activity of (±)-7β,8α-dihydroxy-9α,10α-epoxy-7,8,9,10-tetrahydrobenzo[*a*]pyrene. *Proc Natl Acad Sci USA* 75: 5358–5361

54 Thakker DR, Levin W, Yagi H, Wood AW, Conney AH, Jerina DM (1988) Stereoselective biotransformation of polycyclic aromatic hydrocarbons to ultimate carcinogens. In: AW Wainer, D Dryer (eds): *Stereochemical Aspects of Pharmacologically Active Compounds.* Marcel Dekker, New York, 217–296

55 Yang SK (1988) Stereoselectivity of cytochrome P-450 isozymes and epoxide hydrolase in the metabolism of polycyclic aromatic hydrocarbons. *Biochem Pharmacol* 37: 61–70

56 Ralston SL, Lau HHS, Seidel A, Luch A, Platt KL, Baird WM (1994) The potent carcinogen dibenzo[*a,l*]pyrene is metabolically activated to fjord-region 11,12-diol 13,14-epoxides in human mammary carcinoma MCF-7 cell cultures. *Cancer Res* 54: 887–890

57 Ralston SL, Seidel A, Luch A, Platt KL, Baird WM (1995) Stereoselective activation of dibenzo[*a,l*]pyrene to (–)-*anti*-(11*R*,12*S*,13*S*,14*R*)- and (+)-*syn*-(11*S*,12*R*,13*S*,14*R*)-11,12-diol-13,14-epoxides which bind extensively to deoxyadenosine residues of DNA in the human mammary carcinoma cell line MCF-7. *Carcinogenesis* 16: 2899–2907

58 Devanesan PD, Cremonesi P, Nunnally JE, Rogan EG, Cavalieri EL (1990) Metabolism and mutagenicity of dibenzo[*a,e*]pyrene and the very potent environmental carcinogen dibenzo[*a,l*]pyrene. *Chem Res Toxicol* 3: 580–586

59 Li KM, Todorovic R, Rogan EG, Cavalieri EL, Ariese F, Suh M, Jankowiak R, Small GJ (1995) Identification and quantitation of dibenzo[*a,l*]pyrene-DNA adducts formed by rat liver microsomes *in vitro*: Preponderance of depurinating adducts. *Biochemistry* 34: 8043–8049

60 Arif JM, Gupta RC (1997) Microsome-mediated bioactivation of dibenzo[*a,l*]pyrene and identification of DNA adducts by ^{32}P-postlabeling. *Carcinogenesis* 18: 1999–2007

61 Shou M, Korzekwa KR, Crespi CL, Gonzalez FJ, Gelboin HV (1994) The role of 12 cDNA-expressed human, rodent, and rabbit cytochromes P450 in the metabolism of benzo[*a*]pyrene and benzo[*a*]pyrene *trans*-7,8-dihydrodiol. *Mol Carcinog* 10: 159–168

62 Bauer E, Guo Z, Ueng YF, Bell LC, Zeldin D, Guengerich FP (1995) Oxidation of benzo[*a*]pyrene by recombinant human cytochrome P450 enzymes. *Chem Res Toxicol* 8: 136–142

63 Shou M, Krausz KW, Gonzalez FJ, Gelboin HV (1996) Metabolic activation of the potent carcinogen dibenzo[*a,l*]pyrene by human recombinant cytochromes P450, lung and liver microsomes. *Carcinogenesis* 17: 2429–2433

64 Luch A, Schober W, Soballa VJ, Raab G, Greim H, Jacob J, Doehmer J, Seidel A (1999) Metabolic activation of dibenzo[*a,l*]pyrene by human cytochrome P450 1A1 and 1B1 expressed in V79 Chinese hamster cells. *Chem Res Toxicol* 12: 353–364

65 King LC, Adams L, Allison F, Kohan MJ, Nelson G, Desai D, Amin S, Ross JA (1999) A quantitative comparison of dibenzo[*a,l*]pyrene DNA adduct formation by recombinant human cytochrome P450 microsomes. *Mol Carcinog* 26: 74–82

66 Schober W, Luch A, Soballa VJ, Raab G, Stegeman JJ, Doehmer F, Jacob F, Seidel A (2006) On the species-specific biotransformation of dibenzo[*a,l*]pyrene. *Chem Biol Interact* 161: 37–48

67 Phillips DH, Glatt HR, Seidel A, Bochnitschek W, Oesch F, Grover PL (1986) Mutagenic potential and DNA adducts formed by diol-epoxides, triol-epoxides and K-region epoxide of chrysene in mammalian cells. *Carcinogenesis* 7: 1739–1743

68 Phillips DH, Hewer A, Seidel A, Steinbrecher T, Schrode R, Oesch F, Glatt HR (1991) Relationship between mutagenicity and DNA adduct formation in mammalian cells for fjord-region and bay-region diol-epoxides of polycyclic aromatic hydrocarbons. *Chem Biol Interact* 80: 177–186

69 Glatt H, Piée A, Pauly K, Steinbrecher T, Schrode R, Oesch F, Seidel A (1991) Fjord- and bay-region diol-epoxides investigated for stability, SOS induction in *Escherichia coli* and mutagenicity in *Salmonella typhimurium* and mammalian cells. *Cancer Res* 51: 1659–1667

70 Hughes NC, Phillips DH (1990) Covalent binding of dibenzpyrenes and benzo[*a*]pyrene to DNA: Evidence for synergistic and inhibitory interactions when applied in combination to mouse skin. *Carcinogenesis* 11: 1611–1619

71 Ross JA, Nelson GB, Wilson KH, Rabinowitz JR, Galati A, Stoner GD, Nesnow S, Mass MJ (1995) Adenomas induced by polycyclic aromatic hydrocarbons in strain A/J mouse lung correlate with time-integrated DNA adduct levels. *Cancer Res* 55: 1039–1044

72 Prahalad AK, Ross JA, Nelson GB, Roop BC, King LC, Nesnow S, Mass MJ (1997) Dibenzo[*a,l*]pyrene-induced DNA adduction, tumorigenicity, and Ki-*ras* oncogene mutations in strain A/J mouse lung. *Carcinogenesis* 18: 1955–1963

73 Arif JM, Smith WA, Gupta RC (1997) Tissue distribution of DNA adducts in rats treated by intra-mammillary injection with dibenzo[*a,l*]pyrene, 7,12-dimethylbenz[*a*]anthracene and benzo[*a*]pyrene. *Mutat Res* 378: 31–39

74 Gill HS, Kole PL, Wiley JC, Li KM, Higginbotham S, Rogan EG, Cavalieri EL (1994) Synthesis and tumor-initiating activity in mouse skin of dibenzo[*a,l*]pyrene *syn*- and *anti*-fjord-region diolepoxides. *Carcinogenesis* 15: 2455–2460

75 Amin S, Krzeminski J, Rivenson A, Kurtzke C, Hecht SS, El-Bayoumy K (1995) Mammary carcinogenicity in female CD rats of fjord region diol epoxides of benzo[*c*]phenanthrene, benzo[*g*]chrysene and dibenzo[*a,l*]pyrene. *Carcinogenesis* 16: 1971–1974

76 Amin S, Desai D, Dai W, Harvey RG, Hecht SS (1995) Tumorigenicity in newborn mice of fjord region and other sterically hindered diol epoxides of benzo[*g*]chrysene, dibenzo[*a,l*]pyrene (dibenzo[*def,p*]chrysene), 4*H*-cyclopenta[*def*]chrysene and fluoranthene. *Carcinogenesis* 16: 2813–2817

77 Poirier MC (2004) Chemical-induced DNA damage and human cancer risk. *Nat Rev Cancer* 4: 630–637

78 Xue W, Warshawsky D (2005) Metabolic activation of polycyclic and heterocyclic aromatic hydrocarbons and DNA damage: A review. *Toxicol Appl Pharmacol* 206: 73–93

79 Koreeda M, Moore PD, Wislocki PG, Levin W, Conney AH, Yagi H, Jerina DM (1978) Binding of benzo[*a*]pyrene 7,8-diol 9,10-epoxides to DNA, RNA, and protein of mouse skin occurs with high stereoselectivity. *Science* 199: 778–781

80 Melendez-Colon VJ, Luch A, Seidel A, Baird WM (1999) Cancer initiation by polycyclic aromatic hydrocarbons results from formation of stable DNA adducts rather than apurinic sites. *Carcinogenesis* 20: 1885–1891

81 Melendez-Colon VJ, Luch A, Seidel A, Baird WM (2000) Formation of stable DNA adducts and apurinic sites upon metabolic activation of bay and fjord region polycyclic aromatic hydrocarbons in human cell cultures. *Chem Res Toxicol* 13: 10–17

82 Mahadevan B, Luch A, Bravo CF, Atkin J, Steppan LB, Pereira C, Kerkvliet NI, Baird WM (2005) Dibenzo[*a,l*]pyrene induced DNA adduct formation in lung tissue *in vivo*. *Cancer Lett* 227: 25–32

83 Szeliga J, Dipple A (1998) DNA adduct formation by polycyclic aromatic hydrocarbon dihydrodiol epoxides. *Chem Res Toxicol* 11: 1–11

84 Roberts KP, Lin CH, Jankowiak R, Small GJ (1999) On-line identification of diastereomeric dibenzo[*a,l*]pyrene diol epoxide-derived deoxyadenosine adducts by capillary electrophoresis-fluorescence line-narrowing and non-line narrowing spectroscopy. *J Chromatogr A* 853: 159–170

85 Jankowiak R, Lin CH, Zamzow D, Roberts KP, Li KM, Small GJ (1999) Spectral and conformational analysis of deoxyadenosine adducts derived from *syn*- and *anti*-dibenzo[*a,l*]pyrene diol epoxides: Fluorescence studies. *Chem Res Toxicol* 12: 768–777

86 Luch A, Glatt HR, Platt KL, Oesch F, Seidel A (1994) Synthesis and mutagenicity of the diastereomeric fjord-region 11,12-dihydrodiol 13,14-epoxides of dibenzo[*a,l*]pyrene. *Carcinogenesis* 15: 2507–2516

87 Seidel A, Friedberg T, Löllmann B, Schwierzok A, Funk M, Frank H, Holler R, Oesch F, Glatt HR (1998) Detoxification of optically active bay- and fjord-region polycyclic aromatic hydrocarbon dihydrodiol epoxides by human glutathiol transferase P1-1 expressed in Chinese hamster V79 cells. *Carcinogenesis* 19: 1975–1981

88 Glatt HR (2005) Indicator assays for polycyclic aromatic hydrocarbon-induced genotoxicity. In: A Luch (ed.): *The Carcinogenic Effects of Polycyclic Aromatic Hydrocarbons*. Imperial College Press, London, 283–314

89 Jerina DM, Sayer JM, Agarwal SK, Yagi H, Levin W, Wood AW, Conney AH, Preuss-Schwarz D, Baird WM, Pigott MA, Dipple A (1986) Reactivity and tumorigenicity of bay-region diol epoxides derived from polycyclic aromatic hydrocarbons. In: JJ Kocsis, DJ Jollow, CM Witmer, JO Nelson, R Snyder (eds): *Biological Reactive Intermediates III, Mechanisms of Action in Animal Models and Human Diseases*. Plenum Press, New York, 11–30

90 Jankowiak R, Ariese F, Hewer A, Luch A, Zamzow D, Hughes NC, Phillips DH, Seidel A, Platt KL, Oesch F, Small G (1998) Structure, conformations, and repair of DNA adducts from dibenzo[a,l]pyrene: ^{32}P-postlabeling and fluorescence studies. *Chem Res Toxicol* 11: 674–685

91 Devanesan P, Ariese F, Jankowiak R, Small GJ, Rogan EG, Cavalieri EL (1999) A novel method for the isolation and identification of stable DNA adducts formed by dibenzo[a,l]pyrene and dibenzo[a,l]pyrene 11,12-dihydrodiol 13,14-epoxides *in vitro*. *Chem Res Toxicol* 12: 796–801

92 Wei SJ, Chang RL, Wong CQ, Bhachech N, Cui XX, Hennig E, Yagi Y, Sayer JM, Jerina DM, Preston BD, Conney AH (1991) Dose-dependent differences in the profile of mutations induced by an ultimate carcinogen from benzo[a]pyrene. *Proc Natl Acad Sci USA* 88: 11227–11230

93 Wei SJ, Chang RJ, Hennig E, Cui XX, Merkler KA, Wong CQ, Yagi H, Jerina DM, Conney AH (1994) Mutagenic selectivity at the *HPRT* locus in V79 cells: Comparison of mutations caused by bay-region benzo[a]pyrene 7,8-diol-9,10-epoxide enantiomers with high and low carcinogenic activity. *Carcinogenesis* 15: 1729–1735

94 Chakravarti D, Pelling JC, Cavalieri EL, Rogan EG (1995) Relating aromatic hydrocarbon-induced DNA adducts and c-Ha-*ras* mutations in mouse skin papillomas: The role of apurinic sites. *Proc Natl Acad Sci USA* 92: 10422–10426

95 Chakravarti D, Venugopal D, Mailander PC, Meza JL, Higginbotham S, Cavalieri EL, Rogan EG (2008) The role of polycyclic aromatic hydrocarbon-DNA adducts in inducing mutations in mouse skin. *Mutat Res* 649: 161–178

96 Mahadevan B, Dashwood WM, Luch A, Pecaj A, Doehmer J, Seidel A, Pereira C, Baird WM (2003) Mutations induced by (–)-*anti*-11*R*,12*S*-dihydrodiol 13*S*,14*R*-epoxide of dibenzo-[a,l]pyrene in the coding region of the hypoxanthine phosphoribosyltransferase (*hprt*) gene in Chinese hamster V79 cells. *Environ Mol Mutagen* 41: 131–139

97 Yoon JH, Besaratinia A, Feng Z, Tang MS, Amin S, Luch A, Pfeifer GP (2004) DNA damage, repair, and mutation induction by (+)-*syn*- and (–)-*anti*-dibenzo[a,l]pyrene-11,12-diol-13,14-epoxides in mouse cells. *Cancer Res* 64: 7321–7328

98 Conney AH, Chang RL, Cui XX, Schiltz M, Yagi H, Jerina DM, Wei SJ (2001) Dose-dependent differences in the profile of mutations induced by carcinogenic (R,S,S,R) bay- and fjord-region diol epoxides of polycyclic aromatic hydrocarbons. *Adv Exp Med Biol* 500: 697–707

99 Friedberg EC (2001) How nucleotide excision repair protects against cancer. *Nat Rev Cancer* 1: 22–33

100 Naegeli H, Geacintov NE (2005) Mechanisms of repair of polycylic aromatic hydrocarbon-induced DNA damage. In: A Luch (ed.): *The Carcinogenic Effects of Polycyclic Aromatic Hydrocarbons*. Imperial College Press, London, 211–258

101 Geacintov NE, Broyde S, Buterin T, Naegeli H, Wu M, Yan S, Patel DJ (2002) Thermodynamic and structural factors in the removal of bulky DNA adducts by the nucleotide excision repair machinery. *Biopolymers* 65: 202–210

102 Hess MT, Gunz D, Luneva N, Geacintov NE, Naegeli H (1997) Base pair conformation-dependent excision of benzo[a]pyrene diol epoxide-guanine adducts by human nucleotide excision repair enzymes. *Mol Cell Biol* 17: 7069–7076

103 Geacintov NE, Cosman M, Hingerty BE, Amin S, Broyde S, Patel DJ (1997) NMR solution structures of stereoisomeric covalent polycyclic aromatic carcinogen-DNA adducts: Principles, patterns, and diversity. *Chem Res Toxicol* 10: 111–146

104 Ross JA, Nesnow S (1999) Polycyclic aromatic hydrocarbons: Correlation between DNA adducts and *ras* oncogene mutations. *Mutat Res* 424: 155–166

105 Marshall CJ, Vousden KH, Phillips DH (1984) Activation of c-Ha-*ras*-1 proto-oncogene by *in vitro* modification with a chemical carcinogen, benzo[a]pyrene diol-epoxide. *Nature* 310: 586–589

106 Vousden KH, Bos JL, Marshall CJ, Phillips DH (1986) Mutations activating human c-Ha-*ras1* protooncogene (*HRAS1*) induced by chemical carcinogens and depurination. *Proc Natl Acad Sci USA* 83: 1222–1226

107 Balmain A, Pragnell IB (1983) Mouse skin carcinomas induced *in vivo* by chemical carcinogens have a transforming Harvey-*ras* oncogene. *Nature* 303: 72–74

108 Buterin T, Hess MT, Luneva N, Geacintov NE, Amin S, Kroth H, Seidel A, Naegeli H (2000) Unrepaired fjord region polycyclic aromatic hydrocarbon-DNA adducts in *ras* codon 61 mutational hot spots. *Cancer Res* 60: 1849–1856

109 Lin CH, Huang X, Kolbanovskii A, Hingerty BE, Amin S, Broyde S, Geacintov NE, Patel DJ (2001) Molecular topology of polycyclic aromatic carcinogens determines DNA adduct conformation: A link to tumorigenic activity. *J Mol Biol* 306: 1059–1080

110 Ruan Q, Kolbanovskiy A, Zhuang, P, Chen J, Krzeminski J, Amin S, Geacintov NE (2002) Synthesis and characterization of site-specific and stereoisomeric fjord dibenzo[*a,l*]pyrene diol epoxide-N^6-adenine adducts: Unusual thermal stabilization of modified DNA duplexes. *Chem Res Toxicol* 15: 249–261

111 Geacintov NE, Naegeli H, Dinshaw JP, Broyde S (2005) Structural aspects of polycyclic aromatic carcinogen-damaged DNA and its recognition by NER proteins. In: W Siede, YW Kow, PW Doetsch (eds): *DNA Damage Recognition.* Taylor and Francis, New York

112 Bos JL (1989) *Ras* oncogenes in human cancer: A review. *Cancer Res* 49: 4682–4689

113 Wei SJ, Chang RL, Merkler KA, Gwynne M, Cui XX, Murthy B, Huang MT, Xie JG, Lu YP, Lou YR, Jerina DM, Conney AH (1999) Dose-dependent mutation profile in the c-Ha-*ras* proto-oncogene of skin tumors in mice initiated with benzo[*a*]pyrene. *Carcinogenesis* 20: 1689–1696

114 Chakravarti D, Mailander P, Franzen J, Higginbotham S, Cavalieri EL, Rogan EG (1998) Detection of dibenzo[*a,l*]pyrene-induced H-*ras* codon 61 mutant gene in preneoplastic Sencar mouse skin using a new PCR-RFLP method. *Oncogene* 16: 3203–3210

115 Nesnow S, Ross JA, Mass MJ, Stoner GD (1998) Mechanistic relationships between DNA adducts, oncogene mutations, and lung tumorigenesis in strain A mice. *Exp Lung Res* 24: 395–405

116 Kleiner HE, Suryanarayana V, Vulimiri V, Hatten WB, Reed MJ, Nebert DW, Jefcoate CR, DiGiovanni J (2004) Role of cytochrome P4501 family members in the metabolic activation of polycyclic aromatic hydrocarbons in mouse epidermis. *Chem Res Toxicol* 17: 1667–1674

117 Buters JTM, Mahadevan B, Ouintanilla-Martinez L, Gonzalez FJ, Greim H, Baird WM, Luch A (2002) Cytochrome P450 1B1 determines the susceptibility to dibenzo[*a,l*]pyrene-induced tumor formation. *Chem Res Toxicol* 15: 1127–1135

118 Lau HHS, Baird WM (1991) Separation and characterization of postlabeled DNA adducts of stereoisomers of benzo[*a*]pyrene-7,8-diol-9,10-epoxide by immobilized boronate chromatography and HPLC analysis. *Carcinogenesis* 15: 907–915

119 Mahadevan B, Marston CP, Luch A, Dashwood WM, Brooks E, Pereira C, Doehmer J, Baird WM (2007) Competitive inhibition of carcinogen-activating CYP1A1 and CYP1B1 enzymes by a standardized complex mixture of PAH extracted from coal tar. *Int J Cancer* 120: 1161–1168

120 Nesnow S, Davis C, Nelson G, Ross JA, Allison F, Adams L, King LC (1997) Comparison of the morphological transforming activities of dibenzo[*a,l*]pyrene and benzo[*a*]pyrene in C3H10T1/2CL8 cells and characterization of the dibenzo[*a,l*]pyrene-DNA adducts. *Carcinogenesis* 18: 1973–1978

121 Luch A, Coffing SL, Tang YM, Schneider A, Soballa V, Greim H, Jefcoate CR, Seidel A, Greenlee WF, Baird WM, Doehmer J (1998) Stable expression of human cytochrome P450 1B1 in V79 Chinese hamster cells and metabolically catalyzed DNA adduct formation of dibenzo[*a,l*]pyrene. *Chem Res Toxicol* 11: 686–695

122 Mahadevan B, Luch A, Seidel A, Pelling JC, Baird WM (2001) Effects of the (−)-*anti*-11R,12S-dihydrodiol 13S,14R-epoxide of dibenzo[*a,l*]pyrene on DNA adduct formation and cell cycle arrest in human diploid fibroblasts. *Carcinogenesis* 22: 161–169

123 Sundberg K, Dreij K, Seidel A, Jernström B (2002) Glutathione conjugation and DNA adduct formation of dibenzo[*a,l*]pyrene and benzo[*a*]pyrene diol epoxides in V79 cells stably expressing different human glutathione transferases. *Chem Res Toxicol* 15: 170–179

124 Dreij K, Seidel A, Jerström B (2005) Differential removal of DNA adducts derived from *anti*-diol epoxides of dibenzo[*a,l*]pyrene and benzo[*a*]pyrene in human cells. *Chem Res Toxicol* 18: 655–664

125 Luch A, Kudla K, Seidel A, Doehmer J, Greim H, Baird WM (1999) The level of DNA modification by (+)-*syn*-(11S,12R,13S,14R)- and (−)-*anti*-(11R,12S,13S,14R)-dihydrodiol epoxides of dibenzo[*a,l*]pyrene determined the effect on the proteins p53 and p21^{WAF1} in the human mammary carcinoma cell line MCF-7. *Carcinogenesis* 20: 859–865

126 Sundberg K, Widersten M, Seidel A, Mannervik B, Jernström (1997) Glutathione conjugation of bay- and fjord-region diol epoxides of polycyclic aromatic hydrocarbons by glutathione trans-

ferases M1-1 and P1-1. *Chem Res Toxicol* 10: 1221–1227

127 Hu X, Herzog C, Zimniak P, Singh SV (1999) Differential protection against benzo[*a*] pyrene-7,8-dihydrodiol-9,10-epoxide- induced DNA damage in HepG2 cells stably transfected with allelic variants of pi class human glutathione *S*-transferase. *Cancer Res* 59: 2358–2362

128 Kushman ME, Kabler SL, Fleming MH, Ravoori S, Gupta RC, Doehmer J, Morrow CS, Townsend AJ (2007) Expression of human glutathione *S*-transferase P1 confers resistance to benzo[*a*]pyrene or benzo[*a*]pyrene-7,8-dihydrodiol mutagenesis, macromolecular alkylation and formation of stable N^2-Gua-BPDE adducts in stably transfected V79MZ cells co-expressing hCYP1A1. *Carcinogenesis* 28: 207–214

129 Kushman ME, Kabler SL, Ahmad S, Doehmer J, Morrow CS, Townsend AJ (2007) Cytotoxicity and mutagenicity of dibenzo[*a,l*]pyrene and (±)-dibenzo[*a,l*]pyrene-11,12-dihydrodiol in V79MZ cells co-expressing either hCYP1A1 or hCYP1B1 together with human glutathione *S*-transferase A1. *Mutat Res* 624: 80–87

130 Lagerqvist A, Håkansson D, Prochazka G, Lundin C, Dreij K, Segerbäck D, Jernström B, Törnqvist F, Seidel A, Erixon K, Jenssen D (2008) Both replication bypass fidelity and repair efficiency influence the yield of mutations per target dose in intact mammalian cells induced by benzo[*a*]pyrene-diol-epoxide and dibenzo[*a,l*]pyrene-diol-epoxide. *DNA Rep* 7: 1202–1212

131 Pääjärvi G, Jernström B, Seidel A, Stenius U (2008) *Anti*-diol epoxide of benzo[*a*]pyrene induces transient Mdm2 and p53 Ser15 phosphorylation, while *anti*-diol epoxide of dibenzo[*a,l*]pyrene induces a nontransient p53 Ser15 phosphorylation. *Mol Carinog.* 47: 301–309

132 Bigger CA, Sawicki JT, Blake DM, Raymond LG, Dipple A (1983) Products of binding of 7,12-dimethylbenz[*a*]anthracene to DNA in mouse skin. *Cancer Res* 43: 5647–5651

133 Wei D, Maher VM, McCormick JJ (1995) Site-specific rates of excision repair of benzo[*a*]pyrene diol epoxide adducts in the hypoxanthine phosphoribosyltransferase gene of human fibroblasts: Correlation with mutation spectra. *Proc Natl Acad Sci USA* 92: 2204–2208

134 Luch A, Mahadevan B, Baird WM, Doehmer J, Seidel A, Glatt HR, Greim H, Buters (2002) The role of cytochrome P450 1B1 in dibenzo[*a,l*]pyrene-induced carcinogenesis. *Polycyclic Aromat Compd* 22: 781–789

135 Glatt HR, Pabel U, Muckel E, Meinl W (2002) Activation of polycyclic aromatic compounds by cDNA-expressed phase I and phase II enzymes. *Polycyclic Aromat Compd* 22: 955–967

Molecular, Clinical and Environmental Toxicology. Volume 1: Molecular Toxicology
Edited by A. Luch
© 2009 Birkhäuser Verlag/Switzerland

Chemical induced alterations in p53 signaling

Johan Högberg, Ilona Silins and Ulla Stenius

Institute of Environmental Medicine, Karolinska Institutet, Stockholm, Sweden

Abstract. The p53 protein is one of the most important tumor suppressors. The present review summarizes aspects of p53 function and its role in cancer development. Some of the most well-characterized molecular mechanisms affecting p53 regulation, stabilization, inactivation and downstream events are described. A major focus is on how xenobiotics can interfere with p53 function and on its role in chemical carcinogenesis. In the final section of this chapter we discuss future aspects on how knowledge about p53 can be used in testing of carcinogens and in risk assessment.

Introduction

Experimental and epidemiological/clinical studies performed during the last 25 years show that p53 is one of the most important cellular proteins that protect against cancer. In simplified terms, the p53 tumor suppressor activity protects the integrity of the genome and prevents proliferation of cells with defects in their DNA. The importance of p53 is illustrated by the fact that the majority of all human tumors carry mutations in the p53 gene (*TP53*) or abnormalities in p53 regulation [1–3]. Thus, most tumor types seem to select for a loss of p53 function. The role of p53 in the protection against DNA damage and cancer development has been further demonstrated in a variety of mechanistic studies using animal or *in vitro* models. Furthermore, patients with Li-Fraumeni syndrome and born with one mutated p53 allele have an increased risk for early onset cancers in several organs [4]. This is paralleled by observations in heterozygous p53-deficient (+/–) mice and p53 knockout mice, which exhibit early and frequent tumors [5].

The p53 protein was first described in 1979 [6] and was initially supposed to be an oncogene, since large amounts of mutated p53 was discovered in tumors. Subsequently, p53 was identified as a tumor suppressor gene that normally protects against cancer development [7]. In further studies, p53 was shown to function as a transcription factor transferring DNA damage signals to genes inducing, for example, cell cycle stop [8] or apoptosis [9]. Later studies indicated at least one additional prominent anticancer effect; activation of p53 in rapidly dividing cells, such as those carrying mutated oncogenes [10]. Since its discovery approximately 40 000 papers on p53 have been published. This enormous database indicates a central role for p53 in cell regulation.

However, many open questions remain. One of the key questions is how the p53 pathway interacts with other signaling pathways in the p53 network and which signals determine whether apoptosis or transient cell cycle arrest for DNA repair should be activated.

p53 is a central mediator of the cellular response to DNA damage and protects genome integrity. It facilitates DNA repair or permanently prevents damaged cells from dividing, either by inducing apoptosis or by inducing cellular senescence (permanent cell cycle stop). The p53 network is complex and consists of a large number of genes and proteins that can detect and respond to different types of stresses, including DNA damage. Stress signals activate upstream mediators in the network. They initiate the p53 pathway by activating proteins, which regulate p53 activity mostly *via* post-translational alterations. When activated, the p53 protein induces genes involved in cell cycle arrest, apoptosis, DNA repair and cellular senescence [3]. In other words, when p53 is functional, it prevents cells with defects in their DNA from replicating its genome. The consequence of a non-functional p53 pathway is the lack of protection against DNA-damaging chemicals, radiation and other types of cellular stress. Ultimately, a non-functional p53 pathway can result in uncontrolled growth of cells with unrepaired DNA damage, increased risk for mutations, genomic instability and tumor development. All these aspects are discussed in greater depth later in this chapter.

Xenobiotics or toxicological stress may influence p53 signaling in many ways. Nevertheless, most studies on p53 focus on its basic biological role in cell regulation and tissue homeostasis, or focus on the biological role of p53 mutations. Numerous reviews have been published that summarize different aspects of p53 activity, but in spite of a wealth of data, toxicological aspects of p53 have not often been reviewed. In this chapter we focus on how toxicological stress may evoke transient effects in p53 signaling and how xenobiotics can modify the p53 response. The role of permanent alterations induced by xenobiotics, such as mutations, is also reviewed, as well as how this knowledge can be used in the development of test models and in the risk assessment of carcinogenic chemicals.

p53 regulation

The amount of p53 protein in unstressed cells is low. In a normal cell, p53 is synthesized and degraded continuously and the levels of p53 are mainly regulated by proteolytic (proteasomal) degradation. Usually p53 mRNA levels do not vary and the half-life of the protein is short (6–20 min). Multiple mechanisms exist to control p53 activity including the regulation of protein stability, activity, binding to other proteins and subcellular localization. The ubiquitin ligase murine double minute 2 (Mdm2) protein is a major regulator of p53 and also a transcriptional target of p53. Mdm2 and p53 form an autoregulatory loop [11, 12], in which p53 induces Mdm2 expression and Mdm2 represses p53 activity. Since the discovery of the p53-Mdm2 loop [12] many other loops involving p53 have been characterized.

p53-Mdm2 autoregulatory loop and other genes regulating p53

In unstressed cells p53 and Mdm2 physically interact. The N-terminus of Mdm2 forms a hydrophobic pocket where the transactivation domain of p53 binds and this binding inhibits the transcriptional activity of p53. In complex with p53, Mdm2 also acts as an E3 ubiquitin ligase for p53 and conjugates ubiquitin molecules to p53 (Fig. 1). The formed monoubiquitinated complex is translocated from the nucleus to the cytoplasm where it is polyubiquitinated and thereafter degraded through a proteasomal-dependent mechanism [13, 14]. The activity of Mdm2 is regulated by a variety of post-translational modifications (such as phosphorylations). Phosphorylations of both p53 and Mdm2 affect the binding between the two proteins [15]. Phosphorylation of p53 at the N-terminal domain, e.g., at Ser15, Thr18 and Ser20, prevents Mdm2 binding and leads to p53 accumulation. Stopped ubiquitination and degradation results in increased nuclear levels of p53 and in the activation of the protein [16]. Mdm2 targets p53 for proteasomal degradation and the proteins together form an autoregulatory feedback loop (Fig. 1). Thus, Mdm2 keeps the level of p53

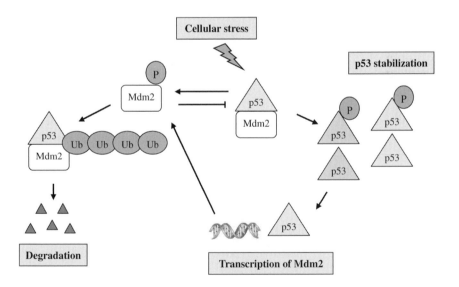

Figure 1. Mdm2-p53 autoregulatory loop. The level of p53 is controlled by Mdm2. During normal conditions p53 binds to Mdm2, and is ubiquitinated and degraded. DNA damage and other stresses induce phosphorylations in both proteins and reduce the binding between them, leading to increased levels of free p53. Subsequently the transcription of Mdm2 is increased by p53. Increased levels of Mdm2 degrade p53 so that p53 is active only for a limited time period. The proteins form an autoregulatory loop.

low in unstressed cells, while an interruption of the loop leads to p53 accumulation. Within minutes or hours this will lead to an increased transcription of Mdm2, which brings p53 levels down to pre-stress levels. This ensures a tight control of p53 activation. The effects of DNA damage on the interplay between p53 and Mdm2 have been studied in several *in vivo* and *in vitro* models [2, 17]. Oscillations in p53-Mdm2 protein levels have been documented by employing *in vivo* imaging [18], but their functions remain to be defined.

A lost control of p53 function may have dramatic consequences, as documented in experiments with knockout mice. Thus, knocking out Mdm2 leads to embryonic lethality, with embryos showing signs of extensive apoptosis. However, knocking out both Mdm2 and p53 gives a viable phenotype [19]. This illustrates that Mdm2 is essential for the control of p53 function during normal development [20]. Certain stress signals can result in a down-regulation of Mdm2, leading to p53 stabilization. Thus, $p14^{ARF}$ ($p19^{ARF}$ in mice) can decrease Mdm2 levels. ARF is activated by oncogenes or by a high cell turnover and binds and inhibits Mdm2 [10]. The anti-apoptotic factor Akt-kinase (PKB) activates Mdm2. Akt can phosphorylate Mdm2 and increase its nuclear localization and its ubiquitin ligase activity [21]. Furthermore, overexpression of Mdm2 by gene amplification can lead to inactivation of p53 so that Mdm2 functions as an oncogene [22]. Xenobiotic-induced alterations in Mdm2 levels are discussed below. Several proteins other than Mdm2 also regulate the stability of p53. Recently, two other ubiquitin ligases involved in p53 degradation, COP-1 and Pirh-2, were described [23, 24]. MdmX is another regulator of p53. It inhibits the p53 transactivation activity, without leading to its ubiquitination [25].

Structure of the p53 protein

The human p53 protein consists of 393 amino acids and constitutes several domains [26]. The major functional domains are shown in Figure 2. The N-terminal domain contains the transactivation domain, which is involved in protein-protein interactions, and a proline-rich domain with a regulatory function. The large central core consists of a DNA-binding domain. The C-terminal regulatory domain contains nuclear localization sequences and a tetramerization domain [27] of importance for the nuclear localization of p53 [28]. As mentioned above, p53 is subjected to numerous post-transcriptional modifications. The p53 sequence contains many serine, threonine and lysine amino acids and post-translational modifications of these residues have a central role in p53 stabilization and activation [3]. Phosphorylations of p53 at the N-terminal domain prevent Mdm2 binding, thus leading to p53 accumulation. Other modifications include acetylation, methylation and ubiquitination. Most mutations in the p53 gene affect the DNA-binding domain [29] and several "hot spots" have been defined. These mutations eliminate or reduce the ability of the p53 protein to bind to DNA and thus prevent the transcriptional activation of target genes [27].

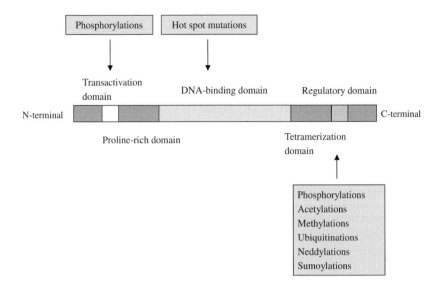

Figure 2. Structure and functional domains of p53 protein. These include the N-terminal transactivation domain, the sequence-specific DNA-binding domain and C-terminal regulatory domain, including the tetramerization domain. Post-transcriptional modifications are mainly induced in N- and C-terminal domains. The majority of point mutations in tumors are found in the central DNA-binding domain.

p53 stabilization and activation induced by stresses

p53 is stabilized and activated as a response to a wide variety of intracellular and external stressors. Among the stress signals that activate the p53 response are DNA damage caused by radiation, chemicals and drugs, activated oncogenes, hypoxia and nutrient deprivation (Fig. 3). In response to stress, the cellular levels of p53 are rapidly elevated through extensive post-translational modifications of the p53 protein. As mentioned above, these modifications include phosphorylation, acetylation, methylation, ubiquitination, neddylation and sumoylation, and they control the stability, activity and subcellular localization of p53 [30, 31]. Activated p53 functions as a sequence-specific transcription factor regulating several target genes [32].

Upstream mediators of p53 activation signals

DNA damage can be induced by many physical and chemical causes such as γ- or ultraviolet (UV) irradiation, reaction with oxidative free radicals, alkylation of DNA, etc. These types of stresses can induce specific pathways and use different signaling molecules for activation of p53 [3, 33]. Cells have complex mechanisms to respond to DNA damage and a common model of p53 regula-

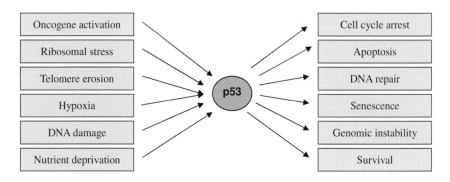

Figure 3. Activation of p53 and outcomes. p53 senses and integrates cellular signals evoked by different types of stressors. Depending on type, severity and interactions between stressors, p53 can trigger different cellular responses.

tion is that DNA damage activates damage-responsive kinases, which phosphorylate Ser15, Ser20 and Thr18 within the transactivation domain of p53. These phosphorylations inhibit p53 binding to Mdm2 and stop its degradation. Many serine and threonine residues, mainly located within the N-terminus of the p53 protein are targets for phosphorylation. The most established damage-responsive kinases phosphorylating p53 are members of the phosphoinositide-3-kinase-like family, namely Ataxia telangiectasia mutated (ATM) and ATM and Rad3 related kinase (ATR) and DNA-dependent protein kinase (DNA-PK, Fig. 4) [34].

These kinases are activated by DNA damage and are involved in sensing damaged DNA. Upon activation, they not only phosphorylate p53 but also other proteins involved in DNA repair, cell cycle arrest and apoptosis [35, 36]. Phosphorylations of the p53 protein after DNA damage and the concurrent stabilization and activation of p53 have been extensively investigated. A range of different kinases, including ATM, ATR, DNA-PK, Chk1, Chk2, CK1, JNK, HIPK2, and DYRK2, have been shown to phosphorylate p53 after DNA damage [37]. Most residues in p53 are phosphorylated by many kinases. One example is Ser15 that can be phosphorylated by at least eight kinases, while some other residues have so far been shown to be phosphorylated by a single kinase [38]. *In vivo* data with genetically modified animals show that even if a mutation in a single phosphorylation site is induced, sufficient p53 stabilization capacity is preserved [38]. These data indicate that p53 stabilization is ensured by a complex signaling pathway and that phosphorylation sites might have overlapping functions (Fig. 4). In the same way, kinases like ATM and ATR have overlapping functions [39]. The picture of phosphorylation patterns is far from clear, and the knowledge of kinases and their targets is continuously growing [26].

ATM is known to mainly respond to double strand breaks (DSBs) produced by ionizing radiation (IR), drugs and reactive oxygen species. In addition, it

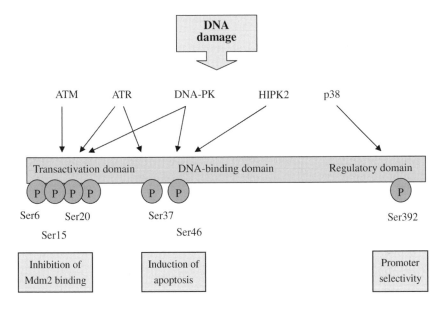

Figure 4. Proteins inducing stabilization and activation of p53 are activated by different types of DNA damage. These proteins include kinases ATM, ATR and DNA-PK. Examples of their effects on p53 are indicated.

responds to DNA strand breaks produced during DNA repair [40]. p53 is extremely sensitive to low levels of DNA damage, a response that might contribute to tumor suppression. Once ATM is activated, it can phosphorylate p53 on Ser15 and Mdm2 on Ser395. Phosphorylations within the N-terminus of the p53 protein, close to the site where Mdm2 binds, influence the stability of p53 due to a changed ability of Mdm2 to bind to p53 [40, 41]. Activation of ATM results in growth arrest, mediated *via* the action of Chk2, a cell cycle checkpoint kinase. Chk2 activation leads to phosphorylation of p53 on Ser20 that in turn results in activation of the G2 checkpoint. Phosphorylation of p53 by ATM also contributes to activation of the checkpoint regulating the entry into S-phase [42]. ATR responds to agents that inhibit DNA replication, such as UV radiation. By monitoring the progress of DNA replication forks during DNA replication, ATR has emerged as an important regulator of the S-phase checkpoint. Chk1, another checkpoint kinase, is mainly responsible for mediating the response of ATR [43].

DNA-PK, another well-studied kinase primarily responds to DSBs and operates as an upstream regulator of the p53-mediated apoptosis pathway. Phosphorylation of p53 on Ser15 by DNA-PK results in apoptosis [44]. The kinases homeodomain interacting protein kinase 2 (HIPK-2) and dual-specificity tyrosine-(Y)-phosphorylation regulated kinase 2 (DYRK-2) are activated by genotoxic stress and phosphorylate p53 on Ser46, a phosphorylation site associated with an induction of apoptosis. Genes important for the apoptotic response are induced, such as the p53-regulated apoptosis-inducing protein 1 (p53AIP) [45].

Unlike the responses upon p53 phosphorylation, the outcome of many of the other post-translational modifications are less clear. Ubiquitination of p53 by Mdm2 targets it for proteasomal degradation. Neddylation by Mdm2 has been shown to inhibit p53 function, while adding of a SUMO-1 molecule to lysines increases p53 activity. Acetylations of p53 in response to DNA damage, caused by IR and UV radiation, have been shown to increase its sequence specific DNA binding and enhance transcription of target genes. Both induction and inhibition of the transcriptional activity have been shown upon methylation of different lysines in the p53 protein [26, 37, 46].

Downstream events in the p53 pathway and their role in tumor suppression

What influences the choice of endpoint induced by p53 stabilization?

Once p53 is activated, it can regulate a variety of genes influencing many cellular endpoints. Three primary responses are cell cycle arrest, apoptosis and cellular senescence (Fig. 5) [14, 26, 47]. p53 acts as a transcription factor and DNA damage-mediated p53 stabilization induces or inhibits the expression of hundreds of genes [48]. Chromatin immunoprecipitation (ChIP) analyses have

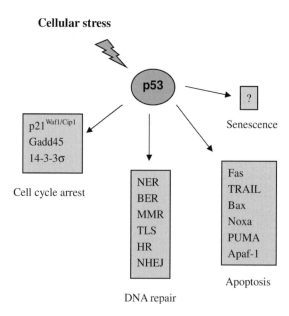

Figure 5. p53-induced signaling pathways and genes involved in regulating them. Four major outcomes of p53 stabilization, cell cycle arrest, DNA repair, apoptosis and senescence are shown.

identified a wide range of p53-regulated genes and there is evidence that p53 is involved in regulation of many processes in addition to those mentioned above, e.g., invasion, angiogenesis, differentiation and glycolysis [30, 49, 50]. p53 downstream proteins are also involved in regulation of DNA repair. It is clear that different stress signals trigger separate transcriptional programs and p53 protein modifications could play a role, but there is still poor understanding about how the cell chooses between these programs. The most studied endpoints in this connection are cell cycle arrest and apoptosis and various posttranslational modifications seem to change the p53 conformation so that it recognizes specific promoters and activate specific genes. Competition between acetylation and ubiquitination of Lys320 has been shown to be important for switching from growth arrest to apoptosis [51]. The regulation of promoter selectivity has been intensely studied to gain an understanding of the regulation of endpoints induced by p53 stabilization [52].

Cell cycle arrest

p53 can have a strong inhibitory effect on cell growth and p53 activation prevents cells from entering or progressing through the cell cycle. It has four distinct steps, i.e., the first gap (G1), synthesis (S), second gap (G2) and mitosis (M). Upon DNA damage, proteins induced by p53 inhibit progression of the cell cycle and give time for repair of damaged DNA. The p53-induced cell cycle block might be reversible or irreversible and, if the damage is mild and easily repaired, p53 induces transcription of genes involved in cell cycle arrest and DNA repair. After the damaged DNA has been successfully restored, the cell may re-enter into a normal cell cycle [53].

The ability of p53 to induce cell cycle arrest depends on three essential target genes: $p21^{Waf1/Cip1}$, 14-3-3σ and the growth arrest and DNA damage inducible gene (Gadd45). Upon p53 activation, $p21^{Waf1/Cip1}$ protein accumulates and causes G1 cell cycle arrest by inhibiting G1 cyclin-dependent kinases. $p21^{Waf1/Cip1}$ is involved in p53-induced tumor suppression and $p21^{Waf1/Cip1}$-deficient cells do not induce cell cycle block after DNA damage [54]. In addition, several studies indicate that $p21^{Waf1/Cip1}$-null animals exhibit increased susceptibility to chemical carcinogenesis [55, 56]. Studies indicate that $p21^{Waf1/Cip1}$-induced tumor suppression is dependent on p53 and that $p21^{Waf1/Cip1}$ plays an important role in p53-dependent tumor suppression [57, 58]. Induction of 14-3-3σ and Gadd45 by p53 results in G2/M arrest in cells that have sustained DNA damage [59–61].

Apoptosis

The most dramatic effect induced by p53 is apoptosis. Apoptosis is a genetically programmed cell death that maintains normal tissue homeostasis and is dependent on the sequential activation of so-called caspases. If the DNA damage is beyond the cell's capacity to repair, p53 induces genes involved in apoptosis. One of the hallmarks of cancer is resistance against apoptosis and lack of p53 activity usually contributes to this resistance.

The p53 target genes that are involved in the apoptotic processes can be divided in two classes, an intrinsic or extrinsic apoptotic pathway. The extrinsic pathway is induced by activation of cell-surface receptors, so-called death receptors, which include members of the tumor necrosis factor receptor superfamily, such as Fas and TRAIL [62]. These receptors are all transcriptionally induced by p53 and contribute to p53-induced apoptosis. In the intrinsic apoptotic pathway p53 also targets genes that are central components of the mitochondrial cell death pathway. The apoptotic protease activating factor 1 (Apaf-1), a major component of the apoptosome and required for pro-caspase-9 activation, is an important apoptosis-associated gene induced by p53 [63]. The phosphorylation of Ser46 of p53 induced by DNA damage selectively enhances the expression of proapoptotic target genes [64]. Also important for the apoptotic p53 response is its transcriptionally independent regulation of a large number of proapoptotic B cell lymphoma 2 (Bcl-2) family members, such as Bax, Noxa, PUMA and Bid. Induction of these so-called BH3-only proteins is essential for p53-induced apoptosis [37, 62].

DNA damage-induced and p53-mediated apoptosis and its importance in the protection against cancer has been studied in *in vitro* and *in vivo* models. High exposure, e.g., to irradiation, kills cells by inducing massive apoptosis in organs sensitive for p53-mediated apoptosis, such as the intestine [54]. However, the role of this apoptotic and toxic p53 response for the protection against cancer is not clear. In a recent study using a mouse model in which p53 activity can be switched on and off, it was shown that the massive early apoptotic response did not protect against radiation-induced cancer to the same extent as p53 switched on later [65]. This finding suggests that p53 still has an important role a long time after DNA damage is repaired, presumably by eliminating preneoplastic/mutated cells [65].

DNA repair

p53 has a central role in regulating the different DNA repair pathways, both by inducing genes involved in the repair process and directly by interacting with repair proteins [66]. Depending on the type of DNA damage, cells employ several DNA repair pathways to restore damaged DNA and with which p53 can interact. These pathways include nucleotide excision repair (NER), which is responsible for repair of pyrimidine dimers and bulky DNA lesions caused, for example, by benzo[*a*]pyrene (BP) adducts, mismatch repair (MMR), which is involved in repair of DNA replication errors, base excision repair (BER), which mainly repairs damage caused by reactive oxygen species, and translesion synthesis (TLS), a process that allows the DNA replication machinery to replicate past DNA lesions. The homologous recombination (HR) pathway and the non-homologous end-joining (NHEJ) pathway are involved in the repair of DSBs and chromosomes [66].

The major substrate for NER is UV-induced DNA damage. p53 is involved in the induction of genes that are important for recognizing damage and for the subsequent repair process. The genes include DNA damage-binding protein 2 (DDB2) and Xeroderma pigmentosum C (XPC). Knockout mice for these genes have an increased risk for developing lung cancer and skin tumors [67]. In addition, p53 interacts with the TFIIH complex (that recognizes and binds to damaged DNA) [68]. p53 has also an important role in induction of MMR genes, e.g., postmeiotic segregation increased 2 (PMS2). PMS2 is involved in sensing severe DNA damage and is able to switch the cellular response from DNA repair to apoptosis. Upon cisplatin treatment, the p53 homologue p73 interacts with the PMS2 protein, resulting in p73-dependent apoptosis [69]. In the BER pathway, p53 is required for enhancing the DNA repair activity of proteins such as the human 8-oxoguanine glycosylase and apurinic/apyrimidinic (AP) endonuclease [70]. p53 also interacts with DNA polymerase beta and helps to stabilize its interaction with abasic DNA [71]. p53 induces genes important in the TLS pathway, such as the DNA polymerase eta (POLH) protein. POLH-knockout mice are more susceptible to UV-induced carcinogenesis [72].

Besides its role in repair of DSBs and chromosomes, the HR pathway is involved in meiotic recombination and V(D)J-recombination (the process that generates diversity in the immune system). A highly regulated HR pathway is necessary given that both a deficit and an excess in HR may result in chromosomal instability. p53 is involved in the negative regulation of the HR pathway, mainly by direct interaction with several repair protein [73]. NHEJ can introduce mutations during repair and p53 has a role in the control of these mutagenic effects to maintain genomic stability. p53 prevents annealing of mismatched DNA and helps in joining of broken DNA ends [74].

Role of p53 in cellular senescence and longevity

DNA damage and, perhaps more importantly, activated oncogenes can trigger cellular senescence. Cells entering senescence are characterized by a permanent cell cycle arrest, an altered transcriptional program and changed morphology [75]. p53 regulates both DNA damage-induced senescence and normal replicative senescence induced by erosion of telomeres. Several studies on mice and humans indicate that cellular senescence prevents the development of many types of cancer [76]. These studies show that early preneoplastic or premalignant lesions, e.g., in lung and bladder, show signs of a DNA damage response by expressing high levels of DNA damage responsive proteins such as ATM, ATR, Chk1, Chk2, γH2AX and p53 binding protein 1 (53BP1), while this DNA damage response is lost in malignant tumors [77, 78]. Subsequent studies implicate oncogene-induced p53 activation as an important factor leading to senescence. Thus, ATM and p53 seem to be triggered by replication stress, characterized by prematurely terminated DNA replication forks and the generation of DNA DSBs [79]. In mice, p53 activation *via* p19^{ARF} might be more important than p53 activation by DNA damage related to replication stress [65, 80].

The finding that a constitutively active and truncated form p53 induced aging in mice [81] has sparked an interest in the question as to whether a slightly increased cellular p53 response, e.g., to DNA damage, would eventually lead to a shortened life span. Studies on mice carrying three alleles of p53 ("super p53") show that these mice are resistant to chemical carcinogenesis but do not age prematurely. On the contrary, they exhibited an extended life span [82]. Two independent epidemiological studies taking advantage of a common polymorphism in the p53 gene (Arg72Pro), which may confer altered apoptotic responses to DNA damage, suggest that the Pro allele is associated with increased longevity in humans [83, 84]. It is possible that the increased longevity seen in humans is related to an increased ability to withstand stresses associated with cancer and other diseases [85].

p53 mutations

p53 mutations commonly result in an inactivated protein, and mutations in the p53 gene are a very common finding in tumors. In 1989, it was reported that p53 was mutated in many sporadic brain, breast, lung and colon tumors [86]. Although the mutation frequencies vary between tumor types, later work indicates that p53 mutations can be found in more than 50% of sporadic human tumors [87] and usually one allele is lost and one is mutated. Most mutations are missense mutations located in the DNA-binding domain [87]. Mutated p53 may act in a dominant negative way, e.g., by disturbing tetramerization. Mutations often inhibit the Mdm2-dependent autoregulatory loop, resulting in the accumulation of mutated p53 protein in cells. Pioneering studies on colorectal cancer pinpoint p53 mutations to relatively advanced morphological stages, and they are preceded by several other mutations found in this tumor type [88]. In studies on non-small cell lung cancer, lung hyperplasia retained wild-type p53, whereas dysplastic lesions often exhibit p53 mutations, perhaps suggesting that p53 mutations will appear earlier in this tobacco-related tumor type [77]. Analysis of other cancers also suggests early p53 mutations [89].

Recent work on preneoplastic lesions in humans indicates that a DNA damage response is a common feature in these lesions, and that functional p53 prevents further development into malignancy. Thus, DNA damage response including activated DNA damage signaling proteins, signs of apoptosis and cellular senescence is seen in several types of preneoplastic lesions in many organs, including chemically induced mouse skin papillomas [79]. Experimental data and data from human specimens indicate that DNA damage response in these lesions is most often induced by activated oncogenes. The activation of DNA damage response may select for mutations in p53 or other genes [76]. Interestingly, these examples mainly concern tumor types often associated with exposure to xenobiotic chemical carcinogens, such as lung and bladder cancer.

Chemical-specific p53 mutations

Relatively few studies have focused on the question whether p53 mutations are caused by endogenous factors or by chemical exposure. Possible endogenous mutants include reactive oxygen species [90]. In the literature, most of p53 mutations seem rather to be regarded as a result of endogenous mutagenic factors and a strong selection pressure, than as a result of exposure to mutagenic xenobiotics. However, given the crucial biological importance of p53 as a tumor suppressor, it is of great interest to understand the role of p53 mutations in chemically induced cancer. Thus, mutation spectra have been analyzed and limited data suggest that certain chemical exposures can be linked to specific mutational alterations (Tab. 1).

A first example was reported 1991 [91] and showed associations between aflatoxin B_1 exposure and G:C to T:A transversions in the third base in codon 249 in hepatocellular carcinoma (HCC). Aflatoxin B_1 is a risk factor for HCC and it induces the same mutation in experimental cell models [92]. However, it has also been shown that hepatitis B virus, another important risk factor in humans, might also influence the codon 249 mutation frequency [93]. In any event, based on this knowledge, the detection of mutant DNA in plasma has been developed into a biomarker for the risk of developing HCC [94]. Hemangiosarcomas, seen after exposure to vinyl chloride, is another tumor type in the liver. A chemical-specific mutation spectrum in the p53 gene has been reported also in this case [95].

The majority of lung cancers are associated with tobacco smoking as a causative factor, and p53 mutations have been analyzed to identify mutations

Table 1. Studies on associations between chemical carcinogen exposure and specific p53 mutations.

Exposure	p53 mutation	Tumor (site or type)	Biomarker
Aflatoxin B_1 [91]	Codon 249	Hepatocellular carcinoma	Plasma DNA
Tobacco smoke [96]	Codon 248, 249	–	Plasma DNA
BPDE [97]	Codon 157, 248, 273	Lung	–
Arylamines [98]	Many codons	Bladder	–
UV radiation [99]	Many codons	Non-melanoma skin	–
Vinyl chloride [95]	Codon 203, 253 and others (in rats)	Liver angiosarcomas	–

BPDE, benzo[a]pyrene diol-epoxide.

characteristic for carcinogens in tobacco smoke. Informative data have been published on "finger prints" in p53 in lung tumors indicating distinct patterns in never, former and current smokers [100]. In other studies, p53 mutations in codons 248 and 249, common in lung tumors, were detected in plasma DNA from healthy smokers. This biomarker correlated to tobacco smoke exposure [97]. Tobacco smoke contains many carcinogenic chemicals, including the very well studied polycyclic aromatic hydrocarbon BP. The ultimate carcinogenic metabolite of BP, BP diol-epoxide (BPDE), has also been shown to selectively form guanine adducts in codon 157, 248 and 273 of the p53 gene, known as mutational hot spots in lung cancer [96]. This indicates a causal link between BP exposure and lung cancer. Later work indicated that acrolein, another component of tobacco smoke, also induced similar adducts as did BPDE [101]. However, this adduct formation may not result in mutagenesis [102].

The high frequency of missense mutations and the observation that mutated p53 often accumulates in tumor cells has raised the question whether some mutations may exhibit gain-of-function characteristics. If so, a p53 mutation may not only lead to a loss of tumor suppressor function, but also to the accumulation of an oncogenic protein. Two of the hot spot mutations in codons 248 and 273, often found in lung tumors and linked to tobacco smoking, have been shown to suppress the binding of the MRN complex (including Mre11, Rad 50 and NBS1) to DNA DSBs. This effect leads to an impaired ATM signaling and may prevent the critical ATM-Chk2-p53 signaling in response to DSBs [103].

Workers exposed to arylamines, such as benzidine and β-naphthylamine, exhibit a specific mutational spectrum in p53 that can be related to exposure to these carcinogens. The frequency of mutation in p53 is closely linked to tumor grade and stage, so an arylamine-induced p53 mutation may be a late event in the development of bladder tumors [98]. Another example is UV radiation, which may induce C:C to T:T transitions in the p53 gene. These mutations are induced by UVB and are not frequently induced by other carcinogens. They are often found in non-melanoma tumors of the skin [104].

The p53 gene exhibits germ-line polymorphisms and some of them have been intensely studied. In a recent study on lung cancer and a polymorphism in codon 72, data suggested that smoking among homozygous carriers of the Arg allele more easily acquire mutations in codon 273, one of the hot spots for BPDE adduct formation [105]. This result suggests that individuals born with this gene variant are more susceptible to tobacco smoke carcinogens, but additional studies confirming this observation are needed.

Epigenetic modifications of p53

Some tumor suppressor genes such as p16(INK4a) are often inactivated by epigenetic silencing, including aberrant cytosine and guanine separated by a phosphate (CpG) island methylation in promoter regions [87]. Reports on promoter hypermethylations in p53 are rare. However, it has been reported that chronic arsenic exposure leads to hypermethylation of DNA in polynuclear leukocytes [106]. Furthermore, an analysis of the p53 promoter has indicated an increased methylation in people exposed to arsenic, even though it was also found that heavily exposed people exhibited hypomethylated p53 promoter regions [107]. In experimental cell studies, arsenic attenuated the p53 response to BP [108] and arsenic hypermethylated the p53 promoter [109]. However, the p53 promoter does not display CpG islands that could regulate p53 expression [87], so a functional impact of arsenic on p53 *via* this mechanism is not to be expected. In addition, arsenic in long-term cell culture experiments inhibited p53, probably by increasing the Mdm2 expression [110].

DNA damage-induced p53 activation by adriamycin and other factors was recently shown to result in microRNA (miRNA) expression in cell models [111, 112] and in mice [113]. These small RNAs may, after processing, suppress many genes by inhibiting translation or by inducing RNA degradation. Downstream consequences of these effects include cell cycle arrest or apoptosis. Due to imperfect base pairing and variable expression in different tissues, these miRNA may affect the p53 response in a tissue-specific manner and in a way that is hard to predict by available data. It has also been observed that cancer cells may loose the ability to express p53-specific miRNA [114].

Metabolic factors affecting p53 signaling

It is well established that glucose availability can regulate cell proliferation, and nutritional stress can affect xenobiotic-induced p53 signaling. Interaction of p53 with the deacetylase, SIRT1, is an example. SIRT1 deacetylates p53 and modifies its activity, and this link has been suggested to adapt the p53 response to nutritional factors. A recent study extends our understanding of the SIRT1-p53 relationship. It was shown that oxidative stress, induced by high doses of H_2O_2, down-regulated SIRT1 by phosphorylating the RNA-binding protein HuR. Interestingly, it was indicated that Chk2 was activated and then phopsphorylated HuR, and that Nijmegen breakage syndrome 1 (NBS1) and ATM mediated the signal to Chk2. NBS1 and ATM might activate p53, so these signaling proteins may be critical for integrating nutritional stress factors with p53 DNA damage signaling [115].

A link to nutritional status on p53 is further illustrated by studies on 5'AMP-activated protein kinase (AMPK) and the Drosophila *par-4* homologue (LKB1). AMPK is activated, e.g., *via* LKB1, by the absence of glucose [116] and can phosphorylate p53 at Ser15 and activate it. This link between AMPK and p53 might be essential for tumor prevention as mice deficient in LKB1 are sensitive to chemical carcinogens and develop skin and lung tumors early by 7,12-dimethylbenz[*a*]anthracene (DMBA) [117]. It has also been observed that lung tumors from smokers often carry mutations in LKB1 [118], so in this case clinical data support a role for nutritional stress signaling in regulating the p53 response to chemical carcinogens.

Drugs affecting cholesterol metabolism may also affect p53 signaling. In recent studies it was found that statins, 3-hydroxy-3-methyl-glutaryl-CoA reductase (HMG-CoA-reductase) inhibitors, commonly used to reduce circulating cholesterol levels and atherosclerosis, activate Mdm2 by phosphorylating Ser166 and attenuated the p53 response on carcinogens [119]. However, a very rapid effect of statins was induced, arguing against an involvement of HMG-CoA-reductase inhibition [120]. Instead, further studies indicate a possible involvement of the P2X7 receptor in these effects of statins [121]. Extracellular ATP is an endogenous ligand for the P2X7 receptor, and ATP is released by damaged cells and may mediate important survival signals to cells downstream in the capillary bed from upstream cells damaged by hypoxia or by oxidative or toxicological stress.

The influence of insulin, insulin-like growth factors (IGFs) and IGF-1R signaling might be an important carcinogenic factor, and high cancer risks associated with body fatness in humans might be explained by circulation levels of endogenous hormones such as insulin and IGFs [122]. In experimental studies connections between the IGF-1–PI3K–Akt–mTOR pathway and the p53 pathway have been indicated in cell models [116]. In response to, for example, IGF-1 or insulin, the anti-apoptotic kinase Akt activates Mdm2 by phosphorylating Ser166 so that p53 is inactivated, and the levels of p53 are decreased in response to, for example, xenobiotic-induced DNA damage [123]. Recent studies also indicate that p53 may regulate metabolic functions in the cell by influencing mitochondrial respiration and the pentose phosphate pathway [49].

Chemically mediated alterations in p53 signaling – Effects of tumor promoters and chemopreventive substances

Xenobiotics can interact with p53 signaling. One example is the adaptation induced by aryl hydrocarbon receptor (AhR) activators. AhR is activated by many xenobiotics, and has been implicated in cell cycle regulation [124]. Early studies suggested that 2,3,7,8-tetrachlorodibenzo-*p*-dioxin (TCDD), a potent liver tumor promoter and AhR ligand, attenuated the p53 response to UV [125]. More recently it was found that TCDD attenuated the p53 response to chemical carcinogens and to DNA-damaging chemicals. The same effect was seen in rat liver *in situ* and in isolated hepatocytes, and the most likely explanation was that TCDD activated Mdm2 by phosphorylating Mdm2 at Ser166. This post-translational modification may facilitate binding of Mdm2 to p53 and activate the E3 ubiquitin ligase activity of Mdm2. This may explain the attenuated p53 response [126].

Another tumor promoter affecting Mdm2 is phenobarbital. It was shown that the constitutive androstane receptor (CAR) agonists phenobarbital and pregnenolone 16α-carbonitrile induced Mdm2 mRNA and protein levels in rats [127]. How these xenobiotics affect Mdm2 is not yet known, but previous studies suggest CAR activation as a possible upstream event, as it has been shown that carcinogenic effects of phenobarbital are mediated by CAR [128]. The significance of these findings is not yet known. However, it can be speculated that liver tumor promoters such as TCDD and phenobarbital inhibit the p53 response to DNA-damaging carcinogens *via* a common mechanism, namely an up-regulation of Mdm2 (Fig. 6). This may lead to increased mutagenesis and activation of oncogenes. Alternatively, the activated Mdm2 may abolish a p53-dependent senescence [75], and thus permit the further development of benign tumors into malignant tumors.

A classical skin tumor promoter and protein kinase C (PKC) activator, 12-*O*-tetradecanylphorbol-13-acetate (TPA), has also been shown to down-regulate the p53 response to BP [129]. In further studies a role for p38 MAP kinase was suggested [130]. Another skin tumor promoter, thapsigargin, can also attenuate the p53 response to DNA adducts [131]. An additional example is the antibiotic anisomycin, which has been shown to activate Mdm2 *via* Ser166 phosphorylation. Interestingly, in this case previous studies have also indicated potential tumor-promoting effects as this antibiotic synergizes with growth factors and phorbol esters to superinduce c-fos and c-jun [132].

Recently, published studies on the chemopreventive food component curcumin indicate that curcumin down-regulates Mdm2 and up-regulates p21$^{Waf1/Cip1}$ in a prostate cancer cell line. An involvement of the PI3K–mTOR pathway was implicated in this effect, and might explain the anticarcinogenic properties of this agent [133].

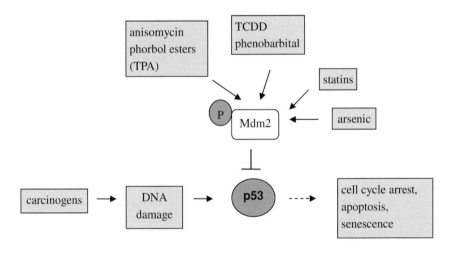

Figure 6. Mdm2 as a focal point for interactions. Xenobiotics that have been shown to activate Mdm2 or increase its levels and activity independently of p53 are shown. Several of these xenobiotics exhibit tumor-promoting activity and their effect on Mdm2 may lead to an attenuated p53 response caused by DNA-damaging carcinogens. In several cases phosphorylation of Ser166 in Mdm2 has been documented (see text).

Taken together these studies indicate that xenobiotics can interact in the p53 network and modulate the p53 response. A common finding is that xenobiotics known as tumor promoters affect Mdm2 and can attenuate the p53 response to DNA-damaging carcinogens. Thus, well-known tumor promoters from two-stage initiation-promotion studies *in vivo* have been shown to affect the response to the genotoxicity of initiators. In these *in vivo* protocols, the promoters and initiators not only exhibit additive effects but often more than additive carcinogenic effects. Future studies may reveal if adaptations of the p53 signaling at least partially can explain some of these interactions. It is also interesting to note that all these tumor promoters interact *via* Mdm2, and that the chemopreventive substance curcumin may have opposite effects. These data suggest an important role for Mdm2 as a focal point for interactions between xenobiotics in the p53 network. As mentioned below, Mdm2 might also mediate hormonal influences.

Sex hormones and alterations in the p53 signaling

Estrogens can have an influence on the chemical carcinogenesis, and long-term menopausal hormone therapy is a risk factor for breast cancer [134, 135]. Studies have demonstrated complex interactions with p53 signaling. For example, a 17β-estradiol-dependent estrogen receptor alpha (ERα) binding to the Mdm2 promoter may result in increased levels of Mdm2 and concomitant p53 translocation to the cytoplasm [136]. This may decrease the p53 response, but exposure to 17β-estradiol and progesterone for 21 days protects against carcinogenic chemicals [137]. Other studies indicate that ERα, Mdm2 and p53 proteins form a ternary complex that may protect p53 from degradation in a ligand-independent way [138]. This ternary complex also prevents ERα degradation [139]. Functional studies indicate that ERα bound to p53 can down-regulate the transcriptional activity of p53, but also inhibit p53-mediated repression of downstream genes such as survivin and multidrug resistance gene 1 (MDR1) [140]. These data thus indicate complex cross-talk between ERα and p53 signaling.

The soy component genistein, on the other hand, down-regulates Mdm2. Both transcriptional and post-translational levels were affected by genistein [141]. These data are in line with the above-mentioned effects of curcumin. They are also in line with many reports on chemopreventive properties of genistein in several animal models and may relate to the fact that genistein is an ERβ agonist. However, it has also been reported that mice neonatally exposed to genistein develop uterine adenocarcinomas [142].

Studies on the Mdm2 promoter single nucleotide polymorphism 309 (SNP309) indicate yet another type of interaction. This SNP affects the binding of ERα to the Mdm2 promoter and can influence the response under carcinogenic stress. Thus, in a cell model it was found that cells carrying the G-allele responded to 17β-estradiol by increasing Mdm2 levels, whereas cells homozygous for the T-allele did not. This effect influenced the p53 response to DNA damage induced by IR [143]. These data support some epidemiological observations. For example, a gender selective association of this polymorphism, e.g., to lung tumors has been reported. Thus female homozygous carriers of the G-allele exhibited significant association to smoke related lung cancer, whereas male homozygous carriers did not [144].

p53 signaling as a biomarker

As presented above and in Table 1, mutations in the p53 gene have been used as endpoints in biomarker studies on humans. Efforts to take advantage of p53 signaling for the development of biomarkers in cellular test models have included p53 [145] and Mdm2. It has been shown that Mdm2 phosphorylations can be a much more sensitive indicator of DNA damage than p53 phosphorylations. Mdm2 phosphorylations were induced by low doses of well-known genotoxic carcinogens, such as diethylnitrosamine (DEN), BP and dibenzo[a,l]pyrene (DBP). The ultimate mutagenic metabolite of BP, BPDE, elicited a clear Mdm2 response even at sub-picomolar concentrations [146]. Studies comparing effects of high and low doses of BP included endpoints such as Mdm2 Ser395, p53 Ser15, ATM Ser1989, histone H2AX phosphorylation (γH2AX) and chromatin binding of Mdm2 and p53. It was concluded that the response to low doses of BP, characterized by a clear Mdm2 phosphorylation without any detectable p53 phosphorylation, reflected easily repaired DNA damages. Other responses, seen at higher concentrations, probably reflected irreparable damage, leading to apoptosis [147].

p53, chemical carcinogenesis and cancer risk assessment

It is by now well established that p53 has a major impact on cancer development, and efforts to use this knowledge in the development of animal models for carcinogen testing have been explored. A p53$^{(+/-)}$ heterozygous mouse model (one copy of the p53 allele is not functional) has been developed and share a similar predisposition to tumors as Li Fraumeni patients [5]. p53$^{(+/-)}$ mice show a shorter latency period for tumors induced by chemicals than wild-type mice [148]. Knockout mouse strains for p53 have also been developed and exhibit spontaneous tumors within 3–6 months [5]. These genetically altered mice, as well as mice modified in many other genetic loci, are commonly used tools in cancer research [149].

Analyses of chemically induced tumors from p53$^{(+/-)}$ mice demonstrate loss of heterozygosity (LOH) and point mutations, mutagenic effects that may underlie the enhanced tumor susceptibility in humans. However, in some chemically induced tumors the wild-type allele was unaffected and, instead, reduction of p53 gene dosage and a haploinsufficient phenotype with reduced p53 function could explain the predisposition for tumors. The distinction between LOH and gene dosage was suggested to depend on the molecular mechanism of the genotoxic agent [148, 150, 151].

Since p53 heterozygous animals have higher risk for both spontaneous and chemically induced tumors, in addition to an earlier onset of tumors, this model has been suggested as an alternative for the conventional 2-year cancer bioassay. A 6-month testing period should be sufficient for tumor induction by genotoxic carcinogens and the model has been evaluated for assessing carcinogenic potentials of chemicals and drugs. It was concluded that the $p53^{(+/-)}$ model was useful in predicting carcinogenicity of genotoxic agents and could be used in risk assessments as a second *in vivo* bioassay, to clarify if carcinogenic effects in long-term bioassays are due to genotoxic mechanisms [152]. The model could also be used for functional analysis of p53 in the carcinogenic process [153] and has both potentials and limitations. Some mechanisms remains to be clarified [154, 155]. For example, experiments using the $p53^{(+/-)}$ mice have not demonstrated an increased sensitivity to certain liver carcinogenic and strongly genotoxic compounds [153, 156]. These findings may depend on the genetic background of mice strains used, but several studies suggest that $p53^{(+/-)}$ mice are not susceptible to chemically induced carcinogenesis in organs where tumors exhibit no or rare p53 mutations [153].

Liver is a major target in cancer bioassay, but rodent HCC rarely carry p53 mutations. Efforts to explain the difference between rodents and humans have not given expected results. For example, HCC in mice with a human p53 knock-in gene did not acquire p53 mutations [157, 158]. Several hypothetical explanations are discussed by the authors [158], but one additional explanation is provided by other studies. Thus, rat and mouse liver enzyme altered foci (EAF), often regarded as preneoplastic lesions, rarely exhibit a p53 response to DNA-damaging carcinogens. Instead, p53 seems to be down-regulated perhaps *via* alterations in Mdm2 expression [17, 159, 160]. Based on the fact that the liver is the main organ for metabolism of xenobiotics entering the body *via* food, and that extensive toxicity can be induced in hepatocytes *via* DNA adduct formation, it may be speculated that the p53 signaling in this organ can attenuate its early apoptotic response. This reasoning is in line with results obtained in mice with a p53 gene that can be turned on or off, and showing no tumor-suppressive effect of such an early toxic response [65]. The rodent liver might thus be exceptional and EAF lesions should perhaps be seen as adaptations and not as *bona fide* preneoplastic lesion.

Future prospects

Although more than 40 000 scientific papers have been published on p53, many questions remain to be answered. p53 is a tumor suppressor that interacts in many ways with basic cellular functions, and uncontrolled p53 activity can have devastating effects. These circumstances indicate a need for a very complex and fine-tuned regulation of p53 activity. The regulation should be able not only to handle stress signals but also to integrate signals related to nutritional status and energy supplies and to integrate mitogenic signals induced by, for example, hormones. It is thus not surprising that many levels of control have been characterized so far, and it can be expected that additional regulatory loops will be reported in the future.

Most studies on p53 concern its basic biology as a tumor suppressor, whereas studies specifically focusing on its role in toxicology are less frequent. For example, ample data indicate that the p53 network is of crucial importance for effects of carcinogens, such as the interactions induced by tumor initiators and promoters. However, the detailed mechanisms behind these interactions remain to be characterized and a more systematic analysis of which chemicals may affect the p53 signaling. Data gathered so far indicate frequent influences *via* Mdm2 alterations, but a more systematic approach may reveal additional influences.

It can also be anticipated that cell test models, monitoring endpoints related to DNA damage signaling and including the p53 pathway, will be developed in the future. It seems feasible to develop cell models that can differentiate easily repaired DNA damage from those that are associated with a high risk of inducing mutations and cancer. The latter type of damage may not only be induced by high doses that overwhelm the repair system, but also by complex mixtures of chemicals that may interact with DNA repair processes and/or in the p53 signaling. This testing approach may also shed new light on the very old controversy on how to perform low-dose extrapolations from high-dose data.

Employing genetically modified mice in long-term cancer bioassays is a promising approach, which may reduce the number of animals needed for testing and the time for testing. However, interpretations of the results might not be as unproblematic as initially thought. Our knowledge today indicates species differences affecting certain organs and a much more complex regulation of p53 than anticipated when these studies were started.

References

1 Oren M, Damalas A, Gottlieb T, Michael D, Taplick J, Leal JF, Maya R, Moas M, Seger R, Taya Y, Ben-Ze'Ev A (2002) Regulation of p53: Intricate loops and delicate balances. *Ann N Y Acad Sci* 973: 374–383
2 Vogelstein B, Lane D, Levine AJ (2000) Surfing the p53 network. *Nature* 408: 307–310
3 Levine AJ, Hu W, Feng Z (2006) The p53 pathway: What questions remain to be explored? *Cell Death Differ* 13: 1027–1036
4 Soussi T, Leblanc T, Baruchel A, Schaison G (1993) Germline mutations of the p53 tumor-suppressor gene in cancer-prone families: A review. *Nouv Rev Fr Hematol* 35: 33–36
5 Donehower LA, Harvey M, Slagle BL, McArthur MJ, Montgomery CA Jr, Butel JS, Bradley A (1992) Mice deficient for p53 are developmentally normal but susceptible to spontaneous tumours. *Nature* 356: 215–221
6 Linzer DI, Levine AJ (1979) Characterization of a 54K dalton cellular SV40 tumor antigen present in SV40-transformed cells and uninfected embryonal carcinoma cells. *Cell* 17: 43–52

7 Baker SJ, Fearon ER, Nigro JM, Hamilton SR, Preisinger AC, Jessup JM, vanTuinen P, Ledbetter DH, Barker DF, Nakamura Y et al (1989) Chromosome 17 deletions and *p53* gene mutations in colorectal carcinomas. *Science* 244: 217–221

8 Kastan MB, Onyekwere O, Sidransky D, Vogelstein B, Craig RW (1991) Participation of p53 protein in the cellular response to DNA damage. *Cancer Res* 51: 6304–6311

9 Yonish-Rouach E, Resnitzky D, Lotem J, Sachs L, Kimchi A, Oren M (1991) Wild-type p53 induces apoptosis of myeloid leukaemic cells that is inhibited by interleukin-6. *Nature* 352: 345–347

10 de Stanchina E, McCurrach ME, Zindy F, Shieh SY, Ferbeyre G, Samuelson AV, Prives C, Roussel MF, Sherr CJ, Lowe SW (1998) E1A signaling to p53 involves the p19(ARF) tumor suppressor. *Genes Dev* 12: 2434–2442

11 Momand J, Zambetti GP, Olson DC, George D, Levine AJ (1992) The mdm-2 oncogene product forms a complex with the p53 protein and inhibits p53-mediated transactivation. *Cell* 69: 1237–1245

12 Wu X, Bayle JH, Olson D, Levine AJ (1993) The p53-mdm-2 autoregulatory feedback loop. *Genes Dev* 7: 1126–1132

13 Haupt Y, Maya R, Kazaz A, Oren M (1997) Mdm2 promotes the rapid degradation of p53. *Nature* 387: 296–299

14 Vousden KH (2000) P53: Death star. *Cell* 103: 691–694

15 Meek DW, Knippschild U (2003) Posttranslational modification of MDM2. *Mol Cancer Res* 1: 1017–1026

16 Banin S, Moyal L, Shieh S, Taya Y, Anderson CW, Chessa L, Smorodinsky NI, Prives C, Reiss Y, Shiloh Y, Ziv Y (1998) Enhanced phosphorylation of p53 by ATM in response to DNA damage. *Science* 281: 1674–1677

17 Finnberg N, Silins I, Stenius U, Högberg J (2004) Characterizing the role of MDM2 in diethylnitrosamine induced acute liver damage and development of pre-neoplastic lesions. *Carcinogenesis* 25: 113–122

18 Hamstra DA, Bhojani MS, Griffin LB, Laxman B, Ross BD, Rehemtulla A (2006) Real-time evaluation of p53 oscillatory behavior *in vivo* using bioluminescent imaging. *Cancer Res* 66: 7482–7489

19 Jones SN, Roe AE, Donehower LA, Bradley A (1995) Rescue of embryonic lethality in Mdm2-deficient mice by absence of p53. *Nature* 378: 206–208

20 Marine JC, Francoz S, Maetens M, Wahl G, Toledo F, Lozano G (2006) Keeping p53 in check: Essential and synergistic functions of Mdm2 and Mdm4. *Cell Death Differ* 13: 927–934

21 Mayo LD, Donner DB (2001) A phosphatidylinositol 3-kinase/Akt pathway promotes translocation of Mdm2 from the cytoplasm to the nucleus. *Proc Natl Acad Sci USA* 98: 11598–11603

22 Momand J, Jung D, Wilczynski S, Niland J (1998) The MDM2 gene amplification database. *Nucleic Acids Res* 26: 3453–3459

23 Dornan D, Wertz I, Shimizu H, Arnott D, Frantz GD, Dowd P, O'Rourke K, Koeppen H, Dixit VM (2004) The ubiquitin ligase COP1 is a critical negative regulator of p53. *Nature* 429: 86–92

24 Leng RP, Lin Y, Ma W, Wu H, Lemmers B, Chung S, Parant JM, Lozano G, Hakem R, Benchimol S (2003) Pirh2, a p53-induced ubiquitin-protein ligase, promotes p53 degradation. *Cell* 112: 779–791

25 Finch RA, Donoviel DB, Potter D, Shi M, Fan A, Freed DD, Wang CY, Zambrowicz BP, Ramirez-Solis R, Sands AT, Zhang N (2002) Mdmx is a negative regulator of p53 activity *in vivo*. *Cancer Res* 62: 3221–3225

26 Bode AM, Dong Z (2004) Post-translational modification of p53 in tumorigenesis. *Nat Rev Cancer* 4: 793–805

27 Joerger AC, Fersht AR (2007) Structural biology of the tumor suppressor p53 and cancer-associated mutants. *Adv Cancer Res* 97: 1–23

28 Foo RS, Nam YJ, Ostreicher MJ, Metzl MD, Whelan RS, Peng CF, Ashton AW, Fu W, Mani K, Chin SF et al (2007) Regulation of p53 tetramerization and nuclear export by ARC. *Proc Natl Acad Sci USA* 104: 20826–20831

29 Hollstein M, Sidransky D, Vogelstein B, Harris CC (1991) P53 mutations in human cancers. *Science* 253: 49–53

30 Vousden KH, Lane DP (2007) P53 in health and disease. *Nat Rev Mol Cell Biol* 8: 275–283

31 Levine AJ, Hu W, Feng Z, Gil G (2007) Reconstructing signal transduction pathways: Challenges and opportunities. *Ann NY Acad Sci* 1115: 32–50

32 Levine AJ, Momand J, Finlay CA (1991) The *p53* tumour suppressor gene. *Nature* 351: 453–456

33 Giaccia AJ, Kastan MB (1998) The complexity of p53 modulation: Emerging patterns from divergent signals. *Genes Dev* 12: 2973–2983

34 Abraham RT (2004) PI 3-kinase related kinases: 'Big' players in stress-induced signaling pathways. *DNA Rep* 3: 883–887

35 Tang X, Hui ZG, Cui XL, Garg R, Kastan MB, Xu B (2008) A novel ATM-dependent pathway regulates protein phosphatase 1 in response to DNA damage. *Mol Cell Biol* 28: 2559–2566

36 Kitagawa R, Kastan MB (2005) The ATM-dependent DNA damage signaling pathway. *Cold Spring Harb Symp Quant Biol* 70: 99–109

37 Olsson A, Manzl C, Strasser A, Villunger A (2007) How important are post-translational modifications in p53 for selectivity in target-gene transcription and tumour suppression? *Cell Death Differ* 14: 1561–1575

38 Toledo F, Wahl GM (2006) Regulating the p53 pathway: *In vitro* hypotheses, *in vivo* veritas. *Nat Rev Cancer* 6: 909–923

39 Matsuoka S, Ballif BA, Smogorzewska A, McDonald ER 3rd, Hurov KE, Luo J, Bakalarski CE, Zhao Z, Solimini N, Lerenthal Y et al (2007) ATM and ATR substrate analysis reveals extensive protein networks responsive to DNA damage. *Science* 316: 1160–1166

40 Shiloh Y, Kastan MB (2001) ATM: Genome stability, neuronal development, and cancer cross paths. *Adv Cancer Res* 83: 209–254

41 Maya R, Balass M, Kim ST, Shkedy D, Leal JF, Shifman O, Moas M, Buschmann T, Ronai Z, Shiloh Y et al (2001) ATM-dependent phosphorylation of Mdm2 on serine 395: Role in p53 activation by DNA damage. *Genes Dev* 15: 1067–1077

42 Hurley PJ, Bunz F (2007) ATM and ATR: Components of an integrated circuit. *Cell Cycle* 6: 414–417

43 Paulsen RD, Cimprich KA (2007) The ATR pathway: Fine-tuning the fork. *DNA Rep* 6: 953–966

44 Burma S, Chen DJ (2004) Role of DNA-PK in the cellular response to DNA double-strand breaks. *DNA Rep* 3: 909–918

45 D'Orazi G, Cecchinelli B, Bruno T, Manni I, Higashimoto Y, Saito S, Gostissa M, Coen S, Marchetti A, Del Sal G et al (2002) Homeodomain-interacting protein kinase-2 phosphorylates p53 at Ser46 and mediates apoptosis. *Nat Cell Biol* 4: 11–19

46 Meek DW (1997) Post-translational modification of p53 and the integration of stress signals. *Pathol Biol* 45: 804–814

47 El-Deiry WS (1998) Regulation of p53 downstream genes. *Semin Cancer Biol* 8: 345–357

48 Sun Y (2006) P53 and its downstream proteins as molecular targets of cancer. *Mol Carcinog* 45: 409–415

49 Bensaad K, Vousden KH (2007) P53: New roles in metabolism. *Trends Cell Biol* 17: 286–291

50 Wei CL, Wu Q, Vega VB, Chiu KP, Ng P, Zhang T, Shahab A, Yong HC, Fu Y, Weng Z et al (2006) A global map of p53 transcription-factor binding sites in the human genome. *Cell* 124: 207–219

51 Le Cam L, Linares LK, Paul C, Julien E, Lacroix M, Hatchi E, Triboulet R, Bossis G, Shmueli A, Rodriguez MS et al (2006) E4F1 is an atypical ubiquitin ligase that modulates p53 effector functions independently of degradation. *Cell* 127: 775–788

52 Das S, Boswell SA, Aaronson SA, Lee SW (2008) P53 promoter selection: Choosing between life and death. *Cell Cycle* 7: 154–157

53 Aylon Y, Oren M (2007) Living with p53, dying of p53. *Cell* 130: 597–600

54 Brugarolas J, Chandrasekaran C, Gordon JI, Beach D, Jacks T, Hannon GJ (1995) Radiation-induced cell cycle arrest compromised by p21 deficiency. *Nature* 377: 552–557

55 Philipp J, Vo K, Gurley KE, Seidel K, Kemp CJ (1999) Tumor suppression by p27[Kip1] and p21[Cip1] during chemically induced skin carcinogenesis. *Oncogene* 18: 4689–4698

56 Weinberg WC, Fernandez-Salas E, Morgan DL, Shalizi A, Mirosh E, Stanulis E, Deng C, Hennings H, Yuspa SH (1999) Genetic deletion of p21[WAF1] enhances papilloma formation but not malignant conversion in experimental mouse skin carcinogenesis. *Cancer Res* 59: 2050–2054

57 De la Cueva E, Garcia-Cao I, Herranz M, Lopez P, Garcia-Palencia P, Flores JM, Serrano M, Fernandez-Piqueras J, Martin-Caballero J (2006) Tumorigenic activity of p21[Waf1/Cip1] in thymic lymphoma. *Oncogene* 25: 4128–4132

58 Efeyan A, Collado M, Velasco-Miguel S, Serrano M (2007) Genetic dissection of the role of p21[Cip1/Waf1] in p53-mediated tumour suppression. *Oncogene* 26: 1645–1649

59 Gupta M, Gupta SK, Balliet AG, Hollander MC, Fornace AJ, Hoffman B, Liebermann DA (2005) Hematopoietic cells from Gadd45a- and Gadd45b-deficient mice are sensitized to genotoxic-

stress-induced apoptosis. *Oncogene* 24: 7170–7179

60 Hermeking H, Benzinger A (2006) 14-3-3 proteins in cell cycle regulation. *Semin Cancer Biol* 16: 183–192

61 Wang XW, Zhan Q, Coursen JD, Khan MA, Kontny HU, Yu L, Hollander MC, O'Connor PM, Fornace AJ Jr, Harris CC (1999) GADD45 induction of a G2/M cell cycle checkpoint. *Proc Natl Acad Sci USA* 96: 3706–3711

62 Jin Z, El-Deiry WS (2005) Overview of cell death signaling pathways. *Cancer Biol Ther* 4: 139–163

63 Robles AI, Bemmels NA, Foraker AB, Harris CC (2001) APAF-1 is a transcriptional target of p53 in DNA damage-induced apoptosis. *Cancer Res* 61: 6660–6664

64 Oda K, Arakawa H, Tanaka T, Matsuda K, Tanikawa C, Mori T, Nishimori H, Tamai K, Tokino T, Nakamura Y, Taya Y (2000) P53AIP1, a potential mediator of p53-dependent apoptosis, and its regulation by Ser-46-phosphorylated p53. *Cell* 102: 849–862

65 Christophorou MA, Ringshausen I, Finch AJ, Swigart LB, Evan GI (2006) The pathological response to DNA damage does not contribute to p53-mediated tumour suppression. *Nature* 443: 214–217

66 Helton ES, Chen X (2007) P53 modulation of the DNA damage response. *J Cell Biochem* 100: 883–896

67 Hollander MC, Philburn RT, Patterson AD, Velasco-Miguel S, Friedberg EC, Linnoila RI, Fornace AJ Jr, (2005) Deletion of XPC leads to lung tumors in mice and is associated with early events in human lung carcinogenesis. *Proc Natl Acad Sci USA* 102: 13200–13205

68 Leveillard T, Andera L, Bissonnette N, Schaeffer L, Bracco L, Egly JM, Wasylyk B (1996) Functional interactions between p53 and the TFIIH complex are affected by tumour-associated mutations. *EMBO J* 15: 1615–1624

69 Shimodaira H, Yoshioka-Yamashita A, Kolodner RD, Wang JY (2003) Interaction of mismatch repair protein PMS2 and the p53-related transcription factor p73 in apoptosis response to cisplatin. *Proc Natl Acad Sci USA* 100: 2420–2425

70 Achanta G, Huang P (2004) Role of p53 in sensing oxidative DNA damage in response to reactive oxygen species-generating agents. *Cancer Res* 64: 6233–6239

71 Zhou J, Ahn J, Wilson SH, Prives C (2001) A role for p53 in base excision repair. *EMBO J* 20: 914–923

72 Lin Q, Clark AB, McCulloch SD, Yuan T, Bronson RT, Kunkel TA, Kucherlapati R (2006) Increased susceptibility to UV-induced skin carcinogenesis in polymerase eta-deficient mice. *Cancer Res* 66: 87–94

73 Romanova LY, Willers H, Blagosklonny MV, Powell SN (2004) The interaction of p53 with replication protein A mediates suppression of homologous recombination. *Oncogene* 23: 9025–9033

74 Dahm-Daphi J, Hubbe P, Horvath F, El-Awady RA, Bouffard KE, Powell SN, Willers H (2005) Nonhomologous end-joining of site-specific but not of radiation-induced DNA double-strand breaks is reduced in the presence of wild-type p53. *Oncogene* 24: 1663–1672

75 Campisi J, d'Adda di Fagagna F (2007) Cellular senescence: When bad things happen to good cells. *Nat Rev Mol Cell Biol* 8: 729–740

76 Bartek J, Bartkova J, Lukas J (2007) DNA damage signalling guards against activated oncogenes and tumour progression. *Oncogene* 26: 7773–7779

77 Gorgoulis VG, Vassiliou LV, Karakaidos P, Zacharatos P, Kotsinas A, Liloglou T, Venere M, Ditullio RA Jr, Kastrinakis NG, Levy B et al (2005) Activation of the DNA damage checkpoint and genomic instability in human precancerous lesions. *Nature* 434: 907–913

78 Bartkova J, Horejsi Z, Koed K, Kramer A, Tort F, Zieger K, Guldberg P, Sehested M, Nesland JM, Lukas C et al (2005) DNA damage response as a candidate anti-cancer barrier in early human tumorigenesis. *Nature* 434: 864–870

79 Bartkova J, Rezaei N, Liontos M, Karakaidos P, Kletsas D, Issaeva N, Vassiliou LV, Kolettas E, Niforou K, Zoumpourlis VC et al (2006) Oncogene-induced senescence is part of the tumorigenesis barrier imposed by DNA damage checkpoints. *Nature* 444: 633–637

80 Matheu A, Maraver A, Klatt P, Flores I, Garcia-Cao I, Borras C, Flores JM, Vina J, Blasco MA, Serrano M (2007) Delayed ageing through damage protection by the Arf/p53 pathway. *Nature* 448: 375–379

81 Tyner SD, Venkatachalam S, Choi J, Jones S, Ghebranious N, Igelmann H, Lu X, Soron G, Cooper B, Brayton C et al (2002) P53 mutant mice that display early ageing-associated phenotypes. *Nature* 415: 45–53

82 Serrano M, Blasco MA (2007) Cancer and ageing: Convergent and divergent mechanisms. *Nat Rev Mol Cell Biol* 8: 715–722

83 Ørsted DD, Bojesen SE, Tybjaerg-Hansen A, Nordestgaard BG (2007) Tumor suppressor p53 Arg72Pro polymorphism and longevity, cancer survival, and risk of cancer in the general population. *J Exp Med* 204: 1295–1301

84 van Heemst D, Mooijaart SP, Beekman M, Schreuder J, de Craen AJ, Brandt BW, Slagboom PE, Westendorp RG (2005) Variation in the human *TP53* gene affects old age survival and cancer mortality. *Exp Gerontol* 40: 11–15

85 Bojesen SE, Nordestgaard BG (2008) The common germline Arg72Pro polymorphism of *p53* and increased longevity in humans. *Cell Cycle* 7: 158–163

86 Nigro JM, Baker SJ, Preisinger AC, Jessup JM, Hostetter R, Cleary K, Bigner SH, Davidson N, Baylin S, Devilee P et al (1989) Mutations in the *p53* gene occur in diverse human tumour types. *Nature* 342: 705–708

87 Soussi T (2007) P53 alterations in human cancer: More questions than answers. *Oncogene* 26: 2145–2156

88 Kinzler KW, Vogelstein B (1996) Lessons from hereditary colorectal cancer. *Cell* 87: 159–170

89 Jenkins GJ, Doak SH, Griffiths AP, Tofazzal N, Shah V, Baxter JN, Parry JM (2003) Early p53 mutations in nondysplastic Barrett's tissue detected by the restriction site mutation (RSM) methodology. *Br J Cancer* 88: 1271–1276

90 Morgan C, Jenkins GJ, Ashton T, Griffiths AP, Baxter JN, Parry EM, Parry JM (2003) Detection of *p53* mutations in precancerous gastric tissue. *Br J Cancer* 89: 1314–1319

91 Ozturk M (1991) P53 mutation in hepatocellular carcinoma after aflatoxin exposure. *Lancet* 338: 1356–1359

92 Aguilar F, Hussain SP, Cerutti P (1993) Aflatoxin B$_1$ induces the transversion of G→T in codon 249 of the p53 tumor suppressor gene in human hepatocytes. *Proc Natl Acad Sci USA* 90: 8586–8590

93 Sohn S, Jaitovitch-Groisman I, Benlimame N, Galipeau J, Batist G, Alaoui-Jamali MA (2000) Retroviral expression of the hepatitis B virus × gene promotes liver cell susceptibility to carcinogen-induced site specific mutagenesis. *Mutat Res* 460: 17–28

94 Hussain SP, Schwank J, Staib F, Wang XW, Harris CC (2007) *TP53* mutations and hepatocellular carcinoma: Insights into the etiology and pathogenesis of liver cancer. *Oncogene* 26: 2166–2176

95 Barbin A, Froment O, Boivin S, Marion M-J, Belpoggi F, Maltoni C, Montesano R (1997) *P53* gene mutation pattern in rat liver tumors induced by vinyl chloride. *Cancer Res* 57: 1695–1698

96 Denissenko MF, Pao A, Tang M, Pfeifer GP (1996) Preferential formation of benzo[*a*]pyrene adducts at lung cancer mutational hotspots in *p53*. *Science* 274: 430–432

97 Hagiwara N, Mechanic LE, Trivers GE, Cawley HL, Taga M, Bowman ED, Kumamoto K, He P, Bernard M, Doja S et al (2006) Quantitative detection of *p53* mutations in plasma DNA from tobacco smokers. *Cancer Res* 66: 8309–8317

98 Taylor JA, Li Y, He M, Mason T, Mettlin C, Vogler WJ, Maygarden S, Liu E (1996) *P53* mutations in bladder tumors from arylamine-exposed workers. *Cancer Res* 56: 294–298

99 Benjamin CL, Ananthaswamy HN (2007) P53 and the pathogenesis of skin cancer. *Toxicol Appl Pharmacol* 224: 241–248

100 Le Calvez F, Mukeria A, Hunt JD, Kelm O, Hung RJ, Taniere P, Brennan P, Boffetta P, Zaridze DG, Hainaut P (2005) *TP53* and *KRAS* mutation load and types in lung cancers in relation to tobacco smoke: Distinct patterns in never, former, and current smokers. *Cancer Res* 65: 5076–5083

101 Feng Z, Hu W, Hu Y, Tang MS (2006) Acrolein is a major cigarette-related lung cancer agent: Preferential binding at p53 mutational hotspots and inhibition of DNA repair. *Proc Natl Acad Sci USA* 103: 15404–15409

102 Kim SI, Pfeifer GP, Besaratinia A (2007) Lack of mutagenicity of acrolein-induced DNA adducts in mouse and human cells. *Cancer Res* 67: 11640–11647

103 Song H, Hollstein M, Xu Y (2007) P53 gain-of-function cancer mutants induce genetic instability by inactivating ATM. *Nat Cell Biol* 9: 573–580

104 Benjamin CL, Ullrich SE, Kripke ML, Ananthaswamy HN (2008) *P53* tumor suppressor gene: A critical molecular target for UV induction and prevention of skin cancer. *Photochem Photobiol* 84: 55–62

105 Lind H, Ekstrom PO, Ryberg D, Skaug V, Andreassen T, Stangeland L, Haugen A, Zienolddiny

S (2007) Frequency of *TP53* mutations in relation to Arg72Pro genotypes in non small cell lung cancer. *Cancer Epidemiol Biomarkers Prev* 16: 2077–2081

106 Pilsner JR, Liu X, Ahsan H, Ilievski V, Slavkovich V, Levy D, Factor-Litvak P, Graziano JH, Gamble MV (2007) Genomic methylation of peripheral blood leukocyte DNA: Influences of arsenic and folate in Bangladeshi adults. *Am J Clin Nutr* 86: 1179–1186

107 Chanda S, Dasgupta UB, Guhamazumder D, Gupta M, Chaudhuri U, Lahiri S, Das S, Ghosh N, Chatterjee D (2006) DNA hypermethylation of promoter of gene *p53* and *p16* in arsenic-exposed people with and without malignancy. *Toxicol Sci* 89: 431–437

108 Shen S, Lee J, Weinfeld M, Le XC (2008) Attenuation of DNA damage-induced p53 expression by arsenic: A possible mechanism for arsenic co-carcinogenesis. *Mol Carcinog* 47: 508–518

109 Mass MJ, Wang L (1997) Arsenic alters cytosine methylation patterns of the promoter of the tumor suppressor gene *p53* in human lung cells: A model for a mechanism of carcinogenesis. *Mutat Res* 386: 263–277

110 Hamadeh HK, Vargas M, Lee E, Menzel DB (1999) Arsenic disrupts cellular levels of p53 and mdm2: A potential mechanism of carcinogenesis. *Biochem Biophys Res Commun* 263: 446–449

111 He L, He X, Lim LP, de Stanchina E, Xuan Z, Liang Y, Xue W, Zender L, Magnus J, Ridzon D et al (2007) A microRNA component of the p53 tumour suppressor network. *Nature* 447: 1130–1134

112 Chang TC, Wentzel EA, Kent OA, Ramachandran K, Mullendore M, Lee KH, Feldmann G, Yamakuchi M, Ferlito M, Lowenstein CJ et al (2007) Transactivation of miR-34a by p53 broadly influences gene expression and promotes apoptosis. *Mol Cell* 26: 745–752

113 Bommer GT, Gerin I, Feng Y, Kaczorowski AJ, Kuick R, Love RE, Zhai Y, Giordano TJ, Qin ZS, Moore BB et al (2007) P53-mediated activation of miRNA34 candidate tumor-suppressor genes. *Curr Biol* 17: 1298–1307

114 Raver-Shapira N, Oren M (2007) Tiny actors, great roles: MicroRNAs in p53's service. *Cell Cycle* 6: 2656–2661

115 Gorospe M, de Cabo R (2008) AsSIRTing the DNA damage response. *Trends Cell Biol* 18: 77–83

116 Levine AJ, Feng Z, Mak TW, You H, Jin S (2006) Coordination and communication between the p53 and IGF-1-AKT-TOR signal transduction pathways. *Genes Dev* 20: 267–275

117 Gurumurthy S, Hezel AF, Berger JH, Bosenberg MW, Bardeesy N (2008) LKB1 deficiency sensitizes mice to carcinogen-induced tumorigenesis. *Cancer Res* 68: 55–63

118 Matsumoto S, Iwakawa R, Takahashi K, Kohno T, Nakanishi Y, Matsuno Y, Suzuki K, Nakamoto M, Shimizu E, Minna JD, Yokota J (2007) Prevalence and specificity of LKB1 genetic alterations in lung cancers. *Oncogene* 26: 5911–5918

119 Pääjärvi G, Roudier E, Crisby M, Högberg J, Stenius U (2005) HMG-CoA reductase inhibitors, statins, induce phosphorylation of Mdm2 and attenuate the p53 response to DNA damage. *FASEB J* 19: 476–478

120 Roudier E, Mistafa O, Stenius U (2006) Statins induce mammalian target of rapamycin (mTOR)-mediated inhibition of Akt signaling and sensitize p53-deficient cells to cytostatic drugs. *Mol Cancer Ther* 5: 2706–2715

121 Mistafa O, Högberg J, Stenius U (2008) Statins and ATP regulate nuclear pAkt *via* the P2X7 purinergic receptor in epithelial cells. *Biochem Biophys Res Commun* 365: 131–136

122 Larsson SC, Wolk A (2008) Excess body fatness: An important cause of most cancers. *Lancet* 371: 536–537

123 Malmlof M, Roudier E, Högberg J, Stenius U (2007) MEK-ERK-mediated phosphorylation of Mdm2 at Ser-166 in hepatocytes. Mdm2 is activated in response to inhibited Akt signaling. *J Biol Chem* 282: 2288–2296

124 Marlowe JL, Puga A (2005) Aryl hydrocarbon receptor, cell cycle regulation, toxicity, and tumorigenesis. *J Cell Biochem* 96: 1174–1184

125 Worner W, Schrenk D (1996) Influence of liver tumor promoters on apoptosis in rat hepatocytes induced by 2-acetylaminofluorene, ultraviolet light, or transforming growth factor beta 1. *Cancer Res* 56: 1272–1278

126 Paajarvi G, Viluksela M, Pohjanvirta R, Stenius U, Högberg J (2005) TCDD activates Mdm2 and attenuates the p53 response to DNA damaging agents. *Carcinogenesis* 26: 201–208

127 Nelson DM, Bhaskaran V, Foster WR, Lehman-McKeeman LD (2006) P53-independent induction of rat hepatic Mdm2 following administration of phenobarbital and pregnenolone 16a-carbonitrile. *Toxicol Sci* 94: 272–280

128 Yamamoto Y, Moore R, Goldsworthy TL, Negishi M, Maronpot RR (2004) The orphan nuclear

receptor constitutive active/androstane receptor is essential for liver tumor promotion by phenobarbital in mice. *Cancer Res* 64: 7197–7200

129 Tapiainen T, Jarvinen K, Paakko P, Bjelogrlic N, Vahakangas K (1996) TPA decreases the p53 response to benzo[*a*]pyrene-DNA adducts *in vivo* in mouse skin. *Carcinogenesis* 17: 1377–1380

130 Mukherjee JJ, Sikka HC (2006) Attenuation of BPDE-induced p53 accumulation by TPA is associated with a decrease in stability and phosphorylation of p53 and downregulation of NFkB activation: Role of p38 MAP kinase. *Carcinogenesis* 27: 631–638

131 Serpi R, Piispala J, Jarvilehto M, Vahakangas K (1999) Thapsigargin has similar effect on p53 protein response to benzo[*a*]pyrene-DNA adducts as TPA in mouse skin. *Carcinogenesis* 20: 1755–1760

132 Kardalinou E, Zhelev N, Hazzalin CA, Mahadevan LC (1994) Anisomycin and rapamycin define an area upstream of p70/85S6k containing a bifurcation to histone H3-HMG-like protein phosphorylation and c-fos-c-jun induction. *Mol Cell Biol* 14: 1066–1074

133 Li M, Zhang Z, Hill DL, Wang H, Zhang R (2007) Curcumin, a dietary component, has anticancer, chemosensitization, and radiosensitization effects by down-regulating the MDM2 oncogene through the PI3K/mTOR/ETS2 pathway. *Cancer Res* 67: 1988–1996

134 Kerlikowske K, Miglioretti DL, Buist DS, Walker R, Carney PA (2007) Declines in invasive breast cancer and use of postmenopausal hormone therapy in a screening mammography population. *J Natl Cancer Inst* 99: 1335–1339

135 Nelson HD, Humphrey LL, Nygren P, Teutsch SM, Allan JD (2002) Postmenopausal hormone replacement therapy: Scientific review. *JAMA* 288: 872–881

136 Kinyamu HK, Archer TK (2003) Estrogen receptor-dependent proteasomal degradation of the glucocorticoid receptor is coupled to an increase in mdm2 protein expression. *Mol Cell Biol* 23: 5867–5881

137 Medina D (2004) Breast cancer: The protective effect of pregnancy. *Clin Cancer Res* 10: 380S–384S

138 Liu G, Schwartz JA, Brooks SC (2000) Estrogen receptor protects p53 from deactivation by human double minute-2. *Cancer Res* 60: 1810–1814

139 Duong V, Boulle N, Daujat S, Chauvet J, Bonnet S, Neel H, Cavailles V (2007) Differential regulation of estrogen receptor a turnover and transactivation by Mdm2 and stress-inducing agents. *Cancer Res* 67: 5513–5521

140 Sayeed A, Konduri SD, Liu W, Bansal S, Li F, Das GM (2007) Estrogen receptor a inhibits p53-mediated transcriptional repression: Implications for the regulation of apoptosis. *Cancer Res* 67: 7746–7755

141 Li M, Zhang Z, Hill DL, Chen X, Wang H, Zhang R (2005) Genistein, a dietary isoflavone, downregulates the MDM2 oncogene at both transcriptional and posttranslational levels. *Cancer Res* 65: 8200–8208

142 Newbold RR, Banks EP, Bullock B, Jefferson WN (2001) Uterine adenocarcinoma in mice treated neonatally with genistein. *Cancer Res* 61: 4325–4328

143 Hu W, Feng Z, Ma L, Wagner J, Rice JJ, Stolovitzky G, Levine AJ (2007) A single nucleotide polymorphism in the MDM2 gene disrupts the oscillation of p53 and MDM2 levels in cells. *Cancer Res* 67: 2757–2765

144 Lind H, Zienolddiny S, Ekstrom PO, Skaug V, Haugen A (2006) Association of a functional polymorphism in the promoter of the *MDM2* gene with risk of nonsmall cell lung cancer. *Int J Cancer* 119: 718–721

145 Saito S, Yamaguchi H, Higashimoto Y, Chao C, Xu Y, Fornace AJ Jr, Appella E, Anderson CW (2003) Phosphorylation site interdependence of human p53 post-translational modifications in response to stress. *J Biol Chem* 278: 37536–37544

146 Pääjärvi G, Jernström B, Seidel A, Stenius U (2008) *Anti*-diol epoxide of benzo[*a*]pyrene induces transient Mdm2 and p53 Ser15 phosphorylation, while *anti*-diol epoxide of dibenzo[*a,l*]pyrene induces a nontransient p53 Ser15 phosphorylation. *Mol Carcinog* 47: 301–309

147 Malmlöf M, Pääjärvi G, Högberg J, Stenius U (2008) Mdm2 as a sensitive and mechanistically informative marker for genotoxicity induced by benzo[*a*]pyrene and dibenzo[*a,l*]pyrene. *Toxicol Sci* 102: 232–240

148 French J, Storer RD, Donehower LA (2001) The nature of the heterozygous *Trp53* knockout model for identification of mutagenic carcinogens. *Toxicol Pathol* 29 Suppl: 24–29

149 Donehower LA, French JE, Hursting SD (2005) The utility of genetically altered mouse models for cancer research. *Mutat Res* 576: 1–3

150 Storer RD, French JE, Haseman J, Hajian G, LeGrand EK, Long GG, Mixson LA, Ochoa R, Sagartz JE, Soper KA (2001) P53$^{(+/-)}$ hemizygous knockout mouse: Overview of available data. *Toxicol Pathol* 29 Suppl: 30–50

151 Venkatachalam S, Tyner SD, Pickering CR, Boley S, Recio L, French JE, Donehower LA (2001) Is p53 haploinsufficient for tumor suppression? Implications for the p53$^{(+/-)}$ mouse model in carcinogenicity testing. *Toxicol Pathol* 29 Suppl: 147–154

152 Storer RD (2000) Current status and use of short/medium term models for carcinogenicity testing of pharmaceuticals – Scientific perspective. *Toxicol Lett* 112–113: 557–566

153 Hirata A, Tsukamoto T, Yamamoto M, Takasu S, Sakai H, Ban H, Yanai T, Masegi T, Donehower LA, Tatematsu M (2007) Organ-dependent susceptibility of p53 knockout mice to 2-amino-3-methylimidazo[4,5-*f*]quinoline (IQ). *Cancer Sci* 98: 1164–1173

154 Mitsumori K (2002) Evaluation on carcinogenicity of chemicals using transgenic mice. *Toxicology* 181–182: 241–244

155 Pritchard JB, French JE, Davis BJ, Haseman JK (2003) The role of transgenic mouse models in carcinogen identification. *Environ Health Perspect* 111: 444–454

156 Finnberg N, Stenius U, Högberg J (2004) Heterozygous p53-deficient (–/–) mice develop fewer p53-negative preneoplastic focal liver lesions in response to treatment with diethylnitrosamine than do wild-type (+/+) mice. *Cancer Lett* 207: 149–155

157 Jaworski M, Hailfinger S, Buchmann A, Hergenhahn M, Hollstein M, Ittrich C, Schwarz M (2005) Human p53 knock-in (hupki) mice do not differ in liver tumor response from their counterparts with murine p53. *Carcinogenesis* 26: 1829–1834

158 Tong WM, Lee MK, Galendo D, Wang ZQ, Sabapathy K (2006) Aflatoxin-B exposure does not lead to p53 mutations but results in enhanced liver cancer of Hupki (human p53 knock-in) mice. *Int J Cancer* 119: 745–749

159 Silins I, Stenius U, Högberg J (2004) Induction of preneoplastic rat liver lesions with an attenuated p53 response by low doses of diethylnitrosamine. *Arch Toxicol* 78: 540–548

160 Van Gijssel HE, Ohlson LC, Torndal UB, Mulder GJ, Eriksson LC, Porsch-Hallstrom I, Meerman JH (2000) Loss of nuclear p53 protein in preneoplastic rat hepatocytes is accompanied by Mdm2 and Bcl-2 overexpression and by defective response to DNA damage *in vivo*. *Hepatology* 32: 701–710

Molecular, Clinical and Environmental Toxicology. Volume 1: Molecular Toxicology
Edited by A. Luch

Molecular pathways involved in cell death after chemically induced DNA damage

Roberto Sánchez-Olea[1], Mónica R. Calera[1], and Alexei Degterev[2]

[1]Instituto de Física, Universidad Autónoma de San Luis Potosí, Mexico
[2]Tufts University School of Medicine, Department of Biochemistry, Boston, MA, USA

Abstract. DNA damage is at the center of the genesis, progression and treatment of cancer. We review here the molecular mechanisms of the DNA damage inducing small molecules most commonly used in cancer therapy. Cell cycle control and DNA repair mechanisms are known to be activated after DNA damage. Here, we revise recent discoveries related to the cell cycle control and DNA repair processes and how these findings are being utilized for the more efficient, powerful and selective therapies for cancer treatment.

Introduction

A large number of agents employed in the clinic as anticancer agents exert their toxic effects by chemically modifying the genetic material or DNA. In addition, accumulating mutations and genetic instability are the hallmarks, and possibly at the core of the genesis of cancer. Therefore, to design rational and effective anticancer treatments, it is essential to unravel the molecular pathways involved in the DNA damage response, including the activation of diverse cellular functions such as cell cycle arrest, DNA repair and cell death. An ultimate goal is the identification of major differences between normal and cancer cells that would allow the high specificity in targeting altered cells and minimizing the secondary effects of therapy. In the current review, we describe the molecular mechanisms of some of the most commonly employed chemotherapies as well as some of manipulations in the DNA damage response that increase the toxicity of these drugs, specifically the inactivation of the G2/M cell cycle checkpoint. Finally, we describe the differences between normal and cancer cell responses to DNA-damaging agents, which are exploited to increase the toxicity of chemical agents towards cancer cells, while, at the same time, decreasing their toxicity to normal cells.

Topoisomerase I inhibitors: Camptothecin

Camptothecin was originally isolated from the bark of the Chinese tree *Campthoteca acuminata* [1]. The cytotoxic effects of campothecin and its potent anticancer derivatives irinotecan and topotecan (Fig. 1) are exerted by inhibiting the enzymatic activity of DNA topoisomerase I (Top I) [2]. Yeast cells lacking Top I are resistant to campothecin [3, 4], demonstrating the remarkable selectivity of this agent. Top I is ubiquitously expressed and its biological function is to release DNA supercoiling produced during basic events of DNA metabolism such as replication and transcription. When the DNA duplex is opened by advancing large macromolecular protein complexes, the rest of the molecule is unable to rotate freely due to its extremely large size. Thus, unwinding of one DNA region produces supercoiling in the neighboring areas. Top I function is to relax this supercoiled DNA by a mechanism that involves a tyrosine-mediated covalent binding of Top I to DNA, breaking one strand, allowing rotation of this strand around the Top I-DNA complex and the final alignment and re-ligation of the nicked strand in the relaxed DNA. During this process Top I encircles DNA as a clamp, allowing for a controlled rotation to relax DNA [5]. The Top I/DNA intermediate is short-lived and is rarely detected in cells under normal conditions. However, it is the target of camptothecin and its derivatives, which act to stabilize these so-called "Top I cleavage complexes". It is interesting to note that campothecin slows down the re-ligation step of the nicked DNA by Top I but do not directly cause DNA damage. Toxicity by Top I inhibitors requires long exposure of cells to the drug

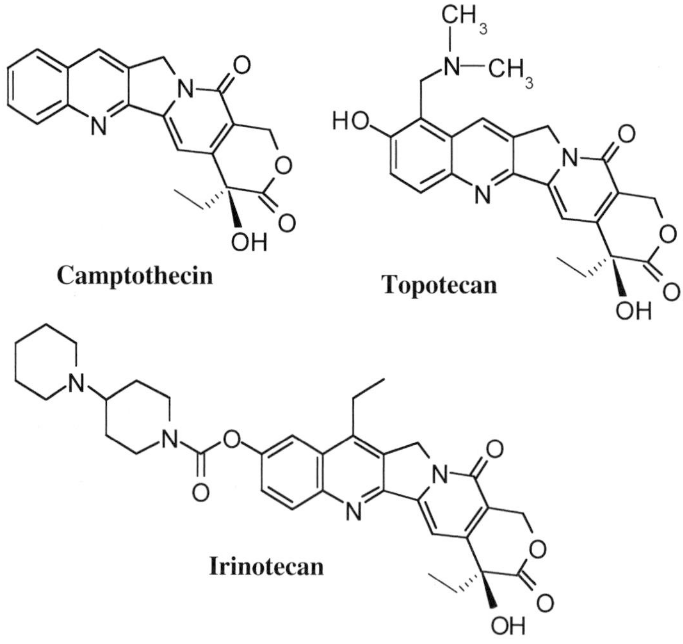

Figure 1. Topoisomerase I (Top I) inhibitors: Camptothecin and its potent derivatives irinotecan and topotecan.

[6], as it is the cellular metabolism that actually converts the reversible cleavage complexes into permanent DNA damage. Cellular transcription and DNA duplication are critical in this process. In culture cells the toxicity of camptothecin is restricted to cells in the S-phase of the cell cycle [7, 8] and is prevented by blocking DNA replication [6, 9]. The advancing of the replication machinery into a stabilized Top I cleavage complex results in the synthesis of a complementary strand up to the last nucleotide of the nicked strand, therefore preventing the possibility of the re-ligation step with the original strand. As cancer cells divide at a significantly higher rate that normal cells, they need to replicate their DNA more often, explaining in part the selectivity of camptothecin and other Top I inhibitors towards cancer cells. However, factors other than DNA replication contribute to the cytotoxicity of camptothecin. The protective effect of aphidicolin, a DNA synthesis blocker, is restricted to doses of camptothecin lower than one micromolar [8, 10–12] and camptothecin is cytotoxic to non-dividing neurons [11]. Replication of double strand breaks (DSBs) as those produced by long exposure to camptothecin in proliferating cells elicit several cellular responses that involve sensors and effectors of the DNA DSB network [13–15].

Camptothecin-induced DSBs activate the ATM-Chk2-p53 pathway [13, 16] and a defect in ATM [13, 17, 18], Chk2 [16, 19, 20] or p53 [21, 22] makes cells more vulnerable to camptothecin. Cockayne syndrome cells, which are defective in the nucleotide excision repair (NER) pathway, display a hypersensitivity to camptothecin [23], reinforcing the idea that DNA repair pathways are activated and limit the toxicity of campothecin. Indeed, a fruitful approach to increase the toxic effects of DNA-inducing agents has been the simultaneous suppression of specific DNA repair pathways (see below).

Topoisomerase II inhibitors

DNA topoisomerase II (Top II) has been identified as the molecular target for a number of anticancer agents, including etoposide, doxorubicin, daunorubicin, mitoxantrone, and amsacrin (m-AMSA) (Fig. 2) [24–26].

These agents poison Top II by stabilizing the Top II-DNA cleavage complex, a normally transient intermediate state of the process. Top IIa is a homodimer of two identical 170-kDa polypeptides [27] and each subunit cleaves one strand of the double helix, leading to the transient formation of a DSB. The ability of a tyrosine in the catalytic site of each subunit of Top II to form a covalent bond with the protruding 5' DNA ends retains the broken DNA together. After the passage of an intact DNA strand through the gap, the same enzymatic activity of Top II is responsible for resealing the DSB. As DSBs are toxic to cells, Top II is an excellent molecular target in chemotherapy. BRCA1- or BRCA2-deficient cells are more sensitive to etoposide [28], indicating that DNA repair by the homologous recombination pathway is a critical determinant for the cytotoxicity caused by Top II inhibiting agents.

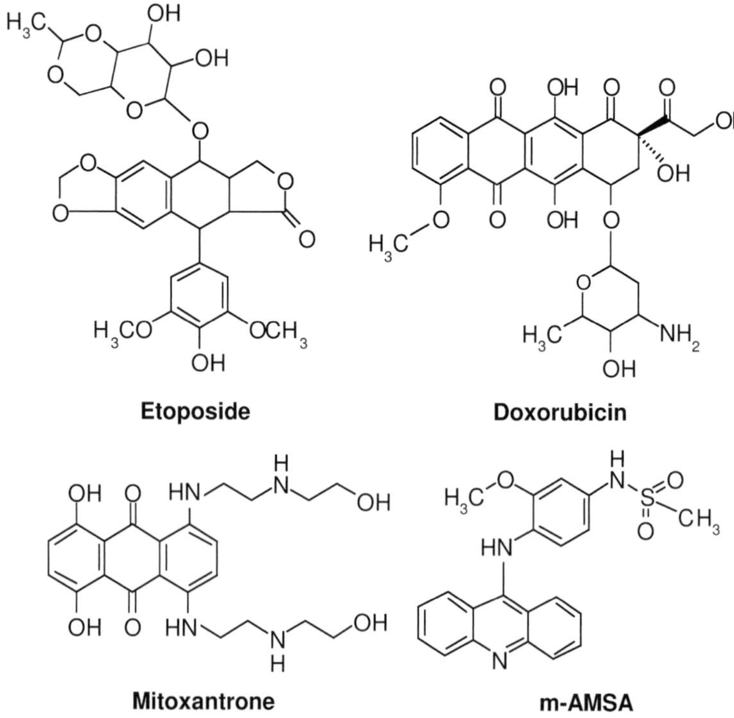

Etoposide

Doxorubicin

Mitoxantrone

m-AMSA

Figure 2. Small molecule inhibitors of topoisomerase II (Top II): etoposide, doxorubicin, mitoxantrone and m-AMSA (amsacrin).

Alkylating agents: Cisplatin

Cisplatin is a potent chemotherapeutic agent derived from the metal platinum (Fig. 3). It is a planar Pt(II) complex (*cis*-dichlorodiamine platinum II) widely used to treat testicular and ovarian cancers [29]. The molecular target of cisplatin is DNA [30, 31]. In solution, cisplatin undergoes a chemical modification where the two chlorides are replaced by water molecules before it can directly interact with DNA. This derivative is a more reactive species towards DNA. The positions originally occupied by chloride are used by cisplatin to form covalent bonds with the N7 atoms of adjacent guanosines in DNA, resulting in the production of intra-strand cross-links [32]. Although cisplatin induces efficient cell death in cancer cells, it has several drawbacks such as kidney toxicity, the need for intravenous administration and development of the resistance over time by some types of cancer. Several cisplatin derivatives have been developed to overcome these limitations. Carboplatin (Fig. 3) is a cisplatin derivative in which the reaction replacing the chloride atoms by water molecules is slowed down, resulting in a reduced reactivity of the compound

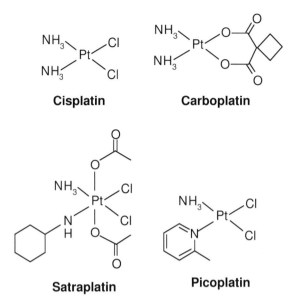

Figure 3. Chemical structure of cisplatin and its derivatives carboplatin, satraplatin and picoplatin. The two chloride atoms in cisplatin are replaced by water molecules before its interaction with DNA. Satraplatin was the first orally available derivative of cisplatin. Picoplatin is orally available and effective in cells that have developed cisplatin resistance by different mechanisms.

towards DNA and other molecules in the cell. Therefore, carboplatin displays a significantly reduced nephrotoxicity compared to cisplatin [33]. Satraplatin (Fig. 3) possesses increased water solubility and was the first orally available derivative of cisplatin. Although its administration had to be divided into five doses a day to compensate for non-linear pharmacokinetics in patients, satraplatin was also active against some acquired cisplatin-resistant cell lines, suggesting the existence of significant difference in the mechanism of action or in the cellular responses to the DNA damage caused by cisplatin and satraplatin. Picoplatin (Fig. 3) is a cisplatin derivative in which the platinum center is less susceptible to the inactivating attack by thiol-containing molecules. In addition to being orally available, the major appeal of picoplatin is its efficiency against cisplatin resistance resulting from the elevated intracellular levels of glutathione, the activation of DNA repair and the reduction of the intracellular drug accumulation [34, 35].

Cellular transport mechanisms for cisplatin

Cisplatin does not freely diffuse through cell membranes, but rather is transported by the high-affinity copper transporter Ctr1. Ctr1 inactivation reduces cisplatin accumulation [36], whereas the overexpression of this protein increases cisplatin uptake [37]. The transporter ATP7B, responsible for copper efflux, is able to confer cisplatin resistance in a cellular model [38] and is over-expressed in some cancers, where it is considered a marker for cisplatin resistance [39–41]. In addition, up-regulation of the ATP-binding cassette transporter ABCC2 is associated with cisplatin resistance [42, 43] and its down-regulation increases sensitivity to cisplatin [44], consistent with a role of this protein in mediating cisplatin efflux from cells.

Mechanism for cisplatin cellular toxicity

The cisplatin-modified DNA is not suitable for translation and transcription as both of these processes are inhibited by cisplatin-DNA adducts [31]. Cisplatin lesions block transcription elongation by RNA polymerase II [45–47] and lead to stable polymerase stalling [48]. The structural basis of RNA polymerase II stalling by cisplatin lesions has been recently determined and it involves the formation of a translocation barrier that prevents the delivery of the lesion to the active site [49].

The cisplatin-DNA adducts are specifically recognized by a number of cellular proteins. High-mobility group (HMG) box proteins are among the best studied and they clearly constitute an important determinant in cisplatin cytotoxicity [50]. The HMG box protein HMGB1 binds to the platinated DNA, shielding the lesion and preventing its repair by the NER pathway [51]. Interestingly, the overexpression of HMGB1 sensitizes ovarian and breast cancer cells to cisplatin [52]. Another HMG domain-containing protein that selectively binds to cisplatin-modified DNA is structure-specific recognition protein (SSR1) [53]. SSRP1 needs to form a complex with SPT16 (suppressor of Ty 16 homologue) to bind to cisplatin-damaged DNA [54].

The kinase activity of nuclear c-Abl, a member of the Src family of non-receptor tyrosine kinases, is increased in response to cisplatin treatment [55], and it has been proposed that c-Abl transmits the DNA damage signal from the nucleus to the cytoplasm. c-Abl clearly plays a role in the induction of apoptosis in response to cisplatin treatment. Mouse embryo fibroblasts deficient in c-Abl are resistant to cisplatin-induced apoptosis and the reintroduction of this protein restored the sensitivity of these cells to the drug [56]. Moreover, suppression of c-Abl by RNAi was associated with cisplatin resistance [57]. The transcription factor p73 seems to function downstream from c-Abl, as cisplatin induces an increase in p73 protein levels in a c-Abl-dependent manner [56]. MAPK/ERK kinase kinase 1 (MEKK1) is among the molecular targets of c-Abl, as cisplatin treatment increases the levels of tyrosine phosphorylation of MEKK1 in wild-type fibroblasts but not in c-Abl-deficient cells [58]. Whereas the activation of p38 MAPK and JNK by ultraviolet light was not affected by c-Abl, the activation by cisplatin was suppressed in c-Abl-deficient cells [59].

Other metal-containing DNA damaging agents

Due to the successful development and applications of cisplatin, other metal-based compounds with chemotherapeutic properties were pursued. Ruthenium-based small molecules have been developed, e.g., RM175 (Fig. 4), and found to target DNA, although using a mechanism different from cisplatin [60]. Indazolium[bis-indazole tetrachlororuthenate] (KP1019, Fig. 4) is a ruthenium-derived compound that is active against colorectal cancer and is transported by the transferrin system [61]. Imidazolium[*trans*-imidazole methylsulfoxide tetrachlororuthenate] (NAMI-A, Fig. 4) displays a remarkable inhibitory effect on tumor metastasis [62].

Selectivity of PARP inhibitors

Poly(ADP-ribose) polymerase (PARP) inhibitors were found to display potent toxic effects and high selectivity towards BRCA-deficient tumors cells but spare normal cells. PARP is an enzyme that cleaves NAD^+ to nicotinamide and ADP-ribose to add long negatively charged (ADP-ribose) polymers to the glutamic acid residues of the PARP protein itself, as well as other nuclear substrates, such as histones. PARP is recruited to the damaged DNA through its

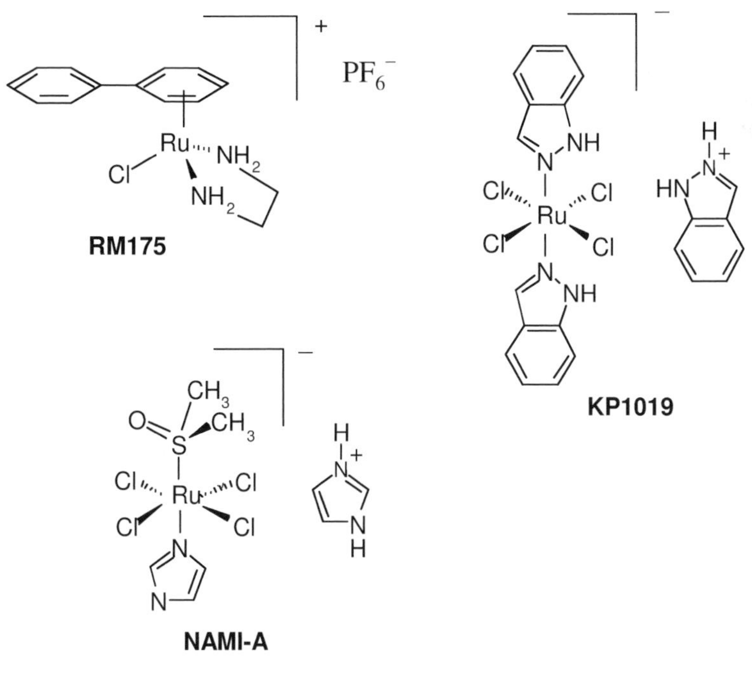

Figure 4. Ruthenium-based small molecules RM175, KP1019 and NAMI-A also target DNA.

two N-terminal zinc fingers and this binding stimulates its enzymatic activity, resulting in the formation of these ADP-ribose polymers [63]. PARP is essential in the base excision repair (BER) pathway, which plays an important role in the repair of single strand breaks (SSBs) [64, 65]. Hence, PARP inhibition leads to permanent SSBs in DNA [66]. PARP is not directly involved in DNA repair but its activity is key for the binding of proteins responsible for the repair of SSBs, such as XRCC1 [67]. Although PARP also binds to DSBs [68], it is not required for the repair of this type of DNA damage [69, 70]. PARP1 is not an essential protein, and PARP1$^{(-/-)}$ knockout mice are healthy and fertile [71, 72]. Furthermore, these animals display reduced tumorigenesis, as PARP knockout mice deficient for p53 show an increase in tumor latency [73]. However, PARP1-deficient cells are unable to repair DNA SSBs and show an increase in homologous recombination, sister chromatid exchange and micronuclei formation [71, 72, 74]. A DNA SSB can lead to the stalling of an advancing replication fork and even to the potential formation of a DNA DSB [75–77], and both SSBs and DSBs are efficiently corrected by the homologous recombination pathway. The activation of homologous recombination suggests that the activation of this alternative DNA repair pathway is a cellular response that reduces the functional consequences of PARP1 deficiency.

BRCA1 and BRCA2 proteins play an important role in the repair of DNA DSBs through homologous recombination [78–80]. A direct interaction of BRCA2 with the recombinase RAD51 [81, 82] is critical for its role in the repair process. In response to DNA damage, BRCA2 recruits RAD51 to the regions containing DSBs to initiate the repair process [83]. The fact that the DNA alterations caused by PARP deficiency are corrected by the homologous recombination pathway led two research groups to explore the possibility that BRCA-deficient cells, which are defective in the homologous recombination pathway, were selectively sensitive to small molecule PARP inhibitors [84, 85]. In support of this hypothesis, the BRCA2-deficient cell line V-C8 [86] was found to be extremely sensitive to low doses of the potent PARP inhibitors 1,5-dihydroxyisoquinoline (ISQ), 8-hydroxy-2-methylquinazolinone (NU1025) and 1-(4-dimethylaminomethyl-phenyl)-8,9-dihydro-7H-2,7,9a-benzo[c,d]azulen-6-one (AG14361) (Fig. 5) when compared with the V-C8 cell line complemented with BRCA2. Importantly, the MCF-7 and MDA-MB-231 breast cancer cell lines also displayed an increase in sensitivity to PARP inhibitors upon BRCA2 depletion using BRCA2 short interfering RNAs (siRNAs) [84]. In addition, mouse embryonic stem (ES) cells lacking Brca1 or Brca2 were highly sensitive to KU0058684 and KU0058948 (Fig. 5), two very potent novel PARP inhibitors [85]. Upon PARP inhibition, ES cells underwent cell cycle arrest in the G2 phase of the cell cycle followed by apoptotic cell death [85]. PARP inhibition alone induced the formation of RAD51 foci in wild-type cells, a hallmark of homologous recombination, confirming the activation of this DNA repair pathway. The finding that PARP activity is so critical for DNA repair even in the absence of genotoxic stress may be explained by high rate of spontaneous DNA damage, i.e., previous studies suggested that

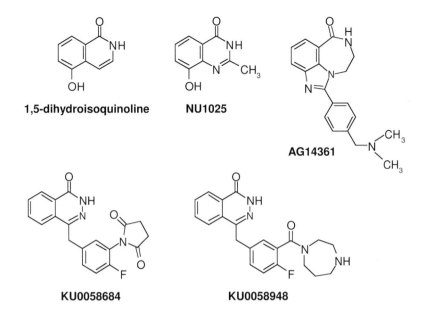

1,5-dihydroisoquinoline   **NU1025**

AG14361

KU0058684   **KU0058948**

Figure 5. Chemical inhibitors of PARP: 1,5-dihydroisoquinoline, NU1025, AG14361, KU0058684, KU0058948.

approximately 10^4 spontaneous SSBs are produced per cell every day [87]. Importantly, RAD51 foci did not form after PARP inhibition in cells lacking BRCA1 or BRCA2 [85]. Thus, the SSBs caused by PARP inhibition are converted into lesions corrected by the BRCA1- and BRCA2-dependent homologous recombination pathway in cells with at least one wild-type allele, but these lesions cannot be repaired in BRCA1- or BRCA2-deficient cells, leading to cell death. Bryant et al. also demonstrated an increase in sensitivity to PARP inhibitors of two cell lines known to be deficient in XRCC2 (irs1 cell line) [88] or XRCC3 (irs1SF) [89], two additional key proteins in the homologous recombination pathway.

Familial breast cancer is often caused by the inactivation of the two BRCA1 or BRCA2 alleles. The first allele is usually inherited defective, while the second one is lost during the lifetime of the individual [90]. The loss of the second allele results in the BRCA$^{(-/-)}$ tumor cells, defective in homologous recombination. At the same time, the BRCA$^{(+/-)}$ surrounding tissue in the patient retains this DNA repair pathway. This raises the intriguing possibility of using small molecule PARP inhibitors as chemotherapeutic agents for breast cancer as these inhibitors would be expected to have a highly cytotoxic, and, importantly, selective effect on tumor cells, producing minimum or no secondary effects. In a tumor model in CD-1 nude mice, xenografts of BRCA2-deficient V-C8 but not of the BRCA2-complimented V-C8-$^+$B2 cell line indeed responded positively to the PARP inhibitor AG14361 [84]. In addition, another PARP inhibitor, KU0058684, prevented the formation of teratocarcinomas in athymic mice by BRCA2-deficient but not wild-type mouse ES cells [85]. The caveat for the therapeutic use of PARP inhibitors is the requirement for their high potency in inhibiting PARP enzymatic activity [91, 92].

The very promising results described above led to the significant effort focused on the use of PARP inhibitors in human patients [93]. The fundamental role of a defective homologous recombination pathway in the sensitivity to PARP inhibitors was recently demonstrated in both, cultured cells and in samples originating from patients with ovarian cancer [93]. CAPAN1 cells express a truncated form of BRCA2 that is defective in DNA repair and the cells are therefore highly sensitive to PARP inhibitors. However, CAPAN1 cells exposed to increasing concentrations of the PARP inhibitor KU0058948 were found to eventually become resistant to this molecule. This acquired resistance was explained by large deletions in the BRCA2 gene that produced a frame shift in the coding region, restoring the expression of the truncated, yet functionally active protein [93]. Transfection of siRNAs for this new form of BRCA2 restored the sensitivity to KU0058948, demonstrating that the short version of the protein was indeed the cause of the resistance [93]. A functionally defective BRCA2 protein also confers high sensitivity to cisplatin [94, 95]. Interestingly, a selection of cisplatin-resistant CAPAN1 clones also yielded cells with diverse intragenic mutations that restored the expression of an almost full-length BRCA2 protein [96]. These cells were also resistant to the PARP inhibitor AG14361, confirming the critical role played by homologous recombination repair in the sensitivity to these two distinct agents.

Regulation of the cell cycle by the DNA damage response

Cell cycle arrest is an essential component of the DNA damage response, as it prevents the expansion of cells with abnormal genetic material. This temporary arrest also provides a window of time to allow one of the numerous DNA repair pathways to correct the alterations before resuming cell proliferation. As we discuss later, one approach to enhance the potency of DNA-targeting chemicals used in chemotherapy is to simultaneously inactivate cell cycle checkpoints.

Overview of the mammalian cell cycle

The increase in cell numbers of normal and cancer cells occurs by mitotic cell division, where several phases can be identified, namely G1 (for "gap 1"), S (synthesis of DNA), G2 (for "gap 2") and M (mitosis) (Fig. 6). A newly originated cell has to gain mass and increase its size during G1, duplicate its genetic material during S, prepare for division in G2, and finally divide all the gained material between two daughter cells during mitosis. These series of

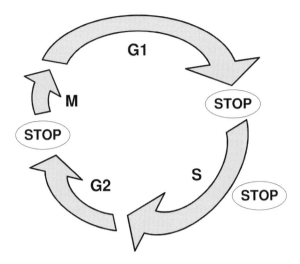

Figure 6. Simplified diagram of the cell cycle. The several phases and main cell cycle checkpoints (represented here by the STOP signals), G1/S, intra-S, and G2/M, are also indicated.

events are repeated in a cyclic manner until the cells withdraw into a quiescent state called G0. Normal cells are driven into the cycle in a regulated manner by the activation of growth factor receptors in the plasma membrane. The signal emanating from growth factors mainly regulate progression through G1, the longest phase of the cell cycle. Cells in early G1 are completely dependent on the presence of growth factors but as they progress through G1 they become independent of the external growth factors. This point is referred to as the "transition point", and molecularly corresponds to the phosphorylation of the retinoblastoma (Rb) protein, a critical regulator of G1 to S transition. Cell cycle progression is driven by the activity of the cyclin-dependent protein kinases (Cdks), which are composed of a catalytic subunit and a regulatory subunit, called cyclin. Cyclins represent the main, although not the only, mechanism controlling the activity of Cdks, as they are synthesized at specific phases of the cell cycle and are degraded by the proteasome after completing their function.

Cyclin D, which pairs with Cdk4 and Cdk6, plays a role in G1, cyclin A and E, which bind to Cdk2, are involved in S phase entry, and cyclin A and B pairing with Cdk1 have a critical role in mitosis. Another level of regulation of Cdk activity is inhibition by protein phosphorylation. The main regulator of mitotic entry, Cdk1, is maintained in an inactive state. Cdk activity in both, G1 to S and G2 to M, requires the involvement of Cdc25, a dual-specificity phosphatase that dephosphorylates tyrosine and threonine residues on Cdks. Three isoforms for Cdc25 phosphatases have been identified in mammalian cells: Cdc25A, B, and C, and all three of them cooperate to regulate normal cell division in mammalian cells [97].

Activation of cell cycle checkpoints by DNA damage

Eukaryotic cells have evolved surveillance mechanisms to ensure that all molecular events corresponding to a specific phase of the cell cycle have been successfully completed before proceeding to the next phase. These mechanisms are called cell cycle checkpoints (Fig. 6) and have the ability to temporarily restrain cell cycle progression to complete the missing steps or to repair any cellular damage [98]. Failure in cell cycle checkpoints would allow the expansion of the population of cells containing permanent genomic alterations. Normal cells possess strong G1, S, and G2 checkpoints and all of them can be activated by DNA damage.

The G1 checkpoint enforces a cell cycle arrest in the G1 phase until the DNA damage is repaired. p53 and Rb proteins play a critical role in this checkpoint. Molecularly, the G1 checkpoint is accomplished by negatively regulating the Cdks CDK4, CDK6 and CDK2, which are essential for G1 progression and the G1-S transition. Mutations in p53 and Rb proteins, which are present in more than half of all human tumors, result in a defective G1 checkpoint. The G1 checkpoint is therefore compromised in cancer cells and they have to depend mainly on the G2 checkpoint (see below). In addition to p53 and Rb, several other tumor suppressor genes are also components of the G1 checkpoint, including p16INK4a, p19ARF, and ATM (Ataxia telangiectasia mutated).

DNA damage and G2 cell cycle arrest

Genotoxic stress also triggers a checkpoint that arrests cells in the G2 phase of the cell cycle. Although multiple mechanisms contribute to this checkpoint, its key aspect is inhibition of the activation of Cdk1 by the Cdc25 protein phosphatases. As the *p53* tumor suppressor gene is mutated in more than half of all human cancers [99], leading to the loss of the G1 checkpoint, the G2 checkpoint is especially critical for the DNA damage response in cancer cells. It has been demonstrated that availability of the intact G1 and G2 checkpoints endow normal cells with the capacity to mount a redundant and more effective response to genotoxic stress, whereas the complete dependency of cancer cells exclusively on the G2 checkpoint makes these cells more vulnerable. This difference between normal and cancer cells has been instrumental in the design of more effective and selective therapies.

The DNA damage triggers a complex protein kinase signaling cascade. In this pathway, the protein kinases ATM and ATR (ATM and Rad-3-related) function as the most upstream components. ATM is activated mainly by DSBs and the recruitment of activated ATM requires the Mre11-Rad50-NBS1 complex. Optimal phosphorylation of ATM substrates, such as Chk2 on Thr68, requires the involvement of several other auxiliary proteins including 53BP1, BRCA1, and MDC1.

Replicative stress caused by stalling of replication forks, as well as DNA cross-linking, nucleotide imbalance and UV light are among the conditions that stimulate the activity of ATR towards Chk1, which in turn phosphorylates and regulates the activity of multiple substrates. The activation of ATR and the phosphorylation of Chk1, observed upon DNA damage, is a complex process, postulated to occur at the sites of the DNA damage and involving a number of additional proteins. The first step in the activation of ATR is mediated by the ATR-interacting protein (ATRIP) that recruits the ATR-ATRIP complex to chromatin through a direct interaction between ATRIP and RPA, a protein that quickly coats exposed single-strand DNA. In turn, activation of Chk1 by ATR requires TopBP1 and claspin, as well as the loading of the sliding clamp Rad9-Hus1-Rad1 by the clamp loader Rad17-RFC (replication factor C). Active Chk1 and Chk2 phosphorylate numerous protein substrates, including all three members of the Cdc25 protein phosphatase family. Cdc25 phosphatases are regulated either through inhibition of their interactions with substrates, by inducing their binding to chaperones, or by induction of their degradation by a proteasome. Cdc25c phosphorylation on Ser216 by Chk1 creates a consensus binding site for the protein 14-3-3, resulting in the sequestration of Cdc25c and inhibition of its interaction with Cdk1-cyclin B [100, 101]. As the Cdc25c-mediated dephosphorylation of Thr14 and Tyr15 in Cdk1 is an absolute requirement for its activation, the DNA damage-induced inactivation of Cdc25c by Chk1 prevents mitotic entry and arrests cells in G2 [102]. Similarly, Chk1 phosphorylates Cdc25a on Ser123, leading to its proteasome-mediated degradation and G2 arrest [103–105].

Abrogation of the G2 checkpoint enhances the cytotoxicity of DNA-damaging agents in cancer but not in normal cells

The defects in p53 observed in many cancer cells confers them with some intrinsic resistance to cytotoxic drugs that induce apoptotic cell death, as p53 is a transcription factor that positively regulates the levels of critical proapoptotic proteins as well as a mediator of apoptosis in a transcription-independent fashion. At the same time, the p53 deficiency constitutes a window of opportunity to design strategies aimed at a synergistic effect between DNA damage and alterations in the G2 checkpoint response. As, in contrast to normal cells, cancer cells lacking p53 are deficient in G1 checkpoint, inhibition of G2 checkpoint represents a very attractive target to improve the efficiency and selectivity of anticancer DNA-damaging agents. The importance of inhibiting the G2 checkpoint to increase drug cytotoxicity was established when it was demonstrated to represent the mechanism behind the synergistic lethal effect of caffeine, an ATM/ATR inhibitor [106], with nitrogen mustard and other DNA-alkylating agents. Caffeine prevented the G2 arrest normally observed in response to the treatment with these agents, allowing cells to enter mitosis without completing DNA repair and resulting in cell death [107, 108]. The key role played by p53 deficiency in sensitization to DNA damage by low concentrations of caffeine was convincingly demonstrated as cell death was observed in p53-deficient but not p53 wild-type mouse embryo fibroblasts [109]. Caffeine also increased the cytotoxicity of etoposide in HeLa cells by abrogating G2 arrest [110]. Pentoxifylline, another methylxanthine similar to caffeine, also showed a selective sensitization of p53 negative breast cancer cells to cisplatin [111]. The p53-deficient prostate tumor cell lines DU145 and BM1604 but not the p53-proficient LNCaP cells showed increased sensitivity to the cytotoxic effects of cisplatin, vinblastin and etoposide when these drugs were applied simultaneously with pentoxifylline [112]. Suppression of the G2 checkpoint with the broad specificity kinase inhibitor UCN-01 (Fig. 7), a blocker of Chk1 [113, 114], Chk2 [115], PDK1 [116], and MK2 [117] also enhanced the toxic effects of DNA damage [117–119].

Involvement of p38 MAPK in G2 arrest induced by DNA damage

Cells possess a second molecular mechanism to enforce a G2 checkpoint in response to DNA damage that functions independently of the ATM/ATR pathway. In addition to more general stresses such as hyperosmotic conditions, the p38 branch of the mitogen-activated protein kinase (MAPK) cascade is activated by DNA-damaging conditions such as UV [120] or doxorubicin [121]. This pathway plays a critical role in the G2 checkpoint caused by these factors, since small molecule p38 inhibitors abrogate G2 arrest [120, 121]. It has been shown that the downstream events regulated by p38 are similar to those for ATM/ATR pathway, i.e., regulation of Ccd25 phosphatases. The UV-induced phosphorylation of Cdc25b on Ser309 is the key event mediating the sequestration of Cdc25b by 14-3-3 proteins and inhibition of Cdk1 dephosphorylation, critical for the mitotic entry [120]. The importance of Cdc25b as a mitotic regulator was demonstrated by the ability of overexpressed Cdc25b but not Cdc25c to induce premature mitotic entry [122]. It has been proposed that the p38 pathway plays an important role in establishing the G2 checkpoint, whereas the Chk1 is more important in the maintenance of G2 arrest upon DNA damage [123].

MAPKAP kinase-2 is essential for the survival of cancer but not normal cells after chemically induced DNA damage

MAPKAP Kinase-2 (MK2) was identified as the kinase functioning downstream of p38 after DNA damage induced by UV [124]. In addition, MK2 is activated by other DNA-damaging agents, including camptothecin, cisplatin and doxorubicin [117]. Using ATM- or ATR-deficient fibroblasts, it was demonstrated that, although UV-mediated activation of p38/MK2 is independent of ATM and ATR, p38/MK2 activation by camptothecin, cisplatin or doxorubicin occurs downstream of ATR, and p38/MK2 activation by doxorubicin also requires ATM [117]. These data suggest complex interplay of p38 and ATM/ATR pathways in DNA damage response. Furthermore, the G2 checkpoint normally induced by doxorubicin in p53-deficient cells was fully dependent on MK2 expression. Abrogation of G2 arrest by inhibiting the expression of MK2 was associated with an increased toxicity of doxorubicin by apoptotic cell death. Curiously, UCN-01 (Fig. 7), typically employed as a Chk1 inhibitor, was found to also block MK2 activity at nanomolar concentrations [117]. This suggests that the beneficial effect of UCN-01 may result from the dual inhibition of Chk1 and MK2. The downstream role of MK2 in both p38 and ATM/ATR pathways following treatment with DNA-damaging agents establish MK2 as a very promising target for the development of DNA damage sensitizers for p53-deficient human cancers. Notably, efficiency of cisplatin and doxorubicin against tumors produced by H-*Ras*-V12-transformed p53$^{(-/-)}$ MEF cells was markedly increased by knocking down MK2 [117]. Equally important, inhibition of MK2 did not have any effect on p53-proficient cells, suggesting potentially high selectivity towards cancer cells.

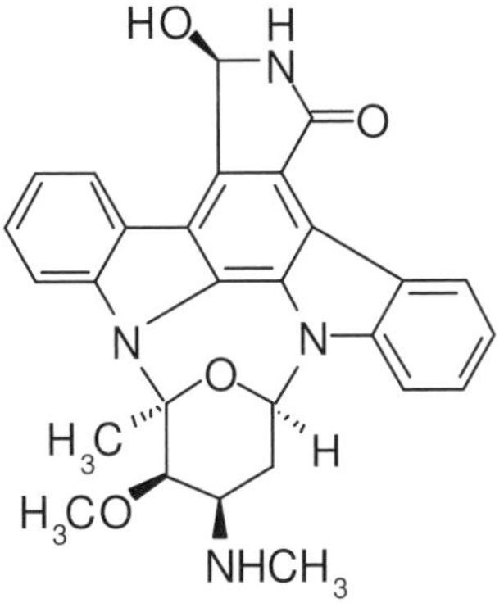

UCN-01

Figure 7. Structure of the G2 checkpoint inhibitor UCN-01. This small molecule has been reported to inhibit the protein kinases Chk1 and more recently, MK2.

DNA damage present in precancerous lesions may constitute a barrier for cancer development

DNA damage is so critical to the origin, maintenance and treatment of cancer, that the unraveling of the molecular mechanisms behind the DNA damage cellular responses is a very active area of research. The activation of the DNA damage checkpoint early on during cancer development [125, 126] points to DNA damage as a direct cause of cancer. This proposal agrees well with the presence of numerous genetic alterations and genetic instability in cancer cells. The inactivation of genes critical to DNA repair pathways, such as BRCA2, has been interpreted as evidence that the accumulation of DNA damage over time is at the core of tumorigenesis. Restoration of BRCA2 in initially BRCA2-defective tumor cells does not induce tumor regression, tentatively because of all the DNA damage already accumulated in cancer cells [96]. Alternatively, the activation of the DNA damage response in early tumorigenesis may constitute a barrier to cancer development [126, 127]. In support of this interpretation, a direct link between the DNA damage response and oncogene-induced senescence has been demonstrated by showing the critical roles played by ATM and Chk2 in this process [128, 129]. As expected, the prevention of oncogene-induced senescence effectively promoted cell transformation [128, 129]. Regardless of the exact role played by DNA damage, the genesis of cancer cells must necessarily be linked to the disabling of this DNA checkpoint, an otherwise very efficient mechanism to limit cell expansion. The most efficient manner by which cancer cells override the DNA checkpoint is by inactivating p53, a transcription factor critical for many of the cellular responses to damaged DNA, including cell cycle arrest, senescence and apoptosis [127]. In addition to directly mutating p53, cancer cells evade the DNA checkpoint by losing the function of one of several upstream activators of p53, such as ATM [130], 53BP1 or MDC [125]. These results are consistent with the idea that the inactivation of the DNA damage checkpoints is an essential prerequisite in the genesis of cancer. Although the precise details are not known, the initial activation of DNA damage checkpoints in precancerous lesions is triggered by DNA damage originating from collapsed replication forks [96].

Understanding the molecular pathways behind the cellular responses to DNA damage and to improve our ability to control these pathways using small molecules is of the utmost importance to understanding not only a fundamental biological process but also to design more rational, effective and selective cancer therapies.

References

1 Wall ME, Wani MC (1995) Camptothecin and taxol: Discovery to clinic – Thirteenth Bruce F. Cain Memorial Award Lecture. *Cancer Res* 55: 753–760
2 Hsiang YH, Hertzberg R, Hecht S, Liu LF (1985) Camptothecin induces protein-linked DNA breaks *via* mammalian DNA topoisomerase I. *J Biol Chem* 260: 14873–14878
3 Eng WK, Faucette L, Johnson RK, Sternglanz R (1988) Evidence that DNA topoisomerase I is necessary for the cytotoxic effects of camptothecin. *Mol Pharmacol* 34: 755–760
4 Nitiss J, Wang JC (1988) DNA topoisomerase-targeting antitumor drugs can be studied in yeast. *Proc Natl Acad Sci USA* 85: 7501–7505
5 Stewart L, Redinbo MR, Qiu X, Hol WG, Champoux JJ (1998) A model for the mechanism of human topoisomerase I. *Science* 279: 1534–1541
6 Holm C, Covey JM, Kerrigan D, Pommier Y (1989) Differential requirement of DNA replication for the cytotoxicity of DNA topoisomerase I and II inhibitors in Chinese hamster DC3F cells. *Cancer Res* 49: 6365–6368
7 Horwitz SB, Horwitz MS (1973) Effects of camptothecin on the breakage and repair of DNA dur-

ing the cell cycle. *Cancer Res* 33: 2834–2836

8 O'Connor PM, Nieves-Neira W, Kerrigan D, Bertrand R, Goldman J, Kohn KW, Pommier Y (1991) S-phase population analysis does not correlate with the cytotoxicity of camptothecin and 10,11-methylenedioxycamptothecin in human colon carcinoma HT-29 cells. *Cancer Commun* 3: 233–240

9 Hsiang YH, Lihou MG, Liu LF (1989) Arrest of replication forks by drug-stabilized topoisomerase I-DNA cleavable complexes as a mechanism of cell killing by camptothecin. *Cancer Res* 49: 5077–5082

10 Wu J, Liu LF (1997) Processing of topoisomerase I cleavable complexes into DNA damage by transcription. *Nucleic Acids Res* 25: 4181–4186

11 Morris EJ, Geller HM (1996) Induction of neuronal apoptosis by camptothecin, an inhibitor of DNA topoisomerase-I: Evidence for cell cycle-independent toxicity. *J Cell Biol* 134: 757–770

12 Borovitskaya AE, D'Arpa P (1998) Replication-dependent and -independent camptothecin cytotoxicity of seven human colon tumor cell lines. *Oncol Res* 10: 271–276

13 Pommier Y, Weinstein JN, Aladjem MI, Kohn KW (2006) Chk2 molecular interaction map and rationale for Chk2 inhibitors. *Clin Cancer Res* 12: 2657–2661

14 Shiloh Y, Lehmann AR (2004) Maintaining integrity. *Nat Cell Biol* 6: 923–928

15 Kastan MB, Bartek J (2004) Cell-cycle checkpoints and cancer. *Nature* 432: 316–323

16 Takemura H, Rao VA, Sordet O, Furuta T, Miao ZH, Meng L, Zhang H, Pommier Y (2006) Defective Mre11-dependent activation of Chk2 by Ataxia telangiectasia mutated in colorectal carcinoma cells in response to replication-dependent DNA double strand breaks. *J Biol Chem* 281: 30814–30823

17 Smith PJ, Makinson TA, Watson JV (1989) Enhanced sensitivity to camptothecin in Ataxia-telangiectasia cells and its relationship with the expression of DNA topoisomerase I. *Int J Radiat Biol* 55: 217–231

18 Johnson MA, Bryant PE, Jones NJ (2000) Isolation of camptothecin-sensitive chinese hamster cell mutants: Phenotypic heterogeneity within the Ataxia telangiectasia-like XRCC8 (irs2) complementation group. *Mutagenesis* 15: 367–374

19 Flatten K, Dai NT, Vroman BT, Loegering D, Erlichman C, Karnitz LM, Kaufmann SH (2005) The role of checkpoint kinase 1 in sensitivity to topoisomerase I poisons. *J Biol Chem* 280: 14349–14355

20 Yu Q, Rose JH, Zhang H, Pommier Y (2001) Antisense inhibition of Chk2/hCds1 expression attenuates DNA damage-induced S and G2 checkpoints and enhances apoptotic activity in HEK-293 cells. *FEBS Lett* 505: 7–12

21 Gupta M, Fan S, Zhan Q, Kohn KW, O'Connor PM, Pommier Y (1997) Inactivation of p53 increases the cytotoxicity of camptothecin in human colon HCT116 and breast MCF-7 cancer cells. *Clin Cancer Res* 3: 1653–1660

22 Han Z, Wei W, Dunaway S, Darnowski JW, Calabresi P, Sedivy J, Hendrickson EA, Balan KV, Pantazis P, Wyche JH (2002) Role of p21 in apoptosis and senescence of human colon cancer cells treated with camptothecin. *J Biol Chem* 277: 17154–17160

23 Squires S, Ryan AJ, Strutt HL, Johnson RT (1993) Hypersensitivity of Cockayne's syndrome cells to camptothecin is associated with the generation of abnormally high levels of double strand breaks in nascent DNA. *Cancer Res* 53: 2012–2019

24 Chen AY, Liu LF (1994) DNA topoisomerases: Essential enzymes and lethal targets. *Annu Rev Pharmacol Toxicol* 34: 191–218

25 Liu LF (1989) DNA topoisomerase poisons as antitumor drugs. *Annu Rev Biochem* 58: 351–375

26 Nitiss JL, Wang JC (1996) Mechanisms of cell killing by drugs that trap covalent complexes between DNA topoisomerases and DNA. *Mol Pharmacol* 50: 1095–1102

27 Tsai-Pflugfelder M, Liu LF, Liu AA, Tewey KM, Whang-Peng J, Knutsen T, Huebner K, Croce CM, Wang JC (1988) Cloning and sequencing of cDNA encoding human DNA topoisomerase II and localization of the gene to chromosome region 17q21–22. *Proc Natl Acad Sci USA* 85: 7177–7181

28 Treszezamsky AD, Kachnic LA, Feng Z, Zhang J, Tokadjian C, Powell SN (2007) BRCA1- and BRCA2-deficient cells are sensitive to etoposide-induced DNA double-strand breaks *via* topoisomerase II. *Cancer Res* 67: 7078–7081

29 Wong E, Giandomenico CM (1999) Current status of platinum-based antitumor drugs. *Chem Rev* 99: 2451–2466

30 Jamieson ER, Lippard SJ (1999) Structure, recognition, and processing of cisplatin-DNA adducts.

Chem Rev 99: 2467–2498

31 Wang D, Lippard SJ (2005) Cellular processing of platinum anticancer drugs. *Nat Rev Drug Discov* 4: 307–320

32 Kartalou M, Essigmann JM (2001) Recognition of cisplatin adducts by cellular proteins. *Mutat Res* 478: 1–21

33 Harrap KR (1985) Preclinical studies identifying carboplatin as a viable cisplatin alternative. *Cancer Treat Rev* 12 Suppl A: 21–33

34 Raynaud FI, Boxall FE, Goddard PM, Valenti M, Jones M, Murrer BA, Abrams M, Kelland LR (1997) cis-Amminedichloro(2-methylpyridine) platinum(II) (AMD473), a novel sterically hindered platinum complex: *In vivo* activity, toxicology, and pharmacokinetics in mice. *Clin Cancer Res* 3: 2063–2074

35 Holford J, Sharp SY, Murrer BA, Abrams M, Kelland LR (1998) *In vitro* circumvention of cisplatin resistance by the novel sterically hindered platinum complex AMD473. *Br J Cancer* 77: 366–373

36 Ishida S, Lee J, Thiele DJ, Herskowitz I (2002) Uptake of the anticancer drug cisplatin mediated by the copper transporter Ctr1 in yeast and mammals. *Proc Natl Acad Sci USA* 99: 14298–14302

37 Holzer AK, Samimi G, Katano K, Naerdemann W, Lin X, Safaei R, Howell SB (2004) The copper influx transporter human copper transport protein 1 regulates the uptake of cisplatin in human ovarian carcinoma cells. *Mol Pharmacol* 66: 817–823

38 Komatsu M, Sumizawa T, Mutoh M, Chen ZS, Terada K, Furukawa T, Yang XL, Gao H, Miura N, Sugiyama T, Akiyama S (2000) Copper-transporting P-type adenosine triphosphatase (ATP7B) is associated with cisplatin resistance. *Cancer Res* 60: 1312–1316

39 Miyashita H, Nitta Y, Mori S, Kanzaki A, Nakayama K, Terada K, Sugiyama T, Kawamura H, Sato A, Morikawa H et al (2003) Expression of copper-transporting P-type adenosine triphosphatase (ATP7B) as a chemoresistance marker in human oral squamous cell carcinoma treated with cisplatin. *Oral Oncol* 39: 157–162

40 Nakayama K, Kanzaki A, Ogawa K, Miyazaki K, Neamati N, Takebayashi Y (2002) Copper-transporting P-type adenosine triphosphatase (ATP7B) as a cisplatin based chemoresistance marker in ovarian carcinoma: Comparative analysis with expression of MDR1, MRP1, MRP2, LRP and BCRP. *Int J Cancer* 101: 488–495

41 Ohbu M, Ogawa K, Konno S, Kanzaki A, Terada K, Sugiyama T, Takebayashi Y (2003) Copper-transporting P-type adenosine triphosphatase (ATP7B) is expressed in human gastric carcinoma. *Cancer Lett* 189: 33–38

42 Cui Y, Konig J, Buchholz JK, Spring H, Leier I, Keppler D (1999) Drug resistance and ATP-dependent conjugate transport mediated by the apical multidrug resistance protein, MRP2, permanently expressed in human and canine cells. *Mol Pharmacol* 55: 929–937

43 Kool M, de Haas M, Scheffer GL, Scheper RJ, van Eijk MJ, Juijn JA, Baas F, Borst P (1997) Analysis of expression of cMOAT (MRP2), MRP3, MRP4, and MRP5, homologues of the multidrug resistance-associated protein gene (MRP1), in human cancer cell lines. *Cancer Res* 57: 3537–3547

44 Koike K, Kawabe T, Tanaka T, Toh S, Uchiumi T, Wada M, Akiyama S, Ono M, Kuwano M (1997) A canalicular multispecific organic anion transporter (cMOAT) antisense cDNA enhances drug sensitivity in human hepatic cancer cells. *Cancer Res* 57: 5475–5479

45 Corda Y, Job C, Anin MF, Leng M, Job D (1991) Transcription by eucaryotic and procaryotic RNA polymerases of DNA modified at a d(GG) or a d(AG) site by the antitumor drug cis-diamminedichloroplatinum(II). *Biochemistry* 30: 222–230

46 Corda Y, Job C, Anin MF, Leng M, Job D (1993) Spectrum of DNA-platinum adduct recognition by prokaryotic and eukaryotic DNA-dependent RNA polymerases. *Biochemistry* 32: 8582–8588

47 Tornaletti S, Patrick SM, Turchi JJ, Hanawalt PC (2003) Behavior of T7 RNA polymerase and mammalian RNA polymerase II at site-specific cisplatin adducts in the template DNA. *J Biol Chem* 278: 35791–35797

48 Jung Y, Lippard SJ (2006) RNA polymerase II blockage by cisplatin-damaged DNA. Stability and polyubiquitylation of stalled polymerase. *J Biol Chem* 281: 1361–1370

49 Damsma GE, Alt A, Brueckner F, Carell T, Cramer P (2007) Mechanism of transcriptional stalling at cisplatin-damaged DNA. *Nat Struct Mol Biol* 14: 1127–1133

50 Zamble DB, Mikata Y, Eng CH, Sandman KE, Lippard SJ (2002) Testis-specific HMG-domain protein alters the responses of cells to cisplatin. *J Inorg Biochem* 91: 451–462

51 Huang JC, Zamble DB, Reardon JT, Lippard SJ, Sancar A (1994) HMG-domain proteins specifi-

cally inhibit the repair of the major DNA adduct of the anticancer drug cisplatin by human excision nuclease. *Proc Natl Acad Sci USA* 91: 10394–10398

52 He Q, Liang CH, Lippard SJ (2000) Steroid hormones induce HMG1 overexpression and sensitize breast cancer cells to cisplatin and carboplatin. *Proc Natl Acad Sci USA* 97: 5768–5772

53 Bruhn SL, Pil PM, Essigmann JM, Housman DE, Lippard SJ (1992) Isolation and characterization of human cDNA clones encoding a high mobility group box protein that recognizes structural distortions to DNA caused by binding of the anticancer agent cisplatin. *Proc Natl Acad Sci USA* 89: 2307–2311

54 Yarnell AT, Oh S, Reinberg D, Lippard SJ (2001) Interaction of FACT, SSRP1, and the high mobility group (HMG) domain of SSRP1 with DNA damaged by the anticancer drug cisplatin. *J Biol Chem* 276: 25736–25741

55 Shaul Y (2000) c-Abl: Activation and nuclear targets. *Cell Death Differ* 7: 10–16

56 Gong JG, Costanzo A, Yang HQ, Melino G, Kaelin WG Jr, Levrero M, Wang JY (1999) The tyrosine kinase c-Abl regulates p73 in apoptotic response to cisplatin-induced DNA damage. *Nature* 399: 806–809

57 Machuy N, Rajalingam K, Rudel T (2004) Requirement of caspase-mediated cleavage of c-Abl during stress-induced apoptosis. *Cell Death Differ* 11: 290–300

58 Kharbanda S, Pandey P, Yamauchi T, Kumar S, Kaneki M, Kumar V, Bharti A, Yuan ZM, Ghanem L, Rana A et al (2000) Activation of MEK kinase 1 by the c-Abl protein tyrosine kinase in response to DNA damage. *Mol Cell Biol* 20: 4979–4989

59 Pandey P, Raingeaud J, Kaneki M, Weichselbaum R, Davis RJ, Kufe D, Kharbanda S (1996) Activation of p38 mitogen-activated protein kinase by c-Abl-dependent and -independent mechanisms. *J Biol Chem* 271: 23775–23779

60 Aird RE, Cummings J, Ritchie AA, Muir M, Morris RE, Chen H, Sadler PJ, Jodrell DI (2002) *In vitro* and *in vivo* activity and cross resistance profiles of novel ruthenium (II) organometallic arene complexes in human ovarian cancer. *Br J Cancer* 86: 1652–1657

61 Hartinger CG, Zorbas-Seifried S, Jakupec MA, Kynast B, Zorbas H, Keppler BK (2006) From bench to bedside – Preclinical and early clinical development of the anticancer agent indazolium trans-[tetrachlorobis(1*H*-indazole)ruthenate(III)] (KP1019 or FFC14A). *J Inorg Biochem* 100: 891–904

62 Bergamo A, Sava G (2007) Ruthenium complexes can target determinants of tumour malignancy. *Dalton Trans* 13: 1267–1272

63 d'Adda di Fagagna F, Hande MP, Tong WM, Lansdorp PM, Wang ZQ, Jackson SP (1999) Functions of poly(ADP-ribose) polymerase in controlling telomere length and chromosomal stability. *Nat Genet* 23: 76–80

64 Jagtap P, Szabo C (2005) Poly(ADP-ribose) polymerase and the therapeutic effects of its inhibitors. *Nat Rev Drug Discov* 4: 421–440

65 Dantzer F, Schreiber V, Niedergang C, Trucco C, Flatter E, De La Rubia G, Oliver J, Rolli V, Menissier-de Murcia J, de Murcia G (1999) Involvement of poly(ADP-ribose) polymerase in base excision repair. *Biochimie* 81: 69–75

66 Boulton S, Kyle S, Durkacz BW (1999) Interactive effects of inhibitors of poly(ADP-ribose) polymerase and DNA-dependent protein kinase on cellular responses to DNA damage. *Carcinogenesis* 20: 199–203

67 El-Khamisy SF, Masutani M, Suzuki H, Caldecott KW (2003) A requirement for PARP-1 for the assembly or stability of XRCC1 nuclear foci at sites of oxidative DNA damage. *Nucleic Acids Res* 31: 5526–5533

68 D'Silva I, Pelletier JD, Lagueux J, D'Amours D, Chaudhry MA, Weinfeld M, Lees-Miller SP, Poirier GG (1999) Relative affinities of poly(ADP-ribose) polymerase and DNA-dependent protein kinase for DNA strand interruptions. *Biochim Biophys Acta* 1430: 119–126

69 Noel G, Giocanti N, Fernet M, Megnin-Chanet F, Favaudon V (2003) Poly(ADP-ribose) polymerase (PARP-1) is not involved in DNA double-strand break recovery. *BMC Cell Biol* 4: 7–17

70 Yang YG, Cortes U, Patnaik S, Jasin M, Wang ZQ (2004) Ablation of PARP-1 does not interfere with the repair of DNA double-strand breaks, but compromises the reactivation of stalled replication forks. *Oncogene* 23: 3872–3882

71 de Murcia JM, Niedergang C, Trucco C, Ricoul M, Dutrillaux B, Mark M, Oliver FJ, Masson M, Dierich A, LeMeur M et al (1997) Requirement of poly(ADP-ribose) polymerase in recovery from DNA damage in mice and in cells. *Proc Natl Acad Sci USA* 94: 7303–7307

72 Wang ZQ, Stingl L, Morrison C, Jantsch M, Los M, Schulze-Osthoff K, Wagner EF (1997) PARP

is important for genomic stability but dispensable in apoptosis. *Genes Dev* 11: 2347–2358

73 Conde C, Mark M, Oliver FJ, Huber A, de Murcia G, Menissier-de Murcia J (2001) Loss of poly(ADP-ribose) polymerase-1 causes increased tumour latency in p53-deficient mice. *EMBO J* 20: 3535–3543

74 Schultz N, Lopez E, Saleh-Gohari N, Helleday T (2003) Poly(ADP-ribose) polymerase (PARP-1) has a controlling role in homologous recombination. *Nucleic Acids Res* 31: 4959–4964

75 Arnaudeau C, Lundin C, Helleday T (2001) DNA double-strand breaks associated with replication forks are predominantly repaired by homologous recombination involving an exchange mechanism in mammalian cells. *J Mol Biol* 307: 1235–1245

76 Haber JE (1999) DNA recombination: The replication connection. *Trends Biochem Sci* 24: 271–275

77 Symington LS (2005) Focus on recombinational DNA repair. *EMBO Rep* 6: 512–517

78 Tutt A, Ashworth A (2002) The relationship between the roles of BRCA genes in DNA repair and cancer predisposition. *Trends Mol Med* 8: 571–576

79 Moynahan ME, Pierce AJ, Jasin M (2001) BRCA2 is required for homology-directed repair of chromosomal breaks. *Mol Cell* 7: 263–272

80 Moynahan ME, Chiu JW, Koller BH, Jasin M (1999) *Brca1* controls homology-directed DNA repair. *Mol Cell* 4: 511–518

81 Davies OR, Pellegrini L (2007) Interaction with the BRCA2 C terminus protects RAD51-DNA filaments from disassembly by BRC repeats. *Nat Struct Mol Biol* 14: 475–483

82 Esashi F, Galkin VE, Yu X, Egelman EH, West SC (2007) Stabilization of RAD51 nucleoprotein filaments by the C-terminal region of BRCA2. *Nat Struct Mol Biol* 14: 468–474

83 Gudmundsdottir K, Ashworth A (2006) The roles of BRCA1 and BRCA2 and associated proteins in the maintenance of genomic stability. *Oncogene* 25: 5864–5874

84 Bryant HE, Schultz N, Thomas HD, Parker KM, Flower D, Lopez E, Kyle S, Meuth M, Curtin NJ, Helleday T (2005) Specific killing of BRCA2-deficient tumours with inhibitors of poly(ADP-ribose) polymerase. *Nature* 434: 913–917

85 Farmer H, McCabe N, Lord CJ, Tutt AN, Johnson DA, Richardson TB, Santarosa M, Dillon KJ, Hickson I, Knights C et al (2005) Targeting the DNA repair defect in BRCA mutant cells as a therapeutic strategy. *Nature* 434: 917–921

86 Kraakman-van der Zwet M, Overkamp WJ, van Lange RE, Essers J, van Duijn-Goedhart A, Wiggers I, Swaminathan S, van Buul PP, Errami A et al (2002) *Brca2* (XRCC11) deficiency results in radioresistant DNA synthesis and a higher frequency of spontaneous deletions. *Mol Cell Biol* 22: 669–679

87 Lindahl T (1993) Instability and decay of the primary structure of DNA. *Nature* 362: 709–715

88 Griffin CS, Simpson PJ, Wilson CR, Thacker J (2000) Mammalian recombination-repair genes XRCC2 and XRCC3 promote correct chromosome segregation. *Nat Cell Biol* 2: 757–761

89 Tebbs RS, Zhao Y, Tucker JD, Scheerer JB, Siciliano MJ, Hwang M, Liu N, Legerski RJ, Thompson LH (1995) Correction of chromosomal instability and sensitivity to diverse mutagens by a cloned cDNA of the XRCC3 DNA repair gene. *Proc Natl Acad Sci USA* 92: 6354–6358

90 Venkitaraman AR (2002) Cancer susceptibility and the functions of BRCA1 and BRCA2. *Cell* 108: 171–182

91 Gallmeier E, Kern SE (2005) Absence of specific cell killing of the BRCA2-deficient human cancer cell line CAPAN1 by poly(ADP-ribose) polymerase inhibition. *Cancer Biol Ther* 4: 703–706

92 McCabe N, Lord CJ, Tutt AN, Martin NM, Smith GC, Ashworth A (2005) BRCA2-deficient CAPAN-1 cells are extremely sensitive to the inhibition of poly (ADP-ribose) polymerase: An issue of potency. *Cancer Biol Ther* 4: 934–936

93 Edwards SL, Brough R, Lord CJ, Natrajan R, Vatcheva R, Levine DA, Boyd J, Reis-Filho JS, Ashworth A (2008) Resistance to therapy caused by intragenic deletion in BRCA2. *Nature* 451: 1111–1115

94 Yuan SS, Lee SY, Chen G, Song M, Tomlinson GE, Lee EY (1999) BRCA2 is required for ionizing radiation-induced assembly of Rad51 complex *in vivo*. *Cancer Res* 59: 3547–3551

95 Bhattacharyya A, Ear US, Koller BH, Weichselbaum RR, Bishop DK (2000) The breast cancer susceptibility gene BRCA1 is required for subnuclear assembly of Rad51 and survival following treatment with the DNA cross-linking agent cisplatin. *J Biol Chem* 275: 23899–23903

96 Sakai W, Swisher EM, Karlan BY, Agarwal MK, Higgins J, Friedman C, Villegas E, Jacquemont C, Farrugia DJ, Couch FJ et al (2008) Secondary mutations as a mechanism of cisplatin resistance in BRCA2-mutated cancers. *Nature* 451: 1116–1120

97 Lindqvist A, Kallstrom H, Lundgren A, Barsoum E, Rosenthal CK (2005) Cdc25B cooperates with Cdc25A to induce mitosis but has a unique role in activating cyclin B1-Cdk1 at the centrosome. *J Cell Biol* 171: 35–45

98 Paulovich AG, Toczyski DP, Hartwell LH (1997) When checkpoints fail. *Cell* 88: 315–321

99 Hollstein M, Sidransky D, Vogelstein B, Harris CC (1991) P53 mutations in human cancers. *Science* 253: 49–53

100 Peng CY, Graves PR, Thoma RS, Wu Z, Shaw AS, Piwnica-Worms H (1997) Mitotic and G2 checkpoint control: Regulation of 14-3-3 protein binding by phosphorylation of Cdc25C on serine-216. *Science* 277: 1501–1505

101 Sanchez Y, Wong C, Thoma RS, Richman R, Wu Z, Piwnica-Worms H, Elledge SJ (1997) Conservation of the Chk1 checkpoint pathway in mammals: Linkage of DNA damage to Cdk regulation through Cdc25. *Science* 277: 1497–1501

102 Abraham RT (2001) Cell cycle checkpoint signaling through the ATM and ATR kinases. *Genes Dev* 15: 2177–2196

103 Mailand N, Podtelejnikov AV, Groth A, Mann M, Bartek J, Lukas J (2002) Regulation of G(2)/M events by Cdc25A through phosphorylation-dependent modulation of its stability. *EMBO J* 21: 5911–5920

104 Zhao H, Watkins JL, Piwnica-Worms H (2002) Disruption of the checkpoint kinase 1/cell division cycle 25A pathway abrogates ionizing radiation-induced S and G2 checkpoints. *Proc Natl Acad Sci USA* 99: 14795–14800

105 Xiao Z, Chen Z, Gunasekera AH, Sowin TJ, Rosenberg SH, Fesik S, Zhang H (2003) Chk1 mediates S and G2 arrests through Cdc25A degradation in response to DNA-damaging agents. *J Biol Chem* 278: 21767–21773

106 Sarkaria JN, Busby EC, Tibbetts RS, Roos P, Taya Y, Karnitz LM, Abraham RT (1999) Inhibition of ATM and ATR kinase activities by the radiosensitizing agent, caffeine. *Cancer Res* 59: 4375–4382

107 Lau CC, Pardee AB (1982) Mechanism by which caffeine potentiates lethality of nitrogen mustard. *Proc Natl Acad Sci USA* 79: 2942–2946

108 Fingert HJ, Chang JD, Pardee AB (1986) Cytotoxic, cell cycle, and chromosomal effects of methylxanthines in human tumor cells treated with alkylating agents. *Cancer Res* 46: 2463–2467

109 Powell SN, DeFrank JS, Connell P, Eogan M, Preffer F, Dombkowski D, Tang W, Friend S (1995) Differential sensitivity of p53$^{(-)}$ and p53$^{(+)}$ cells to caffeine-induced radiosensitization and override of G2 delay. *Cancer Res* 55: 1643–1648

110 Lock RB, Galperina OV, Feldhoff RC, Rhodes LJ (1994) Concentration-dependent differences in the mechanisms by which caffeine potentiates etoposide cytotoxicity in HeLa cells. *Cancer Res* 54: 4933–4939

111 Fan S, Smith ML, Rivet DJ, 2nd, Duba D, Zhan Q, Kohn KW, Fornace AJ Jr, O'Connor PM (1995) Disruption of p53 function sensitizes breast cancer MCF-7 cells to cisplatin and pentoxifylline. *Cancer Res* 55: 1649–1654

112 Serafin AM, Binder AB, Bohm L (2001) Chemosensitivity of prostatic tumour cell lines under conditions of G2 block abrogation. *Urol Res* 29: 221–227

113 Busby EC, Leistritz DF, Abraham RT, Karnitz LM, Sarkaria JN (2000) The radiosensitizing agent 7-hydroxystaurosporine (UCN-01) inhibits the DNA damage checkpoint kinase hChk1. *Cancer Res* 60: 2108–2112

114 Graves PR, Yu L, Schwarz JK, Gales J, Sausville EA, O'Connor PM, Piwnica-Worms H (2000) The Chk1 protein kinase and the Cdc25C regulatory pathways are targets of the anticancer agent UCN-01. *J Biol Chem* 275: 5600–5605

115 Yu Q, La Rose J, Zhang H, Takemura H, Kohn KW, Pommier Y (2002) UCN-01 inhibits p53 upregulation and abrogates gamma-radiation-induced G(2)-M checkpoint independently of p53 by targeting both of the checkpoint kinases, Chk2 and Chk1. *Cancer Res* 62: 5743–5748

116 Sato S, Fujita N, Tsuruo T (2002) Interference with PDK1-Akt survival signaling pathway by UCN-01 (7-hydroxystaurosporine). *Oncogene* 21: 1727–1738

117 Reinhardt HC, Aslanian AS, Lees JA, Yaffe MB (2007) P53-deficient cells rely on ATM- and ATR-mediated checkpoint signaling through the p38MAPK/MK2 pathway for survival after DNA damage. *Cancer Cell* 11: 175–189

118 Wang Q, Fan S, Eastman A, Worland PJ, Sausville EA, O'Connor PM (1996) UCN-01: A potent abrogator of G2 checkpoint function in cancer cells with disrupted p53. *J Natl Cancer Inst* 88: 956–965

119 Yao SL, Akhtar AJ, McKenna KA, Bedi GC, Sidransky D, Mabry M, Ravi R, Collector MI, Jones RJ, Sharkis SJ et al (1996) Selective radiosensitization of p53-deficient cells by caffeine-mediated activation of p34cdc2 kinase. *Nat Med* 2: 1140–1143

120 Bulavin DV, Higashimoto Y, Popoff IJ, Gaarde WA, Basrur V, Potapova O, Appella E, Fornace AJ Jr, (2001) Initiation of a G2/M checkpoint after ultraviolet radiation requires p38 kinase. *Nature* 411: 102–107

121 Mikhailov A, Shinohara M, Rieder CL (2004) Topoisomerase II and histone deacetylase inhibitors delay the G2/M transition by triggering the p38 MAPK checkpoint pathway. *J Cell Biol* 166: 517–526

122 Karlsson C, Katich S, Hagting A, Hoffmann I, Pines J (1999) Cdc25B and Cdc25C differ markedly in their properties as initiators of mitosis. *J Cell Biol* 146: 573–584

123 Bulavin DV, Amundson SA, Fornace AJ (2002) P38 and Chk1 kinases: Different conductors for the G(2)/M checkpoint symphony. *Curr Opin Genet Dev* 12: 92–97

124 Manke IA, Nguyen A, Lim D, Stewart MQ, Elia AE, Yaffe MB (2005) MAPKAP kinase-2 is a cell cycle checkpoint kinase that regulates the G2/M transition and S phase progression in response to UV irradiation. *Mol Cell* 17: 37–48

125 Gorgoulis VG, Vassiliou LV, Karakaidos P, Zacharatos P, Kotsinas A, Liloglou T, Venere M, Ditullio RA Jr, Kastrinakis NG, Levy B et al (2005) Activation of the DNA damage checkpoint and genomic instability in human precancerous lesions. *Nature* 434: 907–913

126 Bartkova J, Horejsi Z, Koed K, Kramer A, Tort F, Zieger K, Guldberg P, Sehested M, Nesland JM, Lukas C et al (2005) DNA damage response as a candidate anti-cancer barrier in early human tumorigenesis. *Nature* 434: 864–870

127 Halazonetis TD, Gorgoulis VG, Bartek J (2008) An oncogene-induced DNA damage model for cancer development. *Science* 319: 1352–1355

128 Bartkova J, Rezaei N, Liontos M, Karakaidos P, Kletsas D, Issaeva N, Vassiliou LV, Kolettas E, Niforou K, Zoumpourlis VC et al (2006) Oncogene-induced senescence is part of the tumorigenesis barrier imposed by DNA damage checkpoints. *Nature* 444: 633–637

129 Di Micco R, Fumagalli M, Cicalese A, Piccinin S, Gasparini P, Luise C, Schurra C, Garre M, Nuciforo PG, Bensimon A et al (2006) Oncogene-induced senescence is a DNA damage response triggered by DNA hyper-replication. *Nature* 444: 638–642

130 Greenman C, Stephens P, Smith R, Dalgliesh GL, Hunter C, Bignell G, Davies H, Teague J, Butler A, Stevens C et al (2007) Patterns of somatic mutation in human cancer genomes. *Nature* 446: 153–158

Molecular, Clinical and Environmental Toxicology. Volume 1: Molecular Toxicology
Edited by A. Luch

The aryl hydrocarbon receptor at the crossroads of multiple signaling pathways

Ci Ma[1], Jennifer L. Marlowe[2] and Alvaro Puga[1]

[1] *Department of Environmental Health and Center for Environmental Genetics, University of Cincinnati College of Medicine, Cincinnati, OH, USA*
[2] *Novartis Pharma AG, Muttenz, Switzerland*

Abstract. The aryl hydrocarbon receptor (AHR) has long been recognized as a ligand-activated transcription factor responsible for the induction of drug-metabolizing enzymes. Its role in the combinatorial matrix of cell functions was established long before the first report of an AHR cDNA sequence was published. It is only recently that other functions of this protein have begun to be recognized, and it is now clear that the AHR also functions in pathways outside of its well-characterized role in xenobiotic enzyme induction. Perturbation of these pathways by xenobiotic ligands may ultimately explain much of the toxicity of these compounds. This chapter focuses on the interactions of the AHR in pathways critical to cell cycle regulation, mitogen-activated protein kinase cascades, differentiation and apoptosis. Ultimately, the effect of a particular AHR ligand on the biology of the organism will depend on the milieu of critical pathways and proteins expressed in specific cells and tissues with which the AHR itself interacts.

Introduction

The aryl hydrocarbon receptor

The aryl hydrocarbon (dioxin) receptor (AHR) is a cytosolic ligand-activated transcription factor known to mediate a large number of toxic and carcinogenic effects in animals and possibly in humans [1, 2]. It is generally accepted that activation of the cytosolic AHR mediates the toxic and carcinogenic effects of a wide variety of environmental contaminants such as dioxin (2,3,7,8-tetrachlorodibenzo-*p*-dioxin, TCDD), coplanar polychlorinated biphenyls (PCBs) and polycyclic or halogenated aromatic hydrocarbons (PAHs, HAHs) in higher organisms. As a consequence of AHR activation by its ligands, many detoxification genes are transcriptionally induced, including those coding for phase I xenobiotic-metabolizing enzymes, such as the cytochromes P450 CYP1A1, CYP1A2, CYP1B1, and CYP2S1, and the phase II enzymes UDP-glucuronosyltransferase UGT1A6, NAD(P)H-dependent quinone oxidoreductase-1 (NQO1), the aldehyde dehydrogenase ALDH3A1, and several glutathione-*S*-transferases. The AHR belongs to the bHLH/PAS family of heterodimeric transcriptional regulators [basic-region helix-loop-helix/period (Per)-aryl hydrocarbon receptor nuclear translocator (ARNT)-Single minded (SIM)] [3] invol-

ved in regulation of development [4] and in the control of diverse physiological processes, including circadian rhythm, neurogenesis, metabolism and
stress response to hypoxia. Evidence from AHR knockout mice, however,
points to functions of the AHR beyond xenobiotic metabolism at several physiological roles that may contribute to AHR-mediated toxic responses. Ablation
of the *Ahr* gene in mice leads to cardiovascular disease, hepatic fibrosis,
reduced liver size, spleen T cell deficiency, dermal fibrosis, liver retinoid accumulation and shortening of life span [5], suggesting that it has biological functions other than xenobiotic detoxification that might contribute to the overall
toxicity resulting from receptor activation.

Mechanism of action of the AHR

The AHR is ubiquitously expressed in mouse tissues [6]. In humans, expression is high in lung, thymus, kidney and liver. In the absence of ligand, the
cytosolic AHR is complexed with two HSP90 chaperone molecules, the
HSP90-interacting protein p23 and the immunophilin-like protein XAP2
(AIP, ARA9) [7–9]. AHR activation by ligand is followed by changes in its
compartmentalization within the cell. The activated receptor complex translocates into the nucleus, dissociates from the chaperones and heterodimerizes
with the AHR nuclear translocator (ARNT). The heterodimer interacts with
several histone acetyltransferases and chromatin remodeling factors [10–14],
and the resulting complex binds to consensus regulatory sequences termed
AHRE (XRE, DRE) located in the promoter of target genes and, by mechanisms still loosely characterized, recruits RNA polymerase II to initiate transcription. The activated AHR is quickly exported to the cytosol where it is
degraded by the 26S proteasome [15], hence preventing its permanent activation. Figure 1 depicts a schematic representation of the main features of lig-
and-mediated AHR activation and the recruitment of the receptor to the promoters of target genes.
    Activation of the AHR by high-affinity HAH or PAH ligands such as
TCDD and benzo[*a*]pyrene (B[*a*]P) results in a wide range of cell cycle perturbations, including G0/G1 and G2/M arrest, diminished capacity for DNA
replication, and inhibition of cell proliferation. Functions of the AHR related
to cell cycle control are often also accomplished in the absence of an exogenous ligand, but their underlying molecular mechanisms remain in part elusive because no definitive endogenous ligands have been identified (reviewed
in [16]). At present, all available evidence indicates that the AHR can trigger
signal transduction pathways involved in proliferation, differentiation or
apoptosis, by mechanisms dependent on xenobiotic ligands or on endogenous
activities that may be ligand mediated or completely ligand independent.
These functions of the AHR coexist with its classical toxicological function
involving the induction of phase I and phase II genes for the detoxification of
foreign compounds.

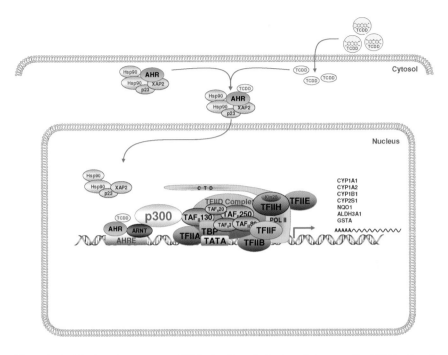

Figure 1. Aryl hydrocarbon receptor (AHR) signaling. Activation of the cytosolic AHR by ligand leads to its nuclear translocation, release from the HSP90 chaperones and binding to specific DNA motifs in a heterodimer with AHR nuclear translocator (ARNT). Interaction with the histone acetyltransferase p300 and various other histone modifying enzymes leads to activation of the transcriptional machinery and recruitment of RNA polymerase II and TFII factors to the promoter of a number of genes in the AHR battery. As a result, transcription of those genes is initiated.

In this chapter, we address novel experimental evidence relating to these less orthodox AHR functions, focusing on new data appearing since our previous review of this subject [16]. The topics discussed encompass the role of the AHR in the activation of mitogen-activated protein kinases (MAPKs), cell cycle regulation, apoptosis and cell differentiation, with a focus on the crosstalk between AHR signaling pathways and the effectors, regulatory events and cell cycle checkpoints responsible for normal cellular functions.

Cross-talk of cellular kinases with the AHR

AHR phosphorylation

Post-translational modifications, such as phosphorylation, play a major role in the regulation of gene expression and protein function in eukaryotic cells. This covalent modification controls intracellular distribution, transcriptional activity and stability of growth hormone receptors and transcription factors, including the AHR, and the physiological activity of a list of proteins too large to be discussed within the scope of this chapter. The reader is directed to a number of previous reviews covering this subject [16, 17].

In silico analysis of the primary structure of the AHR indicates that this protein contains a multiplicity of potential phosphorylation sites (Fig. 2), but evidence of their phosphorylation and of the functional role that phosphorylation of these sites plays in receptor activity is limited. Several studies have shown that the protein kinase C (PKC) pathway is required for AHR activity [18–20], as specific inhibition of PKC blocks ligand-induced DNA binding of AHR/ARNT heterodimers and leads to suppressed *Cyp1* gene expression [18, 21, 22]. In addition, serine/threonine phosphatase inhibitors increase AHR-mediated gene transcription [23]. However, the precise signaling mechanism by which PKC regulates AHR complex activity remains elusive. AHR phosphorylation is required for DNA binding and transcriptional activity of the receptor [18, 24–26]. On the other hand, inhibition of serine/threonine phosphatase activity can remarkably increase AHR-directed gene expression, suggesting the involvement of serine/threonine protein kinases in the activation of the AHR complex [23, 27]. In agreement with the latter findings, recent work from Ikuta and colleagues [28] has identified a set of PKC-dependent phosphorylation events that decrease AHR activity. The AHR protein has both a nuclear localization signal (NLS) and a nuclear export signal (NES), which play important roles in the AHR translocation and intracellular distribution. Ikuta et al. [28] have shown that the NLS, comprised of amino acid residues 13–39, consists of two separate basic amino acid segments, one consisting of residues 13–16 (Arg-Lys-Arg-Arg) and the other spanning residues 37–39 (Lys-Arg-His). Ligand-dependent AHR nuclear import is inhibited by phosphorylation of either of the two serine residues, Ser12 or Ser36, each located one amino acid upstream from either of the two segments. Replacement of these Ser sites with Ala did not affect receptor translocation but their replace-

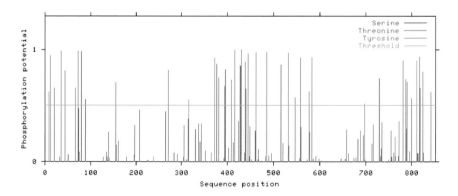

Figure 2. Phosphorylation sites in the AHR. Potential phosphorylation sites in serine, threonine and tyrosine residues were identified by the program NetPhos (http://www.cbs.dtu.dk/services/NetPhos/).

ment with Asp, which mimics the negative charge of phosphorylation, resulted in the retention of the mutant AHR in the cytoplasm. These observations
were supported by *in vitro* nuclear transport assays and a luciferase reporter
assay in which alanine and aspartic acid replacement mutants had much lower
transcriptional activity than the wide-type receptor, suggesting a two-step
mechanism in ligand-dependent nuclear translocation of the AHR.

Phosphorylation also plays an essential role in the transformation of the
unliganded AHR into a functionally active AHR/ARNT heterodimer.
Chemical cleavage patterns have localized phosphorylation sites in two
domains in the C-terminal half of the AHR. These regions include four phosphotyrosine residues (Tyr372, Tyr408, Tyr462, and Tyr532) within residues
368–605, and one more (Tyr698) within residues 639–759 that are highly
phosphorylated *in vivo* [25]. Additionally, an N-terminal tyrosine residue,
Tyr9, although not itself phosphorylated, is essential for proper recognition of
the AHR for PKC-dependent phosphorylation, binding of the AHR to its cognate DNA sequence and for full receptor transcriptional activity [29].

Serine/threonine kinases are found in association with HSP90 complexes in
the cytosol [30], which has prompted the search for the possible consequences
of co-chaperone phosphorylation on receptor functions. Ogiso and coworkers
[31] have shown that the phosphorylation of HSP90 modulates the formation
of the functional cytosolic AHR multiprotein complex. These authors used
mass spectrometry to determine the site-specific phosphorylation of the
steady-state cytosolic AHR complex prepared from Chinese hamster ovary
cells stably expressing mouse AHR. They identified phosphorylation of the
HSP90 subunits at Ser225 and Ser254 of HSP90β and Ser230 of HSP90α.
When these serine residues were substituted with alanine and glutamic acid,
replacement of Ser225 and Ser254 by alanine increased the binding affinity for
AHR, which exhibited more potent transcriptional activity than when the serine was replaced with glutamic acid, suggesting that phosphorylation of the
charged linker region of the HSP90 molecule modulates the formation of the
functional cytosolic AHR complex [31].

Cross-talk of the AHR with MAPK pathways

The three families of MAPKs, extracellular signal-regulated kinases
(ERK1/2), c-Jun N-terminal/stress-activated protein kinases (JNK/SAPK) and
the p38s are important intracellular signal transduction mediators. They are
involved in the control of gene expression and various other events in eukaryotic cells through the phosphorylation of transcription factors and modulation
of their function. MAPKs can phosphorylate a large panel of substrates on serine and threonine residues directly or *via* down-stream MAPK-activated protein kinases. MAPK activities are controlled by the MAPKKK-MAPKK signaling cascades [32] in which the MAPKs are activated by phosphorylation by
MAPK kinases, which in turn are activated by MAPKK kinase-dependent

phosphorylation. As a general rule, ERK1/2 are involved in regulating mitogenic and developmental events and the four p38 kinase isoforms play important roles in inflammatory responses, apoptosis and cell cycle regulation. The
three JNK isoforms are involved in functions ranging from cellular signaling,
the immune system, stress-induced and developmentally programmed apoptosis, to carcinogenesis and the pathogenesis of diabetes [33]. Figure 3 depicts a
simplified overview of the MAPK cascade described above.

TCDD has been shown to activate ERK and JNK [34], but TCDD-stimulated MAPKs do not converge on the induction of the transcription activity of
Elk-1 or c-Jun, the well-known nuclear targets of the ERKs and the JNKs,
respectively. Instead, they are required for AHR activity and AHR-dependent
gene expression [34]. Hence, MAPK activation may represent an alternative
mechanism by which TCDD regulates AHR function, contributing to the
diversity of TCDD-dependent toxic effects in a cell-lineage and gene-specific
manner [35].

Three well-characterized AHR ligands, TCDD, B[a]P and B[a]P diol-epoxide (BPDE), activated JNK in mouse hepatoma Hepa-1 cells, human lung carcinoma A549 cells, AHR-negative CV-1 cells and in both AHR-positive and
AHR-negative mouse embryonic fibroblasts, suggesting that MAPK activation
by TCDD does not require the AHR. However, TCDD-stimulated MAPKs

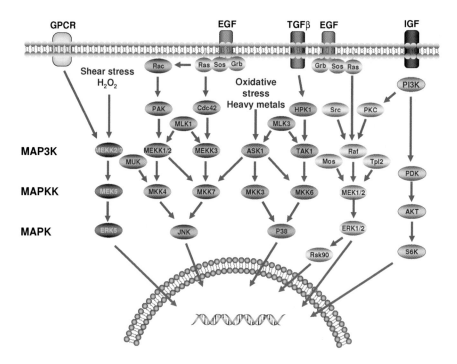

Figure 3. The mitogen-activated protein kinase (MAPK) cascade. Shown is a simplified view of the
interconnecting MAPK cascade and several of its inducers and signaling effectors at the cell surface.

may be critical for the induction of AHR-dependent gene transcription and CYP1A1 expression [34], indicating that AHR ligands elicit AHR-independent epigenetic events that are essential for AHR activation and function.

AHR activation by TCDD or by another AHR ligand, 3-methylcholanthrene, has recently been shown to induce morphological changes that modulate cell plasticity in epithelial cells through a JNK-dependent mechanism [36]. Prolonged treatment with either of these ligands causes these cells to undergo prominent cytoskeletal changes resulting from increased interaction with the extracellular matrix while weakening cell-cell contacts. This promigratory activity of TCDD and 3-methylcholanthrene on epithelial cells correlated with activation of JNK and could be reversed by an inhibitor of JNK activity. Furthermore, all these dioxin-mediated effects were mimicked by constitutive expression and activation of the AHR. Many ligands of the AHR are well-known tumor-promoting agents, and these novel effects of AHR action on cell plasticity point to a possible role for this protein in cancer progression mediated by its ligands.

c-JUN is a down-stream JNK target found to be induced by TCDD [37, 38]. Work with rat hepatoma 5L cells expressing the AHR has shown induction of c-JUN *via* an unconventional AHR-mediated p38-dependent pathway [39] that is unrelated to JNK activity. Weiss and colleagues [39] showed that TCDD causes a persistent increase of c-Jun mRNA levels that leads to an increase in AP-1 levels and activity. This group used an Elk-1-responsive reporter gene as a sensitive assay for p38 activation and showed that TCDD induced p38 phosphorylation. Moreover, Elk-1 activation was undetectable in BP8 cells, an AHR-deficient subclone of the 5L line. None of the kinases known to phosphorylate p38 were activated by the AHR, although phosphorylation did not require transcription activation, suggesting a novel activation mechanism not currently understood. In an extension of this work, Weiss and colleagues have recently reported the induction of JUND by TCDD in rat oval cells [40], an effect that was also reported from our laboratory in mouse hepatoma cells [37, 38].

Several studies have explored the cross-talk between the p38 and AHR signaling pathways and the role of p38 itself in AHR signaling. In human keratinocytes, cell density appears to be critical in determining AHR subcellular distribution, with a predominant cytoplasmic localization at confluence and nuclear localization at sparse cell densities. Ikuta and coworkers [41] found that nuclear accumulation of the activated AHR was associated with p38-dependent phosphorylation of Ser68 in the NES and that nuclear export was suppressed by substitution of Ser68 for aspartic acid. These findings strongly suggest the existence of a functional relationship between cell density, intracellular localization and AHR activity that may have a pivotal role in AHR function.

Activation of p38, and possibly other MAPKs, by AHR ligands seems to be a cell type-specific consequence of exposure. TCDD is responsible for activation of p38 and ERK1/2 in RAW 264.7 murine macrophages by an AHR-independent mechanism [42], and of JNK and ERK, but not p38, in mouse embryonic fibroblasts and African Green Monkey kidney CV-1 cells [34]. In macrophages, TCDD does not cause any apparent changes in JNK activity, although it induces caspase-3 activity, whereas in fibroblasts and CV-1 cells the JNKs are activated. Equilibrium between the ERK and p38 pathways may be critical to cell fate resulting from TCDD exposure.

Studies using MAPK inhibitors have shown a strong connection between the MAPK and AHR signaling pathways, but these studies must be interpreted in the light of the fact that many MAPK inhibitors, particularly those derived from flavonoids, pyridinyl imidazole compounds, and others, are AHR agonists, antagonists or both [34, 35, 43–47]. With this caveat in mind, the two p38 MAPK inhibitors, SB203580 and SB202190 suppressed CYP1A1 induction by TCDD in mouse hepatoma Hepa-1 cells and in human hepatoma HepG2 cells. These inhibitors also suppressed CYP1B1 induction in human breast adenocarcinoma MCF7 cells, although overexpression of a dominant-negative p38 MAPK did not suppress induction of a *Cyp1a1* reporter gene by TCDD in Hepa-1 cells [48]. Hence, suppression of *Cyp1a1* transcription by these pyridinyl imidazole compounds might not be due to p38 inhibition, but to an alternative effect on AHR function. SB203580 did not inhibit AHR transformation by TCDD *in vitro*, indicating that the compound was not acting as a simple AHR antagonist. Instead, it decreased TCDD-induced histone acetylation levels in the TATA box region of the *Cyp1a1* promoter, suggesting the possibility that pyridinyl imidazole compounds might suppress the recruitment of co-activators needed for initiation of *Cyp1a1* transcription [48]. In addition, SB203580 blocked the spontaneous translocation of ectopic AHR into the nucleus of African Green Monkey kidney COS-7 cells and suppressed AHR transcriptional activity [49], pointing at the two alternative conclusions that either the compound is an AHR antagonist that blocks its translocation or that p38-mediated phosphorylation is somehow involved in AHR translocation. These results illustrate the difficulties inherent to the interpretation of inhibitor data.

The ERKs comprise two isoforms, a 44-kDa ERK1 and a 42-kDa ERK2. ERK1 and ERK2 share 83% identity in amino acid sequence and are expressed in virtually all tissues. ERKs are activated by growth factors and mitogens and are involved in the processes of cell growth and differentiation. AHR ligands can activate ERK in many different cell types. TCDD, B[*a*]P and the B[*a*]P metabolite BPDE were reported to activate JNK and ERK in mouse hepatoma Hepa-1 cells, human lung carcinoma A549 cells, AHR-negative CV-1 cells and in AHR-negative and AHR-positive mouse embryonic fibroblasts [34]. Because induction occurred equally well in AHR-negative as in AHR-positive cells in these studies, it was concluded that the induction of these kinases proceed *via* an AHR-independent pathway.

Many recent studies have established a close connection between ERK1/2 function and AHR signaling. Promotion of *N*-nitrosomethylamine-initiated lung adenocarcinomas in mice by TCDD is accompanied by a tumor-suppressive function of K-RAS and a positive role for RAF-1 and ERK1/2 in lung tumorigenesis. TCDD may promote tumors by contributing to the down-regulation of K-RAS and the stimulation of RAF-1 [50]. Characterization of the role of the AHR in ERK1/2 activation was not attempted in these experiments.

Studies in primary human macrophages, African green monkey kidney cells, mouse fibroblasts and mouse hepatoma cells have shown that AHR ligands can induce the activation of an ERK-dependent pathway that, at least in the case of the human macrophages, leads to TNFα induction [51], and in the case of mouse cells culminates in activation of AHR/ARNT transcriptional activity [34, 35]. Along these same lines of evidence, overexpression of constitutively active MEK1, the MAPKK upstream of ERK1/2, reduced total AHR levels and enhanced the TCDD-initiated transactivation potential of the AHR, confirming the direct modulation of the AHR transcriptional response by ERK1/2 activity. Concomitantly, ERK inhibitors delayed TCDD-induced AHR degradation, suggesting that ERK kinase is critically linked to AHR function and expression by facilitating ligand-initiated transcriptional activation while targeting the receptor for degradation. Immunoprecipitation experiments are suggestive of a ligand-independent association of AHR and ERK, suggesting that ERK might be important in the regulation of AHR function, perhaps by targeting receptor phosphorylation or blocking ubiquitinylation [52]. Consistent with the above studies, overexpression of a dominant-negative variant of MEK1 or treatment with a MEK1 inhibitor reduced TCDD-dependent transcription of a reporter gene and inhibited the binding of the AHR to its cognate DNA motif in the promoter of the *Cyp1a1* gene [53].

In summary, a relatively strong body of evidence indicates that there is a two-way cross-talk between MAPK pathways and AHR signaling. For the most part, it appears that AHR ligands activate one or other MAPK, possibly depending on the specific ligand and tissue, and that the kinase in turn activates an ill-defined step in AHR activation, facilitating its binding to DNA and its ability to transactivate target genes. Undoubtedly, as new tools to study the mechanisms of chromatin regulation are developed, we will unravel the deeper mysteries of this signaling cross-talk.

Cross-talk of the AHR with cell cycle progression and apoptosis

Cell cycle progression

The cell cycle is the recurring sequence of events that leads to the duplication of the DNA content and subsequent division of a cell. The cell cycle consists of five distinct phases: G0, G1, S, G2 and M (Fig. 4). During G0, cells are in a quiescent state in which they have temporarily or reversibly stopped dividing. In response to growth factors and mitogens, cells come out of quiescence into the G1 phase, during which cyclins and cyclin-dependent kinases (CDKs) become activated to promote DNA replication. Entry into S phase results from the CDK-mediated phosphorylation of the retinoblastoma (RB) protein, which when hyperphosphorylated can no longer repress the activity of E2F, the main

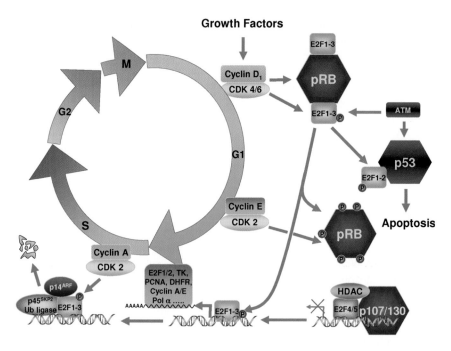

Figure 4. The cell cycle. The figure emphasizes the central role that retinoblastoma protein (RB) and p53 play in regulating entry of the cells into S phase.

transcription factor responsible for the induction of S-phase-specific genes. The G1, S and G2 phases collectively constitute the interphase portion of the cell cycle, during which cells prepare for mitosis. The M phase is itself composed of two tightly coupled processes: mitosis, in which replicated chromosomes are equally divided, and cytokinesis, in which the cytoplasm divides to form two distinct daughter cells. For a more detailed description of the cell cycle, the reader is directed to a number of excellent reviews of the field [54–56].

Several published accounts point to a role for the AHR in cell cycle control, although the precise mechanism remains ill-defined [57–62]. Early studies in this field revealed that an AHR-defective variant of the mouse hepatoma Hepa 1c1c7 cell line exhibited a prolonged doubling time compared with its wild-type counterpart [58]. This effect was attributed to delayed progress through the G1 phase, suggesting that AHR action facilitates G1 cell cycle progression. Mouse embryo fibroblasts (MEF) from AHR-null mice also grow more slowly, but this observation was attributed to an accumulation of cells in the G2/M phase as a result of altered expression of the G2/M kinases Cdc2 and Plk [63]. Using the same AHR-null MEF cells, the AHR was shown to contribute to p300-mediated induction of DNA synthesis (S phase progression) by the adenovirus E1A protein [64]. Collectively, these observations suggested that in the absence of an exogenous ligand, the AHR promotes progression through the cell cycle. In contrast, evidence dating back as far as 1984 indicates that TCDD, the prototypical AHR ligand, can inhibit cell proliferation. Confluent mouse epithelial cell cultures exhibit a diminished capacity for DNA replication in the presence of as little as 10 pM TCDD [65]. TCDD also inhibits DNA synthesis in rat primary hepatocytes [66] and in rat liver following partial hepatectomy [67].

There is extensive evidence in several different cell lines supporting the conclusion that, in the presence of an exogenous ligand such as TCDD, the activated AHR inhibits cell proliferation (reviewed in [16]). Analysis of the molecular basis of this inhibition has shown that interactions between the AHR and the RB/E2F axis control the response of these cells to ligand exposure. As the major target of CDK2 activity, phosphorylation of RB is critical for most cells to enter into S phase. Using yeast two-hybrid assays and co-immunoprecipitation experiments, AHR was found to form complexes with the hypophosphorylated RB protein in G1 and to block its phosphorylation [60, 61]. At least two AHR domains were found to be involved in this interaction. One was the cyclin D LXCXE motif, common to many RB-interacting oncoproteins. The other is present within the glutamine-rich transactivation domain of the AHR (reviewed in [68]). Direct interaction with hypophosphorylated RB constitutes a major G1 checkpoint in cells exposed to AHR ligands [68]. In addition, this interaction appears to regulate the maximal transcriptional levels of AHR activity [62].

Two different molecular mechanisms have been proposed to explain the G1 arrest induced by AHR ligands, with significant supporting evidence for both. One mechanism proposes that AHR activation and its subsequent interaction with RB blocks the phosphorylation of RB and reinforces repression of S-phase-specific gene transcription. In the alternative model, AHR activation is proposed to induce CDK inhibitors that arrest cells in G1. Our group used mouse hepatoma Hepa-1c1c7 cells to study the expression of E2F-regulated S-phase-specific genes and found that the AHR-RB interaction caused the transcriptional repression of several of these genes, including *Cyclin E*, *Cdk2*, *DNA polymerase α*, and *Dhfr*. In addition, we also found a significant increase in the expression of the CDK inhibitor p27[Kip1]. Further analyses using chromatin immunoprecipitation assays showed that TCDD recruited the AHR to E2F-regulated promoters from which it displaced the histone acetyltransferase p300 [69]. This result suggests that the interaction between AHR and RB acts as a negative regulator of cell cycle progression by inhibiting E2F-dependent transcriptional activity, thus preventing the expression of genes required for cell cycle progression through S phase. However, the significant induction of p27[Kip1] indicates that there is also a positive regulatory mechanism at play involving the up-regulation of CDK inhibitors. Similar results were found in a study of the inhibition of androgen-dependent proliferation in prostate cancer LNCaP cells by TCDD, in which TCDD repressed cyclin D1 and blocked RB phosphorylation, but also induced expression of p21[Waf1/Cip1] [70]. Evidence supporting the alternative mechanism suggested that AHR/ARNT dimers interact with RB to play a positive role in the induction of genes encoding CDK inhibitory proteins. Huang and Elferink [71] used dominant-negative DNA-binding-defective AHR and ARNT mutants in rat hepatoma cells and found that TCDD-induced AHR-mediated G1 arrest was only partially regulated by direct AHR transcriptional activity, suggesting that both co-activation and co-repression were responsible for the arrest. Using small interfering RNA (siRNA) to down-regulate ARNT protein expression, it was found that TCDD-induced G1 arrest was dependent on the ARNT protein, a finding that has not been reproduced in other laboratories. Taken together, these results point to the conclusion that the TCDD-activated AHR may inhibit cell cycle progression not only by directly interacting with RB and repressing the expression of genes required for entry into S phase and cell cycle progression, but also by inducing G1 phase regulatory proteins such as $p27^{Kip1}$ that directly inhibit CDK2/4 activity, resulting in RB inactivation.

While these observations provide a plausible mechanistic rationale for the role of the ligand-activated AHR in cell cycle regulation, additional data suggest that other mechanisms may be equally important in the absence of exogenous ligands. Recent work in our laboratory used stably integrated, Tet-OFF-regulated AHR variants in fibroblasts from AHR-null mice to further investigate the role of the AHR in cell cycle regulation. $Ahr^{(+/+)}$ fibroblasts proliferated significantly faster than $Ahr^{(-/-)}$ fibroblasts, and exposure to TCDD or deletion of the ligand-binding domain did not change their proliferation rates, indicating that the AHR functions in the cell cycle independently of ligand activation. Growth-promoting genes, such as cyclins and CDKs, were significantly down-regulated in $Ahr^{(-/-)}$ cells, whereas growth-arresting genes, such as transforming growth factor-β1 (TGFβ1), extracellular matrix-related genes and CDK inhibitors, were up-regulated, again suggesting that the AHR possesses intrinsic cellular functions to regulate cell proliferation that are independent of activation by either exogenous or endogenous ligands [72].

Inhibition of cell proliferation does not seem to be a universal effect of ligand-dependent AHR activation. Weiss and coworkers [40] found that TCDD treatment of rat liver oval cells, presumably a hepatic progenitor cell lineage, leads to AHR-dependent induction of the transcription factor JUND and the transcriptional up-regulation of cyclin A, triggering a release from contact inhibition. Similar results in these cells were obtained by the same authors using benz[a]anthracene, B[a]P and benzo[b]fluoranthene [73], suggesting that an AHR-dependent pathway of promoting cell proliferation through induction of JUND and cyclin A may constitute a novel mechanism in determining AHR-induced deregulation of cell cycle control.

Apoptosis

Apoptosis is a biological process in multicellular organisms leading to morphological changes such as blebbing, cell shrinkage, nuclear fragmentation, chromatin condensation, and chromosomal DNA fragmentation, ultimately ending in cell death. Unlike necrosis, apoptosis does not usually provoke an inflammatory response nor does it cause damage to the organism. Inhibition of apoptosis is a widely accepted mechanism of tumor promotion, and TCDD, by virtue of being one of the most potent tumor promoters known, has been widely studied for its ability to inhibit apoptosis. However, a number of reports indicate that AHR activation by TCDD can induce apoptosis in some cases and inhibit it in others (reviewed in [68]). In studies of liver tumor promotion in a two-stage hepatocarcinogenesis model, TCDD mediates clonal expansion of initiated cells by inhibiting apoptosis and bypassing AHR-dependent cell cycle arrest [74]. In contrast, other reports using thymocytes showed that TCDD induced apoptosis [75–77]. Several explanations have been advanced to explain these paradoxical differences in cell responses. Nebert et al. [78] suggested that AHR-mediated oxidative stress generated by the induction of cytochrome P450 enzymes may be a critical upstream event in the apoptosis cascade. In agreement with this hypothesis, apoptosis initiated by B[a]P in human hepatoma HepG2 cells was found to be linked to the induction of the p38 MAP kinase and to activation of the AHR and induction of CYP1A1, an event that leads to the formation of the ultimate carcinogen BPDE. Confirmatory evidence showed that p38-null mouse embryo fibroblasts were resistant to BPDE-induced apoptosis, indicating that the AHR plays a critical role in B[a]P-induced apoptosis, while p38 links the actions of an electrophilic metabolite like BPDE to the regulation of programmed cell death [79]. Both induction and inhibition of apoptosis by AHR ligands appear to be correct observations depending on the model system examined, pointing at the possibility that the ligand-activated AHR modulates apoptotic cell fate in a tissue- or cell lineage-specific manner.

Evidence for the essential role that the AHR plays in apoptosis is abundant. It has been proposed that the AHR contributes to mammary tumor cell growth by inhibiting apoptosis while promoting the transition to an invasive, metastatic phenotype [80]. If this were correct, it would follow that loss of AHR would be pro-apoptotic. Indeed, in AHR-null mice, liver pathology is associated with an accelerated rate of apoptosis, possibly related to an abnormal accumulation of hepatic retinoic acid that causes activation of TGFβ, resulting in stimulation of apoptosis [81]. Consistent with the anti-apoptotic role of the AHR, tumor promotion in the liver by TCDD treatment is absent in rat strains lacking a functional AHR. Liver cells from these rats also show an increased rate of apoptosis, characterized by activation of MDM2 and attenuation of p53 through increased ubiquitinylation [82, 83]. Similarly, activation of AHR by TCDD in a number of leukemia and lymphoma cells resulted in loss of the apoptotic response, an effect that was associated with an increase in expression of cyclooxygenase-2 and decrease of apoptosis-related Bcl-2 family members Bcl-xl and Mcl-1. Both a cyclooxygenase-2 inhibitor and an AHR antagonist abolished resistance to TCDD-induced apoptosis in vitro, strongly suggesting that AHR activation and the ensuing cyclooxygenase-2 overexpression are directly involved in a mechanism of resistance to apoptosis in lymphoma cell lines [84].

An endogenous role for the AHR in the intrinsic apoptotic process was also uncovered in studies directed at the effect of differing AHR levels on susceptibility to apoptosis. Analyses of a number of different endpoints of apoptosis in murine hepatoma Hepa-1c1c7 cells and in its AHR-deficient derivative LA1 revealed that the LA1 cells were more sensitive to intrinsic apoptosis-induced stresses (UV irradiation, hydrogen peroxide, serum starvation) than wild-type cells, suggesting that the endogenous AHR plays a cytoprotective role in the face of stimuli that initiate the intrinsic apoptotic pathway and that it can regulate cell fate directly. Lack of the AHR appeared to lead to an impaired survival response mediated by PI3 kinase-AKT/PKB and EGFR activation [85].

In contrast to the aforementioned results, other studies have shown that the AHR also modulates susceptibility to pro-apoptotic agents and that it regulates critical players in mitochondrial functions related to apoptosis. Treatment with TNFα plus cycloheximide induced apoptosis in murine hepatoma Hepa-1c1c7 cells, which have high levels of AHR protein, but not in Tao cells, an AHR-deficient variant of the Hepa-1c1c7 line. Extensive characterization of these cells lines led to the conclusion that in the absence of an exogenous ligand, the AHR regulates lysosomal disruption and permeability, an essential step in triggering the apoptosis process [46]. In agreement with this result, analysis of TCDD-induced proteome changes in rat 5L cells showed VDAC2, the voltage-dependent anion channel-selective protein-2, to be an AHR target, and to be dependent on the presence of a functional AHR. VDAC2 plays a central role as an inhibitor of the activation of the pro-apoptotic protein BAK and consequently of the regulation of the mitochondrial apoptotic pathway. Thus, these data point to VDAC2 as a possible effector of AHR-mediated apoptosis [86].

Studies in human fetal testis suggest that developing germ cells may be a target for regulation by AHR ligands and that AHR activation may be one mechanism responsible for the reduction of spermatogenesis in men exposed to environmental toxicants. Expression of the AHR in germ cells was detected in the human testis between 7 and 19 weeks of gestation, treatment with AHR ligands led to increased apoptosis, which was suppressed by AHR antagonists [87]. If confirmed, these results may point at a mechanism to explain a number of male reproductive syndromes.

Cross-talk of the AHR with differentiation pathways

Adipocyte differentiation

A number of studies during the last dozen years have provided strong evidence that the AHR pathway inhibits adipose differentiation in 3T3-L1 and other cells. Early work from Shimba and coworkers [88] found that the AHR was depleted during adipose differentiation of 3T3-L1 cells. The levels of AHR and

ARNT proteins and the binding activity of the AHR/ARNT complex to its cognate DNA motif in response to TCDD were found to decrease as adipose differentiation progressed in these cells. Further experiments suggested a possible negative role of liganded nuclear AHR on adipose differentiation. Down-regulation of the AHR during adipogenesis takes place primarily at the transcriptional level, as determined from nuclear run-on assays and AHR activity on reporter genes during adipose differentiation in 3T3-L1 cells. This work led to the identification of a sequence in the *Ahr* promoter responsible for differentiation-dependent suppression of *Ahr* transcription. It was proposed that down-regulation of a transacting factor binding to this sequence might be responsible for the suppression of *Ahr* transcription during adipose differentiation, but up to now such a factor has not been identified [89].

Jefcoate and coworkers [90] found strong evidence to indicate that the AHR is an inhibitor of lipid synthesis and of commitment to adipogenesis in MEF cells stimulated by an adipogenic hormonal mixture consisting of insulin, dexamethasone and methyisobutylxanthine (IDM). They analyzed constitutive functions of the AHR in primary MEF cells, in spontaneously immortalized MEF cell lines derived from wild-type C57BL/6 mice, and from congenic mice with a targeted disruption of the *Ahr* gene. The results from this work showed that constitutive AHR activity was required for basal expression of *Cyp1b1* and suppressed IDM-induced adipogenesis and lipogenesis in subconfluent cell cultures, an effect that was abrogated in MEF cells from $Ahr^{(-/-)}$ mice. Treatment with TCDD concurrently with hormone suppressed adipogenesis and resulted in limited mitotic expansion, but cells failed to exit the cell cycle, resulting in an increased rate of senescence. Further work along these lines showed that the negative regulation of adipose differentiation in 3T3-L1 cells by the AHR was independent of dioxin treatment. Shimba and colleagues [91] made stable transfectants of 3T3-L1 fibroblasts with a vector expressing high levels of full-length sense or antisense AHR mRNA or a control empty vector. Comparison of the differentiation potency of these clones with that of control cells showed that AHR over-expression suppressed morphological differentiation and induction of adipocyte-related genes, whereas decreased expression of the AHR induced much greater morphological differentiation and expression of adipocyte-related genes. Treatment with PD98059 or U0126, two MEK inhibitors that block activation of ERK1/2 MAPKs, abrogated the inhibitory action of the AHR on adipogenesis and restored the ability of the cells to differentiate, whereas treatment with SB203580, a specific inhibitor of p38 MAPK, inhibited adipogenesis. As indicated before, results with these MAPK inhibitors that are also AHR agonists or antagonists need to be interpreted with caution.

ERK-dependent pathways were found to function synergistically with AHR in mediating TCDD-induced suppression of peroxisome proliferator-activated receptor-γ-1 (PPARγ1) expression and subsequent adipocyte differentiation in C3H10T1/2 cells. Inhibition of MEK-induced ERK phosphorylation had no effect on adipogenesis but prevented TCDD-mediated suppression of adipocyte differentiation, as well as the suppression of the elevation of the key mediator, PPARγ1. This temporal separation of ERK activation from the affected PPARγ1 expression suggests that ERK does not act directly, but that ERK activation and TCDD-mediated activation of the AHR work synergistically to inhibit adipocyte differentiation. Confirmation of these results was obtained by serum depletion, which inhibited ERK activation, and by transfection of a vector expressing constitutively active MEK1, which phosphorylates ERK. Thus, it appears that low levels of activated MEK and ERK cooperate with AHR-induced factor(s) to generate a suppressor that prevents PPARγ1 transcription and differentiation [92, 93]. Microarray analysis to study the cooperative inhibitory effects of growth factors and TCDD on early adipogenesis in C3H10T1/2 cells identified 200 genes that exhibited expression changes of at least twofold after 24 hours of hormonal treatment, encompassing genes associated with cell structure, triglyceride and cholesterol metabolism, oxidative regulation, and secreted proteins. In the absence of growth factors, TCDD inhibited 30% of these hormonal responses without inhibiting the process of differentiation. A combination of EGF and TCDD cooperatively blocks differentiation, as well as additional 44 hormone-responsive genes, most of which have functional links to differentiation, including PPARγ1. EGF and TCDD also cooperatively reversed hormone-induced changes in cell adhesion complexes immediately prior to increases in PPARγ1 expression. These results led to the conclusion that changes in adhesion-linked signaling may play a key role in TCDD effects on differentiation [94].

Interestingly, CYP1B1 expression may represent an early stage of differentiation of pluripotent C3H10T1/2 cells, independent of AHR activity. Stimulation of C3H10T1/2 cells by IDM substantially induces expression of the CYP1B1 along with a subsequent decline in AHR expression. This stimulation represents up to 40% of the level produced by maximum activation of the AHR with TCDD. These responses were unaffected by inhibitors of AHR activity or DNA synthesis and were maintained in AHR-null fibroblasts. CYP1B1 expression partially overlapped with PPARγ1 expression, suggesting that CYP1B1 expression may be critical to an early stage of differentiation that requires factors associated with DNA synthesis to subsequently induce PPARγ1 [95].

The tyrosine kinase cSRC appears to also have a critical role in the effect of TCDD on adipocyte differentiation in MEF cells. Mechanistic studies in MEFs from $Src^{(+/+)}$ and $Src^{(-/-)}$ mice showed that suppression of differentiation by TCDD requires a functional c-SRC activity and maintenance of high levels of induced C/EBPβ and C/EBPδ, rather than an ultimate high level of these C/EBP isoforms [96]. In MCF10A cells, AHR activation by TCDD was accompanied by rapid activation of c-SRC kinase activity, a process mediated by the Cdc37-HSP90 complex, since dissociation of the cytosolic SRC-cdc37-HSP90 complex was sufficient to abolish the action of TCDD in activating c-SRC and cdc37. Blocking cdc37 mRNA translation with an antisense oligonucleotide or by siRNA knockdown confirmed the role of cdc37 in the

TCDD-induced activation of c-SRC. Interestingly, this action of TCDD appears to be limited to activation of c-SRC kinase, but not of kinases associated with the activation of NFκB, C/EBPα, or ERK [97].

More recently, the transcription factor NRF2, which regulates oxidative stress responses through antioxidant response motifs in gene promoters, has been suggested to modulate AHR signaling during adipogenesis. NRF2 was found to regulate expression of AHR and to subsequently modulate several downstream events in the AHR signaling cascade, including inhibition of adipogenesis in MEFs. Reporter analysis and chromatin immunoprecipitation assays revealed that NRF2 directly binds to one antioxidant response motif found in the −230-bp region of the *Ahr* promoter, modulating AHR signaling, and suggesting a novel mechanism mediating the suppression of adipogenesis controlled by the AHR. Consistent with this conclusion, *Nrf2*$^{(-/-)}$ MEFs showed markedly accelerated adipogenesis upon hormone stimulation, which was significantly delayed by ectopic expression of *Ahr* [98].

TGFβ

The TGFβ cytokines (TGFβ1, 2, and 3) belong to a signaling protein family that includes the activins, the nodal subfamily, the bone morphogenetic proteins, growth and differentiation factors, and the Mullerian inhibiting substance subfamily. TGFβ signaling is essential in cellular physiology and development in vertebrates and invertebrates, inhibiting cell proliferation, promoting apoptosis, inducing differentiation, and specifying developmental fates. Much of the evidence of the cross-talk of AHR with TGFβ was reviewed recently [99]. Thus, we review here only recently published results from this field.

Initial work by Greenlee and coworkers [100] showed that TCDD increased TGFα and reduced TGFβ2 mRNA levels in the human keratinocyte cell line SCC-12F, while mRNA levels of TGFβ1 were unchanged. Döhr et al. [101, 102] later described the inhibition of AHR and CYP1A1/1B1 enzyme activity by TGFβ1 in human lung cancer A549 cells in the presence of TCDD, suggesting that TGFβ1 was a direct negative regulator of the expression of these genes. Smad proteins, the transcription factors activated by TGFβ signaling, were shown to be mechanistically responsible for the cell-specific regulation of AHR expression by TGFβ in A549 cells. Transient transfection studies of a reporter gene driven by deletion constructs of the full-length human *AHR* promoter indicated that basal *AHR* promoter activity was cell specific, with human hepatoma HepG2 cells possessing a tenfold higher activity compared with A549 cells. Response to TGFβ treatment was also cell-type specific, resulting in a 50% inhibition of promoter activity in A549 and a threefold induction in HepG2 cells. Deletion mutagenesis identified a TGFβ-responsive sequence containing a functionally conserved Smad-binding element. Transient overexpression of Smad-2, -3, and -4 indicated that these transcription factors regulate human *AHR* promoter activity. Down-regulation of AHR by TGFβ was modulated by 5'-TG-3'-interacting factor (TGIF). Transient overexpression of TGIF in HepG2 cells led to inhibition of promoter activity and a similar decrease in *AHR* mRNA expression [103].

Other studies have shown that AHR can regulate TGFβ expression and function by means of a number of different mechanisms. Livers of *Ahr*[(-/-)] mice exhibited marked increases in active TGFβ1 and TGFβ3 proteins and elevated numbers of hepatocytes undergoing apoptosis and liver fibrosis compared to wild-type mice. Similarly, primary hepatocyte cultures from *Ahr*[(-/-)] mice exhibited a high number of cells in later stages of apoptosis and an elevated secretion of active TGFβ into the media compared with cultures from wild-type mice [104]. Similar results, with an increase in TGFβ2 mRNA, were also observed in MEF cells and smooth muscle cells [105]. Compared to wild-type MEFs, AHR-null cells proliferate slower and have higher rates of apoptosis and higher TGFβ3 mRNA and protein levels [63]. AHR activation by TCDD reduced TGFβ1 expression in epithelial and mesenchymal cells during palatogenesis and reduced *TGFβ2* transcription in human keratinocytes. Furthermore, repression of TGFβ3 signaling played a role in a TCDD-induced cleft palate, as addition of TGFβ3 to an *in vitro* palate culture model prevented the dioxin-induced reduction in filopodial density. Furthermore, TGFβ3 exposure completely prevented the dioxin-induced block of palatal fusion in this system [106], further strengthening the argument that AHR activity inhibits TGFβ function.

Recent work from our laboratory used expression of stably integrated, Tet-Off-regulated AHR variants in fibroblasts from AHR-null mice to investigate the role of the AHR in cell cycle regulation. We found that *Ahr*[(+/+)] fibroblasts proliferated significantly faster than *Ahr*[(-/-)] fibroblasts, and that this function of the AHR was ligand independent. This effect was due to a lengthening of the G1 and S phases of the cell cycle and to the fact that many growth-promoting genes were significantly down-regulated in *Ahr*[(-/-)] cells, whereas growth-arresting genes, such as *Tgfβ1*, ECM-related genes, and CDK inhibitor genes were up-regulated in these cells. *Ahr*[(-/-)] fibroblasts secreted significantly more TGFβ1 into the culture medium than *Ahr*[(+/+)] fibroblasts, and showed increased levels of activated Smad4 and TGFβ1 mRNA. Inhibition of TGFβ1 signaling by overexpression of Smad7 reversed the proliferative and gene expression phenotype of *Ahr*[(-/-)] fibroblasts. Unlike in *Ahr*[(-/-)] cells, TGFβ1 mRNA levels in *Ahr*[(+/+)] cells were down-regulated by mRNA destabilization resulting from the increased activity of the TTP tristetraprolin protein, an AU-rich element-binding protein that destabilizes mRNA. These results demonstrate novel evidence relevant to the mechanisms of AHR-mediated TGFβ1 regulation [72].

TGFβ is secreted as a dimer complexed in a latent form due to its non-covalent association with two latency-associated proteins (LAP). LAP proteins, in turn, are bound to latent TGFβ binding protein (LTBP) by a disulfide bond. The N-terminus of LTBP anchors the complex to the extracellular matrix, and

TGFβ activation is accomplished when TGFβ is released from the LAP-LTBP complex. Several studies of LTBP have provided indirect evidence of interactions between AHR and TGFβ. Analyses of differential patterns of gene expression in wild-type $Ahr^{(+/+)}$ and $Ahr^{(-/-)}$ MEFs have identified LTBP1 as a gene negatively regulated by AHR in the absence of xenobiotic ligands. Both $Ltbp1$ mRNA and protein expression were markedly increased in $Ahr^{(-/-)}$ MEF cells. Secreted LTBP1 was elevated in the culture medium and extracellular matrix of these cells, which also had higher levels of active and total TGFβ. In addition, AHR activation by TCDD down-regulated $Ltbp1$ expression, suggesting a direct AHR-mediated mechanism. Treatment of $Ahr^{(+/+)}$ MEFs with TGFβ down-regulated AHR expression and simultaneously increased LTBP1. High LTBP1 levels correlated with increased TGFβ activity, suggesting that LTBP1 could contribute to the maintenance of TGFβ levels. These data suggest that LTBP1 is negatively regulated by the AHR and, by extension, also support the implication that the AHR can control TGFβ activity [107]. Further analysis of LTBP1 knockdown by siRNA in $Ahr^{(-/-)}$ MEF cells showed a decrease in the amount of active TGFβ, plasminogen activator-1 (PA1)/plasmin, and elastase activities, and thrombospondin-1 (TSP1) expression, and elevation of MMP-2 activity, without significantly affecting their respective mRNA levels. In the absence of AHR, LTBP1 contributes to TGFβ activation in MEF cells, possibly by influencing the activities of various extracellular proteases, such as PA1/plasmin, elastase, TSP-1, and MMP-2, which could also be maintained by TGFβ itself [108]. Additional experiments using collagen protein staining, immunohistochemistry and in $situ$ hybridization were designed to test whether TGFβ and LTBP1 overexpression colocalized within the fibrotic nodules of $Ahr^{(-/-)}$ mouse liver. $Ltbp1$ mRNA and protein were overexpressed in fibrotic regions and colocalized with other indicators of fibrosis. Although TGFβ protein also colocalized with regions of fibrosis, $Tgf\beta$ mRNA was present throughout the hepatic parenchyma and exhibited similar levels in wild-type and $Ahr^{(-/-)}$ mice, suggesting that LTBP1 targets TGFβ to specific areas of the liver and that the AHR could be a negative regulator of liver fibrosis, possibly through the control of LTBP1 and TGFβ activities [109].

NFκB and MYC

The nuclear factor κB (NFκB) family of transcription factors plays a major role in many physiological and pathological processes involved in inflammatory and immune responses, development, apoptosis and cell growth. All members of this family, of which p65 was the first identified and named NFκB, share a common domain termed the Rel homology domain (RHD). Members of this family dimerize through the RHD to form either transcriptionally active or inactive dimers. RelA (p65), RelB, and c-Rel are family members that can form transcriptionally active heterodimers with p52 or p50. The latter lack a transactivation domain of their own but can still form transcriptionally inactive p50/p50 or p52/p52 homodimers. NFκB dimers remain in the cytoplasm in a complex with the inhibitor of κB (IκB). IκB ubiquitinylation, resulting from phosphorylation by inhibitor of κB kinases (IKK), leads to its degradation and release, and ultimately to translocation of NFκB into the nucleus (reviewed in [110]).

In the past several years, evidence has emerged to show that the AHR and NFκB signaling pathways interact and modify each other (reviewed in [111]). Co-immunoprecipitation studies in murine hepatoma cells have shown that RelA/p50 heterodimers physically interact with the activated AHR and that activation of the AHR pathway altered the responsiveness of an NFκB luciferase reporter gene [112]. In fact, RelA and AHR proteins were shown to functionally cooperate to bind to NFκB response elements and induce *c-MYC* gene expression in HS578T breast cancer cells [113]. Interestingly, c-MYC was also found to mediate TCDD-induced telomerase activity in human choriocarcinoma cells. TCDD or a combination of TCDD and 17β-estradiol, increased telomerase activity and the expression of the human telomerase catalytic subunit hTERT in these cells and induced c-MYC, a transcriptional activator of hTERT, but neither of these agents induced telomerase activity in c-MYC-null cells. In addition, TCDD up-regulated telomerase only in cells that expressed c-MYC but lacked estrogen receptor (ER) expression. These findings suggest that AHR signaling induces telomerase activity through ER-independent MYC activation [114]. It appears, however, that whether AHR signaling induces or represses *c-MYC* expression may depend on the specific tissue and on combinatorial interactions with other factors, since when AHR binding sites were deleted from the *c-MYC* promoter or when the AHR repressor was over-expressed in HS578T breast cancer cells, the basal level of MYC expression increased significantly [115], suggesting that AHR represses *MYC* expression.

More recent evidence confirms the strong link between these two pathways in many physiological processes, especially those related to the immune response. Investigation of the role of the AHR in TCDD-induced thymic T cell apoptosis found that AHR regulates FasL and NFκB in stromal cells, playing a critical role in initiating apoptosis in thymic T cells. AHR knockout mice were found to be more resistant to TCDD-induced thymic atrophy and apoptosis when compared with AHR wild-type mice. A search for the mechanisms involved in this effect resulted in the finding that TCDD triggered the expression of several apoptotic genes, including FasL, in AHR wild-type but not AHR-null mice. TCDD treatment both *in vivo* and *in vitro* led to co-localization and translocation of the NFκB subunits p50 and p65 to the nucleus in stromal cells from AHR wild-type mice but not in TCDD-treated AHR-null mice. Mutation of the NFκB binding sites on the FasL promoter showed that TCDD regulates FasL promoter activity through NFκB. Thus, these results suggest that NFκB plays an important role in mediating AHR regulation of FasL in stromal cells, which in turn plays a critical role in initiating apoptosis in thymic T cells [116, 117].

The AHR has also been found to interact with NFκB to regulate the inflammatory response in acute lung inflammation, which was studied using wild-type and AHR-null mice exposed to inhaled cigarette smoke or to bacterial endotoxin. Treated AHR-null mice showed more severe lung inflammation than their wild-type counterparts, suggesting that AHR-deficient mice are more susceptible to inflammation. Smoke-exposed AHR-null mice also showed elevated NFκB DNA-binding activity compared with controls, a correlation that was associated with a rapid loss of RelB. The AHR may function as a regulator of inflammation at least in part through its interactions with NFκB and protection of RelB from premature degradation, resulting in heightened inflammatory responses to multiple proinflammatory stimuli [118].

RelB has emerged as a major partner of the AHR in conditions of deregulated immune cell function. Down-regulation of RelB and eIF3 p170 by B[a]P inhibits the growth and functional differentiation of mouse bone marrow-derived dendritic cells (DCs). Experiments with TCDD and AHR antagonists confirmed that these effects were AHR dependent and that they involved RelB repression, since RelB is necessary for the differentiation and function of mouse DCs [119, 120]. RelB was also found to play an important role in AHR-mediated induction of the expression of inflammatory genes. Induction by TCDD of the expression of several chemokines, including BAFF (B cell-activating factor of the tumor necrosis factor family), BLC (B lymphocyte chemoattractant), CCL1 (CC-chemokine ligand 1), and the transcription factor IFR3 (interferon γ responsive factor) in U937 macrophages is AHR and RelB dependent. Induction is associated with increased binding activity of an AHR/RelB complex on an atypical NFκB binding site (5'-GGGAGATTTG-3') without the participation of ARNT [121, 122].

Conclusions

It is evident that activation of the AHR in response to xenobiotic agonists such as TCDD and B[a]P directly affects multiple cell cycle regulatory pathways, ultimately resulting in the inhibition or promotion of cell proliferation or differentiation, depending on the particular tissue or cell model under investigation. The same is true of AHR activities in the absence of exogenous ligand. Thus, the AHR appears to be involved with regulation or perturbation of a multitude of cellular processes. Given the limited amount of information available, it might seem premature to speculate on the connections between the different signal transduction pathways shown to interact with the AHR and their possible role in adult disease. In this context, it is common to overlook the fact that induction of previously silent genes and repression of previously active genes by AHR ligands, genes that might have little connection with detoxification pathways, are equally likely to derail cellular homeostasis. In the field of developmental toxicology, we lack much needed information – and often the means to collect it – dealing with the consequences that exposure to environmental agents during embryonic life may bring about in the developing embryo and in the adult. When thinking about the possible role of the AHR in environmental injury, it is time for toxicology to synergize with developmental biology.

Acknowledgments
The work in the authors' lab is supported by NIEHS grants P30 ES06096, 2R01 ES06273 and 2R01 ES10708.

References

1 Safe S (2001) Molecular biology of the Ah receptor and its role in carcinogenesis. *Toxicol Lett* 120: 1–7

2 Okey AB (2007) An aryl hydrocarbon receptor odyssey to the shores of toxicology: The Deichmann Lecture, International Congress of Toxicology-XI. *Toxicol Sci* 98: 5–38

3 Hogenesch JB, Chan WK, Jackiw V, Brown RC, Gu Y-Z, Pray-Grant M, Perdew GH, Bradfield CA (1997) Characterization of a subset of the basic-helix-loop-helix-PAS superfamily that interacts with components of the dioxin signaling pathway. *J Biol Chem* 272: 8581–8593

4 Crews ST, Fan CM (1999) Remembrance of things PAS: Regulation of development by bHLH-PAS proteins. *Curr Opin Genet Dev* 9: 580–587

5 Barouki R, Coumoul X, Fernandez-Salguero PM (2007) The aryl hydrocarbon receptor, more than a xenobiotic-interacting protein. *FEBS Lett* 581: 3608–3615

6 Abbott BD, Birnbaum LS, Perdew GH (1995) Developmental expression of two members of a new class of transcription factors: I. Expression of aryl hydrocarbon receptor in the C57BL/6N mouse embryo. *Dev Dyn* 204: 133–143

7 Ma Q, Whitlock JP Jr, (1997) A novel cytoplasmic protein that interacts with the Ah receptor, contains tetratricopeptide repeat motifs, and augments the transcriptional response to 2,3,7,8-tetrachlorodibenzo-*p*-dioxin. *J Biol Chem* 272: 8878–8884

8 Carver LA, Bradfield CA (1997) Ligand-dependent interaction of the aryl hydrocarbon receptor with a novel immunophilin homolog *in vivo*. *J Biol Chem* 272: 11452–11456

9 Petrulis JR, Kusnadi A, Ramadoss P, Hollingshead B, Perdew GH (2003) The hsp90 co-chaperone XAP2 alters importin b recognition of the bipartite nuclear localization signal of the Ah receptor and represses transcriptional activity. *J Biol Chem* 278: 2677–2685

10 Beischlag TV, Wang S, Rose DW, Torchia J, Reisz-Porszasz S, Muhammad K, Nelson WE, Probst MR, Rosenfeld MG, Hankinson O (2002) Recruitment of the NCoA/SRC-1/p160 family of transcriptional coactivators by the aryl hydrocarbon receptor/aryl hydrocarbon receptor nuclear translocator complex. *Mol Cell Biol* 22: 4319–4333

11 Wang S, Hankinson O (2002) Functional involvement of the Brahma/SWI2-related gene 1 protein in cytochrome P4501A1 transcription mediated by the aryl hydrocarbon receptor complex. *J Biol Chem* 277: 11821–11827

12 Hestermann EV, Brown M (2003) Agonist and chemopreventative ligands induce differential transcriptional cofactor recruitment by aryl hydrocarbon receptor. *Mol Cell Biol* 23: 7920–7925

13 Wang S, Ge K, Roeder RG, Hankinson O (2004) Role of mediator in transcriptional activation by the aryl hydrocarbon receptor. *J Biol Chem* 279: 13593–13600

14 Schnekenburger M, Peng L, Puga A (2007) HDAC1 bound to the *Cyp1a1* promoter blocks histone acetylation associated with Ah receptor-mediated trans-activation. *Biochim Biophys Acta* 1769: 569–578

15 Pollenz RS (2002) The mechanism of Ah receptor protein down-regulation (degradation) and its impact on Ah receptor-mediated gene regulation. *Chem Biol Interact* 141: 41–61

16 Puga A, Xia Y, Elferink C (2002) Role of the aryl hydrocarbon receptor in cell cycle regulation. *Chem Biol Interact* 141: 117–130

17 Henklova P, Vrzal R, Ulrichova J, Dvorak Z (2008) Role of mitogen-activated protein kinases in aryl hydrocarbon receptor signaling. *Chem Biol Interact* 172: 93–104

18 Carrier F, Owens RA, Nebert DW, Puga A (1992) Dioxin-dependent activation of murine *Cyp1a-1* gene transcription requires protein kinase C-dependent phosphorylation. *Mol Cell Biol* 12: 1856–1863

19 Chen YH, Tukey RH (1996) Protein kinase C modulates regulation of the *CYP1A1* gene by the

aryl hydrocarbon receptor. *J Biol Chem* 271: 26261–26266

20 Long WP, Pray-Grant M, Tsai JC, Perdew GH (1998) Protein kinase C activity is required for aryl hydrocarbon receptor pathway-mediated signal transduction. *Mol Pharmacol* 53: 691–700

21 Okino ST, Pendurthi UR, Tukey RH (1992) Phorbol esters inhibit the dioxin receptor-mediated transcriptional activation of the mouse *Cyp1a-1* and *Cyp1a-2* genes by 2,3,7,8-tetrachlorodiben-zo-*p*-dioxin. *J Biol Chem* 267: 6991–6998

22 Ikegwuonu FI, Christou M, Jefcoate CR (1999) Regulation of cytochrome P4501B1 (CYP1B1) in mouse embryo fibroblast (C3H10T1/2) cells by protein kinase C (PKC). *Biochem Pharmacol* 57: 619–630

23 Li SY, Dougherty JJ (1997) Inhibitors of serine/threonine-specific protein phosphatases stimulate transcription by the Ah receptor/Arnt dimer by affecting a step subsequent to XRE binding. *Arch Biochem Biophys* 340: 73–82

24 Pongratz I, Strömstedt PE, Mason GG, Poellinger L (1991) Inhibition of the specific DNA bind-ing activity of the dioxin receptor by phosphatase treatment. *J Biol Chem* 266: 16813–16817

25 Mahon MJ, Gasiewicz TA (1995) Ah receptor phosphorylation: Localization of phosphorylation sites to the C-terminal half of the protein. *Arch Biochem Biophys* 318: 166–174

26 Park S, Henry EC, Gasiewicz TA (2000) Regulation of DNA binding activity of the ligand-acti-vated aryl hydrocarbon receptor by tyrosine phosphorylation. *Arch Biochem Biophys* 381: 302–312

27 Dieter MZ, Freshwater SL, Solis WA, Nebert DW, Dalton TP (2001) Tryphostin AG879, a tyro-sine kinase inhibitor: Prevention of transcriptional activation of the electrophile and the aromatic hydrocarbon response elements. *Biochem Pharmacol* 61: 215–225

28 Ikuta T, Kobayashi Y, Kawajiri K (2004) Phosphorylation of nuclear localization signal inhibits the ligand-dependent nuclear import of aryl hydrocarbon receptor. *Biochem Biophys Res Commun* 317: 545–550

29 Minsavage GD, Park SK, Gasiewicz TA (2004) The aryl hydrocarbon receptor (AhR) tyrosine 9, a residue that is essential for AhR DNA binding activity, is not a phosphoresidue but augments AhR phosphorylation. *J Biol Chem* 279: 20582–20593

30 Pratt WB (1997) The role of the hsp90-based chaperone system in signal transduction by nuclear receptors and receptors signaling *via* MAP kinase. *Annu Rev Pharmacol Toxicol* 37: 297–326

31 Ogiso H, Kagi N, Matsumoto E, Nishimoto M, Arai R, Shirouzu M, Mimura J, Fujii-Kuriyama Y, Yokoyama S (2004) Phosphorylation analysis of 90 kDa heat shock protein within the cytosolic arylhydrocarbon receptor complex. *Biochemistry* 43: 15510–15519

32 Cobb MH, Goldsmith EJ (2000) Dimerization in MAP-kinase signaling. *Trends Biochem Sci* 25: 7–9

33 Weston CR, Davis RJ (2007) The JNK signal transduction pathway. *Curr Opin Cell Biol* 19: 142–149

34 Tan Z, Chang X, Puga A, Xia Y (2002) Activation of mitogen-activated protein kinases (MAPKs) by aromatic hydrocarbons: Role in the regulation of aryl hydrocarbon receptor (AHR) function. *Biochem Pharmacol* 64: 771–780

35 Tan Z, Huang M, Puga A, Xia Y (2004) A critical role for MAP kinases in the control of Ah recep-tor complex activity. *Toxicol Sci* 82: 80–87

36 Diry M, Tomkiewicz C, Koehle C, Coumoul X, Bock KW, Barouki R, Transy C (2006) Activation of the dioxin/aryl hydrocarbon receptor (AhR) modulates cell plasticity through a JNK-dependent mechanism. *Oncogene* 25: 5570–5574

37 Puga A, Nebert DW, Carrier F (1992) Dioxin induces expression of c-*fos* and c-*jun* proto-onco-genes and a large increase in transcription factor AP-1. *DNA Cell Biol* 11: 269–281

38 Hoffer A, Chang CY, Puga A (1996) Dioxin induces *fos* and *jun* gene expression by Ah receptor-dependent and -independent pathways. *Toxicol Appl Pharmacol* 141: 238–247

39 Weiss C, Faust D, Durk H, Kolluri SK, Pelzer A, Schneider S, Dietrich C, Oesch F, Göttlicher M (2005) TCDD induces c-jun expression *via* a novel Ah (dioxin) receptor-mediated p38-MAPK-dependent pathway. *Oncogene* 24: 4975–4983

40 Weiss C, Faust D, Schreck I, Ruff A, Farwerck T, Melenberg A, Schneider S, Oesch-Bartlomowicz B, Zatloukalova J, Vondracek J, Oesch F, Dietrich C (2008) TCDD deregulates contact inhibition in rat liver oval cells *via* Ah receptor, JunD and cyclin A. *Oncogene* 27: 2198–2207

41 Ikuta T, Kobayashi Y, Kawajiri K (2004) Cell density regulates intracellular localization of aryl hydrocarbon receptor. *J Biol Chem* 279: 19209–19216

42 Park SJ, Yoon WK, Kim HJ, Son HY, Cho SW, Jeong KS, Kim TH, Kim SH, Kim SR, Ryu SY

(2005) 2,3,7,8-Tetrachlorodibenzo-*p*-dioxin activates ERK and p38 mitogen-activated protein kinases in RAW 264.7 cells. *Anticancer Res* 25: 2831–2836

43 Reiners JJ Jr, Lee JY, Clift RE, Dudley DT, Myrand SP (1998) PD98059 is an equipotent antagonist of the aryl hydrocarbon receptor and inhibitor of mitogen-activated protein kinase kinase. *Mol Pharmacol* 53: 438–445

44 Joiakim A, Mathieu PA, Palermo C, Gasiewicz TA, Reiners JJ Jr, (2003) The Jun N-terminal kinase inhibitor SP600125 is a ligand and antagonist of the aryl hydrocarbon receptor. *Drug Metab Dispos* 31: 1279–1282

45 Andrieux L, Langouet S, Fautrel A, Ezan F, Krauser JA, Savouret JF, Guengerich FP, Baffet G, Guillouzo A (2004) Aryl hydrocarbon receptor activation and cytochrome P450 1A induction by the mitogen-activated protein kinase inhibitor U0126 in hepatocytes. *Mol Pharmacol* 65: 934–943

46 Caruso JA, Mathieu PA, Joiakim A, Zhang H, Reiners JJ Jr, (2006) Aryl hydrocarbon receptor modulation of tumor necrosis factor-α-induced apoptosis and lysosomal disruption in a hepatoma model that is caspase-8-independent. *J Biol Chem* 281: 10954–10967

47 Dvorak Z, Vrzal R, Henklova P, Jancova P, Anzenbacherova E, Maurel P, Svecova L, Pavek P, Ehrmann J, Havlik R, Bednar P, Lemr K, Ulrichova J (2008) JNK inhibitor SP600125 is a partial agonist of human aryl hydrocarbon receptor and induces *CYP1A1* and *CYP1A2* genes in primary human hepatocytes. *Biochem Pharmacol* 75: 580–588

48 Shibazaki M, Takeuchi T, Ahmed S, Kikuchi H (2004) Suppression by p38 MAP kinase inhibitors (pyridinyl imidazole compounds) of Ah receptor target gene activation by 2,3,7,8-tetrachlorodibenzo-*p*-dioxin and the possible mechanism. *J Biol Chem* 279: 3869–3876

49 Shibazaki M, Takeuchi T, Ahmed S, Kikuchi H (2004) Blockade by SB203580 of Cyp1a1 induction by 2,3,7,8-tetrachlorodibenzo-*p*-dioxin, and the possible mechanism: Possible involvement of the p38 mitogen-activated protein kinase pathway in shuttling of Ah receptor overexpressed in COS-7 cells. *Ann NY Acad Sci* 1030: 275–281

50 Ramakrishna G, Perella C, Birely L, Diwan BA, Fornwald LW, Anderson LM (2002) Decrease in K-ras p21 and increase in Raf1 and activated Erk 1 and 2 in murine lung tumors initiated by *N*-nitrosodimethylamine and promoted by 2,3,7,8-tetrachlorodibenzo-*p*-dioxin. *Toxicol Appl Pharmacol* 179: 21–34

51 Lecureur V, Ferrec EL, N'diaye M, Vee ML, Gardyn C, Gilot D, Fardel O (2005) ERK-dependent induction of TNFα expression by the environmental contaminant benzo[*a*]pyrene in primary human macrophages. *FEBS Lett* 579: 1904–1910

52 Chen S, Operana T, Bonzo J, Nguyen N, Tukey RH (2005) ERK kinase inhibition stabilizes the aryl hydrocarbon receptor: Implications for transcriptional activation and protein degradation. *J Biol Chem* 280: 4350–4359

53 Yim S, Oh M, Choi SM, Park H (2004) Inhibition of the MEK-1/p42 MAP kinase reduces aryl hydrocarbon receptor-DNA interactions. *Biochem Biophys Res Commun* 322: 9–16

54 Cheng M, Olivier P, Diehl JA, Fero M, Roussel MF, Roberts JM, Sherr CJ (1999) The p21(Cip1) and p27(Kip1) CDK 'inhibitors' are essential activators of cyclin D-dependent kinases in murine fibroblasts. *EMBO J* 18: 1571–1583

55 Sherr CJ, Roberts JM (1999) CDK inhibitors: Positive and negative regulators of G1-phase progression. *Genes Dev* 13: 1501–1512

56 Smits VA, Medema RH (2001) Checking out the G(2)/M transition. *Biochim Biophys Acta* 1519: 1–12

57 Weiss C, Kolluri SK, Kiefer F, Göttlicher M (1996) Complementation of Ah receptor deficiency in hepatoma cells: Negative feedback regulation and cell cycle control by the Ah receptor. *Exp Cell Res* 226: 154–163

58 Ma Q, Whitlock JPJ (1996) The aromatic hydrocarbon receptor modulates the Hepa 1c1c7 cell cycle and differentiated state independently of dioxin. *Mol Cell Biol* 16: 2144–2150

59 Kolluri SK, Weiss C, Koff A, Göttlicher M (1999) p27^{Kip1} induction and inhibition of proliferation by the intracellular Ah receptor in developing thymus and hepatoma cells. *Genes Dev* 13: 1742–1753

60 Puga A, Barnes SJ, Dalton TP, Chang C, Knudsen ES, Maier MA (2000) Aromatic hydrocarbon receptor interaction with the retinoblastoma protein potentiates repression of E2F-dependent transcription and cell cycle arrest. *J Biol Chem* 275: 2943–2950

61 Ge NL, Elferink CJ (1998) A direct interaction between the aryl hydrocarbon receptor and retinoblastoma protein. *J Biol Chem* 273: 22708–22713

62 Elferink CJ, Ge NL, Levine A (2001) Maximal aryl hydrocarbon receptor activity depends on an

interaction with the retinoblastoma protein. *Mol Pharmacol* 59: 664–673

63 Elizondo G, Fernandez-Salguero P, Sheikh MS, Kim GY, Fornace AJ, Lee KS, Gonzalez FJ (2000) Altered cell cycle control at the G(2)/M phases in aryl hydrocarbon receptor-null embryo fibroblast. *Mol Pharmacol* 57: 1056–1063

64 Tohkin M, Fukuhara M, Elizondo G, Tomita S, Gonzalez FJ (2000) Aryl hydrocarbon receptor is required for p300-mediated induction of DNA synthesis by adenovirus E1A. *Mol Pharmacol* 58: 845–851

65 Gierthy JF, Crane D (1984) Reversible inhibition of *in vitro* epithelial cell proliferation by 2,3,7,8-tetrachlorodibenzo-*p*-dioxin. *Toxicol Appl Pharmacol* 74: 91–98

66 Hushka DR, Greenlee WF (1995) 2,3,7,8-Tetrachlorodibenzo-*p*-dioxin inhibits DNA synthesis in rat primary hepatocytes. *Mutat Res* 333: 89–99

67 Bauman JW, Goldsworthy TL, Dunn CS, Fox TR (1995) Inhibitory effects of 2,3,7,8-tetra-chlorodibenzo-*p*-dioxin on rat hepatocyte proliferation induced by 2/3 partial hepatectomy. *Cell Prolif* 28: 437–451

68 Marlowe JL, Puga A (2005) Aryl hydrocarbon receptor, cell cycle regulation, toxicity, and tumori-genesis. *J Cell Biochem* 96: 1174–1184

69 Marlowe JL, Knudsen ES, Schwemberger S, Puga A (2004) The aryl hydrocarbon receptor dis-places p300 from E2F-dependent promoters and represses S-phase specific gene expression. *J Biol Chem* 279: 29013–29022

70 Barnes-Ellerbe S, Knudsen KE, Puga A (2004) 2,3,7,8-Tetrachlorodibenzo-*p*-dioxin blocks andro-gen-dependent cell proliferation of LNCaP cells through modulation of pRB phosphorylation. *Mol Pharmacol* 66: 502–511

71 Huang G, Elferink CJ (2005) Multiple mechanisms are involved in Ah receptor-mediated cell cycle arrest. *Mol Pharmacol* 67: 88–96

72 Chang X, Fan Y, Karyala S, Schwemberger S, Tomlinson CR, Sartor MA, Puga A (2007) Ligand-independent regulation of transforming growth factor b1 expression and cell cycle progression by the aryl hydrocarbon receptor. *Mol Cell Biol* 27: 6127–6139

73 Andrysik Z, Vondracek J, Machala M, Krcmar P, Svihalkova-Sindlerova L, Kranz A, Weiss C, Faust D, Kozubik A, Dietrich C (2007) The aryl hydrocarbon receptor-dependent deregulation of cell cycle control induced by polycyclic aromatic hydrocarbons in rat liver epithelial cells. *Mutat Res* 615: 87–97

74 Bock KW, Kohle C (2005) Ah receptor- and TCDD-mediated liver tumor promotion: Clonal selec-tion and expansion of cells evading growth arrest and apoptosis. *Biochem Pharmacol* 69: 1403–1408

75 McConkey DJ, Hartzell P, Duddy SK, Hakansson H, Orrenius S (1988) 2,3,7,8-Tetrachloro-dibenzo-*p*-dioxin kills immature thymocytes by Ca^{2+}-mediated endonuclease activation. *Science* 242: 256–259

76 Kurl RN, Abraham M, Olnes MJ (1993) Early effects of 2,3,7,8-tetrachlorodibenzo-*p*-dioxin (TCDD) on rat thymocytes *in vitro*. *Toxicology* 77: 103–114

77 Silverstone AE, Frazier DE Jr, Gasiewicz TA (1994) Alternate immune system targets for TCDD: Lymphocyte stem cells and extrathymic T-cell development. *Exp Clin Immunogenet* 11: 94–101

78 Nebert DW, Roe AL, Dieter MZ, Solis WA, Yang Y, Dalton TP (2000) Role of the aromatic hydro-carbon receptor and [*Ah*] gene battery in the oxidative stress response, cell cycle control, and apoptosis. *Biochem Pharmacol* 59: 65–85

79 Chen S, Nguyen N, Tamura K, Karin M, Tukey RH (2003) The role of the Ah receptor and p38 in benzo[*a*]pyrene-7,8-dihydrodiol and benzo[*a*]pyrene-7,8-dihydrodiol-9,10-epoxide-induced apoptosis. *J Biol Chem* 278: 19526–19533

80 Schlezinger JJ, Liu D, Farago M, Seldin DC, Belguise K, Sonenshein GE, Sherr DH (2006) A role for the aryl hydrocarbon receptor in mammary gland tumorigenesis. *Biol Chem* 387: 1175–1187

81 Gonzalez FJ, Fernandez-Salguero P (1998) The aryl hydrocarbon receptor: Studies using the AHR-null mice. *Drug Metab Dispos* 26: 1194–1198

82 Viluksela M, Bager Y, Tuomisto JT, Scheu G, Unkila M, Pohjanvirta R, Flodstrom S, Kosma VM, Maki-Paakkanen J, Vartiainen T et al (2000) Liver tumor-promoting activity of 2,3,7,8-tetra-chlorodibenzo-*p*-dioxin (TCDD) in TCDD-sensitive and TCDD-resistant rat strains. *Cancer Res* 60: 6911–6920

83 Pääjärvi G, Viluksela M, Pohjanvirta R, Stenius U, Högberg J (2005) TCDD activates Mdm2 and attenuates the p53 response to DNA damaging agents. *Carcinogenesis* 26: 201–208

84 Vogel CF, Li W, Sciullo E, Newman J, Hammock B, Reader JR, Tuscano J, Matsumura F (2007)

Pathogenesis of aryl hydrocarbon receptor-mediated development of lymphoma is associated with increased cyclooxygenase-2 expression. *Am J Pathol* 171: 1538–1548

85 Wu R, Zhang L, Hoagland MS, Swanson HI (2007) Lack of the aryl hydrocarbon receptor leads to impaired activation of AKT/protein kinase B and enhanced sensitivity to apoptosis induced *via* the intrinsic pathway. *J Pharmacol Exp Ther* 320: 448–457

86 Sarioglu H, Brandner S, Haberger M, Jacobsen C, Lichtmannegger J, Wormke M, Andrae U (2008) Analysis of 2,3,7,8-tetrachlorodibenzo-*p*-dioxin-induced proteome changes in 5L rat hepatoma cells reveals novel targets of dioxin action including the mitochondrial apoptosis regulator VDAC2. *Mol Cell Proteomics* 7: 394–410

87 Coutts SM, Fulton N, Anderson RA (2007) Environmental toxicant-induced germ cell apoptosis in the human fetal testis. *Hum Reprod* 22: 2912–2918

88 Shimba S, Todoroki K, Aoyagi T, Tezuka M (1998) Depletion of arylhydrocarbon receptor during adipose differentiation in 3T3-L1 cells. *Biochem Biophys Res Commun* 249: 131–137

89 Shimba S, Hayashi M, Ohno T, Tezuka M (2003) Transcriptional regulation of the *AhR* gene during adipose differentiation. *Biol Pharm Bull* 26: 1266–1271

90 Alexander DL, Ganem LG, Fernandez-Salguero P, Gonzalez F, Jefcoate CR (1998) Aryl-hydrocarbon receptor is an inhibitory regulator of lipid synthesis and of commitment to adipogenesis. *J Cell Sci* 111: 3311–3322

91 Shimba S, Wada T, Tezuka M (2001) Arylhydrocarbon receptor (AhR) is involved in negative regulation of adipose differentiation in 3T3-L1 cells: AhR inhibits adipose differentiation independently of dioxin. *J Cell Sci* 114: 2809–2817

92 Hanlon PR, Ganem LG, Cho YC, Yamamoto M, Jefcoate CR (2003) AhR- and ERK-dependent pathways function synergistically to mediate 2,3,7,8-tetrachlorodibenzo-*p*-dioxin suppression of peroxisome proliferator-activated receptor-γ1 expression and subsequent adipocyte differentiation. *Toxicol Appl Pharmacol* 189: 11–27

93 Cimafranca MA, Hanlon PR, Jefcoate CR (2004) TCDD administration after the pro-adipogenic differentiation stimulus inhibits PPARγ through a MEK-dependent process but less effectively suppresses adipogenesis. *Toxicol Appl Pharmacol* 196: 156–168

94 Hanlon PR, Cimafranca MA, Liu X, Cho YC, Jefcoate CR (2005) Microarray analysis of early adipogenesis in C3H10T1/2 cells: Cooperative inhibitory effects of growth factors and 2,3,7,8-tetrachlorodibenzo-*p*-dioxin. *Toxicol Appl Pharmacol* 207: 39–58

95 Cho YC, Zheng W, Yamamoto M, Liu X, Hanlon PR, Jefcoate CR (2005) Differentiation of pluripotent C3H10T1/2 cells rapidly elevates CYP1B1 through a novel process that overcomes a loss of Ah receptor. *Arch Biochem Biophys* 439: 139–153

96 Vogel CF, Matsumura F (2003) Interaction of 2,3,7,8-tetrachlorodibenzo-*p*-dioxin (TCDD) with induced adipocyte differentiation in mouse embryonic fibroblasts (MEFs) involves tyrosine kinase c-Src. *Biochem Pharmacol* 66: 1231–1244

97 Park S, Dong B, Matsumura F (2007) Rapid activation of c-Src kinase by dioxin is mediated by the Cdc37-HSP90 complex as part of Ah receptor signaling in MCF10A cells. *Biochemistry* 46: 899–908

98 Shin S, Wakabayashi N, Misra V, Biswal S, Lee GH, Agoston ES, Yamamoto M, Kensler TW (2007) NRF2 modulates aryl hydrocarbon receptor signaling: Influence on adipogenesis. *Mol Cell Biol* 27: 7188–7197

99 Puga A, Tomlinson CR, Xia Y (2005) Ah receptor signals cross-talk with multiple developmental pathways. *Biochem Pharmacol* 69: 199–207

100 Gaido KW, Maness SC, Leonard LS, Greenlee WF (1992) 2,3,7,8-Tetrachlorodibenzo-*p*-dioxin-dependent regulation of transforming growth factors-α and -β2 expression in a human keratinocyte cell line involves both transcriptional and post-transcriptional control. *J Biol Chem* 267: 24591–24595

101 Döhr O, Abel J (1997) Transforming growth factor-β1 coregulates mRNA expression of aryl hydrocarbon receptor and cell-cycle-regulating genes in human cancer cell lines. *Biochem Biophys Res Commun* 241: 86–91

102 Döhr O, Sinning R, Vogel C, Münzel P, Abel J (1997) Effect of transforming growth factor-β1 on expression of aryl hydrocarbon receptor and genes of Ah gene battery: Clues for independent down-regulation in A549 cells. *Mol Pharmacol* 51: 703–710

103 Wolff S, Harper PA, Wong JM, Mostert V, Wang Y, Abel J (2001) Cell-specific regulation of human aryl hydrocarbon receptor expression by transforming growth factor-β(1). *Mol Pharmacol* 59: 716–724

104 Zaher H, Fernandez-Salguero PM, Letterio J, Sheikh MS, Fornace AJ Jr, Roberts AB, Gonzalez FJ (1998) The involvement of aryl hydrocarbon receptor in the activation of transforming growth factor-β and apoptosis. *Mol Pharmacol* 54: 313–321

105 Guo J, Sartor M, Karyala S, Medvedovic M, Kann S, Puga A, Ryan P, Tomlinson CR (2004) Expression of genes in the TGF-β signaling pathway is significantly deregulated in smooth muscle cells from aorta of aryl hydrocarbon receptor knockout mice. *Toxicol Appl Pharmacol* 194: 79–89

106 Thomae TL, Stevens EA, Bradfield CA (2005) Transforming growth factor-β3 restores fusion in palatal shelves exposed to 2,3,7,8-tetrachlorodibenzo-*p*-dioxin. *J Biol Chem* 280: 12742–12746

107 Santiago-Josefat B, Mulero-Navarro S, Dallas SL, Fernandez-Salguero PM (2004) Overexpression of latent transforming growth factor-β binding protein 1 (LTBP-1) in dioxin receptor-null mouse embryo fibroblasts. *J Cell Sci* 117: 849–859

108 Gomez-Duran A, Mulero-Navarro S, Chang X, Fernandez-Salguero PM (2006) LTBP-1 blockade in dioxin receptor-null mouse embryo fibroblasts decreases TGF-β activity: Role of extracellular proteases plasmin and elastase. *J Cell Biochem* 97: 380–392

109 Corchero J, Martin-Partido G, Dallas SL, Fernandez-Salguero PM (2004) Liver portal fibrosis in dioxin receptor-null mice that overexpress the latent transforming growth factor-β-binding protein-1. *Int J Exp Pathol* 85: 295–302

110 Liou HC (2002) Regulation of the immune system by NF-κB and IκB. *J Biochem Mol Biol* 35: 537–546

111 Tian Y, Rabson AB, Gallo MA (2002) Ah receptor and NF-kB interactions: Mechanisms and physiological implications. *Chem Biol Interact* 141: 97–115

112 Tian Y, Ke S, Denison MS, Rabson AB, Gallo MA (1999) Ah receptor and NF-κB interactions, a potential mechanism for dioxin toxicity. *J Biol Chem* 274: 510–515

113 Kim DW, Gazourian L, Quadri SA, Romieu-Mourez R, Sherr DH, Sonenshein GE (2000) The RelA NF-κB subunit and the aryl hydrocarbon receptor (AhR) cooperate to transactivate the *c-myc* promoter in mammary cells. *Oncogene* 19: 5498–5506

114 Sarkar P, Shiizaki K, Yonemoto J, Sone H (2006) Activation of telomerase in BeWo cells by estrogen and 2,3,7,8-tetrachlorodibenzo-*p*-dioxin in co-operation with c-Myc. *Int J Oncol* 28: 43–51

115 Yang X, Liu D, Murray TJ, Mitchell GC, Hesterman EV, Karchner SI, Merson RR, Hahn ME, Sherr DH (2005) The aryl hydrocarbon receptor constitutively represses c-myc transcription in human mammary tumor cells. *Oncogene* 24: 7869–7881

116 Camacho IA, Nagarkatti M, Nagarkatti PS (2004) Evidence for induction of apoptosis in T cells from murine fetal thymus following perinatal exposure to 2,3,7,8-tetrachlorodibenzo-*p*-dioxin (TCDD). *Toxicol Sci* 78: 96–106

117 Camacho IA, Singh N, Hegde VL, Nagarkatti M, Nagarkatti PS (2005) Treatment of mice with 2,3,7,8-tetrachlorodibenzo-*p*-dioxin leads to aryl hydrocarbon receptor-dependent nuclear translocation of NF-kB and expression of Fas ligand in thymic stromal cells and consequent apoptosis in T cells. *J Immunol* 175: 90–103

118 Thatcher TH, Maggirwar SB, Baglole CJ, Lakatos HF, Gasiewicz TA, Phipps RP, Sime PJ (2007) Aryl hydrocarbon receptor-deficient mice develop heightened inflammatory responses to cigarette smoke and endotoxin associated with rapid loss of the nuclear factor-κB component RelB. *Am J Pathol* 170: 855–864

119 Hwang JA, Lee JA, Cheong SW, Youn HJ, Park JH (2007) Benzo[*a*]pyrene inhibits growth and functional differentiation of mouse bone marrow-derived dendritic cells. Downregulation of RelB and eIF3 p170 by benzo[*a*]pyrene. *Toxicol Lett* 169: 82–90

120 Lee JA, Hwang JA, Sung HN, Jeon CH, Gill BC, Youn HJ, Park JH (2007) 2,3,7,8-Tetrachlorodibenzo-*p*-dioxin modulates functional differentiation of mouse bone marrow-derived dendritic cells: Downregulation of RelB by 2,3,7,8-tetrachlorodibenzo-*p*-dioxin. *Toxicol Lett* 173: 31–40

121 Vogel CF, Sciullo E, Li W, Wong P, Lazennec G, Matsumura F (2007) RelB, a new partner of aryl hydrocarbon receptor-mediated transcription. *Mol Endocrinol* 21: 2941–2955

122 Vogel CF, Sciullo E, Matsumura F (2007) Involvement of RelB in aryl hydrocarbon receptor-mediated induction of chemokines. *Biochem Biophys Res Commun* 363: 722–726

Molecular, Clinical and Environmental Toxicology. Volume 1: Molecular Toxicology
Edited by A. Luch

Mapping the epigenome – impact for toxicology

Jennifer Marlowe, Soon-Siong Teo, Salah-Dine Chibout, François Pognan and Jonathan Moggs

Novartis Pharma AG, Investigative Toxicology, Preclinical Safety, Basel, Switzerland

Abstract. Recent advances in technological approaches for mapping and characterizing the epigenome are generating a wealth of new opportunities for exploring the relationship between epigenetic modifications, human disease and the therapeutic potential of pharmaceutical drugs. While the best examples for xenobiotic-induced epigenetic perturbations come from the field of non-genotoxic carcinogenesis, there is growing evidence for the relevance of epigenetic mechanisms associated with a wide range of disease areas and drug targets. The application of epigenomic profiling technologies to drug safety sciences has great potential for providing novel insights into the molecular basis of long-lasting cellular perturbations including increased susceptibility to disease and/or toxicity, memory of prior immune stimulation and/or drug exposure, and transgenerational effects.

Introduction

Epigenetics is a burgeoning area of scientific research that encompasses the study of changes in the regulation of gene activity and expression that are not directly dependent on DNA sequence. The principle way in which epigenetic information is stored and propagated is *via* methylation of DNA at cytosine residues, to form the modified base 5-methylcytosine, and through post-translational modification of histone proteins that package genomic DNA into chromatin. Specific patterns of these epigenetic marks form the molecular basis for developmental stage- and cell type-specific gene expression patterns, which are the hallmarks of distinct cellular phenotypes (Tab. 1 and references therein). Thus, while the genetic code represents relatively fixed or stored information, the epigenetic code represents the way in which the genetic code is organized and read, referring to both heritable changes in gene activity and expression as well as stable, long-term alterations in the transcriptional potential of a cell that may or may not be heritable. Recent research has begun to uncover the molecular basis for how cells define, interpret, and modify epigenetic codes. One emerging hypothesis suggests that susceptibility to a wide range of diseases partly results from disregulation of the epigenome (reviewed in [1]). Specifically, the epigenetic mechanisms that control stem cell differentiation and organogenesis also contribute to the biological response to endogenous and exogenous stimuli that influence or cause disease progression. Consequently, the pharmacological inhibition of epigenetic modifications is

Table 1. Key epigenetic factors, modifications, and regulatory proteins.

Epigenetic modification	Enzyme class	Functions	Examples	References
DNA methylation	DNA methyltransferases	methylation maintenance; repression of transcription; embryonic methylation remodeling; genomic imprinting	DNMT1 DNMT2 DNMT3A DNMT3B DNMT3L	[148] [149] [150]
	Methyl-CpG binding proteins	transcriptional repression; DNA repair; repair of deaminated 5-methylcytosine	MeCP2 MBD1 MBD2 MBD3 MBD4	[151] [152] [153] [154] [155]
Histone methylation	Histone methyltransferases	methylation of specific arginines or lysines in histones H3 and H4; transcriptional regulation, RNA metabolism and DNA damage repair	Arginine MTs (PRMTs); SET domain containing lysine MTs (EZH1); Dot1-like lysine MTs (DOT1L)	[156] [157] [158, 159]
	Histone demethylases	activation of genes involved in animal body patterning and inflammation; cell fate determination	H3K27 demethylase (UTX and JMJD3); H3K4 demethylase (LSD1)	[160] [161, 162]
Histone acetylation	Histone acetyltransferases	alter histone-DNA interactions to create open chromatin; neutralize positive charge of lysines on all four core histones	HAT1; CBP/P300; MYST proteins	[163–166]
	Histone deacetylases	remove acetyl groups from histones; transcriptional repression; also target non-histone proteins	HDAC1-10	[167, 168]
Histone phosphorylation	Histone kinases	chromosome condensation; cell-cycle progression; transcriptional control	MSK1-2; CKII; Mst1	[163, 169]

(continued on next page)

Table 1. (continued).

Epigenetic modification	Enzyme class	Functions	Examples	References
Histone ubiquitination	Histone ubiquitinases	regulation of histone methylation	Rad6; Paf1	[163, 170]
Histone sumoylation	Histone sumoylases	regulation of protein-protein interactions, protein function or localization; antagonizes acetylation	E3 SUMO ligases; PIAS family proteins	[171, 172]
Histone variants		regulation of transcription, repair, chromosome assembly and segregation	H2A.Bbd H2A.Z H3.1/H3.3 H1 variants	[173] [174] [175] [176]
Noncoding RNAs		bind to the mRNA of target genes to negatively control their expression; essential roles in development; expression patterns linked to cancer; emerging roles in exposure/disease	microRNAs	[177, 178]
Sirtuins, polycomb, trithorax proteins, etc.	SIR, PcG, TRX	required for proper gene expression through the control of chromatin structures	SIRT1-7	[179–181]

proving to be a promising strategy for drug development. Furthermore, the importance of epigenetic mechanisms in human disease has prompted the initiation of worldwide efforts to map the human epigenome.

It is becoming increasingly evident that some of the earliest events preceding the development of overt pathologies involve perturbations of the epigenome. The best-characterized evidence for mechanistic connections between early epigenetic perturbations and disease comes from the field of intestinal cancer biology and rodent non-genotoxic carcinogenesis. In addition to providing an overview of the key mechanistic links between epigenetic perturbations and carcinogenesis, the focus of this review is to highlight the potential for perturbations of the epigenome by xenobiotics, including pharmaceutical drugs, as well as new opportunities arising from epigenomic profiling technologies for identifying both mechanisms and biomarkers of toxicity.

Mapping the epigenome – Worldwide efforts and resources

Over the course of the past decade, the importance of the epigenome in determining disease susceptibility and the biological response to environmental exposures, including pharmaceutical drugs, has gained worldwide recognition amongst biomedical researchers. The epigenome is increasingly recognized as playing a central role in the development and progression of many, and potentially all, human pathologies. A multitude of national and international scientific efforts, consortia, and publicly available databases have emerged that highlight the importance of this area of biology to the understanding of human development and disease. We outline below some of the most important activities and resources currently available.

The Human Epigenome Project

Following the work of the Human Genome Project (http://www.genome.gov/) to produce a complete DNA sequence of the entire human genome, efforts to map the full human epigenome have been initiated, organized under the auspices of the Human Epigenome Project (HEP) (http://www.epigenome.org/) [2]. The HEP represents a public/private consortium consisting primarily of The Wellcome Trust Sanger Institute, Epigenomics AG, and The Centre National de Génotypage. The ultimate goal of the HEP is to produce a genome-wide map of DNA methylation patterns across all major tissues for all human genes, and to obtain high-resolution methylation profiles through the use of bisulfite sequencing for a limited number of reference epigenomes. The efforts of the consortium have thus far produced a pilot study for the analysis of methylation patterns of the human major histocompatibility complex [3], as well as high-resolution methylation profiles of human chromosomes 6, 20, and

22, consisting of 1.9 million CpG methylation values derived from 12 different tissue sources [4]. Findings from these initial epigenome mapping efforts both underscore the feasibility of large-scale, single-base pair resolution analysis of DNA methylation and characterize important differences in epigenome organization among differing tissue types. Key observations from Eckhardt et al. [4] include the tendency for CpG islands in healthy tissues to remain unmethylated, the existence of tissue-specific regions of differentially methylated DNA that are preferentially located several kilobases from transcription start sites and in orthologous sequences (human and mouse), and the fact that DNA methylation tends to affect the transcription of genes with low CpG density in their 5' UTRs, not just those consisting of *bona fide* CpG islands. These observations are reinforced by additional research showing that genic regions in general are highly methylated, and that at least for colon cancer cells, relatively few genes are hypermethylated compared to cells derived from normal tissues [5]. Research by Weber et al. [6] points to additional evidence indicating the divergence of DNA methylation patterns between humans and chimpanzees, a finding which may eventually impact the preclinical assessment of drug safety in humans.

The Epigenomics Program of the NIH Roadmap for Medical Research

The goal of the NIH Roadmap for Medical Research (http://nihroadmap.nih. gov/), implemented in 2002, is to identify areas of biomedical research that, due to their broad relevance and complexity, warrant the input of multiple NIH institutes, and in doing so make the greatest possible impact on the progress of research in a particular field. The roadmap thus works to stimulate research and/or research resources to transition major program areas out of the auspices of the NIH Roadmap and into the relevant NIH institutes within a 5–10-year timeframe. One of these major program areas is in the field of epigenomics. The working hypothesis of the NIH Roadmap Epigenomics Program (http://nihroadmap.nih.gov/epigenomics) is that the fundamental basis of disease susceptibility and progression lies in the epigenetic regulation of the genetic blueprint. The program aims mainly to drive the development of comprehensive reference epigenome maps as well as new technologies for epigenomic analyses. Specifically, the program has proposed to develop standardized platforms, procedures and reagents for epigenomic research, to conduct pilot projects on the evolution of epigenomes, to develop technologies for *in vivo* imaging and single-cell epigenomic analyses, and to create a public data resource to aid in the application of epigenomic approaches. In addition to the NIH Roadmap Epigenomics Program, the National Institutes of Environmental Health Sciences (NIEHS) (http://www.niehs.nih.gov/) has an ongoing environmental epigenetics program that aims to understand the role of epigenetics in diseases and disorders related to environmental exposures [7]. The program supports research examining the effect of environmental exposures on multiple mechanisms of epigenetic regulation, including DNA methylation and other chromatin modifications, siRNA and gene silencing, and imprinting, among others.

The Epigenome Network of Excellence

The Epigenome Network of Excellence (NoE) [8] (http://www.epigenome-noe.net/index.php) forms the core of the epigenetics research community in Europe, consisting of more than 25 research groups in addition to more than two dozen associate members. At this time, the NoE is working to create a cohesive joint research program for the advancement of scientific discoveries in epigenetics, and towards that end, have established an interactive website including links to member information, tools and web resources, and protocols from leading European research labs.

Publicly available databases

A multitude of publicly accessible databases are currently available for investigating imprinted genes, methylation patterns, and histone protein sequences and functions. We have listed a few of the most recognized and useful here.

DNA Methylation Database (http://www.methdb.de/): The DNA methylation database allows users to search methylation patterns and profiles from a list of sequenced genes. It is searchable by methylation content, species, sex, tissue, gene, sequence, phenotype, and method utilized in the analysis, with links to relevant articles [9].

Histone Database (http://genome.nhgri.nih.gov/histones/): This database is a comprehensive compilation of major public databases representing a curated and searchable collection of full-length sequences and structures of histones and non-histone proteins containing histone-like folds [10].

Imprinted Genes Databases (http://igc.otago.ac.nz/home.html; http://www.mgu.har.mrc.ac.uk/research/imprinting/; http://www.geneimprint.com/): Genes that undergo genomic imprinting are characterized by mono-allelic expression as a result of epigenetic modifications, usually consisting of DNA methylation-mediated silencing of either the paternal or maternal allele [11]. This pattern of mono-allelic expression is transmitted during the process of cell division to the daughter cells. To date, more than fifty human genes are reported to be imprinted [12]. Examples of maternally imprinted genes include *H19* and *CDKN1C*. *IGF2* represents a paternally imprinted gene. The importance of maintaining normal patterns of gene imprinting is highlighted by a number of imprinted gene disorders affecting growth and neurodevelopment, including Prader-Willi and Angelman syndromes (associated with abnormal imprinting on chromosome 15q11–q13) [13] and Beckwith-Weidemann syndrome (associated with loss of imprinting of the *IGF2* gene) [14].

Technological approaches to epigenetic analyses

Approaches to detecting DNA methylation

Technologies for the analysis and quantification of DNA methylation have been available for 25 years or more [15]. Standard methods for analysis of 5-methylcytosine (5-MeC) residues fall into three general categories: differential enzymatic DNA cleavage, selective chemical DNA cleavage, and differential sensitivity to chemical conversion [16]. The first methods utilized for the precise quantification of total 5-MeC included high-performance liquid chromatography (HPLC) and high-performance capillary electrophoresis (HPCE) [17]. However, the analysis of DNA methylation at precise sequences was based solely on the use of enzymes that are sensitive to methylated and unmethylated recognition sites [18]. The many drawbacks of this methodology, including the need for significant amounts of high molecular weight DNA, false positives due to incomplete digestion, and limitations in regions of the genome amenable to analysis, led to the standardization of bisulfite treatment of DNA as a fundamental tool in DNA methylation research [19, 20]. Treatment of genomic DNA with sodium bisulfite followed by alkaline treatment converts unmethylated cytosines to uracil, whereas 5-MeC remains unchanged. Thus, bisulfite conversion combined with the additional methods described below can be used to identify and quantify methylation at single-nucleotide resolution from very small quantities of DNA originating from a variety of sources, including fixed or frozen archived samples [21].

Restriction landmark genomic scanning

While the above techniques are limited to the analysis of DNA methylation at specific gene or DNA loci, multiple approaches are also available for the detection of whole-genome methylation patterns. One of the earliest methods adapted for DNA methylation analysis was restriction landmark genomic scanning (RLGS) [22]. This method involves size fractionation of radioactively labeled regions of unmethylated DNA inside methylation-sensitive restriction enzyme sequences. High-frequency targets are then digested with a second restriction endonuclease and the resulting fragments are again separated to yield scattered hotspots of DNA methylation. RLGS thus results in the quantification of both gene copy number and methylation status, thereby allowing for the simultaneous analysis of genetic and epigenetic changes among different tissue sources. RLGS is limited in scope by the inability to resolve all restriction fragments by electrophoresis, as well as by the use of sequence-specific enzymes that may not target all CpG islands [15].

Arbitrarily primed PCR

A major advantage of the use of arbitrarily primed PCR is the generation of reproducible fingerprints of complex genomes without the prior need for sequence information. This approach requires just two rounds of low stringency amplification followed by a standard high stringency PCR step [23]. The technique has been modified for application to the detection of genome-wide patterns of altered DNA methylation by a number of investigators, all of which result in the preferential amplification of CpG islands and gene promoter regions [15]. Published methods include methylation-sensitive arbitrary primed PCR [24], methylated CpG-island amplification (MCA) [25], and amplification of intermethylated sites (AIMS) [26]. Bisulfite genomic sequencing must be employed for further characterization of results generated by any of the aforementioned techniques.

MethyLight

MethyLight represents a high-throughput quantitative method for DNA methylation analysis with the potential for screening hundreds to thousands of samples at once [27, 28]. The technique utilizes bisulfite treatment followed by fluorescence-based real-time PCR (TaqMan®) using primers designed to overlap with potentially methylated CpG dinucleotides to allow for sequence discrimination. MethyLight is capable of detecting methylated alleles in the presence of a 10 000-fold excess of unmethylated alleles, thus representing a highly specific and sensitive technique. The assay is also highly quantitative, and can determine with great accuracy the prevalence of a particular DNA methylation pattern. As it is a PCR-based method, MethyLight requires only minute amounts of DNA of modest quality, and is thus compatible with small sample sizes and paraffin-embedded tissues.

Pyrosequencing

Bisulfite chemistry-coupled pyrosequencing for locus-specific DNA methylation analysis is an emerging innovative technology, poised to become the gold

---

Figure 1. Methods for genome-wide and locus-specific DNA methylation profiling. The methyl DNA immunoprecipitation (MeDIP) assay (left panel, adapted from [182]), in combination with the use of high-resolution microarrays or high-throughput sequencing (HTS) techniques, is fast becoming a powerful and well-validated method for identifying methylated CpG-rich sequences. The MeDIP technique is performed by randomly fragmenting purified genomic DNA and immunoprecipitating the fragments with a monoclonal antibody against 5-methylcytosine (^mC) to isolate enriched pools of methylated DNA. Total input DNA and immunoprecipitated samples are then co-hybridized to the microarray platform of choice (promoter array, CpG island array, whole genome array, etc.) to identify regions of differential methylation. Pyrosequencing (right panel, adapted from www.biotage.com) was developed as an alternative to dideoxy sequencing, and has been recently adapted for the quanti-

standard for quantitative CpG methylation analyses (Fig. 1). Pyrosequencing represents a DNA sequencing method in which the incorporation of nucleotides into a growing DNA chain is quantitatively monitored in real time [29, 30]. This is made possible through the enzymatic conversion of pyrophos-

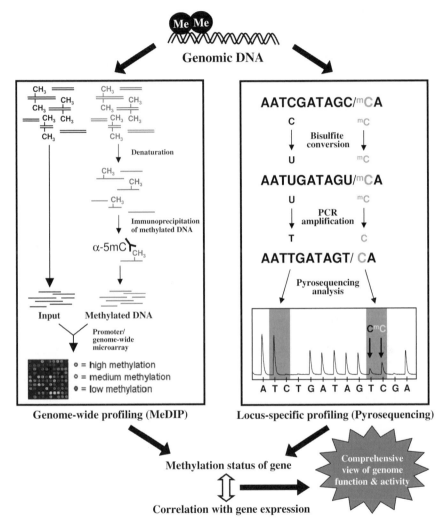

Genome-wide profiling (MeDIP)   **Locus-specific profiling (Pyrosequencing)**

Methylation status of gene

Comprehensive view of genome function & activity

Correlation with gene expression

tative analysis of methylated cytosines. The method is based upon on the detection of pyrophosphate (PPi), which is released from incorporated nucleotides by DNA polymerase during strand elongation. The released pyrophosphate molecules are converted to ATP by ATP sulfurylase, which is then utilized by luciferase to oxidize luciferin and generate light. The relative light signal generated is proportional to the number of base pairs added sequentially during the sequencing reaction. The major advantage of the pyrosequencing method is that the data generated are actual sequences rather than fluorescence data from PCR-based amplification. The combination of the aforementioned techniques provides a powerful discovery tool for evaluating the integrated epigenomic response to environmental exposures or differential states of disease.

phate released during the incorporation of each individual nucleotide, resulting in the production of a proportional light signal. Variations in DNA sequence that are dependent on methylation patterns can be detected using pyrosequencing following an initial bisulfite conversion step and PCR amplification. Thus, the method is similar to conventional SNP typing, where bisulfite-converted cytosines are detected as C-T SNPs [15]. Pyrosequencing also has additional applications in DNA methylation analysis, including monitoring of allele-specific methylation patterns [31], as well as the assessment of global methylation changes through analyzing the degree of methylation at repetitive elements [32] or following restriction digestion [33]. Overall, this method provides a tool for gene-specific hypothesis testing, for example on selected targets identified either *via* genome-wide DNA methylation analysis or *via* gene expression profiling and pathway analysis.

Methyl-DIP (methylated DNA immunoprecipitation)

The application of techniques for enriching methylated genomic sequences to CpG island, promoter, or whole-genome microarrays represents the most efficient method to analyze genome-wide patterns of DNA methylation. Such techniques invariably take advantage of the use of antibodies to methylated DNA [5] or methylated DNA binding proteins [34] to capture methylated sequences on a genome-wide scale. The methylated DNA immunoprecipitation (MeDIP) version of this technology, invented by scientists at the Friedrich Miescher Institute (http://www.fmi.ch/), represents a key method for analyses of the DNA methylome [35]. Hypermethylated regions of DNA are immunoprecipitated with an antibody against 5-MeC and then hybridized to genomic microarrays in a manner similar to the standard Chip-on-chip procedure [36]. This technique provides a powerful discovery tool for evaluating the integrated epigenomic response to environmental exposures or differential states of disease. When combined with locus-specific DNA methylation assays and associated gene expression data, this methodology promises to provide a comprehensive view of genome function and activity (Fig. 1) and examples of the utility of this technology in the investigation of basic biology and disease progression continue to emerge [37–41].

Utilizing technologies for DNA methylation profiling: An example from human cell lines

As outlined above, it is now possible to analyze global alterations in DNA methylation through the use of DNA immunoprecipitation and genomic microarrays. In addition, the pyrosequencing method for quantification of methylation at individual CpG sites provides a tool for gene-specific hypothesis testing, for example on selected targets identified either *via* genome-wide DNA methylation analysis or *via* gene expression profiling and pathway analysis. As outlined in Figure 1, the combination of these two tools could be applied to the analysis of drug-induced perturbations of DNA methylation in preclinical animal models, and as such, provide insight into the mechanisms of the adverse outcomes of drug exposure. Figure 2 represents an illustration of the application of MeDIP and pyrosequencing to a specific region of high DNA methylation in the FoxF1 gene in the human 293FT cell line. The data illustrate the robustness and reproducibility of both assays for analyzing the DNA methylation status of a specified genomic region. In addition, both assays are sensitive to the treatment of cells with the DNA methyltransferase inhibitor 5-azacytosine, illustrating the ability to detect locus-specific alterations in DNA methylation following a drug exposure. Thus, the use of these two robust assays for genome-wide and locus-specific DNA methylation analysis in combination with gene expression data holds great promise to generate a comprehensive view of genome function and activity, including biomarkers for and mechanistic insights into drug-induced toxicity.

Approaches to detecting histone modifications

Available methods for the analysis of histone modifications are limited in comparison to the variety of techniques for DNA methylation analysis. Mass spectrometry represents the gold standard for analysis of post-translational histone modifications, but cannot be used to assess the epigenetic status of gene-specific loci [15]. However, obtaining global measures of histone modifications is possible. Liquid chromatography-electrospray mass spectrometry (LC-ES/MS) has been used to analyze histones isolated by HPLC and potentially provides data on the post-translational modification status of all cellular histone proteins (H3, H4, H2A, H2B, and H1) in one experiment [42–44]. In addition, specific histone modifications can be evaluated using Western blotting, immunohistochemistry [45], or tandem mass spectrometry [44]. Chromatin immunoprecipitation (ChIP) is a very powerful method for the analysis of histone modifications associated with specific genomic DNA sequences (i.e., gene promoters) [46]. Coupled with whole-genome microarray analysis, the ChIP method has allowed for the mapping of histone modifications in a growing number of model species [47–51]. The application of ChIP to tissue samples obtained from preclinical animal models is feasible but technically very challenging.

Relevance of epigenetic perturbations for toxicology

Perturbation of epigenetic status alters the spectrum of genes and proteins expressed in a cell, which in turn leads to alterations in cellular phenotype. It is becoming apparent that a wide range of environmental factors including

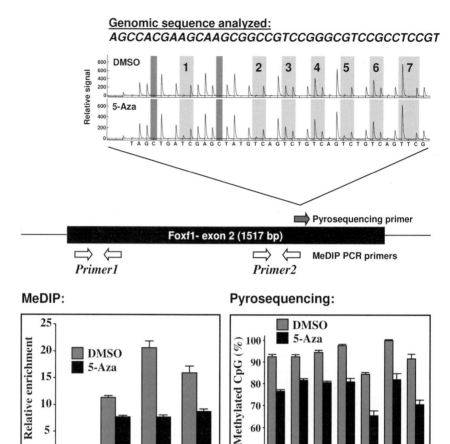

Figure 2. Correlation of results from combined MeDIP and pyrosequencing analyses. In human cells, regions of the FoxF1 gene have been shown to be differentially methylated and responsive to perturbations of methylation status *via* DNA methylation inhibitors such as 5-azacytosine (5-Aza) [6]. 293FT cells were treated with 5 μM 5-Aza or DMSO vehicle control for 96 hours. DNA was subsequently isolated from cell cultures and divided equally for application to either MeDIP [5] or pyrosequencing assays. The relative positions of primer sets in the exon 2 region of the FoxF1 gene (ENSEMBLE reference sequence ENST00000262426) used for real-time PCR of immunoprecipitated samples and for sequencing are depicted by white and gray arrows, respectively. Real-time PCR results (bottom left panel) show that the relative enrichment of two regions of the FoxF1 gene, as well as a positive control (BRDT) [6], is approximately 25-fold higher than that of the unmethylated negative control, UBE2B [6]. This relative enrichment is reduced by roughly 50% following incubation of cells in 5-Aza. The bottom right panel depicts the percentage of methylated cytosines present at seven different CpG sights in the FoxF1 gene. While all seven cytosines are highly methylated in control cells (~93–100% methylated), the percentage of methylated CpGs drops significantly in 5-Aza-treated cells (68–82%). The entirety of the genomic sequence analyzed by pyrosequencing as well as examples of the pyrograms from which the data were generated are presented in the top panel. Dark gray bars represent controls for bisulfite conversion (unmethylated or non-CpG cytosines), and light gray bars represent CpG sites within the genomic sequence analyzed.

diet, stress, behavior and exposure to both natural and synthetic chemicals can influence the epigenome [52, 53]. Importantly, the epigenetic status of the genome can be stably propagated through mitosis and cell division, and therefore toxicants that perturb epigenetic status can potentially have long lasting effects on the phenotype of a somatic cell population. It has also been suggested that certain environmental factors are associated with epigenetic changes and can be transmitted *via* the germline [54–56]. Imprinted genes and metastable epialleles (identical alleles, variably expressed due to early developmental epigenetic modifications [57]) may be especially susceptible to environmental influences, such as nutritional and chemical exposures, as these classes of genes are mainly regulated by epigenetic mechanisms [58] and are currently under intense investigation in the context of fetal development and cellular reprogramming leading to adult disease [59]. Epigenomic profiling technologies will be essential for a more complete understanding of the complexity of the human epigenome, the potential for epigenome perturbation by environmental factors, and how this influences human health and disease. Such efforts will ultimately influence mechanistic toxicology research and the ways in which pharmaceutical drugs are analyzed for both safety and efficacy.

The importance of epigenomics in the understanding of human disease is exemplified by findings in the field of carcinogenesis. Key mechanistic links between early epigenetic perturbations and carcinogenesis have been observed in the field of intestinal cancer biology. Additionally, rodent non-genotoxic carcinogens provide the best examples to date for drug-induced perturbations of epigenome relative to cancer progression. Evidence from both of these fields is outlined below.

Importance of early epigenetic perturbations in carcinogenesis

A wide range of human cancers display aberrations in DNA methylation patterns, leading to the suggestion that changes in the DNA methylation status of certain genes may contribute to the transformed cellular phenotypes. A recent paradigm shift in our understanding of mechanisms of carcinogenesis has led to the classical model of multistage carcinogenesis, involving successive genetic changes that result in initiation, promotion and tumor progression, being refined to include both epigenetic and genetic changes at each stage [60]. Importantly, recent studies indicate that epigenetic alterations (e.g., altered DNA methylation status) may initiate the expansion of pre-malignant cells during the earliest stages of tumorigenesis [61]. The best example of this phenomenon comes from research on intestinal carcinogenesis. A large volume of scientific literature suggests that early epigenetic changes, specifically DNA hypermethylation and the subsequent silencing of tumor suppressor gene expression, are a critical step in the development of intestinal tumors. Evidence suggests that abnormal epigenetic gene silencing through promoter hypermethylation occurs at the earliest stages of tumor development, before specific mutations or chromosomal alterations are detected [61]. For example, promoter hypermethylation has been shown to be present in non-progressed adenomas in which no chromosomal alterations exist, suggesting that epigenetic events are the earliest causes of gene expression changes in tumor progression [62], and that aberrant CpG island methylation tends to accumulate over the course of multistage carcinogenesis [63].

The most well-studied example of this phenomenon is in colon cancer. Studies using the HCT116 human colon cancer cell line have uncovered a large number of both genetic and epigenetic alterations. Significantly, treatment of these cells with DNA-demethylating agents or disruption of genes which encode DNA methyltransferases induces reactivation of a number of epigenetically silenced genes [64–66], resulting in phenotypic changes including reduced proliferation, senescence and apoptosis [67, 68]. One specific example of an epigenetic change associated with colon cancer is the hypermethylation of SFRP gene family members in aberrant crypt foci (ACF), the early-stage, pre-invasive lesions of the colon that are at highest risk for cancer progression. Such cells generally do not contain gene mutations that would activate the Wnt pathway, but methylation of SFRP gene promoters is consistently present in ACFs and persists in primary colon cancer cells as well [69]. SFRP proteins function to antagonize the Wnt pathway, and their aberrant silencing may be required for ultimate progression to malignant phenotype.

There are multiple other examples of genes that are epigenetically silenced early in the progression of colon cancer. HIC1 has been shown to be hypermethylated in cells from early, pre-invasive colon tumors [70], and the loss of its expression appears to predispose cells to the acquisition of mutations in the p53 gene [71]. It is thought that the inactivation of the MLH1 gene plays an initiating role in the pathogenesis of sporadic MSI (microsatellite instability) colon cancers, and that such aberrant methylation, and the subsequent loss of MLH1 gene expression, is the cause rather than the consequence of colon carcinogenesis [72]. Methylation of the MLH1 gene is also strongly associated with loss of MLH1 protein expression and MSI-H phenotype in gastric carcinomas [73]. DAPK promoter hypermethylation has been suggested to play a role in the early steps of tumor progression in colorectal carcinomas [74]. Carcinomas of the colon exhibit DAPK promoter hypermethylation in 81% of cases, whereas intraepithelial neoplasias and normal mucosa exhibit 68% and 25%, respectively. Epigenetic silencing of CXCL12 gene expression appears to enhance the metastatic potential of colon cancer cells [75]. Finally, multiple studies have shown that promoter methylation of multiple tumor-associated genes accumulates in conjunction with tumor progression [76, 77].

Rodent non-genotoxic carcinogenesis as a paradigm for drug-induced perturbations of the epigenome

An emerging body of data suggests that epigenetic perturbations may also be involved in the adverse effects associated with some drugs and toxicants, including certain classes of non-genotoxic carcinogens, specifically phenobarbital and peroxisome proliferators [53, 78–82]. Drug-induced stress (e.g., chronic injury/inflammation/reactive oxygen species) may trigger epigenetic changes that "lock-in" abnormal proliferative states *via* heritable transcriptional repression of key genes/pathways [60]. Thus, epigenomic profiling has significant potential for enhancing our understanding of the molecular basis of the deregulated biological processes that drive tumorigenesis, including aberrant proliferation, apoptosis, intercellular communication, and angiogenesis, among others. Evidence for this potential comes from a number of studies that analyzed the patterns of DNA methylation changes induced by phenobarbital, a classical and well-characterized rodent liver tumor promoter.

Multiple investigations have revealed both gene-specific and global alterations in DNA methylation in response to phenobarbital. These alterations are both time- and target-organ dependent and correlate with the relative susceptibility to tumor formation in a particular mouse strain. When the tumor-susceptible B6C3F1 mouse was given a liver tumor-promoting dose of phenobarbital in drinking water for 14 days, hypomethylation of the raf gene was observed, whereas raf methylation status was unaffected in tumor-resistant C57BL/6 mice [83]. Interestingly, raf gene expression levels were unchanged over the same time course, indicating that changes in methylation status may precede gene expression changes, therefore representing the earliest observable changes in the process of tumorigenesis, as also suggested by the aforementioned data on intestinal carcinogenesis. The raf gene was also found to be hypomethylated in both drug-induced and spontaneous mouse liver tumors following 2 years of phenobarbital administration, with a corresponding increase in raf gene expression in the phenobarbital-induced tumors but not the spontaneous tumors [83].

Additional evidence for the importance of DNA methylation perturbations in xenobiotic-induced carcinogenesis comes from studies comparing the effects of phenobarbital to those of choline-devoid, methionine-deficient diets in tumor-susceptible rodents. Diets devoid of choline and deficient in methionine have been shown to result in increased hepatocyte cell proliferation [84, 85], hypomethylation and increased expression of oncogenes [86]. Such diets also lead to the promotion of liver tumors in rats [87, 88], and choline-deficient diets are carcinogenic in B6C3F1 mice [89]. These observations were subsequently correlated with alterations in DNA methylation. Assessments of methylation changes in phenobarbital-treated mice reveal early (1–4 weeks) global hypomethylation in B6C3F1 (tumor-sensitive) but not in C57BL/6 (tumor-resistant) mice [90]. However, administration of a methionine-deficient diet resulted in global hypomethylation in both strains, indicating that additional analytical precision might be required to identify those alterations with most relevance to the tumorigenic process. Gene-specific analysis of methylation changes induced by phenobarbital and choline-devoid, methionine-deficient diets also indicate hypomethylation of the raf and H-ras genes at most early (1–4 weeks) and late (2 years) time points [91]. Finally, diethanolamine, which inhibits choline uptake and causes biochemical changes similar to choline deficiency, causes similar methylation changes as phenobarbital in cultured hepatocytes from B6C3F1 mice [92].

Not all perturbations in DNA methylation status are hypomethylations. Global examinations of GC-rich regions of DNA reveal predominantly phenobarbital-induced hypermethylation corresponding to the relative tumor sensitivity of the mouse strain examined [93]. Perturbations in DNA methylation following phenobarbital treatment in mice reveal that the changes are reversible following relatively short-term (2–4 weeks) exposure, and are specific to the organ examined. For example, the livers of B6C3F1 mice show extensive hypomethylation and new regions of methylation that are nearly completely dissimilar from the changes observed in the kidney, an organ that is not a target of phenobarbital-induced carcinogenesis [94]. Phenobarbital-mediated liver tumorigenesis is dependent upon the CAR (constitutively active/androstane receptor) protein. CAR is essential for liver tumor promotion by phenobarbital in C3H/He mice following initiation with diethylnitrosamine [95], and chronic activation of CAR by the potent phenobarbital-like inducer 1,4-bis[2-(3,5-dichloropyridyloxy)]benzene results in liver tumor formation in the mouse [96]. Phillips et al. [82] reported patterns of altered DNA methylation that were unique in tumor-sensitive CAR wild-type mice as compared to tumor-resistant CAR knockout mice following exposure to phenobarbital. Pogribny et al. [81] reported a similar finding for the peroxisome proliferator WY14643, which exhibited no effect on DNA and histone methylation status in *Pparα*-null mice, whereas wild-type mice exhibited a distinctive pattern of DNA and histone methylation perturbations over the course of a long-term administration of the compound. While the aforementioned studies suggest an important role for nuclear receptor-mediated mechanisms that are known to exhibit species-specific differences in expression levels, ligand selectivity profiles and/or target genes, the specific mechanisms governing epigenetic perturbations by non-genotoxic carcinogens are generally unknown. Finally, a recent analysis of genomic regions that exhibit phenobarbital-specific changes in DNA methylation status in the livers of phenobarbital-administered mice indicates that such alterations occur at least in part in regulatory regions of genes that are characterized as having a role in the ultimate formation of tumors, including transcription factor 4, transforming growth factor β receptor II and ral guanine nucleotide dissociation stimulator [97]. However, the causal role of such genes in the process of phenobarbital-mediated liver tumorigenesis has not yet been demonstrated.

There is evidence outside of the mouse liver for DNA methylation perturbations related to carcinogenesis. In phenobarbital-treated rats, distinct changes in the methylation pattern of the p16 promoter were observed, and no methylated cytosines were detectable following 14 days of phenobarbital administration [98]. However, in patients with hepatocellular carcinoma, a decrease in the mean detectable $p16^{INK4a}$ methylated sequences in plasma was observed following surgical resection of tumors [99], again illustrating the need for systematic and specific analyses to identify potential early biomarkers of tumorigenesis across specific tumor types. Tamoxifen, another potent rat heptocarcinogen, also induces early, global DNA hypomethylation, concomitant with decreased protein expression of the DNA methyltransferases DNMT1, DNMT3a and DNMT3b, as well as decreased trimethylation of histone H4 lysine 20 [100]. Non-target tissues for tamoxifen-induced carcinogenesis, including mammary gland, pancreas, and spleen, did not exhibit global DNA hypomethylation or decreased activity of key DNMT enzymes. Similar findings were observed in estradiol-17β (E_2)-induced mammary tumors in the rat, including global hypomethylation and loss of H4 Lys20 trimethylation [101]. These alterations in the epigenome were observed after 6 weeks of treatment, whereas hyperplastic lesions were observed only after 12 weeks, again indicating that epigenetic perturbations may be the earliest detectable steps along the path of disease progression. Tumor promotion-dependent changes in normal DNA methylation patterns have also been observed in mouse models of skin carcinogenesis [102].

Evidence for epigenetic mechanisms in inflammation

Inflammation is thought to drive the pathogenesis of a number of diseases, including cancer, and there is growing evidence of mechanistic links between inflammation and DNA methylation changes. For example, inflammation is thought to drive the development of prostate cancer [103], and aberrant silencing of tumor suppressor genes as driven by inflammatory processes may be the earliest events in the development and progression of prostate tumors [104]. Inflammatory bowel diseases such as Crohn's and ulcerative colitis predispose afflicted patients to the development of colon cancer [105], and methylation perturbations also appear to be the earliest changes observed in the progression of these disorders to the eventual formation of cancer [106, 107]. Multiple research labs have identified a link between oxidative stress observed during carcinogenesis and alterations in DNA methylation status [108], and deregulated cysteine and methionine metabolism has been associated with both oxidative stress and inflammation [109]. While the mechanisms governing the interplay between inflammation and altered methylation patterns remain uncertain, recent data using *in vitro* systems indicate that halogenated cytosine damage products that form during inflammatory events can mimic 5-MeC [110]. This could conceivably result in effects on DNA methylation enzymes and methyl-binding proteins, ultimately leading to altered DNA-protein interactions critical for gene regulation, structural maintenance of chromatin, and possibly even the faithful transmission of methylation patterns to gametes. Furthermore, direct control of histone methylation patterns also appears to be effected by the inflammatory process. Polycomb group (PcG) proteins maintain cells in their differentiated state *via* H3 Lys27 trimethylation (H3K27me3)-dependent gene silencing. Macrophages that migrate to regions of inflammation have the ability to transdifferentiate, and this ability is at least partly dependent upon an H3K27me demethylase, Jmjd3, expressed in macrophages in response to inflammatory processes [111]. Jmjd3 binds to PcG target genes to alter H3K27me3 status and transcription, providing a clear mechanistic link between inflammation and epigenomic reprogramming.

It is noteworthy that the link between inflammation and epigenome perturbation might also be relevant to the pathogenesis of drug-induced liver injury (DILI), one of the most common causes for the withdrawal of pharmaceutical drugs from the market. DILI involves the parent drug or metabolites directly affecting liver cell biochemistry or eliciting an immune response, and inflammation is a pathogenic component of a number of types of acute and chronic liver diseases [112]. One recent report shows that changes in DNA methylation patterns occur in pyrazinamide-treated rats, specifically LINE-1 hypomethylation, and GSTP (glutathione-S-transferase placental form) and p16^{INK4a} hypermethylation. These changes occur prior to standard indicators of liver injury (serum bilirubin, aminotransferase activity), and thus might ultimately contribute to the improved prediction of drug-induced hepatotoxicity [113].

Epigenetic mechanisms of non-genotoxic carcinogenesis – Relevance for safety assessment?

Evidence for the epigenetic basis of disease progression as well as perturbations of DNA methylation observed during xenobiotic-induced carcinogenesis emphasizes the potential importance of epigenomic profiling in the assessment of drug safety, both clinically and preclinically. For example, tumor findings are common endpoints in the preclinical testing of drug candidates in rodent models. Such *in vivo* carcinogenesis is rarely genotoxic in nature, as directly genotoxic compounds are excluded at an early point in the drug development process by use of *in vitro* and *in vivo* assays. However, currently there are no sufficiently accurate and well-validated short-term assays to identify non-genotoxic carcinogens, thus necessitating an expensive 2-year rodent bioassay for assessing the carcinogenic risk of such compounds to humans.

Research into the key events that define a particular mode of action for tumor formation have mainly focused on mutational responses resulting from induction of DNA damage and enhanced cell proliferation. However, epigenetic factors are increasingly being considered to play an important role in susceptibility to tumor induction by xenobiotics and as significant modifiers of tumor responses and differential susceptibility to tumor formation. In particular, it is well established that epigenetic modifications can contribute directly to genomic instability (e.g., *via* spontaneous deamination of 5-MeC). Furthermore, it is noteworthy that an emerging mammalian *in vitro* genotoxicity assay relies on the regulation of *GADD45α* [114], a gene that has recently been shown to promote epigenetic gene reactivation by repair-mediated DNA demethylation [115]. Thus, the cancer risk assessment process must also take into consideration the interplay of both genetic and epigenetic events [116]. A key step in cancer risk assessment as applied to pharmaceutical development is to establish the relevance of preclinical toxicity findings to humans. The ability of mammalian cells to retain normal methylation patterns appears to be inversely related to tumor susceptibility [117] and clear differences exist in DNA methylation patterns between mammalian species [6, 80, 118, 119]. Further investigations of these phenomena will be essential for evaluating the utility of incorporating epigenomic-profiling data into future risk assessments.

Opportunities for biomarkers

Given the clear connection between early epigenetic perturbations and carcinogenesis, there is great potential for biomarker identification in the area of rodent non-genotoxic carcinogenesis. The applicability of early biomarkers in the assessment of non-genotoxic carcinogenesis in preclinical rodent toxicity studies has thus far not been established conclusively. Even in those cases where candidate biomarkers have been suggested for specific carcinogenic effects, it is unknown whether the biomarkers involved are mechanistically or causally involved in cancer development. Thus, it would be exceedingly valuable to establish the mechanisms by which early biomarkers are linked to tumor formation, and ultimately demonstrate the concept that early biomarkers can reliably and robustly predict later cancer development, including potential insight into the human relevance of rodent non-genotoxic carcinogens, something that is often called in to question [120, 121]. An additional major long-term benefit of validating robust, early biomarkers of non-genotoxic carcinogenesis is the reduction in the number of animals used for cancer bioassays. Epigenomic profiling represents a powerful approach for evaluating mechanisms of toxicity and biomarker identification. In particular, this approach would complement and build upon existing mechanistic and predictive toxicogenomic studies in which a range of genotoxic and non-genotoxic carcinogens have already been profiled [122, 123]. Together, a combined genomic/epigenomic profiling approach should contribute to the assessment of human cancer risk at a much earlier point in the pharmaceutical development process.

The presence of tumor-specific circulating cells, nucleic acids and methylated DNA in plasma or serum also represent the most promising example of identifying biomarkers for the early detection of cancer in humans [124–126], and could potentially be used in preclinical models as one component of the overall safety assessment of a particular compound. The conceptual and technical feasibility of identifying circulating tumor cells in preclinical rodent models has not yet been explored. In addition, although much work is ongoing in human patients [127–129], the predictive value of tumor-specific circulating nucleic acids and/or methylated DNA in plasma or serum as surrogate biomarkers of non-genotoxic carcinogenesis remains to be assessed in pre-clinical studies.

Molecular classification of tumors

In addition to the identification of early risk factors for non-genotoxic carcinogenesis, epigenomic profiling technologies hold great promise for the molecular classification of tumors. Indeed, the application of both genomic (mRNA and microRNA) and epigenomic profiling to human cancers has delineated molecular subtypes that are associated with distinct biological processes, disease progression and treatment response [130]. Combined gene expression and epigenomic profiling could in principle be applied to archived tissues from long-term rodent carcinogenicity studies. Both tumor and surrounding "normal" tissue could be profiled (isolated by laser capture microdissection, for example) in drug-treated and vehicle control groups. Importantly, baseline data for the molecular profiles of spontaneous tumors observed in aging animals would also need to be established. An integrated systems toxicology approach (e.g., gene ontology, pathway mapping, interactome analysis, transcriptional networks, meta analysis) could subsequently be applied to gain novel insights into the developmental lineage, differentiation state, and ultimately mechanisms of tumor formation. These data should ultimately facilitate human risk assessment during late-phase drug development and would also contribute to an epigenomic profiling strategy for selection of early biomarkers of non-genotoxic carcinogenesis.

Potential limitations of DNA methylation profiling for safety assessment

DNA methylation profiles include a dynamic range of methylation at numerous CpG sites within both strands of the target genomic locus and are thus inherently more complex than mRNA expression data and present significant challenges with respect to bioinformatics analysis. Nevertheless, specific CpG sites are profoundly influenced by drug/chemical exposure [53, 78, 79, 81, 82, 93] and these changes frequently, but not exclusively, correlate with altered genome function. Also, drug-induced epigenetic changes might be limited to a small subset of cells within a target organ or tissue making them difficult to detect. However, growing evidence suggests that aberrant epigenetic events associated with carcinogenesis may occur in numerous cells (i.e., polyclonal or "field" effects).

Not all drug-induced epigenetic changes will be detected using DNA methylation profiling alone (Tab. 1). Indeed, the epigenetic effects of peroxisome proliferators in mouse liver were recently shown to include both altered DNA and histone methylation status [81]. Thus, mechanistic studies of drug-induced changes in epigenetic status using the DNA methylation assays could be further enhanced using complementary assays, such as ChIP, that measure the occupancy of genomic localization of gene regulatory proteins [15].

Emerging opportunities for epigenomic profiling in drug safety sciences

Although relatively few drug-induced perturbations of the epigenome have been characterized to date, the emerging importance of epigenetic mechanisms in a wide range of human diseases emphasizes the need for further research in this area. This is exemplified by the ongoing clinical development of oncology drugs targeting the enzymes responsible for major epigenetic modifications, including histone deacetylases and DNA methyltransferases [131]. Additional epigenetic drug targets being explored in preclinical phases of drug development include histone methyltransferases and sirtuin inhibitors [15]. In addition to the well-established field of cancer epigenetics [132], perturbations of the epigenome have recently been associated with a number of additional disease areas including psychiatric disorders (e.g., depression, drug addiction, schizophrenia [133]) and autoimmune diseases (e.g., drug-induced lupus [134]). It has also been suggested that epigenetic variations may play a more significant role than DNA sequence polymorphisms in determining susceptibility to diseases such as obesity and diabetes [135].

Unique opportunities arising from epigenomic profiling technologies in drug safety sciences include the potential to gain insight into the molecular basis of long-lasting cellular perturbations (e.g., increased susceptibility to disease and/or toxicity; memory of prior immune stimulation and/or drug exposure; transgenerational effects). As outlined above, the identification of early epigenetic biomarkers for increased susceptibility to non-genotoxic carcinogenesis represents a particularly promising area, but the applicability of alterations in the epigenome as biomarkers of toxicity is certainly not limited to this field. For example, evidence suggests that immune cells can be primed by an early exposure for later phenotypic responses to additional exposures *via* alterations in the epigenome. The epigenetic basis of immune cell memory is exemplified by epigenetic changes in IL-2 and IFNγ genes that provide a molecular imprint for enhanced responsiveness and maintenance of a "ready-to-respond" state in memory CD8 T cells [136]. Furthermore, TLR4-mediated LPS stimulation of macrophages during an innate immune response to pathogens triggers differential epigenetic reprogramming of inflammatory genes *versus* antimicrobial genes, thus ensuring avoidance of excessive inflammation upon restimulation without compromising an effective immune response [137, 138]. Similar epigenetic imprints may be associated with specific disease states and/or toxicities and may also provide insight into prior and/or cumulative drug exposure. The latter phenomenon might provide a molecular basis for cumulative, irreversible doxorubicin-induced cardiomyopathy that may appear either shortly after treatment or years subsequent to therapy [139, 140]. It has also been suggested that aberrant DNA methylation can affect the drug-sensitivity of cancers through affecting expression of critical drug response genes. Epigenetic profiling has even been used to identify the molecular mediators of such cancer drug sensitivity [141].

Evidence from epidemiological and clinical studies in human populations suggests that disease susceptibility in adulthood can be influenced by environmental factors occurring early in life, or even during the lifetime of previous generations [59, 142]. It appears that the primary driver of these influences is the epigenome, and a number of epigenetic mechanisms are involved in developmental programming, lifelong stochastic and environmental responses, circadian deteriorations, disease susceptibility, and transgenerational effects [143]. There are now multiple examples in which an adverse phenotypic outcome can be attributed to epigenetic causes [56]. The most notable examples include those involving loss of genomic imprinting at specific gene loci, resulting in increased susceptibility to tumor formation (e.g., loss of imprinting at the insulin-like growth factor 2 gene results in Wilms tumors in children and colorectal neoplasia in adults [144]). In addition, there is now a growing body of evidence that environmental exposures, particularly those that occur during early development, induce epigenetic changes that can be transgenerationally transmitted and thus potentially serve as a basis for later disease susceptibility [55]. For example, the teratogenic effects of valproic acid in differentially susceptible mice were shown to be influenced by genomic imprinting, in that the susceptibility of an exposed fetus was disproportionately influenced by the susceptibility of its grandparents [145]. Another transgenerational study in mice implicates permanent alterations in the epigenome due to *in utero* dietary exposure to genistein, which leads to affects on gene expression and changes in susceptibility to obesity in adulthood [78]. Such evidence has potential implications for the safety assessment of pharmaceuticals as well, particularly in terms of testing strategies for determining reproductive toxicity. Thus, further analysis of the epigenetic modifications that occur in the progeny of drug-exposed parents or even grandparents may offer mechanistic insights into the susceptibility of these progeny to adverse drug-induced outcomes. However, there is little if any evidence as yet to connect drug exposure, general alterations to the epigenome, and detrimental outcomes on reproductive success or long-term disease status of offspring.

The inherently high chemical stability of DNA-based epigenetic modifications greatly enhances the potential for the study of archived (frozen or fixed) tissue samples in both the preclinical and clinical phases of drug development. Furthermore, body fluid-derived methylated DNA has been shown to be a promising surrogate biomarker for a number of human cancers [146, 147], and it is noteworthy that a DNA methylation-based clinical biomarker (PITX2) for progression of prostate cancer is currently being evaluated by the FDA (details at http://www.epigenomics.de/). Blood-based methylated DNA might also represent a useful source of biomarkers for additional toxic endpoints.

In summary, epigenomic profiling technologies have tremendous potential for providing novel mechanistic insights as well as novel candidate biomarkers for safety assessment during both the preclinical and clinical phases of drug development. Continued progress toward mapping the epigenomes of humans and other toxicologically relevant species will allow for mechanistically informed risk assessments concerning individual susceptibility to disease and toxicity, as well as the extrapolation of safety findings in preclinical animal models to humans.

References

1 Robertson KD (2005) DNA methylation and human disease. *Nat Rev Genet* 6: 597–610
2 Eckhardt F, Beck S, Gut IG, Berlin K (2004) Future potential of the Human Epigenome Project. *Expert Rev Mol Diagn* 4: 609–618
3 Rakyan VK, Hildmann T, Novik KL, Lewin J, Tost J, Cox AV, Andrews TD, Howe KL, Otto T, Olek A et al (2004) DNA methylation profiling of the human major histocompatibility complex: A pilot study for the Human Epigenome Project. *PLoS Biology* 2: e405
4 Eckhardt F, Lewin J, Cortese R, Rakyan VK, Attwood J, Burger M, Burton J, Cox TV, Davies R, Down TA et al (2006) DNA methylation profiling of human chromosomes 6, 20 and 22. *Nat Genet* 38: 1378–1385
5 Weber M, Davies JJ, Wittig D, Oakeley EJ, Haase M, Lam WL, Schübeler D (2005) Chromosome-wide and promoter-specific analyses identify sites of differential DNA methylation in normal and transformed human cells. *Nat Genet* 37: 853–862
6 Weber M, Hellmann I, Stadler MB, Ramos L, Paabo S, Rebhan M, Schübeler D (2007) Distribution, silencing potential and evolutionary impact of promoter DNA methylation in the human genome. *Nat Genet* 39: 457–466
7 Tyson FL, Heindel J (2005) Environmental influences on epigenetic regulation. *Environ Health Perspect* 113: A839
8 Akhtar A, Cavalli G (2005) The Epigenome Network of Excellence. *PLoS Biology* 3: e177
9 Grunau C, Renault E, Rosenthal A, Roizes G (2001) MethDB – A public database for DNA methylation data. *Nucleic Acids Res* 29: 270–274
10 Mariño-Ramírez L, Hsu B, Baxevanis AD, Landsman D (2006) The histone database: A comprehensive resource for histones and histone fold-containing proteins. *Proteins* 62: 838–842
11 Nafee TM, Farrell WE, Carroll WD, Fryer AA, Ismail KMK (2008) Epigenetic control of fetal gene expression. *BJOG* 115: 158–168
12 Morison IM, Paton CJ, Cleverley SD (2001) The imprinted gene and parent-of-origin effect database. *Nucleic Acids Res* 29: 275–276
13 Nicholls RD, Knepper JL (2001) Genome organization, function, and imprinting in Prader-Willi and Angelman syndromes. *Annu Rev Genomics Hum Genet* 2: 153–175
14 Tycko B (1997) DNA methylation in genomic imprinting. *Mutat Res* 386: 131–140
15 Esteller M (2007) Cancer epigenomics: DNA methylomes and histone-modification maps. *Nat Rev Genet* 8: 286–298
16 Clark SJ, Statham A, Stirzaker C, Molloy PL, Frommer M (2006) DNA methylation: Bisulphite modification and analysis. *Nat Protoc* 1: 2353–2364
17 Fraga MF, Esteller M (2002) DNA methylation: A profile of methods and applications. *Biotechniques* 33: 632–649
18 Grigg G, Clark S (1994) Sequencing 5-methylcytosine residues in genomic DNA. *Bioessays* 16: 431–436
19 Frommer M, McDonald LE, Millar DS, Collis CM, Watt F, Grigg GW, Molloy PL, Paul CL (1992) A genomic sequencing protocol that yields a positive display of 5-methylcytosine residues in individual DNA strands. *Proc Natl Acad Sci USA* 89: 1827–1831
20 Clark SJ, Harrison J, Paul CL, Frommer M (1994) High sensitivity mapping of methylated cytosines. *Nucleic Acids Res* 22: 2990–2997
21 Clark SJ, Millar DS, Molloy PL (2003) Bisulfite methylation analysis of tumor suppressor genes in prostate cancer from fresh and archival tissue samples. *Methods Mol Med* 81: 219–240
22 Costello JF, Fruhwald MC, Smiraglia DJ, Rush LJ, Robertson GP, Gao X, Wright FA, Feramisco

JD, Peltomaki P, Lang JC et al (2000) Aberrant CpG-island methylation has non-random and tumour-type-specific patterns. *Nat Genet* 24: 132–138

23 Welsh J, McClelland M (1990) Fingerprinting genomes using PCR with arbitrary primers. *Nucleic Acids Res* 18: 7213–7218

24 Gonzalgo ML, Liang G, Spruck CH, III, Zingg JM, Rideout WM III, Jones PA (1997) Identification and characterization of differentially methylated regions of genomic DNA by methylation-sensitive arbitrarily primed PCR. *Cancer Res* 57: 594–599

25 Toyota M, Ho C, Ahuja N, Jair KW, Li Q, Ohe-Toyota M, Baylin SB, Issa JP (1999) Identification of differentially methylated sequences in colorectal cancer by methylated CpG island amplification. *Cancer Res* 59: 2307–2312

26 Frigola J, Ribas M, Risques RA, Peinado MA (2002) Methylome profiling of cancer cells by amplification of inter-methylated sites (AIMS). *Nucleic Acids Res* 30: e28

27 Eads CA, Danenberg KDKK, Saltz LB, Blake C, Shibata D, Danenberg PV, Laird PW (2000) MethyLight: A high-throughput assay to measure DNA methylation. *Nucleic Acids Res* 28: e32

28 Trinh BN, Long TI, Laird PW (2001) DNA Methylation analysis by MethyLight technology. *Methods* 25: 456–462

29 Uhlmann K, Brinckmann A, Toliat MR, Ritter H, Nuernberg P (2002) Evaluation of a potential epigenetic biomarker by quantitative methyl-single nucleotide polymorphism analysis. *Electrophoresis* 23: 4072–4079

30 Tost J, Gut IG (2007) DNA methylation analysis by pyrosequencing. *Nat Protoc* 2: 2265–2275

31 Wong HL, Byun HM, Kwan JM, Campan M, Ingles SA, Laird PW, Yang AS (2006) Rapid and quantitative method of allele-specific DNA methylation analysis. *Biotechniques* 41: 734–739

32 Yang AS, Estecio MRH, Doshi K, Kondo Y, Tajara EH, Issa JP (2004) A simple method for estimating global DNA methylation using bisulfite PCR of repetitive DNA elements. *Nucleic Acids Res* 32: e38

33 Karimi M, Johansson S, Stach D, Corcoran M, Grander D, Schalling M, Bakalkin G, Lyko F, Larsson C, Ekstrom TJ (2006) LUMA (LUminometric Methylation Assay) – A high throughput method to the analysis of genomic DNA methylation. *Exp Cell Res* 312: 1989–1995

34 Matarazzo MR, Lembo F, Angrisano T, Ballestar E, Ferraro M, Pero R, De Bonis ML, Bruni CB, Esteller M, D'Esposito M, Chiariotti L (2004) *In vivo* analysis of DNA methylation patterns recognized by specific proteins: Coupling ChIP and bisulfite analysis. *Biotechniques* 37: 666–673

35 Jacinto FV, Ballestar E, Esteller M (2008) Methyl-DNA immunoprecipitation (MeDIP): Hunting down the DNA methylome. *Biotechniques* 44: 35–43

36 Buck MJ, Lieb JD (2004) ChIP-chip: Considerations for the design, analysis, and application of genome-wide chromatin immunoprecipitation experiments. *Genomics* 83: 349–360

37 Jacinto FV, Ballestar E, Ropero S, Esteller M (2007) Discovery of epigenetically silenced genes by methylated DNA immunoprecipitation in colon cancer cells. *Cancer Res* 67: 11481–11486

38 Milutinovic S, D'Alessio AC, Detich N, Szyf M (2007) Valproate induces widespread epigenetic reprogramming which involves demethylation of specific genes. *Carcinogenesis* 28: 560–571

39 Gebhard C, Schwarzfischer L, Pham TH, Schilling E, Klug M, Andreesen R, Rehli M (2006) Genome-wide profiling of CpG methylation identifies novel targets of aberrant hypermethylation in myeloid leukemia. *Cancer Res* 66: 6118–6128

40 Wardle FC, Odom DT, Bell GW, Yuan B, Danford TW, Wiellette EL, Herbolsheimer E, Sive HL, Young RA, Smith JC (2006) Zebrafish promoter microarrays identify actively transcribed embryonic genes. *Genome Biol* 7: R71

41 Wilson IM, Davies JJ, Weber M, Brown CJ, Alvarez CE, MacAulay C, Schübeler D, Lam WL (2006) Epigenomics: Mapping the methylome. *Cell Cycle* 5: 155–158

42 Galasinski SC, Resing KA, Ahn NG (2003) Protein mass analysis of histones. *Methods* 31: 3–11

43 Smith CM, Haimberger ZW, Johnson CO, Wolf AJ, Gafken PR, Zhang Z, Parthun MR, Gottschling DE (2002) Heritable chromatin structure: Mapping "memory" in histones H3 and H4. *Proc Natl Acad Sci USA* 99: 16454–16461

44 Fraga MF, Ballestar E, Villar-Garea A, Boix-Chornet M, Espada J, Schotta G, Bonaldi T, Haydon C, Ropero S, Petrie K et al (2005) Loss of acetylation at Lys16 and trimethylation at Lys20 of histone H4 is a common hallmark of human cancer. *Nat Genet* 37: 391–400

45 Seligson DB, Horvath S, Shi T, Yu H, Tze S, Grunstein M, Kurdistani SK (2005) Global histone modification patterns predict risk of prostate cancer recurrence. *Nature* 435: 1262–1266

46 Bernstein BE, Humphrey EL, Long Liu C, Schreiber SL (2004) The use of chromatin immunoprecipitation assays in genome-wide analyses of histone modifications. *Methods Enzymol* 376:

349–360

47 Lippman Z, Gendrel AV, Black M, Vaughn MW, Dedhia N, McCombie WR, Lavine K, Mittal V, May B, Kasschau KD et al (2004) Role of transposable elements in heterochromatin and epigenetic control. *Nature* 430: 471–476

48 Bernstein BE, Kamal M, Lindblad-Toh K, Bekiranov S, Bailey DK, Huebert DJ, McMahon S, Karlsson EK, Kulbokas EJ, Gingeras TR et al (2005) Genomic maps and comparative analysis of histone modifications in human and mouse. *Cell* 120: 169–181

49 Martens JHA, O'Sullivan RJ, Braunschweig U, Opravil S, Radolf M, Steinlein P, Jenuwein T (2005) The profile of repeat-associated histone lysine methylation states in the mouse epigenome. *EMBO J* 24: 800–812

50 Schübeler D, MacAlpine DM, Scalzo D, Wirbelauer C, Kooperberg C, van Leeuwen F, Gottschling DE, O'Neill LP, Turner BM, Delrow J et al (2004) The histone modification pattern of active genes revealed through genome-wide chromatin analysis of a higher eukaryote. *Genes Dev* 18: 1263–1271

51 Kurdistani SK, Tavazoie S, Grunstein M (2004) Mapping global histone acetylation patterns to gene expression. *Cell* 117: 721–733

52 Weinhold B (2006) Epigenetics: The science of change. *Environ Health Perspect* 114: A160–A167

53 Bombail V, Moggs JG, Orphanides G (2004) Perturbation of epigenetic status by toxicants. *Toxicol Lett* 149: 51–58

54 Hitchins MP, Wong JJL, Suthers G, Suter CM, Martin DIK, Hawkins NJ, Ward RL (2007) Inheritance of a cancer-associated MLH1 germ-line epimutation. *N Engl J Med* 356: 697–705

55 Jirtle RL, Skinner MK (2007) Environmental epigenomics and disease susceptibility. *Nat Rev Genet* 8: 253–262

56 Reamon-Buettner SM, Borlak J (2007) A new paradigm in toxicology and teratology: Altering gene activity in the absence of DNA sequence variation. *Reprod Toxicol* 24: 20–30

57 Rakyan VK, Blewitt ME, Druker R, Preis JI, Whitelaw E (2002) Metastable epialleles in mammals. *Trends Genet* 18: 348–351

58 Dolinoy DC, Jirtle RL (2008) Environmental epigenomics in human health and disease. *Environ Mol Mutagen* 49: 4–8

59 Dolinoy DC, Weidman JR, Jirtle RL (2007) Epigenetic gene regulation: Linking early developmental environment to adult disease. *Reprod Toxicol* 23: 297–307

60 Feinberg AP, Ohlsson R, Henikoff S (2006) The epigenetic progenitor origin of human cancer. *Nat Rev Genet* 7: 21–33

61 Baylin SB, Ohm JE (2006) Epigenetic gene silencing in cancer – A mechanism for early oncogenic pathway addiction? *Nat Rev Cancer* 6: 107–116

62 Derks S, Postma C, Moerkerk PTM, van den Bosch SM, Carvalho B, Hermsen MAJA, Giaretti W, Herman JG, Weijenberg MP, de Bruïne AP et al (2006) Promoter methylation precedes chromosomal alterations in colorectal cancer development. *Cell Oncol* 28: 247–257

63 Kang GH, Lee S, Kim JS, Jung HY (2003) Profile of aberrant CpG island methylation along the multistep pathway of gastric carcinogenesis. *Lab Invest* 83: 635–641

64 Suzuki H, Gabrielson E, Chen W, Anbazhagan R, van Engeland M, Weijenberg MP, Herman JG, Baylin SB (2002) A genomic screen for genes upregulated by demethylation and histone deacetylase inhibition in human colorectal cancer. *Nat Genet* 31: 141–149

65 Akiyama Y, Watkins N, Suzuki H, Jair KW, van Engeland M, Esteller M, Sakai H, Ren CY, Yuasa Y, Herman JG, Baylin SB (2003) GATA-4 and GATA-5 transcription factor genes and potential downstream antitumor target genes are epigenetically silenced in colorectal and gastric cancer. *Mol Cell Biol* 23: 8429–8439

66 Rhee I, Bachman KE, Park BH, Jair KW, Yen RWC, Schuebel KE, Cui H, Feinberg AP, Lengauer C, Kinzler KW et al (2002) DNMT1 and DNMT3b cooperate to silence genes in human cancer cells. *Nature* 416: 552–556

67 Herman JG, Umar A, Polyak K, Graff JR, Ahuja N, Issa JP, Markowitz S, Willson JK, Hamilton SR, Kinzler KW et al (1998) Incidence and functional consequences of hMLH1 promoter hypermethylation in colorectal carcinoma. *Proc Natl Acad Sci USA* 95: 6870–6875

68 Bachman KE, Park BH, Rhee I, Rajagopalan H, Herman JG, Baylin SB, Kinzler KW, Vogelstein B (2003) Histone modifications and silencing prior to DNA methylation of a tumor suppressor gene. *Cancer Cell* 3: 89–95

69 Suzuki H, Watkins DN, Jair KW, Schuebel KE, Markowitz SD, Dong Chen W, Pretlow TP, Yang B, Akiyama Y, van Engeland M et al (2004) Epigenetic inactivation of SFRP genes allows consti-

tutive WNT signaling in colorectal cancer. *Nat Genet* 36: 417–422

70 Wales MM, Biel MA, Deiry WE, Nelkin BD, Issa JP, Cavenee WK, Kuerbitz SJ, Baylin SB (1995) P53 activates expression of HIC-1, a new candidate tumour suppressor gene on 17p13.3. *Nat Med* 1: 570–577

71 Chen W, Cooper TK, Zahnow CA, Overholtzer M, Zhao Z, Ladanyi M, Karp JE, Gokgoz N, Wunder JS, Andrulis IL et al (2004) Epigenetic and genetic loss of Hic1 function accentuates the role of p53 in tumorigenesis. *Cancer Cell* 6: 387–398

72 Veigl ML, Kasturi L, Olechnowicz J, Ma A, Lutterbaugh JD, Periyasamy S, Li GM, Drummond J, Modrich PL, Sedwick WD, Markowitz SD (1998) Biallelic inactivation of hMLH1 by epigenetic gene silencing, a novel mechanism causing human MSI cancers. *Proc Natl Acad Sci USA* 95: 8698–8702

73 Choi IS, Wu TT (2005) Epigenetic alterations in gastric carcinogenesis. *Cell Res* 15: 247–254

74 Mittag F, Kuester D, Vieth M, Peters B, Stolte B, Roessner A, Schneider-Stock R (2006) DAPK promotor methylation is an early event in colorectal carcinogenesis. *Cancer Lett* 240: 69–75

75 Wendt MK, Johanesen PA, Kang-Decker N, Binion DG, Shah V, Dwinell MB (2006) Silencing of epithelial CXCL12 expression by DNA hypermethylation promotes colonic carcinoma metastasis. *Oncogene* 25: 4986–4997

76 Oue N, Mitani Y, Motoshita J, Matsumura S, Yoshida K, Kuniyasu H, Nakayama H, Yasui W (2006) Accumulation of DNA methylation is associated with tumor stage in gastric cancer. *Cancer* 106: 1250–1259

77 To KF, Leung WK, Lee TL, Yu J, Tong JHM, Chan MWY, Ng EKW, Chung SCS, Sung JJY (2002) Promoter hypermethylation of tumor-related genes in gastric intestinal metaplasia of patients with and without gastric cancer. *Int J Cancer* 102: 623–628

78 Dolinoy DC, Weidman JR, Waterland RA, Jirtle RL (2006) Maternal genistein alters coat color and protects A^{vy} mouse offspring from obesity by modifying the fetal epigenome. *Environ Health Perspect* 114: 567–572

79 Ho SM, Tang WY, Belmonte de Frausto J, Prins GS (2006) Developmental exposure to estradiol and bisphenol A increases susceptibility to prostate carcinogenesis and epigenetically regulates phosphodiesterase type 4 variant 4. *Cancer Res* 66: 5624–5632

80 Watson RE, Goodman JI (2002) Epigenetics and DNA methylation come of age in toxicology. *Toxicol Sci* 67: 11–16

81 Pogribny IP, Tryndyak VP, Woods C, Witt SE, Rusyn I (2007) Epigenetic effects of the continuous exposure to peroxisome proliferator WY-14,643 in mouse liver are dependent upon peroxisome proliferator activated receptor-α. *Mutat Res* 625: 62–71

82 Phillips JM, Yamamoto Y, Negishi M, Maronpot RR, Goodman JI (2007) Orphan nuclear receptor constitutive active/androstane receptor-mediated alterations in DNA methylation during phenobarbital promotion of liver tumorigenesis. *Toxicol Sci* 96: 72–82

83 Ray JS, Harbison ML, McClain RM, Goodman JI (1994) Alterations in the methylation status and expression of the raf oncogene in phenobarbital-induced and spontaneous B6C3F1 mouse live tumors. *Mol Carcinog* 9: 155–166

84 Denda A, Tang Q, Endoh T, Tsujiuchi T, Horiguchi K, Noguchi O, Mizumoto Y, Nakae D, Konishi Y (1994) Prevention by acetylsailcylic acid of liver cirrhosis and carcinogenesis as well as generations of 8-hydroxydeoxyguanosine and thiobarbituric acid-reactive substances caused by a choline-deficient, L-amino acid-defined diet in rats. *Carcinogenesis* 15: 1279–1283

85 Abanobi SE, Lombardi B, Shinozuka H (1982) Stimulation of DNA synthesis and cell proliferation in the liver of rats fed a choline-devoid diet and their suppression by phenobarbital. *Cancer Res* 42: 412–415

86 Wainfan E, Poirier LA (1992) Methyl groups in carcinogenesis: Effects on DNA methylation and gene expression. *Cancer Res* 52: 2071s–2077s

87 Mikol Y, Hoover KL, Creasia D, Poirier LA (1983) Hepatocarcinogenesis in rats fed methyl-deficient, amino acid-defined diets. *Carcinogenesis* 4: 1619–1629

88 Ghoshal AK, Farber E (1984) The induction of liver cancer by dietary deficiency of choline and methionine without added carcinogens. *Carcinogenesis* 5: 1367–1370

89 Newberne PM, de Camargo JLV, Clark AJ (1982) Choline deficiency, partial hepatectomy, and liver tumors in rats and mice. *Toxicol Pathol* 10: 95–106

90 Counts JL, Sarmiento JI, Harbison ML, Downing JC, McClain RM, Goodman JI (1996) Cell proliferation and global methylation status changes in mouse liver after phenobarbital and/or choline-devoid, methionine-deficient diet administration. *Carcinogenesis* 17: 1251–1257

91 Counts JL, McClain RM, Goodman JI (1997) Comparison of effect of tumor promoter treatments on DNA methylation status and gene expression in B6C3F1 and C57BL/6 mouse liver and in B6C3F1 mouse liver tumors. *Mol Carcinog* 18: 97–106

92 Bachman AN, Kamendulis LM, Goodman JI (2006) Diethanolamine and phenobarbital produce an altered pattern of methylation in GC-Rich regions of DNA in B6C3F1 mouse hepatocytes similar to that resulting from choline deficiency. *Toxicol Sci* 90: 317–325

93 Watson RE, Goodman JI (2002) Effects of phenobarbital on DNA methylation in GC-rich regions of hepatic DNA from mice that exhibit different levels of susceptibility to liver tumorigenesis. *Toxicol Sci* 68: 51–58

94 Bachman AN, Phillips JM, Goodman JI (2006) Phenobarbital induces progressive patterns of GC-rich and gene-specific altered DNA methylation in the liver of tumor-prone B6C3F1 mice. *Toxicol Sci* 91: 393–405

95 Yamamoto Y, Moore R, Goldsworthy TL, Negishi M, Maronpot RR (2004) The orphan nuclear receptor constitutive active/androstane receptor is essential for liver tumor promotion by phenobarbital in mice. *Cancer Res* 64: 7197–7200

96 Huang W, Zhang J, Washington M, Liu J, Parant JM, Lozano G, Moore DD (2005) Xenobiotic stress induces hepatomegaly and liver tumors *via* the nuclear receptor constitutive androstane receptor. *Mol Endocrinol* 19: 1646–1653

97 Phillips JM, Goodman JI (2008) Identification of genes that may play critical roles in phenobarbital (PB)-induced liver tumorigenesis due to altered DNA methylation. *Toxicol Sci* 104: 86–99

98 Kostka G, Urbanek K, Ludwicki JK (2007) The effect of phenobarbital on the methylation level of the p16 promoter region in rat liver. *Toxicology* 239: 127–135

99 Wong IH, Zhang J, Lai PB, Lau WY, Lo YM (2003) Quantitative analysis of tumor-derived methylated p16[INK4a] sequences in plasma, serum, and blood cells of hepatocellular carcinoma patients. *Clin Cancer Res* 9: 1047–1052

100 Tryndyak VP, Muskhelishvili L, Kovalchuk O, Rodriguez-Juarez R, Montgomery B, Churchwell MI, Ross SA, Beland FA, Pogribny IP (2006) Effect of long-term tamoxifen exposure on genotoxic and epigenetic changes in rat liver: Implications for tamoxifen-induced hepatocarcinogenesis. *Carcinogenesis* 27: 1713–1720

101 Kovalchuk O, Tryndyak VP, Montgomery B, Boyko A, Kutanzi K, Zemp F, Warbritton AR, Latendresse JR, Kovalchuk I, Beland FA, Pogribny IP (2007) Estrogen-induced rat breast carcinogenesis is characterized by alterations in DNA methylation, histone modifications, and aberrant microRNA expression. *Cell Cycle* 6: 2010–2018

102 Bachman AN, Curtin GM, Doolittle DJ, Goodman JI (2006) Altered methylation in gene-specific and GC-rich regions of DNA is progressive and nonrandom during promotion of skin tumorigenesis. *Toxicol Sci* 91: 406–418

103 Sciarra A, Di Silverio F, Salciccia S, Autran Gomez AM, Gentilucci A, Gentile V (2007) Inflammation and chronic prostatic diseases: Evidence for a link? *Eur Urol* 52: 964–972

104 Nelson WG, Yegnasubramanian S, Agoston AT, Bastian P.J., Lee BH, Nakayama M, De Marzo AM (2008) Abnormal DNA methylation, epigenetics, and prostate cancer. *Front Biosci* 12: 4254–4266

105 Xie J, Itzkowitz SH (2008) Cancer in inflammatory bowel disease. *World J Gastroenterol* 14: 378–389

106 Chan AOO, Rashid A (2006) CpG island methylation in precursors of gastrointestinal malignancies. *Curr Mol Med* 6: 401–408

107 Maeda O, Ando T, Watanabe O, Ishiguro K, Ohmiya N, Niwa Y, Goto H (2006) DNA hypermethylation in colorectal neoplasms and inflammatory bowel disease: A mini review. *Inflamm Pharmacol* 14: 204–206

108 Campos AC, Molognoni F, Melo FH, Galdieri LC, Carneiro CR, D'Almeida V, Correa M, Jasiulionis MG (2007) Oxidative stress modulates DNA methylation during melanocyte anchorage blockade associated with malignant transformation. *Neoplasia* 9: 1111–1121

109 Jiang Y, Sun T, Xiong J, Cao J, Li G, Wang S (2007) Hyperhomocysteinemia-mediated DNA hypomethylation and its potential epigenetic role in rats. *Acta Biochim Biophys Sin* 39: 657–667

110 Valinluck V, Sowers LC (2007) Inflammation-mediated cytosine damage: A mechanistic link between inflammation and the epigenetic alterations in human cancers. *Cancer Res* 67: 5583–5586

111 De Santa F, Totaro MG, Prosperini E, Notarbartolo S, Testa G, Natoli G (2007) The histone H3 lysine-27 demethylase Jmjd3 links inflammation to inhibition of polycomb-mediated gene silenc-

ing. *Cell* 130: 1083–1094
112 Szabo G, Mandrekar P, Dolganiuc A (2007) Innate immune response and hepatic inflammation. *Semin Liver Dis* 339–350
113 Kovalenko VM, Bagnyukova TV, Sergienko OV, Bondarenko LB, Shayakhmetova GM, Matvienko AV, Pogribny IP (2007) Epigenetic changes in the rat livers induced by pyrazinamide treatment. *Toxicol Appl Pharmacol* 225: 293–299
114 Hastwell PW, Chai LL, Roberts KJ, Webster TW, Harvey JS, Rees RW, Walmsley RM (2006) High-specificity and high-sensitivity genotoxicity assessment in a human cell line: Validation of the GreenScreen HC GADD45α-GFP genotoxicity assay. *Mutat Res* 607: 160–175
115 Barreto G, Schäfer A, Marhold J, Stach D, Swaminathan SK, Handa V, Döderlein G, Maltry N, Wu W, Lyko F, Niehrs C (2007) Gadd45α promotes epigenetic gene activation by repair-mediated DNA demethylation. *Nature* 445: 671–675
116 Preston RJ (2007) Epigenetic processes and cancer risk assessment. *Mutat Res* 616: 7–10
117 Rangarajan A, Weinberg RA (2003) Comparative biology of mouse *versus* human cells: Modeling human cancer in mice. *Nat Rev Cancer* 3: 952–959
118 Vu TH, Jirtle RL, Hoffman AR (2007) Cross-species clues of an epigenetic imprinting regulatory code for the *IGF2R* gene. *Cytogenet Genome Res* 113: 202–208
119 Goodman JI, Watson RE (2002) Altered DNA methylation: A secondary mechanism involved in carcinogenesis. *Annu Rev Pharmacol Toxicol* 42: 501–525
120 Knight A, Bailey J, Balcombe J (2006) Animal carcinogenicity studies: 1. Poor human predictivity. *Altern Lab Anim* 34: 19–27
121 Knight A, Bailey J, Balcombe J (2006) Animal carcinogenicity studies: 2. Obstacles to extrapolation of data to humans. *Altern Lab Anim* 34: 29–38
122 Fielden MR, Brennan R, Gollub J (2007) A gene expression biomarker provides early prediction and mechanistic assessment of hepatic tumor induction by nongenotoxic chemicals. *Toxicol Sci* 99: 90–100
123 Predictive Safety Testing Consortium CWG, Fielden MR, Nie A, McMillian M, Elangbam CS, Trela BA, Yang Y, Dunn RT II, Dragan Y, Fransson-Stehen R et al (2008) Inter-laboratory evaluation of genomic signatures for predicting carcinogenicity in the rat. *Toxicol Sci* 103: 28–34
124 Paterlini-Brechot P, Benali NL (2007) Circulating tumor cells (CTC) detection: Clinical impact and future directions. *Cancer Lett* 253: 180–204
125 Jacob K, Sollier C, Jabado N (2007) Circulating tumor cells: Detection, molecular profiling and future prospects. *Expert Rev Proteomics* 4: 741–756
126 Herman JG (2004) Circulating methylated DNA. *Ann NY Acad Sci* 1022: 33–39
127 Pantel K, Alix-Panabières C (2007) The clinical significance of circulating tumor cells. *Nat Clin Pract Oncol* 4: 62–63
128 Nakagawa T, Martinez SR, Goto Y, Koyanagi K, Kitago M, Shingai T, Elashoff DA, Ye X, Singer FR, Giuliano AE, Hoon DSB (2007) Detection of circulating tumor cells in early-stage breast cancer metastasis to axillary lymph nodes. *Clin Cancer Res* 13: 4105–4110
129 He W, Wang H, Hartmann LC, Cheng JX, Low PS (2007) *In vivo* quantitation of rare circulating tumor cells by multiphoton intravital flow cytometry. *Proc Natl Acad Sci USA* 104: 11760–11765
130 Brena R, Huang T, Plass C (2006) Quantitative assessment of DNA methylation: Potential applications for disease diagnosis, classification, and prognosis in clinical settings. *J Mol Med* 84: 365–377
131 Vogiatzi P, Aimola P, Scarano MI, Claudio PP (2007) Epigenome-derived drugs: Recent advances and future perspectives. *Drug News Perspect* 20: 627–633
132 Ballestar E, Esteller M (2008) Epigenetic gene regulation in cancer. *Adv Genet* 61: 247–267
133 Tsankova N, Renthal W, Kumar A, Nestler EJ (2007) Epigenetic regulation in psychiatric disorders. *Nat Rev Neurosci* 8: 355–367
134 Mazari L, Ouarzane M, Zouali M (2007) Subversion of B lymphocyte tolerance by hydralazine, a potential mechanism for drug-induced lupus. *Proc Natl Acad Sci USA* 104: 6317–6322
135 Stöger R (2008) The thrifty epigenotype: An acquired and heritable predisposition for obesity and diabetes? *Bioessays* 30: 156–166
136 Northrop JK, Thomas RM, Wells AD, Shen H (2006) Epigenetic remodeling of the IL-2 and IFN-γ loci in memory CD8 T cells is influenced by CD4 T cells. *J Immunol* 177: 1062–1069
137 Foster SL, Hargreaves DC, Medzhitov R (2007) Gene-specific control of inflammation by TLR-induced chromatin modifications. *Nature* 447: 972–978
138 Arbibe L, Sansonetti PJ (2007) Epigenetic regulation of host response to LPS: Causing tolerance

while avoiding toll errancy. *Cell Host Microbe* 1: 244–246

139 Berthiaume J, Wallace K (2007) Persistent alterations to the gene expression profile of the heart subsequent to chronic doxorubicin treatment. *Cardiovasc Toxicol* 7: 178–191

140 Wallace KB (2003) Doxorubicin-induced cardiac mitochondrionopathy. *Pharmacol Toxicol* 93: 105–115

141 Shen L, Kondo Y, Ahmed S, Boumber Y, Konishi K, Guo Y, Chen X, Vilaythong JN, Issa JP (2007) Drug sensitivity prediction by CpG island methylation profile in the NCI-60 cancer cell line panel. *Cancer Res* 67: 11335–11343

142 Gluckman PD, Hanson MA, Pinal C (2005) The developmental origins of adult disease. *Matern Child Nutr* 1: 130–141

143 Gallou-Kabani C, Vigé, Alexandre, Gross MS, Junien C (2007) Nutri-epigenomics: Lifelong remodeling of our epigenomes by nutritional and metabolic factors and beyond. *Clin Chem Lab Med* 45: 321–327

144 Cui H (2007) Loss of imprinting of IGF2 as an epigenetic marker for the risk of human cancer. *Dis Markers* 23: 105–112

145 Beck SL (2000) Does genomic imprinting contribute to valproic acid teratogenicity? *Reprod Toxicol* 15: 43–48

146 Goebel G, Zitt M, Zitt M, Müller HM (2005) Circulating nucleic acids in plasma or serum (CNAPS) as prognostic and predictive markers in patients with solid neoplasias. *Dis Markers* 21: 105–120

147 Laird PW (2003) The power and the promise of DNA methylation markers. *Nat Rev Cancer* 3: 253–266

148 Svedruzic ZM (2008) Mammalian cytosine DNA methyltransferase Dnmt1: Enzymatic mechanism, novel mechanism-based inhibitors, and RNA-directed DNA methylation. *Curr Med Chem* 15: 92–106

149 Jeltsch A, Nellen W, Lyko F (2006) Two substrates are better than one: Dual specificities for Dnmt2 methyltransferases. *Trends Biochem Sci* 31: 306–308

150 Chen T, Li E (2006) Establishment and maintenance of DNA methylation patterns in mammals. *Curr Top Microbiol Immunol* 301: 179–201

151 LaSalle JM (2007) The odyssey of MeCP2 and parental imprinting. *Epigenetics* 2: 5–10

152 Nakao M, Matsui Si, Yamamoto S, Okumura K, Shirakawa M, Fujita N (2001) Regulation of transcription and chromatin by methyl-CpG binding protein MBD1 *Brain Dev* 23: S174–S176

153 Berger J, Bird A (2005) Role of MBD2 in gene regulation and tumorigenesis. *Biochem Soc Trans* 33: 1537–1540

154 Kaji K, Nichols J, Hendrich B (2007) Mbd3, a component of the NuRD co-repressor complex, is required for development of pluripotent cells. *Development* 134: 1123–1132

155 Abdel-Rahman WM, Knuutila S, Peltomaki P, Harrison DJ, Bader SA (2008) Truncation of MBD4 predisposes to reciprocal chromosomal translocations and alters the response to therapeutic agents in colon cancer cells. *DNA Repair* 7: 321–328

156 Bedford MT (2007) Arginine methylation at a glance. *J Cell Sci* 120: 4243–4246

157 Qian C, Zhou M (2006) SET domain protein lysine methyltransferases: Structure, specificity and catalysis. *Cell Mol Life Sci* 63: 2755–2763

158 Zhang K, Dent SYR (2005) Histone modifying enzymes and cancer: Going beyond histones. *J Cell Biochem* 96: 1137–1148

159 Feng Q, Wang H, Ng HH, Erdjument-Bromage H, Tempst P, Struhl K, Zhang Y (2002) Methylation of H3-lysine 79 is mediated by a new family of HMTases without a SET domain. *Curr Biol* 12: 1052–1058

160 Swigut T, Wysocka J (2007) H3K27 demethylases, at long last. *Cell* 131: 29–32

161 Benevolenskaya EV (2007) Histone H3K4 demethylases are essential in development and differentiation. *Biochem Cell Biol* 85: 435–443

162 Culhane JC, Cole PA (2007) LSD1 and the chemistry of histone demethylation. *Curr Opin Chem Biol* 11: 561–568

163 Kouzarides T (2007) Chromatin modifications and their function. *Cell* 128: 693–705

164 Avvakumov N, Cote J (2007) The MYST family of histone acetyltransferases and their intimate links to cancer. *Oncogene* 26: 5395–5407

165 Bhaumik SR, Smith E, Shilatifard A (2007) Covalent modifications of histones during development and disease pathogenesis. *Nat Struct Mol Biol* 14: 1008–1016

166 Shukla V, Vaissiere T, Herceg Z (2008) Histone acetylation and chromatin signature in stem cell

identity and cancer. *Mutat Res* 637: 1–15

167 Glozak MA, Seto E (2007) Histone deacetylases and cancer. *Oncogene* 26: 5420–5432

168 Smith CL (2008) A shifting paradigm: Histone deacetylases and transcriptional activation. *Bioessays* 30: 15–24

169 Nowak SJ, Corces VG (2004) Phosphorylation of histone H3: A balancing act between chromosome condensation and transcriptional activation *Trends Genet* 20: 214–220

170 Shilatifard A (2006) Chromatin modifications by methylation and ubiquitination: Implications in the regulation of gene expression. *Annu Rev Biochem* 75: 243–269

171 Iñiguez-Lluhí JA (2006) For a healthy histone code, a little SUMO in the tail keeps the acetyl away. *ACS Chem Biol* 1: 204–206

172 Nathan D, Ingvarsdottir K, Sterner DE, Bylebyl GR, Dokmanovic M, Dorsey JA, Whelan KA, Krsmanovic M, Lane WS, Meluh PB et al (2006) Histone sumoylation is a negative regulator in *Saccharomyces cerevisiae* and shows dynamic interplay with positive-acting histone modifications. *Genes Dev* 20: 966–976

173 González-Romero R, Méndez J, Ausió J, Eirín-López JM (2008) Quickly evolving histones, nucleosome stability and chromatin folding: All about histone H2A.Bbd. *Gene* 413: 1–7

174 Zlatanova J, Thakar A (2008) H2A.Z: View from the Top. *Structure* 16: 166–179

175 Loyola A, Almouzni G (2007) Marking histone H3 variants: How, when and why? *Trends Biochem Sci* 32: 425–433

176 Bustin M, Catez F, Lim JH (2005) The dynamics of histone H1 function in chromatin. *Mol Cell* 17: 617–620

177 Hudder A, Novak RF (2008) miRNAs: Effectors of environmental influences on gene expression and disease. *Toxicol Sci* 103: 228–240

178 Taylor EL, Gant TW (2008) Emerging fundamental roles for non-coding RNA species in toxicology. *Toxicology* 246: 34–39

179 Dali-Youcef N, Lagouge M, Froelich S, Koehl C, Schoonjans K, Auwerx J (2007) Sirtuins: The 'magnificent seven', function, metabolism and longevity. *Ann Med* 39: 335–345

180 Rajasekhar VK, Begemann M (2007) Concise review: Roles of polycomb group proteins in development and disease: A stem cell perspective. *Stem Cells* 25: 2498–2510

181 Schuettengruber B, Chourrout D, Vervoort M, Leblanc B, Cavalli G (2007) Genome regulation by polycomb and trithorax proteins. *Cell* 128: 735–745

182 Weber M, Schübeler D (2007) Genomic patterns of DNA methylation: Targets and function of an epigenetic mark. *Curr Opin Cell Biol* 19: 273–280

Molecular, Clinical and Environmental Toxicology. Volume 1: Molecular Toxicology
Edited by A. Luch
© 2009 Birkhäuser Verlag/Switzerland

Receptors mediating toxicity and their involvement in endocrine disruption

Joëlle Rüegg[1], Pauliina Penttinen-Damdimopoulou[2], Sari Mäkelä[2], Ingemar Pongratz[1] and Jan-Åke Gustafsson[1]

[1] *Department of Biosciences and Nutrition, Karolinska Institutet, Huddinge, Sweden*
[2] *Functional Foods Forum & The Department of Biochemistry and Food Chemistry, University of Turku, Turku, Finland*

Abstract. Many toxic compounds exert their harmful effects by activating of certain receptors, which in turn leads to dysregulation of transcription. Some of these receptors are so called xenosensors. They are activated by external chemicals and evoke a cascade of events that lead to the elimination of the chemical from the system. Other receptors that are modulated by toxic substances are hormone receptors, particularly the ones of the nuclear receptor family. Some environmental chemicals resemble endogenous hormones and can falsely activate these receptors, leading to undesired activity in the cell. Furthermore, excessive activation of the xenosensors can lead to disturbances of the integrity of the system as well. In this chapter, the concepts of receptor-mediated toxicity and hormone disruption are introduced. We start by describing environmental chemicals that can bind to xenosensors and nuclear hormone receptors. We then describe the receptors most commonly targeted by environmental chemicals. Finally, the mechanisms by which receptor-mediated events can disrupt the system are depicted.

Introduction

We are constantly exposed to a heavy load of foreign compounds, so-called xenobiotics, both man-made and of natural origin. To survive, the organism has developed systems for tackling these potentially harmful exogenous substances. A number of receptors sense the presence of foreign compounds in the cell and induce a cascade of events that is intended to lead to neutralization and excretion of these compounds. However, in many cases the metabolism of xenobiotic substances can give rise to toxic metabolites or to reactive oxygen species that can harm the cell further. Additionally, the metabolism of foreign compounds can disturb other essential processes in the body, such as production and metabolism of certain hormones. Thus, in some instances the receptors that are meant to induce the elimination of toxic compounds may themselves mediate some of their toxic effects. Some environmental compounds can also bind to receptors other than the xenosensors. In particular the receptors for the female sex steroids (estrogen receptors, ERs) have been identified as targets for many man-made and natural compounds. Interference with hormone receptors by exogenous substances has been termed endocrine disruption. During recent decades, this phenomenon has raised considerable concern. It has been associated with major health threats in the western world, such as development of diabetes and obesity, decreased fertility, and particularly with the growing incidence of hormone-associated cancers.

In this chapter we first give a short overview of environmental chemicals and natural compounds that induce receptor-mediated events. We then describe receptors that mediate toxicity, and finally discuss mechanisms of endocrine disruption.

Xenobiotics that induce receptor-mediated toxicity

Receptor-mediated toxicity is induced by a number of man-made and natural compounds. In contrast to genotoxic substances that are directly carcinogenic by inducing DNA damage, the receptor-mediated effects are versatile and often more subtle, and thus more difficult to identify. In this section, we give examples of xenobiotics inducing receptor-mediated events that can ultimately lead to toxicity to the organism.

Man-made chemicals

As a result of the industrial revolution over the last century, thousands of man-made chemicals are produced and introduced into the environment. A recent European survey reported that the average European consumer is exposed to up to 10 000 different chemicals on a daily basis. They can be divided into compounds that are synthesized and released intentionally and substances that arise as by-products from industrial processes. Most of these chemicals are persistent organic pollutants (POPs), meaning that they are resistant to environmental degradation. They are thus persistent over a long time in the environment, can contaminate drinking water and food, and can biomagnify in the food chain.

Compounds intentionally produced include pesticides and biocides, plasticizers, and additives in food and cosmetics (examples provided in Tab. 1). Many of these substances are structurally related to steroid hormones and may thus act on the respective hormone receptor. Actually, the first man-made chemicals identified as ER disruptors were pesticides. Symptoms in men working with the manufacture of these compounds led to the identification of dichlorodiphenyltrichloroethane (DDT) as an ER agonist. Other examples of pesticides that activate the ER are the DDT metabolite dichlorodiphenyldichloroethylene (DDE), methoxychlor, and dieldrin (e.g., [1]). In addition DDE and the fungicide vinclozolin can also affect the function of other hormone receptors: they both act as androgen receptor (AR) antagonists [2], while vinclozolin can further antagonize the activity of progesterone receptor (PR), glucocorticoid receptor (GR), and mineralocorticoid receptor (MR) in vitro [3]. Phthalates and bisphenol A (BPA) are additives used in the plastics industry. These compounds leach from

Table 1. Overview of man-made chemicals that induce receptor-mediated toxicity. AhR: aryl hydrocarbon receptor, AR: androgen receptor, ER: estrogen receptor, GR: glucocorticoid receptor, MR: mineralocorticoid receptor, PR: progesterone receptor, TR: thyroid hormone receptor.

Substance or class of substances	Examples/structures	Origin	Prevalence	Affinity for
Organochlorines	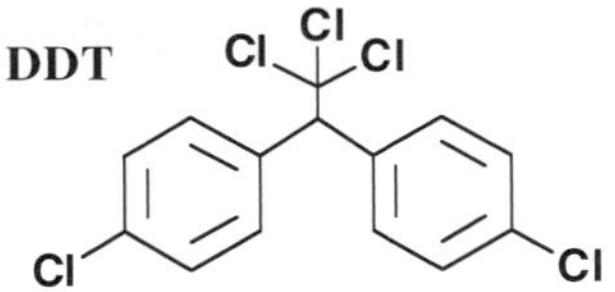 dichlorodiphenyltrichloroethane (DDT) and its metabolites dichloro-diphenyltrichloroethylene (DDE), methoxychlor, aldrin, dieldrin 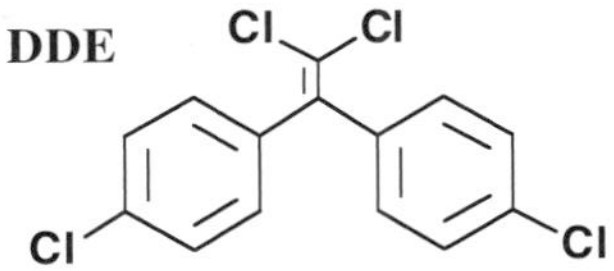	insecticide	banned in agriculture, limited use in malaria combat	**ER, AR**
Organophosphates	dichlorvos, parathion, trichlorphon 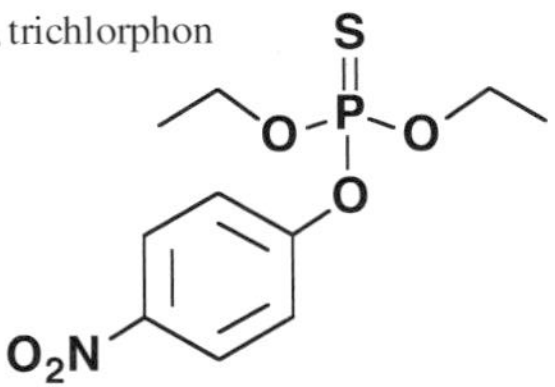	insecticide	widely used in developing countries	**AR**
Vinclozolin	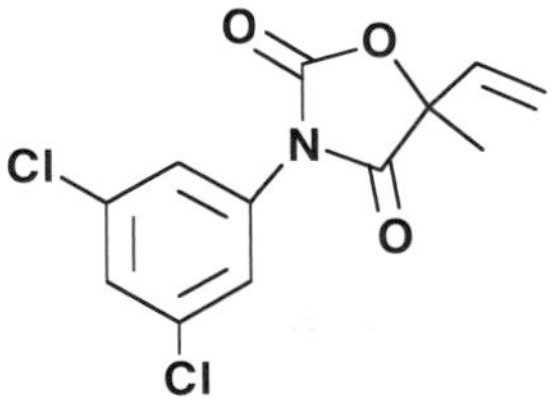	fungicide	commonly used in agriculture e.g. in vineyards	**AR, PR, GR, MR**

DDT   DDE

Parathion

Vinclozolin

(continued on next page)

Table 1. (continued)

Substance or class of substances	Examples/structures	Origin	Prevalence	Affinity for
Phthalates	di-2-ethylhexylphthalate (DEHP), diisodecylphthalate (DIDP) diisononylphthalate (DINP), methoxychlor **DEHP** 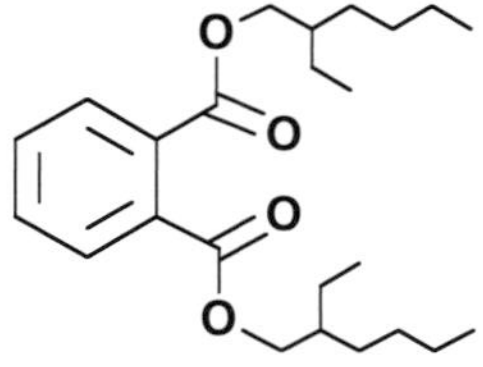	plasticizers	approx. 360 000 tons, produced per year; dominating plasticizers in PVC and soft plastics, for, e.g., cosmetics, adhesives, and electronic devices; partly banned for the use in children's toys	**ER, AR**
Bisphenol A (BPA)	**BPA** 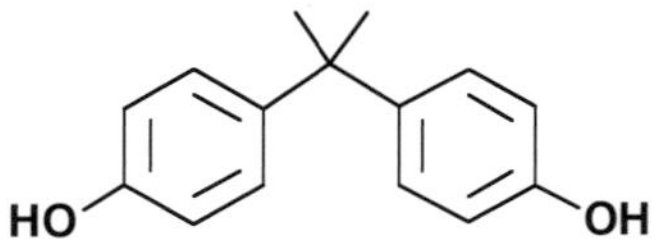	plasticizer, additive in dental fillings	one of the most produced chemical in the world	**ER, AR, TR**
Polychlorinated biphenyls (PCBs)	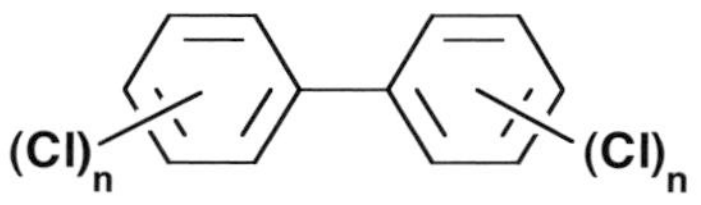 **PCBs**	insulating fluids in transformers, additives in coatings of electric wires	banned in the 1970s for "open" use, widely used for "closed" systems like transformers and hydraulic pumps	**AhR**

(continued on next page)

Table 1. (continued)

Substance or class of substances	Examples/structures	Origin	Prevalence	Affinity for
Halogenated aromatic hydrocarbons (HAHs)	polychlorinated dibenzo-*p*-dioxins (PCDDs), e.g., tetrachlorodibenzo-*p*-dioxin (TCDD) **TCDD**	side-product of waste combustion and herbicide production; part of cigarette smoke	bioaccumulates mostly in fatty fish and meat	**AhR**
Polycyclic aromatic hydrocarbons (PAHs)	e.g., benzo[*a*]pyrene (B[*a*]P), 3-methylcholanthrene (3-MC) **B[*a*]P** **3-MC**	side-product of carbon-containing fuel combustion; part of cigarette smoke	some of the most wide-spread organic pollutants	**AhR** some metabolites bind to ER

plastic packages to food and water, and can be detected in human serum. BPA is also used in dental fillings and in the plastic lining of tin cans. Both phthalates and BPA are ER agonists [4] and AR antagonists [5]. Furthermore, BPA as well as its brominated and chlorinated derivatives, used as flame retardants, have been recognized as thyroid hormone antagonists [6, 7]. Due to their endocrine disruptive nature, the compounds described here are associated with defects of the development and function of the reproductive system, as well as cancer causation in *in vivo* models [8]. Moreover, many pesticides are neurotoxic [9] and can act as strong immunosuppressors [10]. In many cases, unborn and young children are those most susceptible to xenobiotics.

Halogenated aromatic hydrocarbons (HAHs) and polycyclic aromatic hydrocarbons (PAHs) are potent environmental pollutants that arise as by-products from industrial processes. HAHs include polychlorinated dibenzodioxins (PCDD) (also known as dioxins), dibenzofurans (PCDF) and biphenyls (PCB). Examples of PAHs are 3-methylcholanthrene (3-MC), benzo[*a*]pyrene (B[*a*]P) and benzoflavone. Most of these substances are formed by incomplete combustion of carbon-containing materials such as household waste, wood, coal, diesel, and tobacco. Dioxins are also generated as by-products during the production of herbicides. The most toxic dioxin, 2,3,7,8-tetrachlorodibenzo-*p*-dioxin (TCDD), became known as contaminant in Agent Orange, a herbicide used in the Vietnam War. HAHs and PAHs are potent ligands of the aryl hydrocarbon receptor (AhR), one of the main receptors mediating the response to xenobiotics. Even though HAHs and PAHs do not bind hormone receptors as such, some of them (e.g., B[*a*]P and 3-MC) are metabolized into compounds that are potent ER activators [11, 12].

Aromatic hydrocarbons, especially the dioxins, are extremely toxic substances. Humans exposed to high doses of dioxins develop a skin disorder called chloracne and suffer from liver failure. A famous example of dioxin poisoning is the Ukrainian president Viktor Yushchenko. In September 2004, he received a high dose of dioxin that resulted in 1000 times elevated dioxin concentrations in the blood compared to an average individual. He developed severe chloracne and liver damage. Less clear-cut are the effects of chronic exposure to relatively low doses of HAHs and PAHs, i.e., what most of us encounter during our lives. Studies on populations accidentally exposed to elevated levels of dioxins (e.g., Agent Orange during the Vietnam War or after the incidence in a herbicide-producing factory in Seveso, Italy, in 1976) suggest a positive correlation between dioxin exposure and the development of diabetes and cancers including leukemia and lung cancer [13–16].

Nature's own xenobiotics

In addition to the chemicals made by man, our environment contains natural compounds that may regulate receptor-mediated events (examples provided in Tab. 2). These compounds are frequently encountered in plants, where they

Table 2. Overview of natural compounds that induce receptor-mediated cellular signaling. AhR: aryl hydrocarbon receptor, AR: androgen receptor, ER: estrogen receptor.

Class of substances	Examples/structures	Origin	Affinity for
Isoflavones	genistein, daidzein **Genistein**	soy beans, fava beans, chick peas, clover	**ER**
Flavonoids	catechins, quercetin **Quercetin**	green tea, citrus fruits	**ER, AhR**
Lignans	secoisolariciresinol, matairesinol; formed from lignans like enterolactone as metabolites in mammals **Enterolactone**	fiber-rich foods, e.g., cereals	**ER**

(continued on next page)

Table 2. (continued)

Class of substances	Examples/structures	Origin	Affinity for
Stilbenes	resveratrol **Resveratrol**	grapes	**ER, AhR**
Coumestans	coumestrol **Coumestrol**	alfalfa, soy beans, clover	ER
Mycotoxins	zearalenone **Zearalenone**	fungi	ER

often play a role in the self-defense system. To combat herbivores, insects and pests, plants produce compounds that make them less palatable and/or toxic, or compounds that interfere with the invader's capacity to reproduce [17]. Examples of such compounds are glucosinolates in *Brassicaceae* (e.g., 3-indolmethyl glucosinolate), alkaloids in *Solanaceae* (e.g., atropine), lectins in legumes (e.g., ricin), tannins in fruits and berries, and estrogenic polyphenols (e.g., genistein) in various plants. When ingested, some of these compounds are absorbed and are capable of binding to receptors. In particular, compounds possessing a polyphenolic structure can bind to xenosensors and hormone receptors.

Examples of natural activators of AhR are indole-3-carbinole (autolysis product of 3-indolmethyl glucosinolate) from cruciferous vegetables (e.g., cabbage), resveratrol from grapes, some phytoestrogens from soy and citrus fruits, and some flavonoids like chrysin and baicalein [18, 19]. Resveratrol is a stilbene phytoalexin (a compound produced in response to fungal or bacterial attack) produced by grape species, peanuts and lingonberries [20, 21]. Since resveratrol concentrates in the skin of grapes, it is found in significant amounts in red wine, and to a lesser extent in white wines. Indole-3-carbinole is converted to a variety of condensation products, e.g., diindolylmethane and indolocarbazole, which are more potent AhR agonists than indole-3-carbinol itself [22]. Both resveratrol and diindolylmethane are suggested to protect from cancer, the former by inhibiting AhR activity and the latter by inhibiting E_2-induced activity [23–25].

A great number of plant-derived compounds are so-called phytoestrogens, i.e., activators of ER. The potency of biological activity of some phytoestrogens became clear when, some 50 years ago, the red clover flavone formononetin was identified as a reason behind massive infertility in sheep in Australia [26]. Phytoestrogens are characterized by a polyphenolic structure and they are often divided into isoflavones, flavonoids, lignans, coumestanes, and stilbenes [27]. Additionally, mycotoxins produced by certain fungi can act as xenoestrogens [27]. Isoflavones include genistein and daidzein, both mainly found in soy and fava beans, biochanin A found in chick peas, and formononetin found in clover. Catechins in green tea and quercetin in citrus fruits are examples of flavonoids. Plant lignans, such as secoisolariciresinol, matairesinol and lariciresinol, are common constituents of fiber-rich foods, such as cereals. Coumestrol is a coumestan present in alfalfa, soy beans and clover, resveratrol from grapes is a stilbene, and zearalenone is a mycotoxin produced by *Fusarium* fungi, a common plant pathogen in the northern hemisphere. Many phytoestrogens are subjected to metabolism in the gut by microbes when ingested. Daidzein and formononetin are metabolized to equol, and some plant lignans are converted to enterolignans (enterodiol and enterolactone). In many cases, the metabolites are absorbed from the gut and possess more activity than the parental dietary compounds. For example, equol and enterolactone bind to ER, while daidzein, formononetin and many plant lignans are only weak ligands or do not bind the receptor at all [4, 28].

The ER exists as two different subtypes, ERα and ERβ. Many of the phytoestrogens display selectivity towards these subtypes. For example, genistein, coumestrol, equol, the citrus fruit flavonoid naringenin, and tea flavonoids apigenin and kaempferol are ERβ selective in ligand binding and transactivation assays [4, 29]. Zearalenone and enterolactone, on the other hand, preferentially activate ERα [4, 28]. ERβ reduces proliferation of estrogen-dependent cells, whereas ERα enhances proliferation [30]. A popular concept is that consumption of a diet rich in phytoestrogens is associated with beneficial effects for the organism. Most data exist for genistein; its consumption has been associated with decreased risk for breast and prostate cancer, decreased bone loss, and with cardioprotection (reviewed in [31]). However, even though enterolactone is ERα selective, high intake of lignans has also been suggested to be protective against breast and prostate cancer and cardiovascular disease (reviewed in [32]). Hence, the net effect of a phytoestrogen is difficult to predict based on receptor selective activity.

Due to their suggested beneficial effects, phytoestrogen extracts (e.g., soy- and flaxseed-based products) are sold as health-promoting dietary supplements. However, based on current knowledge, it is unclear what risks are involved in excessive intake of phytoestrogens. In addition, the effects of phytoestrogens on males and newborn infants have not been thoroughly studied. Especially the use of soy-based infant formulas, rich in soy phytoestrogens, warrants careful studies.

Receptors involved in toxicity

Receptors mediating toxicity can be roughly divided into two groups: dedicated xenosensors and hormone receptors with no primary role in the defense against xenobiotic insult. Upon binding of a xenobiotic compound, the dedicated xenosensors induce a response intended to metabolize and excrete the compound. Activation of the hormone receptors by xenobiotic substances leads to interference with the hormonal system of the exposed organism, a phenomenon called endocrine disruption (discussed below).

To eliminate the harmful effects of an exogenous chemical, the cell attempts to change the compounds to an inactive state, make them water soluble, and excrete them. Metabolism of xenobiotics occurs in three phases: in phase I the chemical is oxidized; in phase II the oxidized products are conjugated to glutathione, sulfuric acid, or glucuronic acid, resulting in hydrophilic molecules; and finally in phase III these substances are transported out of the cell by ATP-dependent export pumps [33]. In mammals, three different transcription factor superfamilies are responsible for the induction of xenobiotic metabolizing enzymes, basic-helix-loop-helix/Per-ARNT-Sim (bHLH-PAS) proteins, nuclear receptors (NRs) and basic leucine zipper (bZIP) proteins [34, 35]. Although bZIP proteins are important in the detoxification process, they do not bind xenobiotics but rather mediate the cellular response to oxidative stress. The most studied xenosensor is the bHLH-PAS protein AhR, also known as dioxin receptor (DR). Other receptors involved in the xenobiotic recognition are the NRs constitutive androstane receptor (CAR), rodent pregnane X receptor/steroid and its human orthologue human steroid X receptor (PXR/SXR). The NRs farnesoid X receptor (FXR), liver X receptor (LXR), and peroxisome proliferator-activated receptor (PPAR) are also able to induce enzymes involved in the metabolism of xenobiotics. However, as these receptors are not primarily activated by xenobiotics, they are not considered xenosensors. In the following section, we describe the receptors identified as xenosensors, namely AhR, PXR/SXR and CAR, as well as some of the nuclear hormone receptors.

Aryl hydrocarbon receptor

The AhR (Fig. 1A) is a member of the bHLH-PAS protein family, which consists of transcription factors affecting a wide range of physiological functions ranging from circadian regulation to xenobiotic responses. Prominent members in addition to AhR are hypoxia-inducible factors HIF-1α and HIF-2α, the circadian regulatory proteins Clock and Per, and the dimerization partner proteins for these receptors. These partner proteins, AhR nuclear translocators (ARNT), form their own sub-group within the bHLH-PAS family [36]. Three different ARNT proteins have been identified: ARNT-1, ARNT-2, and ARNT-3 (also known as bMAL). ARNT-1 and ARNT-2 display a high degree of sequence homology and are often functionally interchangeable [37]. They both serve as general dimerization partners for AhR as well as for the HIF-1α and HIF-2α. ARNT-3/bMAL on the other hand, is selectively recruited by the circadian regulator Clock and does not dimerize with HIF-1α or AhR [38]. All the members of the bHLH-PAS family share substantial sequence homology over the conserved bHLH and PAS domains (reviewed in [39]). The bHLH domain mediates dimerization and DNA binding of these proteins (Fig. 1A). The HLH part of the bHLH domain provides the dimerization surface for the protein, and the basic region (b) of the domain determines the specificity of DNA binding. The PAS domain is not as thoroughly characterized as the bHLH domain (Fig. 1A). It encompasses two highly conserved hydrophobic repeats, PAS A and PAS B, and a nuclear export sequence (NES). Some studies have indicated that the PAS domain provides specificity for the dimerization between the different bHLH-PAS proteins and is involved in intracellular localization.

The AhR is ubiquitously expressed in most tissues and cell types. According to Northern blot analyses, the highest expression of AhR in humans is found in the lung and placenta, and the lowest in kidney, brain and skeletal muscle. In rats, however, ribonuclease protection assays showed that the receptor was most abundant in lung, thymus, liver and kidney and less abundant in heart and spleen [40]. AhR, like most bHLH-PAS proteins, has been well conserved throughout evolution and is expressed in species as diverse as humans, mice

A. Structure of the AhR

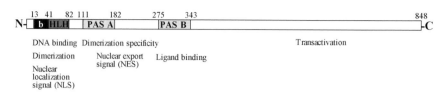

B. General structure of a NR

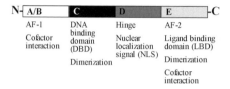

C. Structures of selected NRs

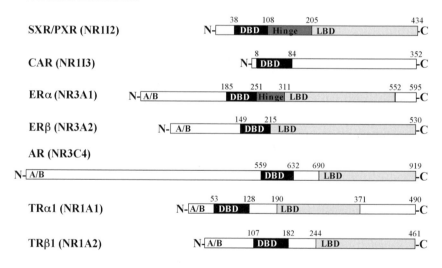

Figure 1. Schematic illustration of the domain arrangements of the receptors involved in xenobiotic response and xenobiotic-induced receptor-mediated toxicity. (A) The structure of human aryl hydrocarbon receptor (AhR) precursor (mature AhR lacks the ten first N-terminal amino acids). The N-terminal basic helix loop helix (bHLH) domain mediates DNA binding, receptor dimerization, and contains a nuclear localization signal (NLS). The PAS domain mediates ligand binding, specificity of receptor dimerization, and contains nuclear export signal (NES). The C-terminus harbors the transactivation function. (B) The general structure of nuclear receptors (NRs) consisting of four separate domains A/B, C, D and E. The A/B domain contains ligand independent activation function 1 (AF-1) and cofactor interaction surface; the C domain contains two zinc fingers, and mediates DNA-binding and receptor dimerization; the D domain functions as a flexible hinge region and contains a nuclear localization signal (NLS); and the C-terminal E domain responsible for ligand binding contains the ligand dependent AF-2 and cofactor interaction surface, and mediates receptor dimerization. (C) The structures of human SXR/PXR, CAR, ERα, ERβ, AR, TRα1 and TRβ1. All the structures are based on the ExPaSy proteomics server of the Swiss Institute of Bioinformatics (http://ca.expasy.org/).

and *Caenorhabditis elegans*. AhR is a highly polymorphic protein. Alleles differing both in ligand-binding activity and in molecular weight have been reported. The differences between the isoforms are primarily due to different positions of the translation termination codon of the receptor, rather than differential splicing or post-translational modifications [41].

The AhR was first characterized as the mediator of toxic responses evoked by HAHs, and soon thereafter PAHs were found to be even stronger activators of this receptor [42]. Even though various other AhR ligands have been identified to date, it seems that the primary activators of the receptor are coplanar aromatic substances (PAH, HAH). Instead of having strict ligand specificity, the ligand-binding cavity of AhR can accommodate a broad range of different chemical structures [18]. This characteristic is crucial for a xenosensor since an important function of such receptors is to recognize a variety of foreign compounds.

The inactive, unliganded AhR resides in the cytoplasm in a complex with chaperone proteins like heat-shock protein (hsp) 90, hepatitis B virus protein X-associated protein (XAP) 2 and immunophilin p23. The chaperone complex localizes AhR in the cytoplasm, stabilizes its conformation, and protects the receptor from degradation [43]. Upon ligand binding, the receptor translocates to the nucleus and dissociates from the chaperone complex. Subsequently, to gain transcriptional activity, AhR forms heterodimers with its partner protein ARNT. Finally, the activated AhR/ARNT complex binds to specific DNA sequences, known as xenobiotic response elements (XREs), in the promoter region of target genes. A simplified illustration of receptor-mediated response is presented in Figure 2.

Binding of the AhR/ARNT dimer to the promoter region of xenobiotic-responsive genes attracts co-activators to the promoter, and finally recruits the basal transcription machinery, leading to activated gene expression [39]. The mechanistic details behind the regulation of gene expression by AhR have been extensively studied using the promoter of the cytochrome P450 CYP1A1 gene (one of the most prominent AhR target genes) in response to dioxin. The structure of the CYP1A1 promoter is well characterized; it consists of an enhancer region containing several XREs approximately 1000 bp upstream of the transcription start site, and a TATA-box region with a TATA-box and a CCAAT-box in proximity to the transcription initiation site [44]. It has been proposed that dioxin-activated AhR/ARNT can bind to XREs located in the enhancer region, leading to recruitment of a number of coactivators and transcription factors to both the enhancer and transcription initiation sites. This results in the formation of a large multiprotein complex that bridges the two promoter regions, which in turn leads to nucleosomal disruption over the transcribed region and subsequent mRNA synthesis [44, 45]. Several coactivators from different families have been shown to physically interact with the AhR, including CBP/p300, the p160 coactivators, BRG1, thyroid receptor-associated protein TRAP220, and NR coactivator 4 [44, 46]. Most of these coactivators can bind to the promoter region of the CYP1A1 gene in a defined order: binding

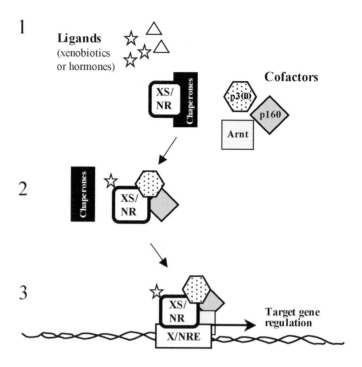

Figure 2. Simplified illustration of receptor-mediated gene regulation in response to ligand. (1) The xenosensor (XS) or nuclear receptor (NR) is bound to chaperone proteins in the absence of ligand. (2) Upon ligand (depicted by starts and triangles) binding, the chaperone complex dissociates from the receptor, leading to increased affinity of the receptor for cofactors and xenobiotic/nuclear receptor response elements (X/NRE) on DNA. (3) The activated receptor complex binds to the response elements and regulates the expression of the target gene (activation or inhibition).

of the p160 coactivators precedes recruitment of CBP/p300, TRAP220 and polymerase II [44, 45]. Transcriptional activity of the AhR is not only enhanced by transcriptional coactivators. For example, the female sex hormone 17β-estradiol (E$_2$) enhances AhR function in different cell lines [47, 48]. A recent study demonstrated that ERα is recruited to the promoter of the AhR target gene CYP1A1 after TCDD treatment, and that this recruitment is enhanced in the presence of E$_2$ [45]. This recruitment coincides with increased transcription of CYP1A1 mRNA after treatment with TCDD and E$_2$ compared to TCDD only. On the other hand, E$_2$ does not increase AhR transcriptional activity after down-regulation of ERα by small inhibitory RNA. These findings suggest that ERα is recruited to promoters of AhR target genes, thereby enhancing AhR-mediated gene expression.

In addition to the positive coregulation of AhR, there are factors that negatively regulate AhR function. For instance, AhR repressor (AhRR), a protein belonging to the bHLH-PAS family, represses AhR function. It has been suggested that AhRR competes with AhR for the common dimerization partner

ARNT. Like AhR/ARNT, the AhRR/ARNT complex binds to XRE-containing promoters but is transcriptionally inactive. Hence, the formation of transcriptionally inactive AhRR/ARNT competes with the formation of transcriptionally active AhR/ARNT and with the binding of AhR/ARNT to XREs, thereby reducing AhR activity [49]. However, this mechanism has recently been challenged [50]. Interestingly, AhRR is a target gene of AhR. Thus, AhR and AhRR form a regulatory circuit in the xenobiotic signaling pathway. This property of controlling transcription *via* induction of negative signal transduction factors within the same family is a common feature in the bHLH-PAS family. Another example is the modulation of the circadian rhythm repressor protein Per by the members of the clock gene family [51]. Besides the repressor AhRR, several negative cofactors are known to down-regulate AhR activity. These include corepressors like receptor interacting protein RIP140 and small heterodimer partner SHP [52, 53]. Additionally, ARNT interacting protein AINT is known to bind ARNT and thereby decrease its nuclear localization, leading to reduced AhR activity [54].

In addition to the XRE-mediated AhR pathway, other mechanisms can lead to AhR dependent gene activation. For instance, AhR can regulate gene activation through interaction with other transcription factors like retinoblastoma protein (Rb) and nuclear factor (NF)-κB [55]. AhR can also activate phosphorylation cascades *via* tyrosine kinase c-Src and p38-mitogen-activated protein kinase [56]. Some ligands can induce phosphorylation of AhR, leading to facilitated nuclear translocation without dimerization with ARNT [57].

Many AhR target genes code for enzymes involved in metabolism of xenobiotics, primarily phase I drug-metabolizing monooxygenases such as CYP1A1, CYP1A2, and CYP1B1, but also phase II enzymes like glutathione-*S*-transferase, UDP-glucuronosyltransferase, NADPH-quinone oxidoreductase and xanthine oxidase [58, 59]. However, AhR also induces a variety of genes other than the drug-metabolizing enzymes. For example, many genes involved in growth, differentiation and cellular homeostasis are induced by TCDD-activated AhR, including transforming growth factors (TGFs) α and β, plasminogen activator inhibitor-2, c-fos and c-jun [60].

Although AhR has been extensively studied in the context of the xenobiotic response, its endogenous ligands and functions are poorly understood. Creation of AhR-knockout mice has shed some light on the physiological role of this receptor, and it has become clear that AhR possesses other roles than recognition of xenobiotics in living organisms. The AhR$^{(-/-)}$ mice are viable, but have reduced liver size at birth, perhaps due to reduced blood supply during fetal development. They also have slowed early growth, reduced female fertility and disturbed immune system [61]. In addition, the AhR-knockout mice differ from wild-type animals upon TCDD exposure. The knock-out mice are unaffected by acute exposure to this dioxin at doses ten times that needed to cause severe toxic effects in their wild-type littermates [62]. Usual responses to TCDD, including teratogenic effects, wasting syndrome, thymic atrophy, and decreased prostate weight are missing in the AhR$^{(-/-)}$ mice [63, 64]. On the other hand, transgenic mice expressing constitutively active AhR develop gastric tumors, suggesting that AhR is involved in some aspects of the regulation of growth and proliferation [65].

Endogenous ligands for AhR include several prostaglandins that transiently induce CYP1A1 [66]. Furthermore, it has recently been shown that AhR is strongly induced by photoproducts of tryptophan, such as 6-formylindolo-[3,2-*b*]carbazole (FICZ) *in vitro* [67, 68] and *in vivo* [69]. Tryptophan photoproducts are suggested to act as light sensors and they induce changes in the expression of genes regulating the circadian rhythm [69]. This puts forward a role for AhR as mediator of circadian adaptation to light and darkness.

The nuclear receptor family

General features of the nuclear receptors

NRs are a large family of nuclear transcription factors involved in a plethora of vital functions ranging from fetal development to homeostasis, reproduction, immune function and metabolism. In mammals, 48 structurally and functionally conserved NRs have been identified. These receptors serve as sensors for numerous endogenous, but also exogenous compounds, such as sex hormones, cholesterol metabolites, bile acids, fatty acids and vitamins. However, only half of the NRs have a recognized ligand; the rest are so-called orphan receptors.

The NRs share an evolutionarily conserved domain arrangement consisting of an N-terminal variable A/B domain, followed by a C domain harboring two zinc fingers responsible for the DNA binding (DNA-binding domain or DBD) and a C-terminal ligand binding domain (LBD) (Fig. 1B). NRs have two distinct activation functions (AF), a ligand independent AF-1 within the A/B domain and a ligand-dependent AF-2 within the LBD [70] (Fig. 1B). Depending on the ligand bound to AF-2, there is interaction and synergism between AF-1 and AF-2.

In the inactive state, the NRs reside either in the cytoplasm or in the cell nucleus bound to chaperone proteins like hsp90 and p23 and/or co-repressors. This complex represses NR function and keeps it in a conformation that allows binding of ligand (reviewed in [71]). Ligand binding triggers conformation changes that lead to dissociation of the repressor proteins and to dimerization of the receptors. The most prominent conformational change is a switch of helix 12 (H12) to an agonist bound conformation, where H12 is aligned across the ligand binding cavity, forming a docking site for coactivator binding together with H3 and H5 (Fig. 3). If the NR is bound to an antagonist, the alignment of H12 into the agonist position is prevented due to steric hindrance and the coactivator docking site is not formed properly (Fig. 3) (reviewed in [72]). The agonist-activated NR recruits coactivators that enable the transcriptional activity of the receptor. Most is known about the AF-2 interacting coactivators, the primary role of which is in chromatin remodeling, in bridging NRs

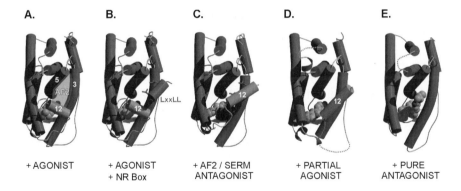

Figure 3. Depiction of conformations of ERα ligand binding domain (LBD) in response to: (A) the full agonist E$_2$, (B) E$_2$ and the cofactor NR-Box with the LxxLL motif, (C) the selective ER modulator raloxifene, (D) the phytoestrogen genistein and (D) the full antagonist ICI. AF: activation function, H12: helix 12. Used with permission by Physiological Reviews [72].

to the basal transcription machinery, and in promoting the assembly of RNA polymerase II transcription complexes. The p160 family of co-activators, including steroid receptor co-activators SRC-1 and SRC-3 and transcriptional intermediary factor TIF-2, interact with the coactivator docking site of the NR through a conserved LxxLL motif (NR-box). The p160 co-activators in turn recruit other transcription coactivators and histone acetyltransferases such as p300, CREB-binding protein (CBP) and p300/CBP-associated factor (pCAF). As a consequence of coactivator recruitment, a large multiprotein complex forms on responsive areas of DNA, and histones become acetylated, which leads to de-condensation of local histone structures. The opened DNA structure attracts the basal transcriptional machinery and facilitates transcription (reviewed in, e.g., [73]). In an antagonist-bound NR conformation, corepressors such as SHP and RIP140 are recruited to the LxxLL docking site, leading to repression of transcriptional activity. A simplified illustration of receptor mediated signaling is presented in Figure 2.

The xenosensors PXR/SXR and CAR

PXR and its human orthologue SXR as well as CAR share the general NR structures but lack the A/B domain and hence AF-1 [74] (Fig. 1C). PXR/SXR is primarily expressed in the liver with some expression in testis and embryonic tissues in humans and in intestine in mice. CAR is primarily found in the liver and kidney with some expression in heart and intestine [74].

PXR/SXR responds to a large number of ligands, including many endogenous substances like pregnanes, bile acids and hormones as well as pharmaceuticals such as the anticancer drugs tamoxifen and taxol and the antibiotic rifampicin. Environmental contaminants that activate PXR/SXR include many pesticides (e.g., DDT, dieldrin, and methoxychlor), vinclozolin and BPA [33]. Even though SXR is the human orthologue of rodent PXR, these receptor isoforms do not respond similarly to all xenobiotics. Certain highly chlorinated PCBs activate PXR and inactivate SXR [75]. This is important as rodent models are widely used to predict the toxicity of a certain compound in humans. CAR is less promiscuous in ligand binding than PXR/SXR. Its prototypical activator is the anticonvulsant drug phenobarbital (PB). However, the activity of CAR is also modulated by several environmental contaminants, mostly pesticides like DDT and its metabolite DDE as well as methoxychlor [76, 77]. Some ligands are species specific; the hydrocarbons 1,4-bis-[2-(3,5-dichloropyridyl-oxy)]benzene (TCPOBOP) and 6-(4-chlorophenyl)imidazo[2,1-b][1,3]thi-azole-5-carbaldehyde O-3,4-dichlorobenzyl)oxime (CITCO) are, for example, ligands for mouse CAR and human CAR, respectively [74]. Additionally, CAR is activated by endogenous compounds such as steroids and retinoic acid.

In the absence of ligand, both PXR/SXR and CAR reside in the cytoplasm. They are bound in a complex with the co-chaperones cytoplasmic CAR retention protein (CCRP) and Hsp90 [78, 79]. Upon ligand binding, the receptors dissociate from this complex and translocate to the nucleus where they form heterodimers with the retinoid X receptor-α (RXRα) [80]. This leads to the regulation of target gene expression by binding to xenobiotic-responsive enhancer modules (XREM) in the case of PXR, and PB-responsive enhancer modules (PBREM) in the case of CAR. The activity of the receptors is further modulated by recruitment of coactivators including those of the p160 family, e.g., SRC, PPARγ coactivator PGC-1α, TRAP220 and PPAR-binding protein (PBP) [74]. Furthermore, the activity of PXR is potentiated by the activation of the protein kinase A (PKA) signaling pathway induced by forskolin [81]. In contrast, PXR activity is repressed by PKC. Unlike other NRs, CAR exhibits constitutive activity, at least *in vitro* [82]. It is thought that even without a ligand, a fraction of receptor molecules is able to translocate to the nucleus and interact with coactivators. The presence of agonist strengthens this interaction, thereby increasing the activity of the receptor, whereas an antagonist (also called inverse agonist) abolishes the receptor binding to coactivators thus repressing receptor activity [76]. Additionally, nuclear translocation of CAR is a highly regulated process that involves many other proteins [74], which suggests that nuclear translocation of the receptor might also play a role in regulating its constitutive activity.

The target genes of PXR/SXR and CAR activity are involved in all phases of the xenobiotic response. They induce phase I factors like CYP3A, CYP2B, and CYP2C, phase II conjugation enzymes like glutathione-S-transferases and UDP-glucuronosyltransferases, and phase III transporters such as multi-drug resistance-associated proteins (MRPs) [33]. Although PXR/SXR and CAR have overlapping target genes, some enzymes are exclusively regulated by either PXR/SXR or CAR [83].

In addition to their pivotal role in the xenobiotic response, both PXR/SXR and CAR are important players in the metabolism of steroids [33]. As a result, activation of these receptors by xenobiotics can change steroid hormone levels, leading to altered hormone responses and thus endocrine disruption [84].

Conversely, the activity of PXR/SXR and CAR is also regulated by endocrine signals. For example, PXR/SXR is activated by pregnanes and glucocorticoids. More recently, CAR and PXR were found to be involved in gluconeogenesis and homeostasis of cholesterol and bile acids in the liver [74]. Thus, CAR and PXR seem to form a complex regulatory network with other NRs such as LXR, FXR, and PPAR [85].

Hormone receptors of the NR family
Estrogen receptors:
In mammals, two distinct functional ERs are recognized, namely ERα (first described in 1962 by Jensen and Jacobsen [86]) and ERβ (discovered in 1996 by Kuiper et al. [87]) (Fig. 1C). ERα and ERβ are encoded by separate genes on separate chromosomes (in human chromosomes 6 and 14, respectively), and share a high degree of homology (overall 55%, ranging from 17% in A/B domain to 97% in DBD) [88] (Fig. 1C). Several alternative splice variants have been recognized for both ERs; however, it is unclear if all identified transcripts are expressed as functional proteins *in vivo*. For ERα, one shorter splice variant has been characterized that is expressed in several cell lines. This truncated variant lacks exon 1 that harbors the N-terminal AF-1 and has been termed ERα 46. ERα 46 can dimerize with full-length ERα and repress its AF-1-mediated activity [89]. Of the several ERβ isoforms, at least six splicing variants have been found in the human in addition to the first characterized wild-type receptor. Most functional studies have focused on ERβcx, which is expressed both in normal and in cancerous tissues. ERβcx has a low affinity for E_2 due to amino acid deletions and replacements in the C-terminus, and it can inhibit ERα activity through preferential dimerization with this receptor subtype. The full-length ERs are expressed in several tissues, e.g., the central nervous system, cardiovascular system, bone, gastrointestinal tract, reproductive tissues and urogenital tract. Specifically, ERα is preferentially expressed in uterus, ovaries, breast, liver, and kidney, and ERβ in prostate, ovaries, colon, lung and ovaries. Even though both receptors are present in the same tissues their expression is often limited to distinct cell types. For example, in ovaries ERα is expressed in the theca cells and ERβ in the granulosa cells [90].

Estradiol (E_2) is the most potent ER ligand among the endogenous estrogens. In premenopausal women, E_2 is produced in the ovaries by aromatization from androgens. At menopause, the ovarian E_2 production seizes and peripheral CYP19 (aromatase) activity (e.g., in adipose tissue), also present in premenopausal women, becomes the main source of circulating estrogens. The ligand-binding cavities of ERα and ERβ differ by two amino acids, allowing selective ligand binding by the two receptors. A wide variety of ligands other than endogenous estrogens binds to and regulates the activity of the ERs.

The ligand-activated ER complex binds to target promoters directly or indirectly through tethering to other transcription factors. The direct binding occurs through estrogen-responsive elements (EREs) resembling the palindromic consensus sequence GGTCAnnnTGACC. Genes that are regulated through EREs include trefoil factor 1 (TFF1, also known as pS2), oxytocin, angiotensinogen, cathepsin D, complement C3 and vascular endothelium growth factor (VEGF) [91]. ER tethering to promoters may occur through transcription factors such as activating protein AP-1 and specificity protein SP-1. Genes that respond to ERs through AP-1 and SP-1 sites include c-fos, and hsp27 [91]. In addition to ligand-dependent transcriptional activity of ERs, activation of unliganded ER through phosphorylation has also been described, as well as signaling through membrane-bound ERs in response to estrogens.

Estrogens and ERs play important roles in regulation of development and maintenance of reproductive functions. Additionally, estrogens affect, for example, bone, adipose tissue and central nervous and cardiovascular systems as well as metabolism. Estrogens are mitogenic and are involved in carcino-genesis in hormone-dependent tissues. Even though the two ERs are highly homologous, have similar affinity for E_2, and often share target genes, they generally appear to have different functions. To elucidate the specific roles of ERs *in vivo*, knockout mice have been created with targeted disruption of either the ERα or the ERβ gene, termed αERKO and βERKO, respectively. The phenotypes of these mice reveal distinct roles in physiology for the two receptors (extensively reviewed in [92–94]). Both αERKO and βERKO mice are viable (heterozygous mating produces expected Mendelian distribution of the genotypes), indicating that the ERs are not essential for embryonic devel-opment [95–97]. However, the phenotypes of the mice clearly indicate that ERs are crucial, e.g., for normal growth, development, and homeostasis. The reproductive capacity of both αERKO and βERKO mice is either severely impaired or completely abolished. The αERKO females are infertile and dis-play altered mating behavior. αERKO males have drastically reduced sperm counts and are infertile [95, 97]. The reproductive phenotype of the βERKO mice is milder than that of αERKO mice. βERKO females are subfertile or infertile, yet their mating behavior is normal. βERKO males are normal in terms of fertility [96, 97]. Hence, it can be concluded that the role of ERα in reproduction is indispensable, while the effect of ERβ seems less profound. Instead of playing a vital role in classical estrogen target tissues, ERβ is impor-tant in non-reproductive tissues. For instance, βERKO mice develop hyperten-sion as they age [98], indicating a possible role for ERβ in the cardiovascular system. Furthermore, βERKO mice display changes in the colon and prostate that indicate that ERβ is anti-proliferative in these tissues [96, 99, 100]. Additionally, βERKO females show increased anxiety and decreased serotonin concentration in certain brain areas [101, 102]. In cell cultures, ERβ acts as negative regulator of ERα-mediated activity on the cyclin D1 promoter [103]. The two isoforms have opposing effects on cell growth: ERα promotes prolif-eration, whereas ERβ is anti-proliferative and promotes differentiation [30]. The opposing activities of the ER isoforms suggest that ERβ and ERα balance each other's effects. Dysregulation of one ER isoform would lead to inappro-priate activity of the other. The anti-proliferative principle of ERβ may prove useful in development of antitumor drugs [104].

Androgen receptor:
Two AR isoforms have been described, AR-A and AR-B (Fig. 1C). They are encoded by the same gene; however, AR-A is truncated at the N-terminus, lakking the first 187 amino acids. The function of AR-A is not clear, but it has been suggested that it can act as repressor of AR-B [105]. Both receptors are ubiquitously expressed but most prominently in male reproductive tissues [106, 107]. In most tissues, AR-B is more abundant than AR-A [107].

The two main physiological androgens are testosterone and dihydrotestosterone (DHT), the latter being a metabolite of testosterone that exhibits a higher affinity for the AR than its mother compound. In males, testosterone is produced by the testes; the remaining androgens, including DHT, are either produced by the adrenal cortex or are derived from peripheral conversion of testosterone. In females, androgens are produced in the adrenal cortex and the ovaries.

In its unliganded state, the AR resides predominantly in the cytoplasm. Upon ligand binding, it translocates to the nucleus, forms homodimers and regulates gene expression by binding to androgen response elements (ARE) in the promoter region of target genes. Apart from the actions of AR as transcription factor, non-genomic actions of a plasma membrane-bound androgen-binding receptor have been proposed. This receptor has not yet been identified, however, it has been suggested to increase intracellular calcium levels, thereby activating kinase signaling cascades [108].

Androgens and their receptor are crucial for the development of the male reproductive tissues and their maturation in puberty. In adulthood, they are essential for maintaining male reproductive behavior and function. Furthermore, they affect a number of non-productive tissues including adipose tissue, skin, bone, and muscle.

Thyroid hormone receptors:
There are two thyroid hormone receptors (TRs), TRα and TRβ, encoded by two separate genes. The TRα gene gives rise to two splice variants, the fully functional TRα1 (Fig. 1C) as well as TRα2 that is unable to bind ligand or activate genes. The TRβ gene encodes three splicing isoforms, TRβ1–3 (Fig. 1C), that vary in their N-terminal sequence [109]. Recently, novel truncated variants for both TRα and TRβ have been identified, with dominant negative or unknown function [110, 111]. Most of the TRs are ubiquitously expressed; however, TRα1 levels are highest in muscle and brown adipose tissue, whereas TRβ1 expression peaks in brain, liver, and kidney. TRβ2 expression is restricted to the anterior pituitary gland and specific areas of the hypothalamus as well as the developing brain and inner ear [109].

In contrast to the other hormone receptors of the NR family, TRs are activated by an amino acid derivative rather than a steroid hormone. Their main ligand triiodothyronine (T_3) is produced as a prohormone, 3,5,3',5'-tetraiodothyronine or thyroxine (T_4), in the thyroid gland. Synthesis of T_4 is stimulated by thyroid-stimulating hormone (TSH), which is secreted by the pituitary. Upon secretion, T_4 is transported to the target tissues and converted to T_3 by the iodothyronine deiodinase enzymes D1 and D2 (reviewed in [112]). The activation process of TRs varies slightly from the principle introduced in the section 'General features of the NRs' (see above). In the absence of ligand, TR is already bound to thyroid hormone response elements (TREs) in promoter regions of target genes as a homodimer or heterodimer with RXR and in complex with corepressors like silencing mediator for retinoic acid receptor and TR (SMRT) and nuclear receptor corepressor (NCoR). This DNA binding represses the transcription of the target gene or allows basal transcription. Upon ligand binding the corepressor complex dissociates from the TR/RXR heterodimer, and coactivators are recruited. This leads to induction or enhancement of gene transcription [109]. Several genes have also been identified that are activated by unliganded TR and repressed upon hormone binding to the receptor. Thus far, the mechanism of this negative regulation is not fully understood; however, it is clear that TRs bind to negative TREs in the promoter region of these genes [113].

Thyroid hormones and receptors play a crucial role in a multitude of important physiological processes including homeostasis, differentiation, growth and development. In fact, they are necessary for the normal function of virtually all tissues, and dysfunction of the thyroid gland is one of the most common endocrine disorders.

Other hormone receptors belonging to the NR family:
The three other steroid hormone receptors of the NR family are the progesterone receptor (PR), glucocorticoid receptor (GR), and mineralocorticoid receptor (MR). There are two PR isoforms, PR-A and PR-B, which mediate the response to progesterone. PRs play a crucial role in female reproductive functions like ovulation, and uterine as well as mammary gland development. Furthermore, they are believed to have functions in the cardiovascular and central nervous system, in bone maintenance, and in thymic involution [114]. Both GR and MR mediate the response to glucocorticoids such as cortisol and corticosterone. MR has a tenfold higher affinity for the glucocorticoids and is additionally responsible for the effects of mineralocorticoids like aldosterone. Glucocorticoids play important roles in metabolic homeostasis, inflammatory processes and in particular the response to stress, whereas mineralocorticoids are crucial for sodium homeostasis. GR is abundantly expressed, whereas MR expression is mainly restricted to kidney, intestine, pancreas, and defined regions in the central nervous system.

Endocrine disruption

The human endocrine system is a body-wide network of signaling pathways in which hormones deliver messages, directly or indirectly affecting all aspects of physiology. The phenomenon of exogenous compounds causing dysregulation of the endocrine system has been termed endocrine disruption, and the compounds are collectively referred to as endocrine disrupting chemicals (EDCs). Although xenobiotics could theoretically disturb the function of all hormone receptors, research to date has mainly focused on their effects on ER, AR and TR. In particular, the ER is recognized as a target for many EDCs. There are two main pathways by which a xenobiotic can disrupt hormone signaling through receptor-mediated mechanisms: (1) by directly binding to the hormone receptors (Fig. 4), or (2) by binding to xenosensors (AhR, PXR, CAR) and indirectly affecting hormone signaling by various mechanisms (Fig. 5). Direct binding to hormone receptors is by far the most studied mechanism of endocrine disruption. However, there are several other mechanisms by which EDCs may indirectly affect hormonal pathways, such as inhibition of steroidogenic enzymes and binding to steroid transport proteins, reflecting how complex and interwoven the endocrine and EDC detoxification systems

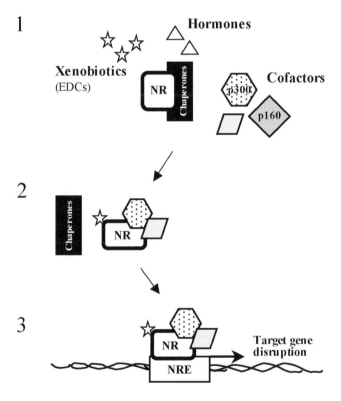

Figure 4. Mechanistic representation of receptor-mediated endocrine disruption. (1) Endocrine disruptive chemicals (EDCs) of suitable structure and affinity (illustrated with stars) bind to a nuclear receptor (NR) instead of the cognate hormone (illustrated with triangles). (2) Depending on the receptor conformation induced by the ligand, different cofactors are recruited to the ligand-NR complex. (3) The transcriptional activity of the receptor is governed through the ligand and cofactors, and may lead to inappropriate activation or repression of genes containing NR response elements (NRE) in their regulatory regions.

A. Competition for cofactors

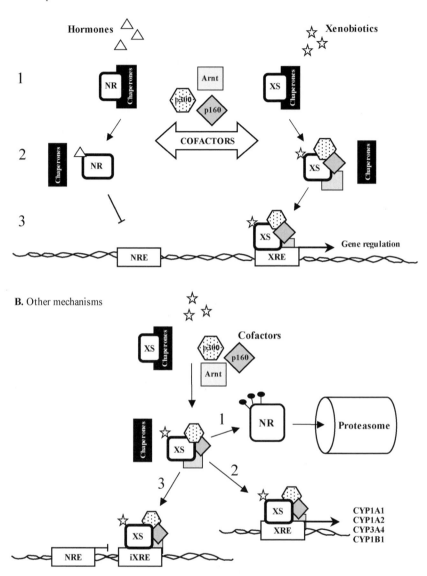

Figure 5. Indirect mechanisms of receptor-mediated endocrine disruption. (A) Xenobiotics can disrupt nuclear receptor (NR) activity indirectly through competition for common cofactors: (1) When
both NR and xenosensor (XS)-mediated signaling occurs in the same cell, common cofactors (e.g.,
p160, p300 and ARNT) may become limiting factors in the signaling event. (2) When the cofactors
are recruited to XS pathway, they are not sufficiently available for NR signaling. (3) NR signaling is
hampered, while XS-mediated activity occurs normally. (B) More indirect mechanisms of endocrine
disruption include: (1) Targeted degradation of NR in proteasome as a consequence of XS-induced
ubiquitination (depicted with pins) of the receptor; (2) XS-induced transcription of enzymes involved
in hormone metabolism; and (3) Binding of XS to inhibitory xenobiotic response elements (iXRE) in
close proximity of NREs on DNA, blocking gene regulation through the NRE.

are. In this section we describe the mechanisms by which xenobiotics can disturb the hormone signaling directly and indirectly through xenosensors.

Direct binding to the nuclear hormone receptors

The nuclear hormone receptors regulate gene transcription in response to their cognate ligands. As many xenobiotics structurally resemble these cognate ligands, they too can bind to the receptors causing inappropriate signals in the cell. The activation of NRs by xenobiotics follows the general mechanism introduced in the section 'General features of the NRs' (see above); a xenobiotic compound with suitable structural features binds to the hormone-binding cavity, or in some instances to some other part of the receptor, causing a conformational change of the receptor. This change determines the activity of the receptor: an agonistic conformation leads to recruitment of coactivators and thus increased transcriptional activity of the receptor, whereas an antagonistic conformation prevents coactivator recruitment and/or attracts corepressors, leading to decreased transcriptional activity of the receptor (Fig. 3).

There are many alternative receptor conformations between the "full agonist" and "full antagonist" conformations (Fig. 3) resulting in differential receptor activities. For example, ERα adopts distinct conformations when bound to E_2 or to phytoestrogens like genistein and coumestrol as compared to mono- or bisphenolic substances like methoxychlor [115]. E_2 and some phytoestrogens are coplanar, whereas methoxychlor is nonplanar; this latter structure results in an incomplete sealing of the ligand-binding pocket by H12. This in turn prevents synergy between AF-1 and AF-2, and thus leads to differences in ER function when ER is bound to nonplanar compounds compared to planar ones [115].

This type of ligand-specific effects makes the identification of EDCs and their specific effects challenging. It has become evident that the traditional EDC tests, which are based on physiological changes such as the rodent uterotrophic tests for ER activity, do not detect all compounds that are capable of interfering with the hormone receptors. For example, BPA and enterolactone are recognized as ERα agonists in *in vitro* transactivation and ligand-binding assays, but neither of the compounds induces consistent uterine weight increase in rodents [28, 116]. This is most probably due to selective regulation of gene expression by the xenobiotic-ER complexes in question.

Proper identification of xenobiotic-induced hormone receptor modulation thus requires a combination of different assays. At present, reporter gene assays (e.g., hormone-responsive promoters coupled to reporter genes such as luciferase and GFP) in different cell lines and *in vivo* models such as *Xenopus,* zebra fish and mouse are used for screening purposes. Positive hits are then further characterized in more detail with reference to ligand binding, target gene expression, promoter binding and cofactor recruitment as well as *in vivo* effects. These mechanistic studies help to understand why different EDCs have

varying effects on human health (e.g., phytoestrogens may protect from cancer, while man-made ER ligands may increase the risk of cancer).

Xenosensor-mediated endocrine disruption

Competition for common cofactors

The biological function of transcription factors, including the NRs and AhR, is
dependent on the availability of transcriptional cofactors. Many of these cofactors are not receptor specific but are used in several different signaling pathways in the cell. Hence, the cofactors are a potential target for endocrine disruption. If one receptor is persistently activated, the continuous recruitment of
cofactors to this receptor may reduce the availability of the cofactors to other
receptors, thereby impairing their activity (Fig. 5A). Both the bHLH-PAS proteins and NRs use p160/p300 class cofactors [117, 118], and this overlap creates a means for interaction between the two protein families.

The best-described example of endocrine disruptive cofactor sequestering is
the apparent anti-estrogenicity of TCDD. TCDD decreases ER-mediated estrogen signaling by binding to AhR thereby leading to recruitment of common
cofactors, such as SRC-1, RIP140, and CBP/p300 [119, 120]. Thus, if the
coactivator-binding domain of AhR is overexpressed in cells, ER-mediated
activity is impaired. Similarly, overexpression of the coactivator binding site
of ERα results in decreased signaling *via* AhR [58].

The obligatory heterodimerization partner of AhR, ARNT, is not considered
as a classical cofactor of transcription. Still, many features of ARNT resemble
those of coactivators of the SRC family: both ARNT and SRC proteins have a
bHLH-PAS domain and they are both constitutively active nuclear proteins.
These characteristics may suggest that the functions of ARNT and SRC could
overlap. Indeed, recent reports show that ARNT may coactivate both ERα and
ERβ. Overexpression of ARNT in cells leads to increased ER activity, whereas
down-regulation of ARNT leads to impairment of ER function [121]. Promoter
studies indicate that, following E_2 treatment, ARNT is recruited to estrogen-
responsive promoters [121]. Thus, ARNT represents one more means of cross-
talk between ER and AhR. Moreover, recent studies demonstrate that the effects
of AhR-mediated anti-estrogenicity differ between ER subtypes; upon AhR activation, the function of ERβ is more severely impaired than that of ERα [122].

Cross-talk between xenosensors and NRs in terms of cofactor recruitment
is not limited to AhR and ER. For example, activation of CAR inhibits ER
activity by squelching the p160 coactivator GRIP-1 [123].

Dysregulation of hormone metabolism

Steroid hormones are small lipophilic molecules that are synthesized from
cholesterol. They have a cyclopentano-perhydrophenanthrene four-ring hydrocarbon nucleus, the so-called steroid nucleus, as a core structure. The synthesis of steroid hormones occurs primarily in the adrenal cortex, gonads (testes

and ovaries) and placenta. The catabolism of sex steroids occurs in the liver in a process that closely resembles the metabolism of xenobiotics. To render the steroid hormones hydrophilic, they are hydroxylated and conjugated. Hydroxylation reactions are carried out by steroid hydroxylases of the CYP family, whereas sulfotransferases and the UDP-glucuronosyltransferases are responsible for steroid conjugation.

Many of the xenobiotic metabolizing enzymes are also involved in the catabolism of endogenous hormones. For example, the phase I P450 enzymes CYP1A2, CYP3A4, CYP1A1, and CYP1B1 are responsible for the hydroxylation of E_2 [124]. Interestingly, CYP1B1 has been reported to be activated by estrogens [125]. Other NRs can also activate members of the P450 enzyme family. PPARs activate CYP4A, and liver X receptor (LXR) and FXR activate CYP7A. Hence, even hormone receptors possibly activated by EDCs may be involved in creation of the xenobiotic response in an organism.

As a consequence of the similarities between xenobiotic and steroid metabolism, activation of the xenobiotic response can affect metabolism of hormones (Fig. 5B). Enzyme activation upon exposure to xenobiotics can lead to alterations in the endogenous levels of hormones in the organism, and subsequently compromise hormone signaling. The most important regulators of the activity of these enzymes are PXR and CAR. For example, DDE and PCBs can induce both sulfotransferases and CYP enzymes, whereas 3-MC and PB may up-regulate UDP-glucuronosyltransferases [126].

Inhibitory xenobiotic response elements
As described earlier in this chapter, ligand-activated xenosensors bind to XREs in the DNA and induce the expression of target genes. Interestingly, many ER-regulated genes have also XRE-like elements in their promoters. These inhibitory XREs (iXREs) differ slightly in base composition from the response elements on xenobiotic-responsive genes, rendering the sequence capable of binding xenosensors but not of activating the downstream gene (reviewed in [119]). Binding of the AhR to iXREs results in a decrease of ER activity by disturbing either the assembly of the preinitiation complex [127, 128] or the binding of additional transcription factors necessary for full activity of the ER, e.g., AP-1 [129] (Fig. 5B). Recently, it has also been found that the presence of unliganded AhR on XREs close to ER binding sites on certain genes, instead of being inhibitory, is in fact necessary for the transcriptional activation of the respective gene by ER [130]. Binding of liganded AhR to the same XRE, on the other hand, leads to abrogation of gene induction by ER. These findings suggest that a complex interaction between AhR and ER occurs on the promoters that contain both XREs and EREs (or other transcription factor binding sites through which ER can act).

Dysregulation of receptor stability
The stability of hormone receptors is an integral feature of receptor biology. Since cells need to rapidly respond to fluctuating hormone levels, the amount

of the receptors has to be tightly and rapidly regulated. For example, in the absence of estrogens the ER levels are up-regulated and upon estrogen treatment, the levels quickly decrease. The degradation of several NRs occurs in the ubiquitin-proteasome pathway. Receptors are targeted for degradation with ubiquitin, a 76-amino acid protein. Polyubiquitinated proteins are transported to the proteasome, a multi-protein complex, where they are degraded. Inhibition of this pathway affects different hormone receptors differently. For instance, upon ubiquitination the transcriptional activity of AR [131] or GR [132] is increased, whereas the activity of ER is decreased [133]. Recent findings indicate that both AhR and ARNT are a part of cullin 4B ubiquitin ligase, a multi-protein complex involved in targeting proteins to the proteasome [134]. This complex seems to specifically target ER and AR for degradation. Furthermore, the assembly of this complex is suggested to be AhR ligand dependent [134]. Therefore, activation of AhR by environmental chemicals may alter the amount of steroid hormone receptors in the cell, leading to disrupted estrogen and androgen signaling (Fig. 5B). For instance, TCDD treatment has been shown to decrease ERα levels, and this effect can be blocked by drugs inhibiting proteasomal degradation [135].

Prospects

Through food, water and air, humans are exposed daily to a diversity of chemicals. Some of the xenobiotic substances have accompanied human existence for a long time, while a plethora of chemicals has been introduced recently, mainly during the last century. The former compounds, natural plant-derived substances, most likely do not pose a threat to humans when consumed as part of a normal healthy diet. However, the latter group of compounds, man-made chemicals, are new to the cellular detoxification system and may present a hazard to health. As described in this chapter, many problems may arise upon exposure to man-made chemicals, for example endocrine disruption. Indeed, it is alarming that, while the manufacture and use of chemicals has drastically increased during the last decades, the incidence of hormone-related diseases, such as cancers of breast and prostate, has increased markedly in the industrialized countries. Even though the link between increased use of chemicals and human disease may seem evident, it has yet to be causally proven.

At present, over 100 000 different chemicals, untested for their effects on human health are in use in the European Community. Recently, a new regulation concerning safer management of chemicals was passed in the EU. The REACH Regulation (Registration, Evaluation, Authorization and restriction of Chemicals), which entered into force on 1 July 2007, aims to protect humans and the environment through earlier identification of potential risks associated with chemicals. The manufacturers and importers of the chemicals rather than public health authorities are responsible for the risk identification. During the implementation period of 11 years, over 30 000 chemicals are to be registered with information on their health hazards. In the hazard identification, REACH encourages the use of alternative testing methods to reduce the use of experimental animals to a strict minimum. Hence, REACH creates a strong need for reliable, cost- and time-efficient *in vitro* test methods that can be used to predict effects *in vivo*. At present, there are a wide variety of *in vitro* test models available that can be used to characterize effects of various xenobiotics in different settings. However, at present, a major limitation of these approaches is the interpretation of the results in terms of human health.

The difficulty of identifying receptor-mediated human health hazards based on *in vitro* methodology depends partly on poor understanding of the function of the receptors with respect to disease development. It is not clear which end-points should be considered alarming, or which end-points should be used to identify all possible biological effects of chemicals. To tackle these problems, basic research on the function of xenosensors and hormone receptors is needed. The growing field of genomics, proteomics (analysis of protein expression and interactions), and structural biology will contribute to a more detailed understanding of the complexity of the signaling of these receptors. Research should also consider the action of xenobiotics on other hormonal systems other than steroid and thyroid hormone signaling. To date, only little is known about how xenobiotic substances can affect the pathways of peptide hormones and fatty acid derivatives.

It is obvious that one approach cannot cover all aspects needed for risk assessment of chemicals, but a combination of different methods and model systems has to be used. It is clear that this cannot be achieved by the research of single laboratories or institutes. In an effort to create collaborations between scientists from various fields related to chemical risk characterization to tackle the problems associated with risk assessment, European Commission has funded multi-disciplinary projects within Europe. One of these is the CASCADE Network of Excellence (www.cascadenet.org), which is devoted to risk assessment of endocrine disruptive chemicals in food, and is a good example of an EC-funded joint attempt to characterize food-related health concerns. Such inter-disciplinary collaborations are needed to create reliable and sustainable routines for chemical hazard identification in the future.

Acknowledgements
The authors would like to thank Dr. Nina Heldring for the contribution of Figure 3 and Dr. Krista Power for proofreading this chapter. The authors are supported by the Swiss National Science Foundation (SNF) (J.R.), the European Union funded CASCADE Network of Excellence (FOOD-CT-2004-506319), the Swedish Cancer Foundation, and the Swedish Research Council.

References

1 Lemaire G, Mnif W, Mauvais P, Balaguer P, Rahmani R (2006) Activation of α- and β-estrogen receptors by persistent pesticides in reporter cell lines. *Life Sci* 79: 1160–1169
2 Kavlock R, Cummings A (2005) Mode of action: Inhibition of androgen receptor function – Vinclozolin-induced malformations in reproductive development. *Crit Rev Toxicol* 35: 721–726

3 Molina-Molina JM, Hillenweck A, Jouanin I, Zalko D, Cravedi JP, Fernandez MF, Pillon A, Nicolas JC, Olea N, Balaguer P (2006) Steroid receptor profiling of vinclozolin and its primary metabolites. *Toxicol Appl Pharmacol* 216: 44–54

4 Kuiper GG, Lemmen JG, Carlsson B, Corton JC, Safe SH, van der Saag PT, van der Burg B, Gustafsson JA (1998) Interaction of estrogenic chemicals and phytoestrogens with estrogen receptor β. *Endocrinology* 139: 4252–4263

5 Bonefeld-Jorgensen EC, Long M, Hofmeister MV, Vinggaard AM (2007) Endocrine-disrupting potential of bisphenol A, bisphenol A dimethacrylate, 4-n-nonylphenol, and 4-n-octylphenol *in vitro*: New data and a brief review. *Environ Health Perspect* 115 Suppl 1: 69–76

6 Kitamura S, Suzuki T, Sanoh S, Kohta R, Jinno N, Sugihara K, Yoshihara S, Fujimoto N, Watanabe H, Ohta S (2005) Comparative study of the endocrine-disrupting activity of bisphenol A and 19 related compounds. *Toxicol Sci* 84: 249–259

7 Fini JB, Le Mevel S, Turque N, Palmier K, Zalko D, Cravedi JP, Demeneix BA (2007) An *in vivo* multiwell-based fluorescent screen for monitoring vertebrate thyroid hormone disruption. *Environ Sci Technol* 41: 5908–5914

8 Belpomme D, Irigaray P, Hardell L, Clapp R, Montagnier L, Epstein S, Sasco AJ (2007) The multitude and diversity of environmental carcinogens. *Environ Res* 105: 414–429

9 Costa LG, Giordano G, Guizzetti M, Vitalone A (2008) Neurotoxicity of pesticides: A brief review. *Front Biosci* 13: 1240–1249

10 Repetto R, Baliga SS (1997) Pesticides and immunosuppression: The risks to public health. *Health Policy Plan* 12: 97–106

11 Arcaro KF, O'Keefe PW, Yang Y, Clayton W, Gierthy JF (1999) Antiestrogenicity of environmental polycyclic aromatic hydrocarbons in human breast cancer cells. *Toxicology* 133: 115–127

12 Swedenborg E, Ruegg J, Hillenweck A, Rehnmark S, Faulds MH, Zalko D, Pongratz I, Pettersson K (2008) 3-Methylcholanthrene displays dual effects on estrogen receptor (ER) α and ER β signaling in a cell-type specific fashion. *Mol Pharmacol* 73: 575–586

13 Bock KW, Kohle C (2006) Ah receptor: Dioxin-mediated toxic responses as hints to deregulated physiologic functions. *Biochem Pharmacol* 72: 393–404

14 IARC (1997) Working Group on the Evaluation of Carcinogenic Risks to Humans: Polychlorinated dibenzo-*p*-dioxins and polychlorinated dibenzofurans. *IARC Monogr Eval Carcinog Risks Hum* 69: 1–631

15 Remillard RB, Bunce NJ (2002) Linking dioxins to diabetes: Epidemiology and biologic plausibility. *Environ Health Perspect* 110: 853–858

16 Warner M, Eskenazi B, Mocarelli P, Gerthoux PM, Samuels S, Needham L, Patterson D, Brambilla P (2002) Serum dioxin concentrations and breast cancer risk in the Seveso Women's Health Study. *Environ Health Perspect* 110: 625–628

17 Wynne-Edwards KE (2001) Evolutionary biology of plant defenses against herbivory and their predictive implications for endocrine disruptor susceptibility in vertebrates. *Environ Health Perspect* 109: 443–448

18 Denison MS, Nagy SR (2003) Activation of the aryl hydrocarbon receptor by structurally diverse exogenous and endogenous chemicals. *Annu Rev Pharmacol Toxicol* 43: 309–334

19 Amakura Y, Tsutsumi T, Sasaki K, Nakamura M, Yoshida T, Maitani T (2008) Influence of food polyphenols on aryl hydrocarbon receptor-signaling pathway estimated by *in vitro* bioassay. *Phytochemistry*; *in press*

20 Sanders TH, McMichael RW Jr, Hendrix KW (2000) Occurrence of resveratrol in edible peanuts. *J Agric Food Chem* 48: 1243–1246

21 Rimando AM, Kalt W, Magee JB, Dewey J, Ballington JR (2004) Resveratrol, pterostilbene, and piceatannol in vaccinium berries. *J Agric Food Chem* 52: 4713–4719

22 Gillner M, Bergman J, Cambillau C, Fernstrom B, Gustafsson JA (1985) Interactions of indoles with specific binding sites for 2,3,7,8-tetrachlorodibenzo-*p*-dioxin in rat liver. *Mol Pharmacol* 28: 357–363

23 Ciolino HP, Daschner PJ, Yeh GC (1998) Resveratrol inhibits transcription of CYP1A1 *in vitro* by preventing activation of the aryl hydrocarbon receptor. *Cancer Res* 58: 5707–5712

24 Casper RF, Quesne M, Rogers IM, Shirota T, Jolivet A, Milgrom E, Savouret JF (1999) Resveratrol has antagonist activity on the aryl hydrocarbon receptor: Implications for prevention of dioxin toxicity. *Mol Pharmacol* 56: 784–790

25 Chen ZH, Hurh YJ, Na HK, Kim JH, Chun YJ, Kim DH, Kang KS, Cho MH, Surh YJ (2004) Resveratrol inhibits TCDD-induced expression of CYP1A1 and CYP1B1 and catechol estrogen-

mediated oxidative DNA damage in cultured human mammary epithelial cells. *Carcinogenesis* 25: 2005–2013

26 Bradbury RB, White DE (1954) Estrogens and related substances in plants. *Vitam Horm* 12: 207–233

27 Benassayag C, Perrot-Applanat M, Ferre F (2002) Phytoestrogens as modulators of steroid action in target cells. *J Chromatogr B Analyt Technol Biomed Life Sci* 777: 233–248

28 Penttinen P, Jaehrling J, Damdimopoulos AE, Inzunza J, Lemmen JG, van der Saag P, Pettersson K, Gauglitz G, Makela S, Pongratz I (2007) Diet-derived polyphenol metabolite enterolactone is a tissue-specific estrogen receptor activator. *Endocrinology* 148: 4875–4886

29 Mueller SO, Simon S, Chae K, Metzler M, Korach KS (2004) Phytoestrogens and their human metabolites show distinct agonistic and antagonistic properties on estrogen receptor α (ERα) and ERβ in human cells. *Toxicol Sci* 80: 14–25

30 Helguero LA, Faulds MH, Gustafsson JA, Haldosen LA (2005) Estrogen receptors α (ERα) and β (ERβ) differentially regulate proliferation and apoptosis of the normal murine mammary epithelial cell line HC11. *Oncogene* 24: 6605–6616

31 Moutsatsou P (2007) The spectrum of phytoestrogens in nature: Our knowledge is expanding. *Hormones (Athens)* 6: 173–193

32 Adlercreutz H (2002) Phyto-oestrogens and cancer. *Lancet Oncol* 3: 364–373

33 Nakata K, Tanaka Y, Nakano T, Adachi T, Tanaka H, Kaminuma T, Ishikawa T (2006) Nuclear receptor-mediated transcriptional regulation in phase I, II, and III xenobiotic metabolizing systems. *Drug Metab Pharmacokinet* 21: 437–457

34 Kliewer SA, Willson TM (2002) Regulation of xenobiotic and bile acid metabolism by the nuclear pregnane X receptor. *J Lipid Res* 43: 359–364

35 Pascussi JM, Gerbal-Chaloin S, Drocourt L, Assenat E, Larrey D, Pichard-Garcia L, Vilarem MJ, Maurel P (2004) Cross-talk between xenobiotic detoxication and other signalling pathways: Clinical and toxicological consequences. *Xenobiotica* 34: 633–664

36 Gu YZ, Hogenesch JB, Bradfield CA (2000) The PAS superfamily: Sensors of environmental and developmental signals. *Annu Rev Pharmacol Toxicol* 40: 519–561

37 Keith B, Adelman DM, Simon MC (2001) Targeted mutation of the murine arylhydrocarbon receptor nuclear translocator 2 (*Arnt2*) gene reveals partial redundancy with Arnt. *Proc Natl Acad Sci USA* 98: 6692–6697

38 Takahata S, Sogawa K, Kobayashi A, Ema M, Mimura J, Ozaki N, Fujii-Kuriyama Y (1998) Transcriptionally active heterodimer formation of an Arnt-like PAS protein, Arnt3, with HIF-1α, HLF, and clock. *Biochem Biophys Res Commun* 248: 789–794

39 Rowlands JC, Gustafsson JA (1997) Aryl hydrocarbon receptor-mediated signal transduction. *Crit Rev Toxicol* 27: 109–134

40 Wilson CL, Safe S (1998) Mechanisms of ligand-induced aryl hydrocarbon receptor-mediated biochemical and toxic responses. *Toxicol Pathol* 26: 657–671

41 Schmidt JV, Bradfield CA (1996) Ah receptor signaling pathways. *Annu Rev Cell Dev Biol* 12: 55–89

42 Poland A, Knutson JC (1982) 2,3,7,8-Tetrachlorodibenzo-*p*-dioxin and related halogenated aromatic hydrocarbons: Examination of the mechanism of toxicity. *Annu Rev Pharmacol Toxicol* 22: 517–554

43 Petrulis JR, Perdew GH (2002) The role of chaperone proteins in the aryl hydrocarbon receptor core complex. *Chem Biol Interact* 141: 25–40

44 Hankinson O (2005) Role of coactivators in transcriptional activation by the aryl hydrocarbon receptor. *Arch Biochem Biophys* 433: 379–386

45 Matthews J, Wihlen B, Thomsen J, Gustafsson JA (2005) Aryl hydrocarbon receptor-mediated transcription: Ligand-dependent recruitment of estrogen receptor α to 2,3,7,8-tetrachlorodibenzo-*p*-dioxin-responsive promoters. *Mol Cell Biol* 25: 5317–5328

46 Kollara A, Brown TJ (2006) Functional interaction of nuclear receptor coactivator 4 with aryl hydrocarbon receptor. *Biochem Biophys Res Commun* 346: 526–534

47 Son DS, Roby KF, Rozman KK, Terranova PF (2002) Estradiol enhances and estriol inhibits the expression of CYP1A1 induced by 2,3,7,8-tetrachlorodibenzo-*p*-dioxin in a mouse ovarian cancer cell line. *Toxicology* 176: 229–243

48 Spink DC, Katz BH, Hussain MM, Pentecost BT, Cao Z, Spink BC (2003) Estrogen regulates Ah responsiveness in MCF-7 breast cancer cells. *Carcinogenesis* 24: 1941–1950

49 Mimura J, Ema M, Sogawa K, Fujii-Kuriyama Y (1999) Identification of a novel mechanism of

regulation of Ah (dioxin) receptor function. *Genes Dev* 13: 20–25

50 Evans BR, Karchner SI, Allan LL, Pollenz RS, Tanguay RL, Jenny MJ, Sherr DH, Hahn ME (2008) Repression of aryl hydrocarbon receptor (AHR) signaling by AHR repressor: Role of DNA binding and competition for AHR nuclear translocator. *Mol Pharmacol* 73: 387–398

51 Dunlap JC, Loros JJ, Liu Y, Crosthwaite SK (1999) Eukaryotic circadian systems: Cycles in common. *Genes Cells* 4: 1–10

52 Klinge CM, Jernigan SC, Risinger KE, Lee JE, Tyulmenkov VV, Falkner KC, Prough RA (2001) Short heterodimer partner (SHP) orphan nuclear receptor inhibits the transcriptional activity of aryl hydrocarbon receptor (AHR)/AHR nuclear translocator (ARNT). *Arch Biochem Biophys* 390: 64–70

53 Kumar MB, Tarpey RW, Perdew GH (1999) Differential recruitment of coactivator RIP140 by Ah and estrogen receptors. Absence of a role for LXXLL motifs. *J Biol Chem* 274: 22155–22164

54 Sadek CM, Jalaguier S, Feeney EP, Aitola M, Damdimopoulos AE, Pelto-Huikko M, Gustafsson JA (2000) Isolation and characterization of AINT: A novel ARNT interacting protein expressed during murine embryonic development. *Mech Dev* 97: 13–26

55 Puga A, Xia Y, Elferink C (2002) Role of the aryl hydrocarbon receptor in cell cycle regulation. *Chem Biol Interact* 141: 117–130

56 Weiss C, Faust D, Durk H, Kolluri SK, Pelzer A, Schneider S, Dietrich C, Oesch F, Göttlicher M (2005) TCDD induces c-jun expression *via* a novel Ah (dioxin) receptor-mediated p38-MAPK-dependent pathway. *Oncogene* 24: 4975–4983

57 Oesch-Bartlomowicz B, Huelster A, Wiss O, Antoniou-Lipfert P, Dietrich C, Arand M, Weiss C, Bockamp E, Oesch F (2005) Aryl hydrocarbon receptor activation by cAMP *versus* dioxin: Divergent signaling pathways. *Proc Natl Acad Sci USA* 102: 9218–9223

58 Reen RK, Cadwallader A, Perdew GH (2002) The subdomains of the transactivation domain of the aryl hydrocarbon receptor (AhR) inhibit AhR and estrogen receptor transcriptional activity. *Arch Biochem Biophys* 408: 93–102

59 Kohle C, Bock KW (2007) Coordinate regulation of phase I and II xenobiotic metabolisms by the Ah receptor and Nrf2. *Biochem Pharmacol* 73: 1853–1862

60 Nebert DW, Roe AL, Dieter MZ, Solis WA, Yang Y, Dalton TP (2000) Role of the aromatic hydrocarbon receptor and [Ah] gene battery in the oxidative stress response, cell cycle control, and apoptosis. *Biochem Pharmacol* 59: 65–85

61 Lahvis GP, Bradfield CA (1998) Ahr null alleles: Distinctive or different? *Biochem Pharmacol* 56: 781–787

62 Fernandez-Salguero PM, Hilbert DM, Rudikoff S, Ward JM, Gonzalez FJ (1996) Aryl-hydrocarbon receptor-deficient mice are resistant to 2,3,7,8-tetrachlorodibenzo-*p*-dioxin-induced toxicity. *Toxicol Appl Pharmacol* 140: 173–179

63 Peters JM, Narotsky MG, Elizondo G, Fernandez-Salguero PM, Gonzalez FJ, Abbott BD (1999) Amelioration of TCDD-induced teratogenesis in aryl hydrocarbon receptor (AhR)-null mice. *Toxicol Sci* 47: 86–92

64 Lin TM, Ko K, Moore RW, Simanainen U, Oberley TD, Peterson RE (2002) Effects of aryl hydrocarbon receptor null mutation and *in utero* and lactational 2,3,7,8-tetrachlorodibenzo-*p*-dioxin exposure on prostate and seminal vesicle development in C57BL/6 mice. *Toxicol Sci* 68: 479–487

65 Andersson P, McGuire J, Rubio C, Gradin K, Whitelaw ML, Pettersson S, Hanberg A, Poellinger L (2002) A constitutively active dioxin/aryl hydrocarbon receptor induces stomach tumors. *Proc Natl Acad Sci U S A* 99: 9990–9995

66 Seidel SD, Winters GM, Rogers WJ, Ziccardi MH, Li V, Keser B, Denison MS (2001) Activation of the Ah receptor signaling pathway by prostaglandins. *J Biochem Mol Toxicol* 15: 187–196

67 Rannug U, Rannug A, Sjoberg U, Li H, Westerholm R, Bergman J (1995) Structure elucidation of two tryptophan-derived, high affinity Ah receptor ligands. *Chem Biol* 2: 841–845

68 Adachi J, Mori Y, Matsui S, Takigami H, Fujino J, Kitagawa H, Miller CA 3rd, Kato T, Saeki K, Matsuda T (2001) Indirubin and indigo are potent aryl hydrocarbon receptor ligands present in human urine. *J Biol Chem* 276: 31475–31478

69 Mukai M, Tischkau SA (2007) Effects of tryptophan photoproducts in the circadian timing system: Searching for a physiological role for aryl hydrocarbon receptor. *Toxicol Sci* 95: 172–181

70 Gronemeyer H, Gustafsson JA, Laudet V (2004) Principles for modulation of the nuclear receptor superfamily. *Nat Rev Drug Discov* 3: 950–964

71 Picard D (2006) Chaperoning steroid hormone action. *Trends Endocrinol Metab* 17: 229–235

72 Heldring N, Pike A, Andersson S, Matthews J, Cheng G, Hartman J, Tujague M, Strom A, Treuter

E, Warner M, Gustafsson JA (2007) Estrogen receptors: How do they signal and what are their targets. *Physiol Rev* 87: 905–931

73 Lee KC, Lee Kraus W (2001) Nuclear receptors, coactivators and chromatin: New approaches, new insights. *Trends Endocrinol Metab* 12: 191–197

74 Timsit YE, Negishi M (2007) CAR and PXR: The xenobiotic-sensing receptors. *Steroids* 72: 231–246

75 Tabb MM, Kholodovych V, Grun F, Zhou C, Welsh WJ, Blumberg B (2004) Highly chlorinated PCBs inhibit the human xenobiotic response mediated by the steroid and xenobiotic receptor (SXR). *Environ Health Perspect* 112: 163–169

76 Tzameli I, Moore DD (2001) Role reversal: New insights from new ligands for the xenobiotic receptor CAR. *Trends Endocrinol Metab* 12: 7–10

77 Moore LB, Parks DJ, Jones SA, Bledsoe RK, Consler TG, Stimmel JB, Goodwin B, Liddle C, Blanchard SG, Willson TM et al (2000) Orphan nuclear receptors constitutive androstane receptor and pregnane X receptor share xenobiotic and steroid ligands. *J Biol Chem* 275: 15122–15127

78 Kobayashi K, Sueyoshi T, Inoue K, Moore R, Negishi M (2003) Cytoplasmic accumulation of the nuclear receptor CAR by a tetratricopeptide repeat protein in HepG2 cells. *Mol Pharmacol* 64: 1069–1075

79 Squires EJ, Sueyoshi T, Negishi M (2004) Cytoplasmic localization of pregnane X receptor and ligand-dependent nuclear translocation in mouse liver. *J Biol Chem* 279: 49307–49314

80 Wang K, Mendy AJ, Dai G, Luo HR, He L, Wan YJ (2006) Retinoids activate the RXR/SXR-mediated pathway and induce the endogenous CYP3A4 activity in Huh7 human hepatoma cells. *Toxicol Sci* 92: 51–60

81 Ding X, Staudinger JL (2005) Induction of drug metabolism by forskolin: The role of the pregnane X receptor and the protein kinase a signal transduction pathway. *J Pharmacol Exp Ther* 312: 849–856

82 Baes M, Gulick T, Choi HS, Martinoli MG, Simha D, Moore DD (1994) A new orphan member of the nuclear hormone receptor superfamily that interacts with a subset of retinoic acid response elements. *Mol Cell Biol* 14: 1544–1552

83 Maglich JM, Stoltz CM, Goodwin B, Hawkins-Brown D, Moore JT, Kliewer SA (2002) Nuclear pregnane x receptor and constitutive androstane receptor regulate overlapping but distinct sets of genes involved in xenobiotic detoxification. *Mol Pharmacol* 62: 638–646

84 Kretschmer XC, Baldwin WS (2005) CAR and PXR: Xenosensors of endocrine disrupters? *Chem Biol Interact* 155: 111–128

85 Handschin C, Meyer UA (2005) Regulatory network of lipid-sensing nuclear receptors: Roles for CAR, PXR, LXR, and FXR. *Arch Biochem Biophys* 433: 387–396

86 Jensen EV, Jacobson HJ (1962) Basic guides to the mechanism of estrogen action. *Recent Prog Horm Res* 18: 318–414

87 Kuiper GG, Enmark E, Pelto-Huikko M, Nilsson S, Gustafsson JA (1996) Cloning of a novel receptor expressed in rat prostate and ovary. *Proc Natl Acad Sci USA* 93: 5925–5930

88 Enmark E, Pelto-Huikko M, Grandien K, Lagercrantz S, Lagercrantz J, Fried G, Nordenskjold M, Gustafsson JA (1997) Human estrogen receptor β-gene structure, chromosomal localization, and expression pattern. *J Clin Endocrinol Metab* 82: 4258–4265

89 Flouriot G, Brand H, Denger S, Metivier R, Kos M, Reid G, Sonntag-Buck V, Gannon F (2000) Identification of a new isoform of the human estrogen receptor-α (hER-α) that is encoded by distinct transcripts and that is able to repress hER-α activation function 1. *EMBO J* 19: 4688–4700

90 Saunders PT, Millar MR, Williams K, Macpherson S, Harkiss D, Anderson RA, Orr B, Groome NP, Scobie G, Fraser HM (2000) Differential expression of estrogen receptor-α and -β and androgen receptor in the ovaries of marmosets and humans. *Biol Reprod* 63: 1098–1105

91 Gruber CJ, Gruber DM, Gruber IM, Wieser F, Huber JC (2004) Anatomy of the estrogen response element. *Trends Endocrinol Metab* 15: 73–78

92 Couse JF, Korach KS (1999) Estrogen receptor null mice: What have we learned and where will they lead us? *Endocr Rev* 20: 358–417

93 Hewitt SC, Harrell JC, Korach KS (2005) Lessons in estrogen biology from knockout and transgenic animals. *Annu Rev Physiol* 67: 285–308

94 Harris HA (2007) Estrogen receptor-β: Recent lessons from *in vivo* studies. *Mol Endocrinol* 21: 1–13

95 Lubahn DB, Moyer JS, Golding TS, Couse JF, Korach KS, Smithies O (1993) Alteration of reproductive function but not prenatal sexual development after insertional disruption of the mouse

estrogen receptor gene. *Proc Natl Acad Sci USA* 90: 11162–11166

96 Krege JH, Hodgin JB, Couse JF, Enmark E, Warner M, Mahler JF, Sar M, Korach KS, Gustafsson JA, Smithies O (1998) Generation and reproductive phenotypes of mice lacking estrogen receptor β. *Proc Natl Acad Sci USA* 95: 15677–15682

97 Dupont S, Krust A, Gansmuller A, Dierich A, Chambon P, Mark M (2000) Effect of single and compound knockouts of estrogen receptors α (ERα) and β (ERβ) on mouse reproductive phenotypes. *Development* 127: 4277–4291

98 Zhu Y, Bian Z, Lu P, Karas RH, Bao L, Cox D, Hodgin J, Shaul PW, Thoren P, Smithies O, Gustafsson JA, Mendelsohn ME (2002) Abnormal vascular function and hypertension in mice deficient in estrogen receptor β. *Science* 295: 505–508

99 Imamov O, Morani A, Shim GJ, Omoto Y, Thulin-Andersson C, Warner M, Gustafsson JA (2004) Estrogen receptor β regulates epithelial cellular differentiation in the mouse ventral prostate. *Proc Natl Acad Sci USA* 101: 9375–9380

100 Wada-Hiraike O, Imamov O, Hiraike H, Hultenby K, Schwend T, Omoto Y, Warner M, Gustafsson JA (2006) Role of estrogen receptor β in colonic epithelium. *Proc Natl Acad Sci USA* 103: 2959–2964

101 Krezel W, Dupont S, Krust A, Chambon P, Chapman PF (2001) Increased anxiety and synaptic plasticity in estrogen receptor β-deficient mice. *Proc Natl Acad Sci USA* 98: 12278–12282

102 Imwalle DB, Gustafsson JA, Rissman EF (2005) Lack of functional estrogen receptor β influences anxiety behavior and serotonin content in female mice. *Physiol Behav* 84: 157–163

103 Liu MM, Albanese C, Anderson CM, Hilty K, Webb P, Uht RM, Price RH Jr, Pestell RG, Kushner PJ (2002) Opposing action of estrogen receptors α and β on cyclin D1 gene expression. *J Biol Chem* 277: 24353–24360

104 Matthews J, Gustafsson JA (2003) Estrogen signaling: A subtle balance between ER α and ER β. *Mol Interv* 3: 281–292

105 Wilson CM, McPhaul MJ (1994) A and B forms of the androgen receptor are present in human genital skin fibroblasts. *Proc Natl Acad Sci USA* 91: 1234–1238

106 Takeda H, Chodak G, Mutchnik S, Nakamoto T, Chang C (1990) Immunohistochemical localization of androgen receptors with mono- and polyclonal antibodies to androgen receptor. *J Endocrinol* 126: 17–25

107 Wilson CM, McPhaul MJ (1996) A and B forms of the androgen receptor are expressed in a variety of human tissues. *Mol Cell Endocrinol* 120: 51–57

108 Heinlein CA, Chang C (2002) The roles of androgen receptors and androgen-binding proteins in nongenomic androgen actions. *Mol Endocrinol* 16: 2181–2187

109 Yen PM (2001) Physiological and molecular basis of thyroid hormone action. *Physiol Rev* 81: 1097–1142

110 Chassande O, Fraichard A, Gauthier K, Flamant F, Legrand C, Savatier P, Laudet V, Samarut J (1997) Identification of transcripts initiated from an internal promoter in the c-erbA α locus that encode inhibitors of retinoic acid receptor-α and triiodothyronine receptor activities. *Mol Endocrinol* 11: 1278–1290

111 Plateroti M, Gauthier K, Domon-Dell C, Freund JN, Samarut J, Chassande O (2001) Functional interference between thyroid hormone receptor α (TRα) and natural truncated TRδα isoforms in the control of intestine development. *Mol Cell Biol* 21: 4761–4772

112 Boas M, Feldt-Rasmussen U, Skakkebaek NE, Main KM (2006) Environmental chemicals and thyroid function. *Eur J Endocrinol* 154: 599–611

113 Bassett JH, Harvey CB, Williams GR (2003) Mechanisms of thyroid hormone receptor-specific nuclear and extra nuclear actions. *Mol Cell Endocrinol* 213: 1–11

114 Li X, Lonard DM, O'Malley BW (2004) A contemporary understanding of progesterone receptor function. *Mech Ageing Dev* 125: 669–678

115 Bentrem D, Fox JE, Pearce ST, Liu H, Pappas S, Kupfer D, Zapf JW, Jordan VC (2003) Distinct molecular conformations of the estrogen receptor α complex exploited by environmental estrogens. *Cancer Res* 63: 7490–7496

116 Nagel SC, Hagelbarger JL, McDonnell DP (2001) Development of an ER action indicator mouse for the study of estrogens, selective ER modulators (SERMs), and xenobiotics. *Endocrinology* 142: 4721–4728

117 Hall JM, McDonnell DP (2005) Coregulators in nuclear estrogen receptor action: From concept to therapeutic targeting. *Mol Interv* 5: 343–357

118 Hestermann EV, Brown M (2003) Agonist and chemopreventative ligands induce differential

transcriptional cofactor recruitment by aryl hydrocarbon receptor. *Mol Cell Biol* 23: 7920–7925

119 Safe S, Wormke M (2003) Inhibitory aryl hydrocarbon receptor-estrogen receptor α cross-talk and mechanisms of action. *Chem Res Toxicol* 16: 807–816

120 Safe S, Wormke M, Samudio I (2000) Mechanisms of inhibitory aryl hydrocarbon receptor-estrogen receptor crosstalk in human breast cancer cells. *J Mammary Gland Biol Neoplasia* 5: 295–306

121 Brunnberg S, Pettersson K, Rydin E, Matthews J, Hanberg A, Pongratz I (2003) The basic helix-loop-helix-PAS protein ARNT functions as a potent coactivator of estrogen receptor-dependent transcription. *Proc Natl Acad Sci USA* 100: 6517–6522

122 Ruegg J, Swedenborg E, Wahlström D, Escande A, Balaguer P, Pettersson K, Pongratz I (2008) The transcription factor aryl hydrocarbon receptor nuclear translocator functions as an estrogen receptor β-selective co-activator, and its recruitment to alternative pathways mediates antiestrogenic effects of dioxin. *Mol Endocrinol* 22: 304–316

123 Min G, Kim H, Bae Y, Petz L, Kemper JK (2002) Inhibitory cross-talk between estrogen receptor (ER) and constitutively activated androstane receptor (CAR). CAR inhibits ER-mediated signaling pathway by squelching p160 coactivators. *J Biol Chem* 277: 34626–34633

124 Tsuchiya Y, Nakajima M, Yokoi T (2005) Cytochrome P450-mediated metabolism of estrogens and its regulation in human. *Cancer Lett* 227: 115–124

125 Tsuchiya Y, Nakajima M, Kyo S, Kanaya T, Inoue M, Yokoi T (2004) Human CYP1B1 is regulated by estradiol *via* estrogen receptor. *Cancer Res* 64: 3119–3125

126 You L (2004) Steroid hormone biotransformation and xenobiotic induction of hepatic steroid metabolizing enzymes. *Chem Biol Interact* 147: 233–246

127 Porter W, Wang F, Duan R, Qin C, Castro-Rivera E, Kim K, Safe S (2001) Transcriptional activation of heat shock protein 27 gene expression by 17β-estradiol and modulation by antiestrogens and aryl hydrocarbon receptor agonists. *J Mol Endocrinol* 26: 31–42

128 Wang F, Samudio I, Safe S (2001) Transcriptional activation of cathepsin D gene expression by 17β-estradiol: Mechanism of aryl hydrocarbon receptor-mediated inhibition. *Mol Cell Endocrinol* 172: 91–103

129 Gillesby BE, Stanostefano M, Porter W, Safe S, Wu ZF, Zacharewski TR (1997) Identification of a motif within the 5' regulatory region of pS2 which is responsible for AP-1 binding and TCDD-mediated suppression. *Biochemistry* 36: 6080–6089

130 Hockings JK, Thorne PA, Kemp MQ, Morgan SS, Selmin O, Romagnolo DF (2006) The ligand status of the aromatic hydrocarbon receptor modulates transcriptional activation of BRCA-1 promoter by estrogen. *Cancer Res* 66: 2224–2232

131 Lin HK, Altuwaijri S, Lin WJ, Kan PY, Collins LL, Chang C (2002) Proteasome activity is required for androgen receptor transcriptional activity *via* regulation of androgen receptor nuclear translocation and interaction with coregulators in prostate cancer cells. *J Biol Chem* 277: 36570–36576

132 Deroo BJ, Rentsch C, Sampath S, Young J, DeFranco DB, Archer TK (2002) Proteasomal inhibition enhances glucocorticoid receptor transactivation and alters its subnuclear trafficking. *Mol Cell Biol* 22: 4113–4123

133 Wijayaratne AL, McDonnell DP (2001) The human estrogen receptor-α is a ubiquitinated protein whose stability is affected differentially by agonists, antagonists, and selective estrogen receptor modulators. *J Biol Chem* 276: 35684–35692

134 Ohtake F, Baba A, Takada I, Okada M, Iwasaki K, Miki H, Takahashi S, Kouzmenko A, Nohara K, Chiba T et al (2007) Dioxin receptor is a ligand-dependent E3 ubiquitin ligase. *Nature* 446: 562–566

135 Wang X, Porter W, Krishnan V, Narasimhan TR, Safe S (1993) Mechanism of 2,3,7,8-tetrachlorodibenzo-*p*-dioxin (TCDD)-mediated decrease of the nuclear estrogen receptor in MCF-7 human breast cancer cells. *Mol Cell Endocrinol* 96: 159–166

Molecular, Clinical and Environmental Toxicology. Volume 1: Molecular Toxicology
Edited by A. Luch

Toxicogenomics: transcription profiling for toxicology assessment

Tong Zhou[1], Jeff Chou[2], Paul B. Watkins[1] and William K. Kaufmann[3]

[1] *Center for Drug Safety Sciences, The Hamner Institutes for Health Sciences, University of North Carolina at Chapel Hill, Research Triangle Park, NC, USA*
[2] *Microarray Group, National Institute for Environmental Health Sciences (NIEHS), Research Triangle Park, NC, USA*
[3] *Department of Pathology and Laboratory Medicine, Center for Environmental Health and Susceptibility, University of North Carolina at Chapel Hill, Chapel Hill, NC, USA*

Abstract. Toxicogenomics, the application of transcription profiling to toxicology, has been widely used for elucidating the molecular and cellular actions of chemicals and other environmental stressors on biological systems, predicting toxicity before any functional damages, and classification of known or new toxicants based on signatures of gene expression. The success of a toxicogenomics study depends upon close collaboration among experts in different fields, including a toxicologist or biologist, a bioinformatician, statistician, physician and, sometimes, mathematician. This review is focused on toxicogenomics studies, including transcription profiling technology, experimental design, significant gene extraction, toxicological results interpretation, potential pathway identification, database input and the applications of toxicogenomics in various fields of toxicological study.

Introduction

The rapid evolution of genome-based technologies has greatly accelerated the application of gene expression profiling in toxicology studies. These technological advances have led to the development of the field of toxicogenomics, which proposes to apply mRNA expression technologies to study effects of hazards in biological systems [1, 2]. Toxicogenomics combines traditional toxicology using appropriate pharmacological and toxicological models with global "omics" technologies to provide a comprehensive view of the functioning of the genetic and biochemical machinery in organisms under stress [3]. Applications of these technologies help in predicting the potential toxicity of a drug or chemical before functional damages are recognized, in classification of toxicants, and in screening human susceptibility to diseases, drugs or environmental hazards.

Microarray analysis

Experimental design in microarray studies

DNA microarrays enable researchers to simultaneously determine changes in the levels of expression of thousands of genes. Experimental design is very important in microarray studies due to the unique characteristics in experimental performance and the complexities of large amounts of data for analysis and interpretation. In studies on mechanisms of toxicity, toxicity prediction, toxicant classification and biomarker identification, clear objectives are essential for effective experimental design. The statistical power for identifying differentially expressed genes or for developing classifiers is generally determined by the number of biological replicates in each class. To achieve a variety of experimental goals, optimization in determination of sample sizes and replications can reduce costs effectively and still retain statistical and biological significance [4–6]. A replicate is an independent repeat of an experiment. Generally, there are two types of replication in microarray analysis: biological and technical. A biological replicate is a sample from a different individual (cell line, animal or person) and is used to estimate biological variations, since a particular sample might not be representative of the whole affected sample group. A technical replicate refers to a sample from the same animal or cell culture. Due to the large size of samples, microarray experiments are often performed in different batches (labs, platforms, time, etc.) and give rise to the so-called "batch effect" that may dominate the biological/toxicological signals in data analysis and yield information that is completely unrelated to the research goals. To be able to consolidate data from different batches, the experimental design should be certain that the batch effect can be minimized or corrected [7].

Technologies of transcription profiling

There are two main platforms of microarrays used in gene expression analyses, oligonucleotide arrays and cDNA arrays. The oligonucleotide arrays apply high-density arrays of short oligonucleotide (20–80 bases) probes spotted by either robotic deposition or *in situ* synthesis on a solid substrate to examine up to 44 000 signals. cDNA arrays are made from long double-stranded DNA molecules (0.5–2 kb in length) generated by PCR and are usually spotted onto the surface of treated glass slides or membranes [1].

For microarray analysis, RNAs are isolated from biological samples with or without treatment to determine and compare differential gene expression profiles. In toxicogenomics studies, the gene expression profiles can be used to compare dose-dependent and/or time-dependent responses after toxicant exposure, to compare toxicity patterns caused by two or more toxicants in the same organ, tissue or cells, and to compare susceptibility to a specific toxicant or disease among different individuals. Expression profiling may provide information about early responses, subtle changes and potential mechanisms that cannot be obtained from traditional toxicology studies.

The quality of all RNA samples is usually determined to warranty success in further array analysis. A Bioanalyzer is commonly used for RNA quality analysis prior to microarray assay (Agilent). After quality confirmation, isolated RNA is converted to cDNA by a reverse transcriptase reaction and cDNA then is further labeled in different ways according protocols developed by different companies. There are two typical labeling technologies: single-color and dual-color. Examples of widely used single-color and dual-color labeling are Affymetrix and Agilent protocols, respectively. In the Affymetrix protocol, cDNA that is converted from each RNA sample is transcribed to biotin-labeled cRNA that is hybridized to an Affymetrix array chip. The array is then scanned and digitalized. The intensity data extracted for the scanned images reflect the gene expression levels and are directly used for further microarray data analysis [8]. In the Agilent protocol, two cDNAs converted from two different RNA samples (one control and one treated, or one global reference RNA and one sample RNA) when being transcribed to cDNAs are labeled with different fluorescent tags, either Cy3 or Cy5, then the two fluorescence-labeled targets are mixed and hybridized to an Agilent array chip. The array is scanned at two wavelengths, usually at 632 and 532 nm for the red (Cy5) and green (Cy3) labels. The ratios of the intensities of the treated (Cy5) *versus* control (Cy3) samples or the sample (Cy5) *versus* global reference (Cy3) from the scanned images are used to determine gene expression levels and for further microarray data analysis. Dye swap labeling is sometimes advised to eliminate bias from Cy5 and Cy3 labeling [9]. Real-time PCR has been used to validate gene expression levels determined by microarray analysis. With improvement in the technologies of microarray analysis, microarray-extracted genes with significant changes in expression have been shown to have nearly identical results in real-time PCR determination in recent studies [10].

Microarray data analysis

Microarray data analysis includes image analysis [11, 12], quality assurance analysis, pre-processing, normalization [13], differentially expressed gene identification, data mining, ontology analysis and pathway analysis. Statistical methods such as Student's *t*-test and its variants [14], ANOVA [15, 16] and Bayesian method [17, 18] are widely used in identifying differentially expressed genes. To balance the false positives and the false negatives, a common practice is to control the false discovery rate (FDR) [19]. The statistical analysis of microarray (SAM) [20] and BRB-ArrayTools (http://linus.nci.nih.gov/BRB-ArrayTools.html) utilize this FDR concept to assist in determining a cutoff after performing adjusted *t*-tests. Some commonly used methods in data mining include principal component analysis (PCA), hierarchical clustering and k-mean clustering [21, 22].

In a toxicological microarray experiment, samples are often treated with a series of different conditions, such as agents, doses, time points, tissues and cell lines. The formed gene expression profiles may appear relatively similar or co-expressed. In this review, we discuss a newly developed profile-based method for extracting microarray gene expression patterns and identifying co-expressed genes, designated as EPIG [23]. In extracting gene expression patterns, EPIG uses a filtering process where all profiles initially are considered as pattern candidates. In pair-wise correlation calculation, each candidate is given a correlation score, which is the number of genes with which the candidate is highly correlated (larger than a given threshold). The candidate then is removed from the pattern candidate pool if its correlation score is less than a predefined value or its correlation with another candidate is higher. A pattern candidate will also be filtered if its signal-to-noise ratio (S/N) and signal intensity (S) values are lower than given thresholds. After this filtering processing, the remaining candidates consist of the extracted patterns, which are designated as representatives of each of the local clusters. Subsequently, EPIG categorizes each gene whose expression profile passed the S and S/N tests, to the one pattern for which its correlation value is the highest and larger than a given threshold. EPIG has shown a unique ability to extract more biologically informative patterns and co-expressed genes in its applications to biological data [24, 25] in comparison with the patterns given by other techniques, such as CLICK (cluster identification *via* connectivity kernels) [26, 27].

Once significant genes and expression patterns are extracted, the most important interpretation of microarray data is to link the results to biological phenomena. A general method is to look for pathway or function enrichment when a subset of regulated genes is compared to the entire array. Such analyses based on the gene ontology annotation (http://www.geneontology.org) can be performed by several web-based tools, e.g., EASE (http://apps1.niaid.nih.gov/david/), or GoStat (http://gostat.wehi.edu.au/). Within a given pathway individual gene expression can be visualized by tools such as GenMAPP (www.genmapp.org), KEGG (http://www.genome.ad.jp/kegg/), Ingenuity (http://www.ingenuity.com/), or MetaCore (www.GeneGO.com). For identification of transcription factors within a group of co-regulated genes, the useful tools and databases include: TRANSFAC (http://www.gene-regulation.com/), Ensembl (http://www.ensembl.org), UCSC Genome Bioinformatics (http://genome.ucsc.edu/), PAINT (http://www.dbi.tju.edu/dbi/tools/paint/) [28], and PRIMA [29].

The microarray data, when published, is usually requested to be deposited into a public database, so other researchers are able to access it for further analysis and comparison. There are five extensive public databases that house toxicologically relevant microarray data; (1) Comparative Toxicogenomics Database (CTD) [30], (2) Environment, Drugs and Gene Expression database (EDGE) [31], (3) Chemical Effects in Biological System (CEBS) knowledge-base [32, 33], (4) ArrayExpress [34], and (5) Gene Expression Omnibus (GEO) [35]. The last two databases include microarray data from many kinds of biological/toxicological studies, and therefore are not toxicology-specific databases. These databases offer a flexible and convenient tool in the management and analyses of information obtained from a microarray, and facilitate pattern recognition. Furthermore, they allow comparison of molecular expression data sets from "omics" technologies and conventional toxicological approaches with toxicological pathway and gene regulatory network information relevant to environmental toxicology and human disease [36].

An example in microarray analysis: Gene expression profiling in response to IR- and UV-induced DNA damage

Gene expression profiles in response to ionizing radiation (IR)- and ultraviolet (UV)-induced DNA damage were studied using microarray analysis in three human fibroblast cell lines with two research goals: (1) to find similar and dissimilar responses between treatments, and (2) to reveal differences in gene regulation upon DNA damage caused by two different toxic radiations, IR or UV. Cells were treated with IR or UV at doses reducing colony formation by about 40%. Microarray assays were performed using Agilent Human 22k arrays. In each of the two treatments, the data consisted of four biological states, i.e., sham-treated controls, 2, 6, and 24 h post UV or IR treatment. The gene expression profiles consisted of eight inter-groups, corresponding to four states from each of the two treatments. Each of the intra-groups contains six data points from three biological replicates and two technical replicates (dye-swap pairs) for a given treatment at a given time point. As such, each gene expression profile consisted of 48 data points. EPIG analysis using the whole data as its input was performed and extracted a total of 18 patterns, as shown in Figure 1, with a total of 2661 co-expressed genes being identified. Each of the co-expressed genes was categorized to a particular pattern. The over-represented biological processes based on Gene Ontology (GO) categories in each expression pattern were analyzed using EASE [37].

As indicted in Figure 1 and Table 1, patterns 1–8 showed UV-specific expression (either up- or down-regulation) with little or no changes in gene expression in IR-treated cells. UV-specific regulation of gene expression happened only at early time points, 2 and/or 6 h after the treatment, and fully recovered to baseline levels at 24 h. Patterns 1–4 showed gene induction with modest variations in kinetics. GO categories that were over-represented in these profiles were regulation of transcription and RNA metabolism. Patterns 5–8 showed repression of gene expression also with modest variations in kinetics. Over-represented GO categories were associated with transcription regulation and protein kinase activity. There were over 600 genes in pattern 8, which were substantially repressed at 6 h post UV, and this profile was associated with about 30 biological processes including purine nucleotide binding, protein modification, ubiquitin cycle, kinase activity, and cell growth. Thus, in response to a dose of UV for which most cells retained the ability to expand

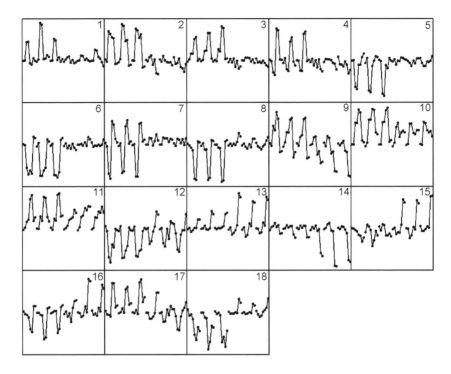

Figure 1. The patterns extracted by EPIG from the combined UV- and IR-treatment data sets. In each of these patterns, 1–18, the first half were UV treated and the second half were IR treated. For each treatment, there were three individual cell lines, F1-HTERT, F3-HTERT and F10-HTERT, positioned from left to right. Each cell line consisted of eight data points with four different treatment conditions, i.e., sham-treatment and 2, 6, and 24 h post treatment, and two dye-swap (Cy3 and Cy5) replicates for each treatment condition, respectively. The vertical axes with zero at the middle are the changes in gene expression (log2 intensity) relative to the sham-treated controls (Chou et al. [23]).

clonally, transcription factors were induced or repressed. The reasons why some transcription factors are induced and others repressed remain to be determined. While GO can point to broad categories with altered transcript levels, a more focused gene mining or pathway analysis will be required to determine the biological significance of the UV-specific signatures. Genes in patterns 9–14 responded to both UV and IR treatment but with different time dependencies, which may reflect different mechanisms of biological response to IR- and UV-induced DNA damage. Among the 172 genes in pattern 9 were genes that are maximally expressed in S phase during a normal cell cycle, such as *CDC6, FEN1, MSH6, ORC6L, PCNA, POLG, RBM14, CCNE2* and *TOP3A*. The changes in expression of these genes were coincidental with changes in the S-phase compartment of the cell cycle. S-phase cells were increased over control at 2 and 6 h post-UV (data not shown), but were moderately reduced relative to control at 6 h and markedly reduced at 24 h post-IR. Patterns 10 and 11 contained a total of 42 genes, many of which are p53 targets and are high-

Table 1. Pattern response trends and selected over represented Gene Ontology (GO) categories.

Pattern	No. of genes	UV response trends	IR response trends	Selected over represented GO categories[*]
1	21	up-regulated at 2 h post UV	insignificant response	transcription factor complex, development, nucleoplasm, regulation of transcription from polymerase II promoter, morphogenesis
2	57	up-regulated at 2 and 6 h post UV	insignificant response	nucleus, nucleic acid binding
3	173	moderately up-regulated at 2 h and peak 6 h post UV	insignificant response	RNA metabolism, RNA processing, methyltransferase activity, nucleolus
4	73	moderately down-regulated at 2 h and up-regulated at 6 h post UV	insignificant response	metabolism, nucleobase/nucleoside/nucleotide and nucleic acid metabolism, regulation of transcription, DNA-dependent
5	100	down-regulated at 2 h post UV	insignificant response	regulation of transcription, transcription/DNA-dependent, nucleic acid binding, nucleus
6	79	down-regulated at 2 and 6 h post UV	insignificant response	protein serine/threonine kinase activity, nucleus
7	52	down-regulated at 2 h and moderately up-regulated 6 h post UV	insignificant response	transcription regulator activity
8	616	down-regulated at 6 h post UV	insignificant response	purine nucleotide binding, protein modification, protein amino acid phosphorylation, ubiquitin cycle, kinase activity, cell growth and/or maintenance
9	172	moderately up-regulated at 2 h and peak 6 h post UV	moderately up-regulated at 2 h and down-regulated at 24 h post IR	nucleolus, ribosome biogenesis and assembly, mitotic cell cycle, rRNA processing, DNA replication, S phase of mitotic cell cycle
10	23	moderately up-regulated at 2 h and peak 6 h post UV	up-regulated at 2 h post IR, then decrease at 6 h and remained stable through 24 h post IR	cell proliferation

(continued on next page)

Table 1. (continued)

Pattern	No. of genes	UV response trends	IR response trends	Selected over represented GO categories[*]
11	19	moderately up-regulated at 2 h and peak 6 h post UV	progressively up-regulated from 2 to 24 h post IR	
12	47	down-regulated at 2 h and moderately down-regulated 6 h post UV	down-regulated at 2 h post IR	protein binding
13	385	moderately up-regulated at 24 h post UV	up-regulated at 24 h post IR	lysosome, lytic vacuole, complement activation
14	563	moderately down-regulated at 24 h post UV	down-regulated at 24 h post IR	mitotic cell cycle, DNA replication and chromosome cycle, M phase, nuclear division, cell growth and/or maintenance, RNA processing, RNA metabolism, response to DNA damage stimulus, DNA repair, cell cycle checkpoint, cell growth and/or maintenance, G1/S transition of mitotic cell cycle, G2/M transition of mitotic cell cycle
15	49	insignificant response	up-regulated at 24 h post IR	cell adhesion
16	115	down-regulated at 6 h post UV	up-regulated at 24 h post IR	catalytic activity
17	61	up-regulated at 6 h post UV	moderately down-regulated at 2 h post IR	DNA binding
18	56	down-regulated at 6 h and 24 h post UV	moderately up-regulated at 24 h post IR	plasma membrane, morphogenesis

[*] Includes biological processes, molecular function and cellular component (Chou et al. [23]).

ly induced by DNA-damaging agents, including *GDF15*, *BTG2*, *PLK2*, *BAK1*, *PLK3* and *CDKN1A* (pattern 10) and *TP53INP1*, *SESN1*, *DDB2* and *FDXR* (pattern 11). Patterns 13 and 14 contained nearly one thousand genes in total that were regulated only at the late time point after treatment (24 h). There were more than 500 genes in pattern 14 that were down-regulated at 24 h post IR or UV treatment, and these included many cell cycle regulators functioning in the G1/S and G2/M transitions. These genes were down-regulated as proliferating cells were transiently synchronized to a G0-like state of quiescence behind the p53-dependent G1 checkpoint [25]. Genes that were up-regulated in pattern 13 are associated with complement activation with lysosome and lytic vacuole being the main cellular components. Finally, groups of 50–100 genes exhibited co-expression in patterns 15–18. Pattern 15 was related to cell adhesion genes, while pattern 18 was related to the plasma membrane and morphogenesis. Thus, we see that in response to even modestly toxic doses of radiation, human fibroblasts display stereotypic patterns of change in gene expression that may affect a wide variety of cellular processes and functions. Some responses are radiation specific and some are induced by both types of radiation.

Computational promoter analysis of the genes in each expression pattern was done using the PRIMA method [29] that analyses both strands in a region spanning from 2000 bp upstream to 2000 bp downstream of a gene's transcription start site. Table 2 summarizes the list of transcription factor binding sites that were enriched in genes in many of the expression patterns. p53 binding sites were, not surprisingly, over-represented in patterns 10 and 11 that contain many p53 target genes. Identification of E2F binding sites in patterns 9 and 14 confirmed its function in transcriptional regulation of cell cycle-regulated genes that were prominent in these two patterns. We have discussed a potential mechanism of gene repression in pattern 14 secondary to p53-de-

Table 2. Identification of transcription factors in each expression pattern.

Pattern	No. of genes	Enriched transcription factor(s)
1	18	ZF5
2	35	ZF5
3	108	E2F, Sp-1
5	76	RSRFC4
7	36	MEF-2
8	472	SRY, HNF-1, Oct-1, FOXD3, HFH-4, MEF-2, E2F, HMG_IY, ZF5
9	111	E2F, Tax/CREB
10	18	AP-2γ, p53
11	15	p53
13	301	p53
14	435	E2F, Nrf-1, NF-Y, ZF5, Elk-1, Sp1

pendent G1 arrest [24]. Interestingly, ZF5 was identified in four expression patterns (1, 2, 8 and 14), suggesting an important function in regulating gene expression in response to UV- and IR-induced DNA damage.

Applications of gene expression profiles in toxicological studies

When an organism is exposed to a toxic stimulus, the immediate cellular responses are usually post-translational modification of a few key proteins, such as phosphorylation of ATM and p53 upon DNA damage. These signals are further amplified in very short time by altering the expression patterns of genes that function in mechanisms of protecting cells from adverse effects of the exposure. Later, if the toxicant dose is high, irreversible damages in cells may generate changes in gene regulation related to apoptosis or cell lysis [38]. Changes in gene expression profiles can reflect a variety of biological reactions to a toxic stimulus. Toxicogenomics has proved to be a powerful tool in monitoring the patterns of gene expression that result in and/or from pathological alterations within cells and tissues following exposure to a toxin. The characteristic gene expression profile is termed a 'molecular signature' or 'fingerprint' of exposure or toxicological response to specific classes of toxic compounds [36]. Three common uses of the molecular signatures in current toxicology research include: (1) mechanistic studies of toxicity, (2) prediction of toxicity and classification of compounds, and (3) identification of early biomarkers of toxic responses.

Active, but not passive, transcriptional regulation of gene expression is of importance in the responses of cells or tissues to toxicant exposure. Alterations in gene expression profiles provide a better understanding of mechanisms of toxicity and the biological responses that protect the organism from toxic effects. For example, induction of the expression factors associated with apoptosis, such as *Puma*, *Bax* and *Casp3* (caspase 3), is a strong indicator of toxic effect in cellular killing through apoptosis, the pathway of cell suicide upon exposure to a variety of toxicants. When exposed to DNA-damaging agents, up-regulation of *Cdkn1a* (p21^{Waf1}) and down-regulation of cell cycle-regulated genes are closely associated with cell cycle arrest to allow more time for DNA repair [25]. However, not all alterations in gene expression should be viewed as direct responses to a toxin. Only the genes that are actively induced by xenobiotic stimulation and/or by biological responses are of real interest in mechanistic studies; those with changes in expression that follow biological dysfunction or histopathological change passively are less important in both mechanistic studies or biomarker identification [24]. Accurate and exhaustive analysis of the type of gene expression pattern is crucial to distinguish between harmful toxic responses and responses representing benign homeostatic adjustments [39]. Gene expression results obtained from exposure of model systems to known compounds are analyzed, and a common set of changes in gene expression in cells and tissues unique to the specific class of compound (molecular signature) is identified. This molecular signature is characteristic of a specific mechanism of induced toxicity [36].

Molecular signatures have been used in prediction of toxicity and carcinogenicity and classification of compounds. Chemicals or toxicants are traditionally classified by similar toxic endpoints, mechanisms of action, chemical structure, or target organ specificity. Toxicogenomics allows the use of gene expression profiles of known toxins from representative toxicological classes (model compounds) to predict the toxicological effects of unknown compounds based on similarities between gene expression profiles [40]. Gene expression profiles induced by candidate compounds with unknown toxicity can be compared with the established and validated signatures from known toxicants. A positive correlation with an archived molecular signature provides an indication of the potential adverse effects of a new compound similar to those of known toxicants [36]. Another practical application is the use of toxicogenomics for toxicity prediction in complex mixtures such as environmental samples or product impurities. Complex mixtures can be tested to determine if they produce gene expression profiles that indicate adverse effects and then the observed profiles can be associated with a particular compound class, providing valuable mechanistic information about the nature of the unexpected effects [36, 41].

Identification of the earliest biological/toxicological responses from a toxicant exposure is one of the main goals of toxicological studies. The molecular signatures of gene expression, when used as biomarkers, can provide an early indication of toxicity because toxin-mediated changes in gene expression are often detectable before chemistry, histopathology, or clinical observations demonstrate toxicity. Molecular signatures are applied as biomarkers in two ways. The first is to determine gene expression patterns in a time course and observe the relationship between early changes in gene expression and later pathological changes or biological dysfunctions in a specific organ or tissue. The target genes and specific expression patterns (fingerprint) altered at early time points prior to pathological changes or biological dysfunctions that are the common endpoints in conventional toxicological studies are used to predict the potential toxic effects at later times. The second, which is more useful in clinical practice, is to use molecular signatures obtained from blood samples to predict potential toxic damage or dysfunction of internal body organs or tissues that are difficult or dangerous to examine directly [42, 43]. Molecular signatures of gene expression have proven to be useful biomarkers in many recent toxicological studies as discussed in detail below.

Toxicogenomics in studies of genotoxicity and carcinogenicity

The genotoxicity of a chemical is often discussed together with its potential carcinogenicity. Based on their properties in genotoxicity and carcinogenicity studies compounds have been classified as genotoxic carcinogens, genotoxic non-carcinogens, non-genotoxic carcinogens and non-genotoxic non-carcinogens. Gene expression profiles obtained from toxicogenomic studies in either *in vitro* cell culture experiments or short-term animal studies have shown the ability to distinguish genotoxicants from non-genotoxicants [44–47], and carcinogens from non-carcinogens [48–51]. An *in vitro* study indicated that changes in expression of genes that are usually involved in DNA repair or cell cycle checkpoint control can provide valuable evaluation of genotoxicity but have limitations in predicting carcinogenicity [45]. A set of genes was used to predict the carcinogenicity of 39 chemicals at non-toxic doses, including 25 carcinogens and 14 non-carcinogens, in cultured rat hepatoma cells (MH1C1). The correct prediction rate of carcinogens and non-carcinogens was 89% [52]. Gene expression profiles have also been shown to be able to distinguish carcinogenic and non-carcinogenic isomers and tissue-specific carcinogenicity of a chemical in a short-term rat study [50, 53]. The genes that are significantly altered by carcinogens and/or genotoxins are usually associated with defense response, apoptosis, DNA repair, cell cycle control, cell proliferation and immune response in target tissues [53].

A unique advantage of using gene expression patterns to predict the carcinogenicity of non-genotoxic compounds, which usually requires an expensive 2-year rodent bioassay, was demonstrated in a recent study that analyzed 147 non-genotoxic hepatocarcinogens and non-hepatocarcinogens. When 25 hepatocarcinogens and 75 non-hepatocarcinogens were used as training set to extract a gene expression signature, independent validation of the signature on 47 test chemicals indicated an assay sensitivity and specificity of 86% and 81%, respectively [54, 55].

Toxicogenomics in studies of hepatotoxicity

Liver is the main organ that metabolizes and excretes exogenous chemicals. Therefore, liver cells are exposed to various chemicals at significant concentrations, which can result in liver damage. In the pharmaceutical industry, hepatotoxicity is one of the important reasons for discontinuing the development of a drug or withdrawing a drug from the market [56]. Prediction of hepatotoxicity at early stage of drug development or early detection of hepatotoxicity of a clinically used drug can not only save money but also prevent clinical organ damage and, sometimes, the patient's death. Gene expression profiles have successfully been used as early biomarkers to detect the potential adverse effects of hepatotoxicants before any liver dysfunction or histological changes [57, 58]. A good example for the application of toxicogenomics in hepatoxicity studies is acetaminophen (*N*-acetyl-*p*-aminophenol, APAP) [39, 43, 57–59]. APAP-induced hepatotoxicity at subtoxic doses, below the limits of detection of conventional toxicological parameters, was clearly identified by gene expression profiling and patterns of gene expression indicated cellular energy loss and oxidative stress as a consequence of APAP toxicity [39, 59]. In a recent APAP study performed at the National Institute for Environmental Health Sciences (NIEHS), a blood gene expression data set from rats exposed to APAP was used to train classifiers in two prediction algorithms and to extract patterns for prediction of APAP exposure levels and liver toxicity using a profiling algorithm. Prediction accuracy was tested on a blinded, independent rat blood test data set and ranged from 89% to 96%. Genomic markers outperformed predictions based on traditional clinical parameters. Results from this study indicate that gene expression patterns derived from peripheral white blood cells can provide valuable information about exposure levels, well before liver damage is detected by classical parameters. It also supports the potential use of genomic markers in the blood as surrogates for clinical markers of potential acute liver damage [43].

Gene expression profiles were used for discrimination of hepatotoxicants from non-hepatotoxicants. Three genes, *Gckr* (glucokinase regulatory protein), *Oat* (ornithine aminotransferase) and *Cyp2c29* (cytochrome P450 2C29) that were extracted from gene expression profiles of rat liver successfully predicted hepatotoxicity of six compounds, α-naphthylisothiocyanate, dimethylformamide, *N*-methylformamide, caerulein, dinitrophenol and rosiglitazone, with a classification rate of 96% [60]. Gene expression profiles from *in vitro*-cultured hepatocytes indicated that genes involved in necrotic, apoptotic and cell proliferation pathways were able to separate troglitazone and ciglitazone from rosiglitazone and piolitazone, four agonists of peroxisome proliferator-activated receptor γ that are a new class of oral drugs for treatment of diabetes. The first two compounds had more severe hepatotoxicity [61]. Microarray analysis of the effects of three hepatotoxins including two peroxisome proliferators, clofibrate and gefibrozil, and an antiepileptic drug, phenytoin, in Sprague Dawley rats showed that expression of many genes associated with cell proliferation, metabolism and cellular component were significantly changed. Clustering of gene expression was found to have similarity to the hepatotoxic phenotypes of the three compounds [62].

Gene expression profiles can also help to understand the mechanisms of hepatotoxicity [63–65]. Transcripts related to ribosomes, apoptosis and proteasomal degradation were associated with diethylhexylphthalate (DEHP)-induced liver toxicity, while valproic acid toxicity in the liver was more associated with oxidative stress and cell cycle transcripts [63]. Oxidative stress and energy usage-related genes were changed by methapyrilene [65]. 2,3,7,8-Tetrachlorodibenzo-*p*-dioxin (TCDD) induces a variety of changes in expression of genes associated with oxidative stress, differentiation, apoptosis, gluconeogenesis and fatty acid uptake that may relate to its hepatotoxicity [66]. An *in vitro* study of eight drugs with different mechanisms of hepatotoxicity was performed using rat primary hepatocytes, including drugs that are directly toxic to the liver *in vivo*, APAP and cyclophosphamide, drugs that are indirectly toxic to the liver *in vivo*, clofibrate, chlorpromazine and lithocholic acid and drugs that are rarely toxic to the liver *in vivo*, cisplatin, diclofenac and disulfiram. APAP, cyclophosphamide and clofibrate induced a higher number of genes with changes in their expression, including stress response genes and metabolism-related genes, while the number of transcripts affected by chole-static hepatotoxicants (lithocholic acid and chlorpromazine) or drugs that rarely cause hepatotoxicity (cisplatin, diclofenac and disulfiram) were limited. In the hierarchical cluster analysis, drugs formed clusters depending on their mode of toxicity against cells. Microarray results from this *in vitro* study con-firmed consistency in previously reported *in vivo* data, indicating *in vitro* gene expression analysis of hepatocytes is a useful tool for evaluating the toxico-logical profiles of drugs and in screening for hepatotoxicants [64]. A unique liver-specific microarray containing 550 genes has been developed for hepato-toxicity studies. In one of these studies 64 potential marker genes were identi-fied to classify hepatotoxicants and reflect the common pathway specifically deregulated by hepatotoxicants but not by non-hepatotoxicants, including car-bohydrate metabolism, oxidative stress and xenobiotic metabolism [67].

Toxicogenomics in studies of nephrotoxicity

The kidney plays a principal role in the excretion of metabolic waste, regula-tion of fluid volume and electrolyte and acid-base balance. Important hor-mones, such as renin, erythropoietin, and active vitamin D3 are synthesized and released from the kidney [68]. The kidney is especially vulnerable to toxic insult by various drugs and xenobiotics because of its unique functions in transporting, concentrating and, sometimes, metabolizing a variety of toxic substances within its parenchyma [69, 70]. Nephrotoxicity caused by toxicant exposure not only damages the kidney itself, the renal dysfunction that follows may greatly increase the plasma concentrations of drugs or chemicals and therefore cause additional organ damage. A major challenge in the study of nephrotoxicity is to provide measurements, both qualitative and quantitative, that give an early indication of the initial site of renal damage before gross deterioration in kidney function has occurred [71]. Traditional markers of kid-ney toxicity such as blood urea nitrogen and creatinine have their limitations for toxicological evaluation because they are not region specific, and signifi-cant changes may not occur until 30–50% damage has occurred [72]. Thus, it is important to determine whether genomics evaluation can lead to the identi-fication of biomarkers that can provide increased sensitivity or earlier detec-tion of renal dysfunction [71].

Changes in the expression of mRNA specifically expressed in injured kid-ney cells were shown to be some of the earliest events that accompany renal injury [73]. These changes contribute either to cellular repair or recovery of renal function or mediate fibrosis and further pathology of the kidney [74–76]. The gene-based markers of nephrotoxicity may improve current sensitivity for assessing xenobiotic-induced nephrotoxicity. In a toxicogenomics analysis of the nephrotoxicity of cisplatin, gentamicin, and puromycin, several potential biomarkers of renal toxicity/repair, including kidney injury molecule-1 (KIM-1), osteopontin, and vimentin, were found to change at transcript levels earlier than alteration of creatinine, blood urea nitrogen excretion, and histological change [73]. KIM-1 is a good example of the identification of a biomarker of nephrotoxicity based on changes in gene expression. It was originally discovered in 1998 using representational difference analysis (RDA) to identify differences in gene expression between normal and post-ischemic kidney [77]. KIM-1 is an epithelial cell adhesion molecule and the mRNA and protein product are expressed at low levels in normal kidney but are increased dramatically in post-ischemic kidney. Many groups have reported its usefulness as a gene and/or protein biomarker of renal injury in rats and humans [71, 73, 78, 79]. In the summer of 2007 the Predictive Safety Testing Consortium (PSTC) submitted a set of urinary protein biomarkers for validation to the U.S. Federal Drug Administration (FDA) for their use in detecting various forms of nephrotoxicity in rats during preclinical testing. KIM-1 was one of the biomarkers submitted in this nephrotoxicity panel [42].

Gene expression changes related to nephrotoxicity can be used not only as biomarkers for early renal damage, but can also help to identify patterns of gene expression that reflect different types of nephrotoxicity and predict potential nephrotoxicants. Amin et al. [71] reported a number of region-specific changes in gene expression including changes in a group of proximal tubule-expressed genes that were induced by cisplatin and gentamycin, two proximal tubule-specific toxicants. Changes in transcription profiles that were induced by puromycin, a glomerular toxicant, indicated only a mild secondary tubular lesion [73, 80]. The gene expression profiles in the kidney of rats that were treated with the haloalkene fluoromethyl-2,2-difluoro-1-(trifluoromethyl)vinyl ether (FDVE), the major degradation product of the volatile anesthetic sevoflurane, showed that genes associated with apoptosis, oxidative stress, inflammatory response, regeneration and repair were up-regulated; down-regulated genes were associated with transporters and intermediary metabolism. The rapid and brisk changes in gene expression were thought to be potential biomarkers for FDVE nephrotoxicity and more sensitive than conventional measures of renal function [81]. Nedaplatin (NDP), an anti-neoplastic platinum complex, was reported to cause renal lesions, such as necrosis and regeneration/hyperplasia of the epithelial cells in both cortex and papilla of rat kidneys. Global gene expression analysis of NDP-induced nephrotoxicity revealed that genes involved in various functional categories, including apoptosis, cell cycle regulation, DNA metabolism, cell migration/adhesion and cytoskeleton organization were down-regulated, while genes involved in perturbation of oxidative status and calcium homeostasis were induced. Cytokeratins 14 and 19 were identified in this study as potential biomarkers of renal papillary toxicity [82, 83].

Toxicogenomics in studies of endocrine and reproductive toxicity

The reproductive cycles of both males and females can be interrupted at any stage by toxicant exposure leading to either infertility or developmental defects [84, 85]. The endocrine system, besides its unique regulation of hormones, has great effect on maintaining normal reproductive functions. An endocrine disrupter can often disrupt both endocrine and reproductive functions. Gene expression profiling has been applied in studies of male and female reproductive toxicity, developmental toxicity and endocrine disruption for mechanism exploration, biomarker identification and toxicity prediction.

Male reproductive toxicity involves a broad range of targets and mechanisms including disruption on the seminiferous epithelium, Leydig and Sertoli cells, spermatogenesis, epididymal sperm maturation and endocrine interruptions [86]. Identification of gene expression signatures as predictive biomarkers of testicular toxicity in rats was performed with four known reproductive toxicants, 2,5-hexanedione (a Sertoli cell toxicant), ethylene glycol monomethyl ether (a spermatocyte toxicant), cyclophosphamide (a spermatogonia toxicant) and sulfasalazine [87]. In this study, differential gene expression profiles based on the characteristic action of the compounds were identified before histopathological abnormality was detected in the testes. Three spermatogenesis-related genes that were affected by all compounds, *Hsp70-2* (heat shock protein 70-2), *Igfbp-3* (insulin growth factor binding protein 3) and *Gst-pi* (glutathione-*S*-transferase P), may be used as potential biomarkers of testicular toxicity [58]. Microarray analysis on the effects of the phosphorothioate insecticide profenofos, a known testicular toxicant, on gene expression revealed that genes related to steroidogenesis including *Cyp17a1*, *Cyp11a1* and *Star* (steroidogenic acute regulatory protein) were significantly induced in the testes of treated adult rats. The three steroidogenic-related genes were considered as potential biomarkers of testicular toxicity [88]. Mechanistic studies on male reproductive toxicity using microarray technology have been done with many compounds. Gene expression profiles from a study of cadmium, a well-known testicular toxicant and carcinogen, at a single dose of 5 µmol/kg revealed that down-regulation of *Ccnb1* (cyclin B1) and *Cdc2* (cell division cycle 2), two key regulators of the G2/M transition, may be the cause of meiosis disturbance, and decreased expression of the pro-apoptotic gene *Casp3*, and DNA repair genes possibly contribute to the carcinogenicity of cadmium [89]. Exposure of rats at post-natal day 28 to mono-(2-ethylhexyl)phthalate, a phthalate plasticizer, at 1000 mg/kg caused significant changes in 1675 genes in the rat testes, of *Thbs1* (thrombospondin-1) showed 114-fold induction. Gene profiling data were phenotypically anchored to increased germ cell apoptosis and an interstitial neutrophil infiltrate [90]. An *in vitro* study on an endocrine-disrupting chemical, bisphenol A (BPA), with cultured mouse Leydig tumor cells (mLTC-1) revealed 8 up-regulated and 16 down-regulated genes associating with steroid/cholesterol metabolism/transport and cell cycle regulation. However, BPA has only subtle modulating effects on gene expression in gonadotrophin-stimulated mLTC-1 cells [91].

Female reproductive toxicity can be caused by interruption of oogenesis, and oviduct, uterus, vagina and endocrine functions. The application of gene expression profiling to the rodent uterotrophic assay is a powerful approach for identification of genes targeted *in vivo* by estrogens [92]. Diethylstilbestrol (DES) is a well-known carcinogen with estrogenic activity causing uterine adenocarcinoma following neonatal treatment, and *in utero* exposure to DES was found to induce various abnormalities in the Mullerian duct of mice [93]. DES has been studied for its effects on gene expression in female reproductive tissues. Uterine gene expression profiles after DES treatment indicated that 3% of more than 20 000 transcripts were differentially expressed compared to the control group. GO annotation revealed alterations in genes associated with cell growth, differentiation and adhesion [94]. Hundreds of genes were up- or down-regulated by *in utero* exposure to DES in the oviduct, uterus and vagina of mice. DES-induced changes in the expression of *Dkk2, Nkd2, sFrp1, Hox, Wnt* and *Eph* could be the basis for various abnormalities in the Mullerian ducts [93]. Another microarray study on *in utero* exposure to DES and octyl-phenol (OP) revealed that the expression levels of known estrogen-responsive genes, i.e., *complement component 3, epidermal growth factor receptor, c-fos* and *calbindin 3*, as well as *transcription factor IIa, transcription factor 4*, and *lymphocyte specific 1 (Lsp1)*, were increased by OP and/or DES treatment in the uteri of both maternal and neonate groups. However, the magnitude of these alterations in gene expression differed markedly between dams and neonates, most likely reflecting the temporal susceptibility of the reproductive tract to estrogenic chemicals [95]. Estrogen exposure was reported to cause changes in gene expression in the uterus of immature rats. Over 3000 genes were altered in a time-course study after a single dose of 17α-ethinyl estradi-ol (EE), which involved various biological processes based on GO analysis, such as transcriptional regulation, signal transduction, protein synthesis, ener-gy utilization, solute transport, cell proliferation, differentiation and immuno-logical responses [96]. Comparison of the alterations in gene expression induced by estrogen (17β-estradiol, E_2) and other endocrine disrupters, includ-ing DES, OP, nonyl-phenol, BPA or genistein in the uterus of immature rats revealed that 7–10% of the 7636 genes were changed by more than twofold by estrogen and/or endocrine disrupter exposure. Representative genes included *calbindin-D9k, oxytocin, adipocyte complement-related protein, lactate dehy-drogenase A* and *calcium-binding protein A6*. The results demonstrated dis-tinct alterations in expression of responsive genes following exposure to E_2 and other estrogenic compounds, and implicated distinct effects of endogenous E_2 and environmental endocrine-disrupting chemicals in the uterus of imma-ture rats [97].

Some chemicals can produce both male and female toxicities. An example is BPA, which, as discussed above, can disrupt the function of cultured mouse Leydig tumor cells and the function of the uterus of immature rats with corre-spondent alterations in gene expression [91, 97]. Another example is methyl-mercury (MeHg), which was reported to cause changes in expression of genes associated with egg fertilization and development, sugar metabolism, apopto-sis and electron transport in both male and female fathead minnows. Vitellogenin was markedly up-regulated in male fish but was significantly down-regulated in female fish with increasing MeHg concentrations, indicating a potential physiological difference in the reaction of males and females to MeHg [98].

Environmental endocrine-disrupting chemicals include a wide range of compounds, such as organochlorines, pesticides, plasticizers, pharmaceutics and natural hormones. These compounds interact with various receptors, such as those for estrogen (ER), androgen (AR) and aryl hydrocarbons (AhR), and therefore interrupt normal endocrine and/or reproductive functions. Disruptor-induced changes in gene expression may be involved in the underlying molecular mechanisms of their toxic effects [92, 99, 100]. TCDD is a well-studied endocrine disrupter that acts on multiple components of the endocrine axis. TCDD exposure alters the levels of many hormones and growth factors, and their receptors. Exposure to very low doses of TCDD produces infertility, puberty delay and a number of unusual reproductive alterations in male and female rodent progeny. Much, if not all, of the toxicity and changes in gene expression mediated by TCDD appears to result from activation of the AhR [101, 102]. Miyamoto et al. [103] reported more than 100 TCDD-sensitive genes in cultured rat granulosa cells, rat placentas and ovaries that contain the xenobiotic-responsive element (XRE) in their promoter region. TCDD modulated a small portion (133/2753) of the EE-mediated differential gene expression in the mouse uterus. Inhibition of some of the estrogen-response genes, functioning in cell proliferation, water and ion transport and maintenance of cellular structure and integrity, may contribute to the anti-uterotrophic effects of TCDD [102, 104].

Toxicogenomics in studies of pulmonary toxicity

Exposure to chemicals by inhalation can have effects on lung tissue and on distant organs that are reached after chemicals are inhaled and absorbed. Numerous lung diseases are known to be induced or promoted by occupational exposure [105]. Gene expression profiling has been widely used to study toxicant-induced pulmonary disorders for understanding mechanisms of lung disease, screening for sensitive individuals and identifying potential biomarkers.

Genomic biomarkers were recently used to predict lung tumor incidence in a 2-year rodent cancer bioassay [48, 49]. A set of 13 chemicals tested by the U.S. National Toxicology Program were used in this study. Seven chemicals were positive for increased lung tumor incidence: 1,5-naphthalenediamine, 2,3-benzofuran, 2,2-bis(bromomethyl)-1,3-propanediol, *N*-methylolacrylamide, 1,2-dibromoethane, coumarin, and benzene; and 6 chemicals were negative: *N*-(1-naphthyl)ethylenediamine, pentachloronitrobenzene, *N*-nitroanthranilic acid, 2-chloromethylpyridine, diazinone, and malathione. Microarray analysis of lung tissues after 90 days of chemical exposure identified six highly discriminating genes, *Ugt1a* (UDP-glucuronosyltransferase 1A), *Ces1* (carboxylesterase 1), *Fgfr2* (fibroblast growth factor receptor 2), *Ephx1* (microsomal epoxide hydrolase), *Au018778* and *Gst-m1*, as biomarkers to predict an increase in lung tumor incidence with 94% accuracy. The sensitivity and specificity were 95% and 92%, respectively [49].

Exposure to diesel exhaust particles (DEP) has been associated with asthma, chronic bronchitis, lung inflammation and lung cancer. The underlying mechanisms of DEP toxicity have been studied in cultured cells *in vitro* and in rodents *in vivo* using microarray analysis [106–110]. Alveolar epithelial cells are an important target of toxic insult in the lung. Genes associated with drug metabolism, antioxidation, cell cycle/proliferation/apoptosis, and coagulation/fibrinolysis in cultured rat alveolar type II epithelial cells were changed by exposure to DEP extracts, of which *Ho-1* (heme oxygenase-1) showed the maximum increase and *Tgm-2* (type II transglutaminase) showed the maximum decrease. Changes in the transcript levels of these two genes might be good markers of the biological responses to organic compounds [108]. Alveolar macrophages are another important target of DEP. Six genes, *Ho-1*, *Ho-2*, *Tdpx-2* (thioredoxin peroxidase 2), *Gst-pi*, *NADPH dehydrogenase* and *Pcna* (proliferating cell nuclear antigen), were induced in primary cultured rat alveolar macrophages by exposure to DEP, indicating the pulmonary defense against oxidative stress caused by DEP [109]. Another experiment with human THP-1 cells that were differentiated to macrophages showed modest changes in expression of over 100 genes after DEP exposure reflecting alterations in cytoskeletal rearrangement, motility and filopodia formation [107]. Microarray analysis of global gene expression in the lungs of rats exposed to DEP displayed changes in only a small number of genes representing signaling hormone/G protein-coupled receptors, growth factor/cytokines/hormones and intracellular kinases. Induction of *Il-2* (interleukin 2) and *Pcna* are associated with proliferation of a wide range of cell types, including epithelial type II cells, and might contribute to pulmonary carcinogenesis in rats [106, 110]. Other pollution particles, such as residual oil fly ash (ROFA) and Asian sand dust particles (ASDP), were also studied for their effects on gene expression in rodent lungs [111, 112]. ROFA exposure increased expression of phospho-ERK and phospho-IKB proteins, suggesting changes in cell growth, transformation and inflammation within the airway [111]. ASDP exposure markedly enhanced inflammatory response-related genes, which could be attributed to lipopolysaccharides (LPS) and β-glucan absorbed in ASDP [112].

LPS-induced acute lung injury causes high mortality in humans. Understanding the pathogenesis of LPS-induced acute lung injury is therefore important for designing better treatments. Microarray analysis of transcriptional responses to LPS in the lungs of mice revealed that inflammatory response genes and cell proliferation-related genes were up-regulated by LPS exposure, while heat shock response and chaperone activity related genes were downregulated [113–115]. Up-regulation of *Lix* (CXC chemokine), *Tnf-α* (tumor necrosis factor-α), *macrophage inflammatory protein 2*, *Noxo1*, *serum amyloid A1*, and macrophage receptor *Marco* was thought to play an important role in the pathogenesis of LPS-induced lung injuries. These genes may be useful as biomarkers to monitor the lung injury process [113, 114]. Anti-LIX antibodies were able to attenuate the LPS-induced accumulation of neutrophils in the lungs, suggesting that LIX may serve as a novel therapeutic target [114].

Heavy metals are another group of toxicants causing lung injury. Nickel, a known lung carcinogen, has been widely used in an acute lung injury model in mice [116–118]. After acute exposure to nickel, changes in the expression of genes that are associated with a wide range of biological and pathological processes have been observed. Genes found at increased expression levels are associated with oxidative stress, inflammatory mediators, matrix injury-repair, cell proliferation, and TGFβ signaling, and genes with decreased expression levels are associated with pulmonary lung surfactant proteins, lung fluid absorption and epithelial and endothelial cell injury [116–119]. In one of the above studies, the nitric oxide inhibitor N^G-nitro-L-arginine methyl ester (L-NAME) was observed to decrease expression of cytokine genes and restore expression of surfactant protein genes, and these effects were associated with attenuation of the lung injury caused by nickel exposure [117]. In another study, TGFβ was identified as a central mediator of nickel-induced acute lung injury based on pathway analysis of gene expression profiles [119]. Macrophage stimulating 1 (Mst1) tyrosine kinase-deficient [TK$^{(-/-)}$] mice, compared to normal mice, were found to have higher basal expression of *granzymes D, E, F* and *G*, and to show, after nickel exposure, enhanced expression of genes associated with inflammatory mediators such as *Il-6*, *Illrn* (IL-1 receptor antagonist) and some chemokines. Considering that Mst1 TK$^{(-/-)}$ mice have an earlier onset of pulmonary pathology and a decreased survival time after nickel challenge, Mst1 signaling may be involved in the regulation of macrophage and T lymphocyte activation [118]. An *in vitro* study with human peripheral lung epithelial HPL1D cells that were exposed to a non-toxic dose of nickel revealed changes in some potential cancer-related genes, such as *RhoA, dyskerin, interferon regulatory factor 1* and *Rad21*, that may relate to its carcinogenicity in the lung [120]. Mercury (Hg) is another highly toxic metal raising many public health concerns. Effects of Hg exposure on gene expression in the lung of rats included increased expression levels of genes involved in inflammatory responses, such as *chemokines, Tnf-α* and *interleukins*, in adaptive responses, such as *Gst* (glutathione-*S*-transferase) and *metallothioneins*, and some transporter genes, such as *Mrp, P-glycoprotein* and *Znt1*. The expression of the transcription factor *c-jun/AP-1* and *PI3-kinase* was suppressed. As no pathological alteration was observed in the lung, the changes in gene expression may reflect pulmonary adaptation to Hg exposure [121]. Cadmium-induced changes in the gene expression in cultured lung WI38-VA13 fibroblasts showed interruption of cell cycle, immunity and defense, nucleoside metabolism and signal transduction. Down-regulation of *Eno1* (enolase 1) may cause proto-oncogene expression that is possibly related to cadmium-induced carcinogenicity in the lung [122]. Another metal, titanium, induced differential expression of hundreds of genes in the lungs of mice, including activation of pathways involved in cell cycle, apoptosis, chemokines and complement cascades, which mainly reflected its toxicity in causing pulmonary emphysema, macrophage accumulation and extensive disruption of alveolar septa [123].

Microarray studies on therapeutic drug-induced pulmonary toxicity have also been reported. Bleomycin is a widely used anti-tumor drug with the severe adverse effect of pulmonary fibrosis. Analysis of bleomycin-induced fibrosis in murine models demonstrated early changes in expression of genes associated with inflammation and later changes in genes involved in fibrosis development. Further, discrete changes in expression patterns contributing to TGFβ1-associated pulmonary fibrosis were identified [124, 125].

Effects on global gene expression of many other pulmonary toxicants have also been studied. Paraquat exposure was associated with changes in expression of *Cftr* (cystic fibrosis transmembrane conductance regulator) and *Nf1* (neurofibromatosis type 1) genes, which are important in development of pulmonary fibrosis [126, 127]. Bitumen fumes exposure was reported to highly induce expression of genes associated with polycyclic aromatic hydrocarbon metabolism and detoxification [128]. Hexamethylene diisocyanate exposure was related to changes in expression of genes involved in stress responses, apoptosis, growth arrest and inflammation in the lung of mice [129]. Finally, jet propulsion fuel (JP-8) exposure was related to induction of genes of antioxidant responses and detoxification, including *Gst* and *Cyp* [130]. Many antioxidant response genes that were induced by pulmonary toxicants were reported to be regulated by nuclear factor erythroid-2 related factor 2 (Nrf2). Nrf2 may play a key role in gene regulation responding to toxicant-induced acute lung injury [131].

Toxicogenomics in studies of cardiotoxicity

Cardiovascular functions can be adversely affected by exposure to extrinsic and intrinsic stresses on the heart and vascular system. Extrinsic stress, which is the focus in this discussion, includes exposure to therapeutic drugs, natural products and environmental toxicants [132]. Several drugs used for cancer treatment have shown cardiotoxicity. Doxorubicin-induced cardiotoxicity has often been reported and the underlying mechanism has been studied with microarray technology in recent years. Two models of doxorubicin-mediated cardiotoxicity in mice, acute and chronic, were established to investigate the relationship of changes in gene expression and cardiomyopathy. In the acute model where mice received one dose of 15 mg/kg doxorubicin, up-regulated genes included those involved in oxidative stress, signal transduction and apoptosis, indicating an acute nonspecific response. In the chronic model where mice received 3 mg/kg doxorubicin weekly for 12 weeks, some dysregulated genes implicated specific mechanisms of cardiotoxicity, including the hypertrophy-responsive gene *Stars* (striated muscle activator of Rho signaling), the potential modulator of ATP levels *Snf1-kinase*, and *Axud1* (AXIN1-up-regulated protein 1) [133]. Dystrophin deficiency was reported to enhance doxorubicin-induced cardiotoxicity. Dystrophin-deficient mice (MDX) exhibited pronounced mortality, cardiac insufficiency and cardiac interstitial fibrosis when compared with wild-type mice after doxorubicin treatment, and more genes were changed in MDX mice, including those involved in cell adhesion, oxidative stress, cytoskeleton organization, inflammation, immune response and cell death [134]. In another global analysis of doxorubicin cardiotoxicity in mice, phospholipase c-δ1 was identified as a critical target for TNF receptor-mediated protection against doxorubicin cardiotoxicity [135]. Anti-erbB2 monoclonal antibodies were shown to benefit patients with Her-2-overexpressing breast cancer, either as a single agent or in combination with chemotherapy. However, cardiac dysfunction is one of the serious side effects of this therapy [136]. Blocking erbB2 with the anti-erbB2 monoclonal antibody B-10 in cultured rat cardiomyocytes led to changes in the expression of heat shock proteins and genes involved in cell cycle and calcium exchange. B-10 seemed to inhibit pro-survival pathways and to reduce cellular contractility that may impair the stress response of the heart [137]. Cocaine, a commonly abused drug, is associated with heart ischemia and heart failure. Cocaine administration was shown to activate genes including *angiotensinogen*, *Adrb1* (β1 adrenergic receptor) and *Crp* (C-reactive protein) in rabbit ventricular tissue and cultured H9C2 cells. In H9C2 cells, *N*-acetylcysteine pretreatment blocked the cocaine-mediated increase of expression in *Crp*, *Fas*, *Fas ligand* and *Crlf1* (cytokine receptor-like factor 1), which may relate to its protective effects on cocaine-induced mitochondrial membrane depolarization. These data suggest that acute cocaine administration initiates cellular and genetic changes that, if chronically manifested, could cause cardiac deficits similar to those seen in heart failure and ischemia [138].

Several environmental toxicants also elicit cardiotoxicity. TCDD has been demonstrated to target the heart during fetal development in several species. Exposure to TCDD of pregnant mice on gestational day 14.5 led to significant changes in expression of genes in the fetal heart at gestational day 17.5. Altered genes are associated with xenobiotic metabolism, cardiac homeostasis, extracellular matrix production/remodeling, cardiac hypertrophy, cell cycle regulation and AhR activation [139, 140]. Adult male offspring also showed similar changes in the TCDD-induced genes that were observed fetally [140]. AhR was required for TCDD-induced changes in gene expression since gene expression was not changed in AhR$^{(-/-)}$ fetuses [140]. TCDD induced changes in expression of genes that regulate xenobiotic metabolism, proliferation, contractility and development in the heart of zebrafish larvae [141]. Bis-(2-chloroethoxy)methane (CEM), an organic compound of which polysulfide polymers are generated for the use in sealant applications, has been shown to cause cardiotoxicity in rodents, characterized by cytoplasmic vacuolation of myocytes, necrosis, and inflammation [142]. Transcript changes occurred within 2 days of CEM exposure, including genes involved in energy production (*Atp5j*, *Atp5k*), control of calcium levels (*ucp1*, *calb3*) and maintenance of heart function [143]. The response of the heart to oxidative stress was studied with rat neonatal cardiac myocytes exposed to apoptotic or non-toxic concentrations of H_2O_2. A ubiquitously expressed, novel H_2O_2-response gene, *Perit 1*, was identified and many transcriptional regulators were changed by H_2O_2 exposure, which may potentially influence the progression of the apoptotic response [144].

Toxicogenomics in studies of neurotoxicity

Neurotoxicity may be defined as any adverse effect on the structure or function of the central and/or peripheral nervous system by biological, chemical or physical agents. Gene expression assays can help define neurotoxic mechanisms and be used as biomarkers of effect. The site-selective nature of gene expression in the nervous system may mandate assessment of selected cell populations [145].

Agents with abuse potential have been studied for their effects on gene expression in the human and animal central neural systems. Amphetamine (AMPH) and methamphetamine (METH) are psychomotor stimulants and have high drug abuse potential. Their toxicological effects include hyperthermia, tremor, panic states, paranoid hallucinations and mental confusion [146]. However, microarray analysis of gene expression in several brain regions of rats after 5–30 mg/kg AMPH treatment showed a few changes in gene expression. *Npy-p* (neuropeptide Y precursor protein) was increased twofold in the parietal cortex, *Igfbp-1* (insulin-like growth factor binding protein 1) was increased twofold in the cortical amygdaloid nucleus, *Dat* (plasma membrane-associated dopamine transporter) and *Dad2* (dopamine D2 receptor) were decreased by 40% in the substania nigra, and no significant changes in gene expression were noted in the striatum [145, 147]. METH intoxication leads to persistent damage of dopamine nerve endings of the striatum [148]. Studies on METH neurotoxicity with dopamine neuron-specific cDNA microarrays in adult mice indicated *CoxI* (cytochrome c oxidase polypeptide I) was induced, while *Nadh2* (NADH dehydrogenase chain 2) and *Ndr3* (N-myc downstream regulated 3) were decreased [149, 150]. The mRNA synthesis inhibitor, actinomycin-D, and the protein synthesis inhibitor, cycloheximide, both completely protected against METH-induced dopamine neuron toxicity, indicating that dysregulations at gene transcription and translation levels are involved in the mechanisms of METH-induced neurotoxicity [150]. A later global gene expression analysis of METH-induced dopamine neuron toxicity in the striatum of wild-type C57BL/6 mice identified 152 significant genes grouped into functional categories with inflammatory/immune response elements, receptor/signal transduction components and ion channel/transport proteins. Many genes could be linked to ion regulation and apoptosis, and numerous factors associated with microglial activation emerged with significant changes in expression [151]. In a subsequent study, *Cox-2* (cyclooxygenase-2) and its transcription factor *CCAAT/enhancer-binding protein*, but not *Cox-1*, were found to be induced. Mice bearing a null mutation of the gene for *Cox-2* were resistant to METH-induced neurotoxicity, while *Cox-1* knockouts, like wild-type mice, showed extensive dopamine neuron damage [152]. Dopamine-quinones (DAQ) formed after METH administration caused an early activation of microglial cells that was manifested by an alteration in the expression of a broad biomarker panel of genes [153].

Long-term alcohol abuse has been reported to result in changes in gene expression in the brain due to activation or repression of alcohol-responsive transcription factors, and these changes are responsible, at least partly, for alcohol tolerance, dependence and neurotoxicity [154, 155]. Gene expression in human alcoholism was studied in the superior frontal cortex and/or motor cortex of postmortem human brain. Genes involved in myelination, ubiquiti-nation, apoptosis, cell adhesion, neurogenesis, and neural disease showed altered expression levels when alcoholic cases were compared with non-alco-holic controls [154–158]. Importantly, genes involved in neurodegenerative diseases such as Alzheimer's disease were significantly altered, suggesting a link between alcoholism and other neurodegenerative conditions [158]. Gene expression profiles in the frontal cortex accurately distinguished not only alco-holics from controls, but also cirrhotic alcoholics from non-cirrhotic alcoholics [159]. Animal studies showed significant changes in expression of genes asso-ciated with oxidative stress, membrane trafficking, cell signaling and homeo-stasis/stress response, in which new ethanol-responsive genes were identified and potential pathways in ethanol-induced neurotoxicity were explored [160, 161]. While chronic alcohol abuse produces neurotoxicity in the brain, with-drawal from alcohol abuse could also cause brain damage. Microarray analy-sis of gene expression in the brain of mice after withdrawal from chronic alco-hol intoxication identified cell death and DNA/RNA binding as targeted class-es of genes in females, which may be related to the more severe neurotoxicity of alcohol withdrawal in females. In males, protein degradation and calcium ion binding pathways were altered upon withdrawal of alcohol [162].

The molecular mechanisms of neurotoxicity from environmental neurotox-in exposure have been widely studied using microarray analysis. *N*-Methyl-4-phenyl-1,2,3,6-tetrahydropyridine (MPTP) and 6-hydroxydopamine (6-OHDA) are dopaminergic neuron agonists that have been used to produce animal models for Parkinson's disease. Neurodegeneration by MPTP or 6-OH-DA has been related to changes in expression of genes associated with oxida-tive stress, synthesis and modification of proteins, protein degradation and cell death [163–165]. Iron deposition in the substantia nigra pars compacta (SNPC) of parkinsonian brains was thought to be involved in oxidative stress-induced neurodegeneration [164]. Rapomorphine, an iron chelator, can protect against MPTP-mediated neurotoxicity by preventing the loss of iron-regulato-ry protein IRP2, the increase in α-synuclein and the overexpression of genes involved in cell death [163, 164]. Toki-to, a Japanese/Chinese herbal remedy, was reported to have a protective effect on MPTP neurotoxicity by reverting MPTP-suppressed expression of *Th* (tyrosine hydroxylase) and *Dat* (dopamine transporter) genes and by suppressing the expression of *Sgk* (serum- and glucocorticorid-regulated kinase) that is believed to drive the pathogenesis of Parkinson's disease [166].

Organophosphates, such as insecticides, and heavy metals are two major groups of environmental neurotoxins. Chlorpyrifos, cyfuthrin and diazinon were studied for their neurotoxicity and effects on gene expression in primary human fetal astrocytes and the brains of neonatal rats [167, 168]. The target genes changed by chlorpyrifos and cyfuthrin in fetal astrocytes included molecular chaperons, signal transducers, transcriptional regulators, transporters and those involved in behavior and development. The up-regulation of genes functioning in interferon-γ and insulin-signaling pathways suggested that inflammatory activation of astrocytes might be an important mechanism of their neurotoxicity [167]. Effects of chlorpyrifos and diazinon on gene expression during neural cell development elicited changes in genes involved in neural cell growth, development of glia and myelin, and transcription factors for neural cell differentiation [168].

Lead, mercury and manganese are among the heavy metals well known for their neurotoxicity. Although neurotoxicities caused by heavy metal exposure have been extensively studied, there are surprisingly few reports on changes in global gene expression induced by heavy metals in the nervous system. Immortalized human fetal astrocytes (SV-FHA) were used to assess lead-induced changes in gene expression. *VEGF* (vascular endothelial growth factor) was the most sensitive gene that was induced in SV-FHA cells by lead *via* a protein kinase C (PKC)/AP-1-dependent and hypoxia-inducible factor 1 (HIF-1)-independent signaling pathway [169]. Differentially expressed genes in the brains of rats exposed to mercury were studied, and the results showed that genes involved in many biological processes, including immune response, detoxification, transfer and expression of genetic information, cell signaling, neurotransduction, cell proliferation, cell differentiation, and apoptosis were changed [170]. The molecular targets of manganese in primary human astrocytes were identified using microarray gene expression analysis. Manganese was shown to selectively up-regulate expression of genes encoding pro-inflammatory chemokines, cytokines and related functions, and to down-regulate genes encoding functions involved in DNA replication and repair and cell cycle checkpoint control [171].

Toxicogenomics in studies of dermal toxicity

Skin is the body's first line of defense against external insult and is routinely exposed to various environmental or occupational chemicals. Skin disease was placed in the top ten work-related diseases in 1982 [172]. Differential gene expression has been used to understand the potential mechanisms of allergic contact dermatitis, inflammatory response and carcinogenesis in the skin. A low-density cDNA array with 165 genes related to dendritic cell (DC) biology was used to study the mechanism of interaction of immature DC (iDC) with four haptens representative of strong (2,4-dinitrobenzenesulfonic acid, DNBS), moderate (isoeugenol) and weak (eugenol, hydroxycitronellal) contact sensitizers and with one irritant, sodium dodecyl sulfate (SDS). A total of 21 genes were significantly modulated. The four haptens consistently modulated the chemokine receptor (CCR) and ligand (CCL) genes *Ccr5*, *Ccl27*, *Ccl2*, and *Ccr7*, whereas SDS regulated *Cxcl10* (chemokine ligand 10). The 21 target genes fell into four groups associated with a particular type of chemical endowed with distinct sensitizing or irritant properties. Gene profiling of iDC may be useful to identify chemicals with weak sensitizing properties [173]. Jet propulsion fuel (JP-8) can cause an inflammatory response in skin characterized by erythema, edema and hyperplasia. Gene expression was studied in the epidermis of rats after 1-h cutaneous exposure to JP-8. Consistent increases in expression of genes associated with structural proteins, cell signaling, inflammatory mediators, growth factors and enzymes were observed as early as 1 h after the exposure [174]. Comparison of gene expression in response to JP-8 and its aromatic or aliphatic components, undecane, tetradecane, trimethylbenzene, and dimethylnaphthalene indicated that signaling pathways changed by JP-8 were nearly all activated by the components, but to different extents. While no single component mimicked the gene expression resulting from the JP-8 exposure, undecane produced the most similar responses [175].

To distinguish the genes associated with neoplastic transformation from those linked with proliferation and differentiation in the skin, differential gene expression was determined in two established initiation and promotion skin carcinogenesis models. The first model utilized 7,12-dimethylbenz[*a*]anthracene (DMBA) initiation and 12-*O*-tetradecanoylphorbol 13-acetate (TPA) promotion in the FVB/N mouse, and the second model was TPA promotion of the Tg.Ac mouse, which is endogenously initiated by virtue of an activated H-*ras* transgene. Genes associated with differentiation, such as *Cryαβ* (αβ crystallin), *Eno3* (enolase 3), *Flg* (filaggrin) and *Cstβ* (cystatin β), and genes associated with tumor formation, such as *Gst-o1*, *Txn1* (thioredoxin 1) and *Car2* (carbonic anhydrase 2), were identified by comparing changes in gene expression profiles in TPA-promoted skin, DMBA-initiated skin and DMBA-initiated/TPA-promoted skin in FVB/N mice and in TPA-promoted skin in Tg.Ac mice [176].

Toxicogenomics in studies of hematotoxicity

Hematotoxicity is the adverse effects of drugs, chemicals, and other agents in the environment on blood and blood-forming tissues [177]. Hemolytic anemia is a serious adverse effect of many therapeutic drugs that is caused by increased destruction of drug-damaged erythrocytes by macrophages in the liver and spleen [178, 179]. Studies searching for potential marker genes for drug-induced hematotoxicity were performed with two hemolytic anemia-inducing drugs, phenylhydrazine and phenacetin, in the liver and spleen of rats. Microarray analysis of gene expression in the liver revealed changes in genes involved in hepatic events characteristic of hemolytic anemia, such as hemoglobin biosynthesis, heme metabolism, and phagocytosis. Six up-regulated genes, *Alas2* (aminolevulinic acid synthase), *beta-glo*, *Eraf* (erythroid-associated factor), *Hmox1* (heme oxygenase 1), *Lgals3* (lectin, galactose-binding, soluble 3), and *Rhced* (Rhesus blood group CE and D), were selected as putative biomarkers for hemolytic anemia [178]. Similar analysis in the spleen revealed hundreds of genes commonly deregulated under severe hemolytic conditions, which included genes related to splenic events characteristic of the hematotoxicity, such as proteolysis and iron metabolism. Eleven up-regulated genes were selected as biomarker candidates that can sensitively reflect the erythrocyte damage even under a condition that caused no decrease in erythrocyte counts. Among the selected genes, *Hmox1* was one of the most promising biomarker candidates that could be used to predict hematotoxicity [178].

Benzene is a well-known toxic solvent that causes a progressive decline of hematopoietic function and various hematological disorders, including aplastic anemia, myelodysplastic syndrome, and leukemia. Toxicogenomics studies of benzene have been performed in both occupationally exposed workers and experimental animals. More than 100 genes were found to be differentially expressed in the blood of workers occupationally exposed to benzene. Genes related to apoptosis and immune function were the most significantly affected. Four genes, *CXCL16* [chemokine (C-X-C) motif ligand 16], *ZNF331* (zinc finger protein 331), *JUN*, and *PF4* (platelet factor 4), with the most significant changes are potential biomarkers of benzene exposure [180, 181]. Benzene exposure in animals was found to result in changes in expression of genes that may be related to its hematotoxicity and leukemogenicity. Induction in the bone marrow of the expression of the cyclin-dependent kinase inhibitor *Cdkn1a* (p21^{Waf1}) *via* the p53-dependent pathway was found to be an important regulator in cell cycle arrest associated with the suppression of hematopoiesis. Changes in expression of genes related to apoptosis, DNA repair, cell cycle and growth control were also observed, consistent with the established genotoxicity of benzene [182–185].

Toxicogenomics in studies of immunotoxicity

The immune system can be adversely affected by exposure to occupational or therapeutic drugs, environmental chemicals and, in some instance, biological materials. The immune system has been shown to be compromised (decreased lymphoid cellularity, alternation in lymphocyte subpopulations, decreased host resistance and altered specific immune function responses) in the absence of observed toxicity in other organs [186]. A recent method to study immunotoxicity is examination of changes in gene expression profiles in immunological-ly relevant organs, also referred to as immuno-toxicogenomics [187, 188]. Examples of the application of toxicogenomics in immunotoxicity studies have been reviewed by Baken et al. [188].

Hexachlorobenzene (HCB) is a persistent environmental pollutant with toxic effects in man and rat. Reported adverse effects include hepatic porphyria, toxic effects on the immune and reproductive system, and neurotoxicity. Microarray analysis of gene expression in the blood, thymus, kidney, liver, spleen, and mesenteric lymph nodes (MLN) of brown Norway rats exposed to HCB sub-chronically has been performed [189]. Statistically significant ($p < 0.001$) changes in expression of genes in the spleen and MLN were observed, includ-ing those associated with granulocytes, chemokines, cytokines, immunoglobu-lins, and genes involved in drug metabolism and acute-phase responses. Genes in which expression was significantly changed in the liver included CYP enzymes, genes involved in estrogen and porphyrin metabolism, and genes involved in immune function. In the blood, genes encoding major histocompat-ibility complex class II and T cell markers were affected. Alterations in the kid-ney included genes for cytokines and complement components, and also CYP enzymes. As expected, the thymus was only weakly affected since it is not a tar-get organ of HCB. New findings in this study were the up-regulation of genes encoding pro-inflammatory cytokines, antioxidants, acute-phase proteins, mast cell markers, complement, chemokines, and cell adhesion molecules. The gene expression data indicated that HCB induces a systemic inflammatory response, accompanied by oxidative stress and the acute-phase response.

The xenobiotic TCDD produces a variety of toxic effects, one of the target sites being the immune system. Immunotoxic effects include induction of thy-mus atrophy, suppression of cytotoxic T cell activity, and reduction of humoral immunity [190]. Zeytun et al. [191] studied the effects of TCDD in mice at the transcriptome level. Pathway-specific microarray analysis interrogating 83 genes involved in apoptosis, cytokine production, and angiogenesis revealed up-regulation of expression of apoptosis-related genes in the thymus and spleen and to a lesser extent in the liver after TCDD administration. The results suggested that TCDD-induced apoptosis is mediated by the death receptor pathway in thymocytes [191]. Gene expression changes that underlie suppres-sion of antibody production by TCDD were investigated in lymphocytes of mice immunized with ovalbumin and adjuvant. Exposure to TCDD resulted in down-regulation of gene expression in the T cells that were up-regulated by immunization alone. In B cells, TCDD mainly up-regulated gene expression, indicating that TCDD causes cell type-specific effects [192].

Several studies were performed to determine the gene expression changes induced by various sensitizers. Two allergens, oxazolone and toluene diiso-cyanate, as well as the non-sensitizing irritant nonanoic acid, were applied on the ears of mice on four consecutive days, and the draining (auricular) lymph nodes (LN) were sampled on day 5. Differentially expressed genes included those that are involved in immune response, transcription factors and signal transduction [193]. Betts et al. [194] measured effects on global gene expression in the draining LN after a single topical exposure to the contact allergen dinitrofluorobenzene (DNFB) on the ears of mice. Three genes were significantly affected, *GlyCAM-1* (glycosylation-dependent cell adhesion molecule 1, down-regulated), *guanylate binding protein 2* (up-regulated), and *onzin* (up-regulated). The reduced *GlyCAM-1* expression might be linked to increased LN cellularity in the absence of LN cell proliferation. The up-regulation of *guanylate-binding protein 2* and *onzin* expression may be in keeping with induction of T-helper 1 cell (T_H1 cell) responses, which is generally accepted to occur after contact allergen exposure [194]. Ryan et al. [195] have demonstrated dose-dependent changes in gene expression in blood-derived DC induced by the contact allergen dinitrobenzenesulfonic acid (DNBS) that are associated with DC maturation, a process that is proposed to occur during DC migration to LN after activation by encountering a chemical allergen. Microarray analysis revealed up-regulation of 60 genes and down-regulation of 58 genes after exposure representing several cellular processes such as transcription, signal transduction, protein modification and small molecule transport. A number of gene expression changes were consistent with known features of DC maturation, and some of those were recently reported to also occur after exposure of human DC to the contact allergen nickel sulfate. A list of target genes from this study was derived that could serve to predict skin sensitization by chemicals [195, 196]. Hansen et al. [197] were able to identify 26 differentially expressed ($p < 0.01$) genes by microarray analysis in chromium-stimulated peripheral blood mononuclear cells from allergic patients compared to healthy controls.

Besides chemical compounds and metals, biological toxins can also produce immunomodulatory effects. The mycotoxin deoxynivalenol was reported to change expression of genes involved in immunity, inflammation and chemotaxis in the spleen of exposed mice [198]. Cholera toxin is not only responsible for the clinical symptoms of cholera but also is a powerful mucosal adjuvant that enhances the production of various cytokines and suppression of IL-12 expression, thereby stimulating the development of T_H2 cells. A study with a cDNA microarray containing 800 selected genes indicated that genes associated with immunomodulation, inflammation, and oxidative stress in cultured human lymphocytes and monocytes responded to cholera toxin. As expected, expression of T_H1 cell markers were down-regulated, whereas T_H2 cell markers were up-regulated. Many genes affected by cholera toxin were regulated through the NF-κB pathway. Expression of pro-inflammatory genes such as *IL-8*, *IL-1β*, *IL-6* and *VEGF* was found to be affected the most. Since the proteins encoded by the regulated genes are known to mediate the recruitment, migration, activation, proliferation, and development of lymphocytes, they are likely to contribute to the adjuvanticity of cholera toxin [199–201].

The immunotoxic effects of a set of model compounds [bis(tri-n-butyltin)oxide (TBTO), cyclosporin A, and benzo[*a*]pyrene], as well as acetaminophen (APAP, see above), a compound with reported immunomodulating properties [202, 203], were investigated in mice. Gene expression profiles were determined in the spleens of mice exposed to the four compounds. TBTO exposure caused the most pronounced effect on gene expression and also resulted in the most severe reduction of body weight gain and induction of splenic irregularities. All compounds caused inhibition of cell division in the spleen as shown by microarray analysis as well as by suppression of lympho-cyte proliferation. The immuno-toxicogenomics approach applied in this study pointed to immunosuppression through cell cycle arrest as a common mecha-nism of action of immuntoxicants, including APAP. Genes related to cell divi-sion such as *Ccna2* (cyclin A2), *Brca1* (breast cancer antigen 1), *Birc5* (bac-uloviral IAP repeat-containing 5), *Incenp* (inner centromer protein) and *Cdkn1a* (p21^{Waf1}) were identified as candidate genes to indicate anti-prolifera-tive effects of xenobiotics in immune cells for future screening assays [204].

Novel applications of microarrays in toxicological research

While the toxicological community has been working to apply transcriptomics (mRNA analysis) in toxicology, technology has moved beyond this application into new arenas. An expert opinion about the new genomics technologies in toxicological study was given by Gant in 2007 [205]. In the opinion of this author there are three potentially major applications: (1) array comparative genome hybridization (arrayCGH) in assessment and recognition of genotox-icity; (2) epigenetic assessment in developmental and transgenerational toxi-cology; and (3) miRNA assessment in all toxicology types, but particularly developmental toxicology [205].

ArrayCGH is one of the first alternative applications of microarray technol-ogy in determination of copy number alterations (amplification and deletion) in the genome [206]. The process of determining chromosomal changes is essentially the same as that of measuring mRNA transcript levels, except that the probe is genomic DNA and not mRNA [207, 208]. Hybridization and data collection occur in a similar manner to that used for expression microarrays (for a two-color system) with probes from a control and test system hybridized to the same microarray to produce the familiar red and green spot image. The ratio of fluorescent dyes in the hybridized spot indicates either an amplifica-tion or deletion in the test genome. These data can be plotted against the chro-mosomal location of the probe to produce a map of copy number alterations across the genome. Use of the arrayCGH technique has demonstrated the diversity that exists in DNA copy numbers in the human genome [209]. Such genetic change affects resistance and susceptibility to toxicity, and under-standing the effect of genome variation could have a fundamentally important application in the assessment of drug efficacy and safety [205].

The determination of epigenetic modification in toxicology can help in: (1) understanding how drugs and chemicals alter DNA methylation patterns in cells and how these might affect toxicity, and (2) elucidating how epigenetic changes affect gene expression and, in turn, susceptibility and resistance to toxicity [210]. One means of epigenetic modulation is *via* cytosine methylation. The methylated DNA is first collected by immunoprecipitation using an antibody against 5-methylcytosine. A microarray-based comparison of the differences in DNA sequence content between the immunoprecipitated fractions then indicates those regions of the chromosome where differential methylation has occurred [211]. Epigenetic assessment has played an important role in understanding the mechanisms of transgenerational toxicology indicating genome alteration that may give rise to a phenotype being present in progeny as a result of germline transmission from the exposed parents [212]. Transgenerational genome instability arising from chemical exposure has been indicated by minisatellite mutation in both individual sperm and somatic cells of first generation mice where the father has been exposed to radiation or genotoxins [213, 214]. Modification of epigenetic status in the embryo through maternal exposure may have important long-term effects for the fetus and subsequent generations. An effect of DES on DNA methylation patterns may be a plausible mechanism of the increased incidence of uterine adenocarcinoma in females [215, 216].

Chromatin immunoprecipitation (ChIP) analysis, similar to that used for epigenetic analysis, is a methodology to determine the binding of transcription factors to the promoter regions of target genes. ChIP analysis requires an antibody against the transcription factor of interest, and it is necessary to chemically cross-link the transcription factor to the DNA before immunoprecipitation [217]. A microarray that contains gene promoter region target sequences also is required. Data from the ChIP assay is able to show the gene promoter regions that bind the transcription factor in one condition relative to another, and, therefore, help to identify the target genes of a specific transcription factor. Single nucleotide polymorphism (SNP) analysis in the gene promoter regions may explain the difference in binding efficiency of a transcription factor among different individuals, which may be the underlying mechanism of the susceptibility or resistance to a specific toxicity [218]. ChIP assay and microarray analysis can be complementary to each other for understanding the mechanisms of toxicity and identification of potential biomarkers. First, co-regulated genes that show the same or very similar expression patterns after toxic agent exposure can be extracted in microarray analysis. Genes showing similar expression patterns may share the same transcription factor. Bioinformatics analysis of the promoter regions of these genes may identify enriched binding sites for one or several transcription factors. As noted earlier in our example of radiation-induced changes in gene expression in human fibroblasts, *cis*-acting target sequences for the transcription factor ZF5 were enriched in UV-responsive transcripts. A ChIP assay that is focused on transcription factors and their target genes can be performed to confirm the role of the identified transcription factors in response to the toxicant and the biological significance of genes extracted from the microarray analysis. Activation of transcription factors usually precedes changes in transcription. So transcription factor binding to a promoter can be an early biomarker of response to a toxicant. A good example is the phosphorylation of p53 upon DNA damage, which happens prior to any transcriptional changes of its target genes, such as *CDKN1A* (p21$^{\text{Waf1}}$), *GADD45α* (growth-arrest and DNA damage inducible gene), *BTG2* (B cell translocation gene 2), and *CCNG1* (cyclin G1) [25].

One of the main control methods of gene translation occurs through the recently recognized miRNA species [219]. For toxicology research, there are some interesting features of these miRNA species. First, they are transcribed from polycistronic regions and the control of their transcription appears to be very similar to that used for protein coding genes [220]. This implies that miRNAs may be differentially regulated under the influence of chemicals, which could then substantially change the translational profile of the cell. Second, each miRNA species has an effect on the translation of many mRNA species and so a change in its level of expression could substantially affect the protein complement of the cell, leading to profound responses to chemical exposure. Finally, because of the potential control of the polycistronic regions at the transcriptional level, they form an expression profile in a similar manner to mRNA species. Therefore, the pattern of their expression could potentially be used to identify the type of toxicity that may be associated with the xenobiotic exposure [205]. Presently there are some technical challenges in the use of microarrays to assess the expression of the miRNA species. First, it is necessary to use an RNA tailing method for the labeling because the miRNA species are too short for conventional labeling techniques. Second, it is often necessary to place targets on the microarray that contain modified nucleotides called locked nucleic acid nucleotides [221]. The expression of miRNA species is fundamental to the control of the protein complement of the cells and this may be material both to the understanding of mechanisms of toxicity as well as differential susceptibility [205].

References

1 Hamadeh HK, Amin RP, Paules RS, Afshari CA (2002) An overview of toxicogenomics. *Curr Issues Mol Biol* 4: 45–56
2 Tennant RW (2002) The National Center for Toxicogenomics: Using new technologies to inform mechanistic toxicology. *Environ Health Perspect* 110: A8–10
3 Waters M, Yauk C (2007) Consensus recommendations to promote and advance predictive systems toxicology and toxicogenomics. *Environ Mol Mutagen* 48: 400–403
4 Dobbin K, Simon R (2005) Sample size determination in microarray experiments for class comparison and prognostic classification. *Biostatistics* 6: 27–38
5 Dobbin KK, Simon RM (2007) Sample size planning for developing classifiers using high-dimensional DNA microarray data. *Biostatistics* 8: 101–117
6 Simon R (2008) Microarray-based expression profiling and informatics. *Curr Opin Biotechnol* 19: 26–29
7 Johnson WE, Li C, Rabinovic A (2007) Adjusting batch effects in microarray expression data using empirical Bayes methods. *Biostatistics* 8: 118–127
8 Dalma-Weiszhausz DD, Warrington J, Tanimoto EY, Miyada CG (2006) The affymetrix GeneChip platform: An overview. *Methods Enzymol* 410: 3–28

9   Wolber PK, Collins PJ, Lucas AB, De Witte A, Shannon KW (2006) The Agilent *in situ*-synthesized microarray platform. *Methods Enzymol* 410: 28–57

10 Pedotti P, 't Hoen PA, Vreugdenhil E, Schenk GJ, Vossen RH, Ariyurek Y, de Hollander M, Kuiper R, van Ommen GJ, den Dunnen JT et al (2008) Can subtle changes in gene expression be consistently detected with different microarray platforms? *BMC Genomics* 9: 124

11 Jain AN, Tokuyasu TA, Snijders AM, Segraves R, Albertson DG, Pinkel D (2002) Fully automatic quantification of microarray image data. *Genome Res* 12: 325–332

12 Yang YH, Buckley MJ, Speed TP (2001) Analysis of cDNA microarray images. *Brief Bioinform* 2: 341–349

13 Quackenbush J (2002) Microarray data normalization and transformation. *Nat Genet* 32 Suppl: 496–501

14 Baggerly KA, Coombes KR, Hess KR, Stivers DN, Abruzzo LV, Zhang W (2001) Identifying differentially expressed genes in cDNA microarray experiments. *J Comput Biol* 8: 639–659

15 Kerr MK, Martin M, Churchill GA (2000) Analysis of variance for gene expression microarray data. *J Comput Biol* 7: 819–837

16 Kerr MK, Churchill GA (2001) Experimental design for gene expression microarrays. *Biostatistics* 2: 183–201

17 Baldi P, Long AD (2001) A Bayesian framework for the analysis of microarray expression data: Regularized *t*-test and statistical inferences of gene changes. *Bioinformatics* 17: 509–519

18 Long AD, Mangalam HJ, Chan BY, Tolleri L, Hatfield GW, Baldi P (2001) Improved statistical inference from DNA microarray data using analysis of variance and a Bayesian statistical framework. Analysis of global gene expression in *Escherichia coli* K12. *J Biol Chem* 276: 19937–19944

19 Reiner A, Yekutieli D, Benjamini Y (2003) Identifying differentially expressed genes using false discovery rate controlling procedures. *Bioinformatics* 19: 368–375

20 Tusher VG, Tibshirani R, Chu G (2001) Significance analysis of microarrays applied to the ionizing radiation response. *Proc Natl Acad Sci USA* 98: 5116–5121

21 Quackenbush J (2001) Computational analysis of microarray data. *Nat Rev Genet* 2: 418–427

22 Valafar F (2002) Pattern recognition techniques in microarray data analysis: A survey. *Ann NY Acad Sci* 980: 41–64

23 Chou JW, Zhou T, Kaufmann WK, Paules RS, Bushel PR (2007) Extracting gene expression patterns and identifying co-expressed genes from microarray data reveals biologically responsive processes. *BMC Bioinformatics* 8: 427

24 Zhou T, Chou J, Mullen TE, Elkon R, Zhou Y, Simpson DA, Bushel PR, Paules RS, Lobenhofer EK, Hurban P, Kaufmann WK (2007) Identification of primary transcriptional regulation of cell cycle-regulated genes upon DNA damage. *Cell Cycle* 6: 972–981

25 Zhou T, Chou JW, Simpson DA, Zhou Y, Mullen TE, Medeiros M, Bushel PR, Paules RS, Yang X, Hurban P et al (2006) Profiles of global gene expression in ionizing-radiation-damaged human diploid fibroblasts reveal synchronization behind the G1 checkpoint in a G0-like state of quiescence. *Environ Health Perspect* 114: 553–559

26 Sharan R, Maron-Katz A, Shamir R (2003) CLICK and EXPANDER: A system for clustering and visualizing gene expression data. *Bioinformatics* 19: 1787–1799

27 Sharan R, Shamir R (2000) CLICK: A clustering algorithm with applications to gene expression analysis. *Proc Int Conf Intell Syst Mol Biol* 8: 307–316

28 Wasserman WW, Sandelin A (2004) Applied bioinformatics for the identification of regulatory elements. *Nat Rev Genet* 5: 276–287

29 Elkon R, Linhart C, Sharan R, Shamir R, Shiloh Y (2003) Genome-wide *in silico* identification of transcriptional regulators controlling the cell cycle in human cells. *Genome Res* 13: 773–780

30 Mattingly CJ, Rosenstein MC, Davis AP, Colby GT, Forrest JN Jr, Boyer JL (2006) The comparative toxicogenomics database: A cross-species resource for building chemical-gene interaction networks. *Toxicol Sci* 92: 587–595

31 Hayes KR, Vollrath AL, Zastrow GM, McMillan BJ, Craven M, Jovanovich S, Rank DR, Penn S, Walisser JA, Reddy JK et al (2005) EDGE: A centralized resource for the comparison, analysis, and distribution of toxicogenomic information. *Mol Pharmacol* 67: 1360–1368

32 Waters M, Boorman G, Bushel P, Cunningham M, Irwin R, Merrick A, Olden K, Paules R, Selkirk J, Stasiewicz S et al (2003) Systems toxicology and the Chemical Effects in Biological Systems (CEBS) knowledge base. *EHP Toxicogenomics* 111: 15–28

33 Waters M, Stasiewicz S, Merrick BA, Tomer K, Bushel P, Paules R, Stegman N, Nehls G, Yost KJ, Johnson CH et al (2008) CEBS – Chemical Effects in Biological Systems: A public data reposi-

tory integrating study design and toxicity data with microarray and proteomics data. *Nucleic Acids Res* 36: 892–900

34 Parkinson H, Kapushesky M, Shojatalab M, Abeygunawardena N, Coulson R, Farne A, Holloway E, Kolesnykov N, Lilja P, Lukk M et al (2007) ArrayExpress – A public database of microarray experiments and gene expression profiles. *Nucleic Acids Res* 35: D747–750

35 Barrett T, Troup DB, Wilhite SE, Ledoux P, Rudnev D, Evangelista C, Kim IF, Soboleva A, Tomashevsky M, Edgar R (2007) NCBI GEO: Mining tens of millions of expression profiles – Database and tools update. *Nucleic Acids Res* 35: D760–765

36 Gatzidou ET, Zira AN, Theocharis SE (2007) Toxicogenomics: A pivotal piece in the puzzle of toxicological research. *J Appl Toxicol* 27: 302–309

37 Hosack DA, Dennis G Jr, Sherman BT, Lane HC, Lempicki RA (2003) Identifying biological themes within lists of genes with EASE. *Genome Biol* 4: R70

38 Zhou T, Chou J, Zhou Y, Simpson DA, Cao F, Bushel PR, Paules RS, Kaufmann WK (2007) *Ataxia telangiectasia*-mutated dependent DNA damage checkpoint functions regulate gene expression in human fibroblasts. *Mol Cancer Res* 5: 813–822

39 Heinloth AN, Irwin RD, Boorman GA, Nettesheim P, Fannin RD, Sieber SO, Snell ML, Tucker CJ, Li L, Travlos GS et al (2004) Gene expression profiling of rat livers reveals indicators of potential adverse effects. *Toxicol Sci* 80: 193–202

40 Maggioli J, Hoover A, Weng L (2006) Toxicogenomic analysis methods for predictive toxicology. *J Pharmacol Toxicol Methods* 53: 31–37

41 Feron VJ, Groten JP (2002) Toxicological evaluation of chemical mixtures. *Food Chem Toxicol* 40: 825–839

42 Mendrick DL (2008) Genomic and genetic biomarkers of toxicity. *Toxicology* 245: 175–181

43 Bushel PR, Heinloth AN, Li J, Huang L, Chou JW, Boorman GA, Malarkey DE, Houle CD, Ward SM, Wilson RE et al (2007) Blood gene expression signatures predict exposure levels. *Proc Natl Acad Sci USA* 104: 18211–18216

44 Newton RK, Aardema M, Aubrecht J (2004) The utility of DNA microarrays for characterizing genotoxicity. *Environ Health Perspect* 112: 420–422

45 Kim JY, Kwon J, Kim JE, Koh WS, Chung MK, Yoon S, Song CW, Lee M (2005) Identification of potential biomarkers of genotoxicity and carcinogenicity in L5178Y mouse lymphoma cells by cDNA microarray analysis. *Environ Mol Mutagen* 45: 80–89

46 Lee M, Kwon J, Kim SN, Kim JE, Koh WS, Kim EJ, Chung MK, Han SS, Song CW (2003) cDNA microarray gene expression profiling of hydroxyurea, paclitaxel, and *p*-anisidine, genotoxic compounds with differing tumorigenicity results. *Environ Mol Mutagen* 42: 91–97

47 Seidel SD, Kan HL, Stott WT, Schisler MR, Gollapudi BB (2003) Identification of transcriptome profiles for the DNA-damaging agents bleomycin and hydrogen peroxide in L5178Y mouse lymphoma cells. *Environ Mol Mutagen* 42: 19–25

48 Thomas RS, O'Connell TM, Pluta L, Wolfinger RD, Yang L, Page TJ (2007) A comparison of transcriptomic and metabonomic technologies for identifying biomarkers predictive of two-year rodent cancer bioassays. *Toxicol Sci* 96: 40–46

49 Thomas RS, Pluta L, Yang L, Halsey TA (2007) Application of genomic biomarkers to predict increased lung tumor incidence in 2-year rodent cancer bioassays. *Toxicol Sci* 97: 55–64

50 Nakayama K, Kawano Y, Kawakami Y, Moriwaki N, Sekijima M, Otsuka M, Yakabe Y, Miyaura H, Saito K, Sumida K, Shirai T (2006) Differences in gene expression profiles in the liver between carcinogenic and non-carcinogenic isomers of compounds given to rats in a 28-day repeat-dose toxicity study. *Toxicol Appl Pharmacol* 217: 299–307

51 Sumida K, Saito K, Oeda K, Yakabe Y, Otsuka M, Matsumoto H, Sekijima M, Nakayama K, Kawano Y, Shirai T (2007) A comparative study of gene expression profiles in rat liver after administration of α-hexachlorocyclohexane and lindane. *J Toxicol Sci* 32: 261–288

52 Tsujimura K, Asamoto M, Suzuki S, Hokaiwado N, Ogawa K, Shirai T (2006) Prediction of carcinogenic potential by a toxicogenomic approach using rat hepatoma cells. *Cancer Sci* 97: 1002–1010

53 Chen T, Guo L, Zhang L, Shi L, Fang H, Sun Y, Fuscoe JC, Mei N (2006) Gene expression profiles distinguish the carcinogenic effects of aristolochic acid in target (kidney) and non-target (liver) tissues in rats. *BMC Bioinformatics* 7 Suppl 2: S20

54 Fielden MR, Brennan R, Gollub J (2007) A gene expression biomarker provides early prediction and mechanistic assessment of hepatic tumor induction by nongenotoxic chemicals. *Toxicol Sci* 99: 90–100

55 Fielden MR, Nie A, McMillian M, Elangbam CS, Trela BA, Yang Y, Dunn RT 2nd, Dragan Y, Fransson-Stehen R, Bogdanffy M et al (2008) Interlaboratory evaluation of genomic signatures for predicting carcinogenicity in the rat. *Toxicol Sci* 103: 28–34

56 Jaeschke H (2008) Toxic responses of the liver. In: C Klaassen (ed.): *Casarett & Doull's Toxicology: The Basic Sciences of Poisons*, 7th edn. McGraw-Hill, New York, 557–582

57 Minami K, Saito T, Narahara M, Tomita H, Kato H, Sugiyama H, Katoh M, Nakajima M, Yokoi T (2005) Relationship between hepatic gene expression profiles and hepatotoxicity in five typical hepatotoxicant-administered rats. *Toxicol Sci* 87: 296–305

58 Fukushima T, Kikkawa R, Hamada Y, Horii I (2006) Genomic cluster and network analysis for predictive screening for hepatotoxicity. *J Toxicol Sci* 31: 419–432

59 Powell CL, Kosyk O, Ross PK, Schoonhoven R, Boysen G, Swenberg JA, Heinloth AN, Boorman GA, Cunningham ML, Paules RS, Rusyn I (2006) Phenotypic anchoring of acetaminophen-induced oxidative stress with gene expression profiles in rat liver. *Toxicol Sci* 93: 213–222

60 Spicker JS, Pedersen HT, Nielsen HB, Brunak S (2007) Analysis of cell death inducing compounds. *Arch Toxicol* 81: 803–811

61 Guo L, Zhang L, Sun Y, Muskhelishvili L, Blann E, Dial S, Shi L, Schroth G, Dragan YP (2006) Differences in hepatotoxicity and gene expression profiles by anti-diabetic PPAR γ agonists on rat primary hepatocytes and human HepG2 cells. *Mol Divers* 10: 349–360

62 Jung JW, Park JS, Hwang JW, Kang KS, Lee YS, Song BS, Lee GJ, Yeo CD, Kang JS, Lee WS et al (2004) Gene expression analysis of peroxisome proliferators- and phenytoin-induced hepatotoxicity using cDNA microarray. *J Vet Med Sci* 66: 1329–1333

63 Jolly RA, Goldstein KM, Wei T, Gao H, Chen P, Huang S, Colet JM, Ryan TP, Thomas CE, Estrem ST (2005) Pooling samples within microarray studies: A comparative analysis of rat liver transcription response to prototypical toxicants. *Physiol Genomics* 22: 346–355

64 Suzuki H, Inoue T, Matsushita T, Kobayashi K, Horii I, Hirabayashi Y (2008) *In vitro* gene expression analysis of hepatotoxic drugs in rat primary hepatocytes. *J Appl Toxicol* 28: 227–236

65 Craig A, Sidaway J, Holmes E, Orton T, Jackson D, Rowlinson R, Nickson J, Tonge R, Wilson I, Nicholson J (2006) Systems toxicology: Integrated genomic, proteomic and metabonomic analysis of methapyrilene induced hepatotoxicity in the rat. *J Proteome Res* 5: 1586–1601

66 Boverhof DR, Burgoon LD, Tashiro C, Chittim B, Harkema JR, Jump DB, Zacharewski TR (2005) Temporal and dose-dependent hepatic gene expression patterns in mice provide new insights into TCDD-Mediated hepatotoxicity. *Toxicol Sci* 85: 1048–1063

67 Zidek N, Hellmann J, Kramer PJ, Hewitt PG (2007) Acute hepatotoxicity: A predictive model based on focused illumina microarrays. *Toxicol Sci* 99: 289–302

68 Schnellmann RG (2008) Toxic responses of the kidney. In: C Klaassen (ed.): *Casarett & Doull's Toxicology: The Basic Sciences of Poisons*, 7th edn. McGraw-Hill, New York, 583–608

69 Toback FG (1992) Regeneration after acute tubular necrosis. *Kidney Int* 41: 226–246

70 Bennett WM (1997) Drug nephrotoxicity: An overview. *Ren Fail* 19: 221–224

71 Thukral SK, Nordone PJ, Hu R, Sullivan L, Galambos E, Fitzpatrick VD, Healy L, Bass MB, Cosenza ME, Afshari CA (2005) Prediction of nephrotoxicant action and identification of candidate toxicity-related biomarkers. *Toxicol Pathol* 33: 343–355

72 Duarte CG, Preuss HG (1993) Assessment of renal function – Glomerular and tubular. *Clin Lab Med* 13: 33–52

73 Amin RP, Vickers AE, Sistare F, Thompson KL, Roman RJ, Lawton M, Kramer J, Hamadeh HK, Collins J, Grissom S et al (2004) Identification of putative gene based markers of renal toxicity. *Environ Health Perspect* 112: 465–479

74 Matejka GL (1998) Expression of GH receptor, IGF-I receptor and IGF-I mRNA in the kidney and liver of rats recovering from unilateral renal ischemia. *Growth Horm IGF Res* 8: 77–82

75 Norman JT, Bohman RE, Fischmann G, Bowen JW, McDonough A, Slamon D, Fine LG (1988) Patterns of mRNA expression during early cell growth differ in kidney epithelial cells destined to undergo compensatory hypertrophy *versus* regenerative hyperplasia. *Proc Natl Acad Sci USA* 85: 6768–6772

76 Safirstein R, Price PM, Saggi SJ, Harris RC (1990) Changes in gene expression after temporary renal ischemia. *Kidney Int* 37: 1515–1521

77 Ichimura T, Bonventre JV, Bailly V, Wei H, Hession CA, Cate RL, Sanicola M (1998) Kidney injury molecule-1 (KIM-1), a putative epithelial cell adhesion molecule containing a novel immunoglobulin domain, is up-regulated in renal cells after injury. *J Biol Chem* 273: 4135–4142

78 Chen G, Bridenbaugh EA, Akintola AD, Catania JM, Vaidya VS, Bonventre JV, Dearman AC,

Sampson HW, Zawieja DC, Burghardt RC, Parrish AR (2007) Increased susceptibility of aging kidney to ischemic injury: Identification of candidate genes changed during aging, but corrected by caloric restriction. *Am J Physiol Renal Physiol* 293: F1272–1281

79 Tsuji M, Monkawa T, Yoshino J, Asai M, Fukuda S, Kawachi H, Shimizu F, Hayashi M, Saruta T (2006) Microarray analysis of a reversible model and an irreversible model of anti-Thy-1 nephritis. *Kidney Int* 69: 996–1004

80 Kramer JA, Pettit SD, Amin RP, Bertram TA, Car B, Cunningham M, Curtiss SW, Davis JW, Kind C, Lawton M et al (2004) Overview on the application of transcription profiling using selected nephrotoxicants for toxicology assessment. *Environ Health Perspect* 112: 460–464

81 Kharasch ED, Schroeder JL, Bammler T, Beyer R, Srinouanprachanh S (2006) Gene expression profiling of nephrotoxicity from the sevoflurane degradation product fluoromethyl-2,2-difluoro-1-(trifluoromethyl)vinyl ether ("compound A") in rats. *Toxicol Sci* 90: 419–431

82 Uehara T, Miyoshi T, Tsuchiya N, Masuno K, Okada M, Inoue S, Torii M, Yamate J, Maruyama T (2007) Comparative analysis of gene expression between renal cortex and papilla in nedaplatin-induced nephrotoxicity in rats. *Hum Exp Toxicol* 26: 767–780

83 Uehara T, Tsuchiya N, Masuda A, Torii M, Nakamura M, Yamate J, Maruyama T (2008) Time course of the change and amelioration of nedaplatin-induced nephrotoxicity in rats. *J Appl Toxicol* 28: 388–398

84 Foster PM, Earl Gary L Jr, (2008) Toxic responses of the reproductive system. In: C Klaassen (ed.): *Casarett & Doull's Toxicology: The Basic Sciences of Poisons*, 7th edn. McGraw-Hill, New York, 761–806

85 Capen CC (2008) Toxic responses of the endocrine system. In: C Klaassen (ed.): *Casarett & Doull's Toxicology: The Basic Sciences of Poisons*, 7th edn. McGraw-Hill, New York, 807–879

86 Mantovani A, Maranghi F (2005) Risk assessment of chemicals potentially affecting male fertility. *Contraception* 72: 308–313

87 Fukushima T, Yamamoto T, Kikkawa R, Hamada Y, Komiyama M, Mori C, Horii I (2005) Effects of male reproductive toxicants on gene expression in rat testes. *J Toxicol Sci* 30: 195–206

88 Moustafa GG, Ibrahim ZS, Hashimoto Y, Alkelch AM, Sakamoto KQ, Ishizuka M, Fujita S (2007) Testicular toxicity of profenofos in matured male rats. *Arch Toxicol* 81: 875–881

89 Zhou T, Jia X, Chapin RE, Maronpot RR, Harris MW, Liu J, Waalkes MP, Eddy EM (2004) Cadmium at a non-toxic dose alters gene expression in mouse testes. *Toxicol Lett* 154: 191–200

90 Lahousse SA, Wallace DG, Liu D, Gaido KW, Johnson KJ (2006) Testicular gene expression profiling following prepubertal rat mono-(2-ethylhexyl)phthalate exposure suggests a common initial genetic response at fetal and prepubertal ages. *Toxicol Sci* 93: 369–381

91 Takamiya M, Lambard S, Huhtaniemi IT (2007) Effect of bisphenol A on human chorionic gonadotrophin-stimulated gene expression of cultured mouse Leydig tumour cells. *Reprod Toxicol* 24: 265–275

92 Moggs JG (2005) Molecular responses to xenoestrogens: Mechanistic insights from toxicogenomics. *Toxicology* 213: 177–193

93 Suzuki A, Urushitani H, Sato T, Kobayashi T, Watanabe H, Ohta Y, Iguchi T (2007) Gene expression change in the Müllerian duct of the mouse fetus exposed to diethylstilbestrol *in utero*. *Exp Biol Med* 232: 503–514

94 Newbold RR, Jefferson WN, Grissom SF, Padilla-Banks E, Snyder RJ, Lobenhofer EK (2007) Developmental exposure to diethylstilbestrol alters uterine gene expression that may be associated with uterine neoplasia later in life. *Mol Carcinog* 46: 783–796

95 Dang VH, Choi KC, Hyun SH, Jeung EB (2007) Analysis of gene expression profiles in the offspring of rats following maternal exposure to xenoestrogens. *Reprod Toxicol* 23: 42–54

96 Naciff JM, Overmann GJ, Torontali SM, Carr GJ, Khambatta ZS, Tiesman JP, Richardson BD, Daston GP (2007) Uterine temporal response to acute exposure to 17α-ethinyl estradiol in the immature rat. *Toxicol Sci* 97: 467–490

97 Hong EJ, Park SH, Choi KC, Leung PC, Jeung EB (2006) Identification of estrogen-regulated genes by microarray analysis of the uterus of immature rats exposed to endocrine disrupting chemicals. *Reprod Biol Endocrinol* 4: 49

98 Klaper R, Rees CB, Drevnick P, Weber D, Sandheinrich M, Carvan MJ (2006) Gene expression changes related to endocrine function and decline in reproduction in fathead minnow (*Pimephales promelas*) after dietary methylmercury exposure. *Environ Health Perspect* 114: 1337–1343

99 Iguchi T, Sumi M, Tanabe S (2002) Endocrine disruptor issues in Japan. *Congenit Anom* 42: 106–119

100 Iguchi T, Watanabe H, Katsu Y (2007) Toxicogenomics and ecotoxicogenomics for studying endocrine disruption and basic biology. *Gen Comp Endocrinol* 153: 25–29

101 Hotchkiss AK, Rider CV, Blystone CR, Wilson VS, Hartig PC, Ankley GT, Foster PM, Gray CL, Gray LE (2008) Fifteen years after "Wingspread"- Environmental Endocrine Disrupters and human and wildlife health: Where we are today and where we need to go. *Toxicol Sci* 105: 235–259

102 Boverhof DR, Kwekel JC, Humes DG, Burgoon LD, Zacharewski TR (2006) Dioxin induces an estrogen-like, estrogen receptor-dependent gene expression response in the murine uterus. *Mol Pharmacol* 69: 1599–1606

103 Miyamoto K (2004) Effects of dioxin on gene expression in female reproductive system in the rat. *Environ Sci* 11: 47–55

104 Boverhof DR, Burgoon LD, Williams KJ, Zacharewski TR (2008) Inhibition of estrogen-mediated uterine gene expression responses by dioxin. *Mol Pharmacol* 73: 82–93

105 Witschi HR, Pinkerton KE, Van Winkle LS, Last JA (2008) Toxic responses of the respiratory system. In: C Klaassen (ed.): *Casarett & Doull's Toxicology: The Basic Sciences of Poisons*, 7th edn. McGraw-Hill, New York, 609–630

106 Sato H, Sagai M, Suzuki KT, Aoki Y (1999) Identification, by cDNA microarray, of A-raf and proliferating cell nuclear antigen as genes induced in rat lung by exposure to diesel exhaust. *Res Commun Mol Pathol Pharmacol* 105: 77–86

107 Verheyen GR, Nuijten JM, Van Hummelen P, Schoeters GR (2004) Microarray analysis of the effect of diesel exhaust particles on *in vitro* cultured macrophages. *Toxicol In Vitro* 18: 377–391

108 Koike E, Hirano S, Furuyama A, Kobayashi T (2004) cDNA microarray analysis of rat alveolar epithelial cells following exposure to organic extract of diesel exhaust particles. *Toxicol Appl Pharmacol* 201: 178–185

109 Koike E, Hirano S, Shimojo N, Kobayashi T (2002) cDNA microarray analysis of gene expression in rat alveolar macrophages in response to organic extract of diesel exhaust particles. *Toxicol Sci* 67: 241–246

110 Wise H, Balharry D, Reynolds LJ, Sexton K, Richards RJ (2006) Conventional and toxicogenomic assessment of the acute pulmonary damage induced by the instillation of Cardiff PM10 into the rat lung. *Sci Total Environ* 360: 60–67

111 Roberts E, Charboneau L, Espina V, Liotta L, Petricoin E, Dreher K (2004) Application of laser capture microdissection and protein microarray technologies in the molecular analysis of airway injury following pollution particle exposure. *J Toxicol Environ Health A* 67: 851–861

112 Yanagisawa R, Takano H, Ichinose T, Mizushima K, Nishikawa M, Mori I, Inoue K, Sadakane K, Yoshikawa T (2007) Gene expression analysis of murine lungs following pulmonary exposure to Asian sand dust particles. *Exp Biol Med* 232: 1109–1118

113 Meng QR, Gideon KM, Harbo SJ, Renne RA, Lee MK, Brys AM, Jones R (2006) Gene expression profiling in lung tissues from mice exposed to cigarette smoke, lipopolysaccharide, or smoke plus lipopolysaccharide by inhalation. *Inhal Toxicol* 18: 555–568

114 Jeyaseelan S, Chu HW, Young SK, Worthen GS (2004) Transcriptional profiling of lipopolysaccharide-induced acute lung injury. *Infect Immun* 72: 7247–7256

115 Cook DN, Wang S, Wang Y, Howles GP, Whitehead GS, Berman KG, Church TD, Frank BC, Gaspard RM, Yu Y et al (2004) Genetic regulation of endotoxin-induced airway disease. *Genomics* 83: 961–969

116 McDowell SA, Gammon K, Bachurski CJ, Wiest JS, Leikauf JE, Prows DR, Leikauf GD (2000) Differential gene expression in the initiation and progression of nickel-induced acute lung injury. *Am J Respir Cell Mol Biol* 23: 466–474

117 McDowell SA, Gammon K, Zingarelli B, Bachurski CJ, Aronow BJ, Prows DR, Leikauf GD (2003) Inhibition of nitric oxide restores surfactant gene expression following nickel-induced acute lung injury. *Am J Respir Cell Mol Biol* 28: 188–198

118 Mallakin A, Kutcher LW, McDowell SA, Kong S, Schuster R, Lentsch AB, Aronow BJ, Leikauf GD, Waltz SE (2006) Gene expression profiles of Mst1r-deficient mice during nickel-induced acute lung injury. *Am J Respir Cell Mol Biol* 34: 15–27

119 Wesselkamper SC, Case LM, Henning LN, Borchers MT, Tichelaar JW, Mason JM, Dragin N, Medvedovic M, Sartor MA, Tomlinson CR, Leikauf GD (2005) Gene expression changes during the development of acute lung injury: Role of transforming growth factor β. *Am J Respir Crit Care Med* 172: 1399–1411

120 Cheng RY, Zhao A, Alvord WG, Powell DA, Bare RM, Masuda A, Takahashi T, Anderson LM,

Kasprzak KS (2003) Gene expression dose-response changes in microarrays after exposure of human peripheral lung epithelial cells to nickel(II). *Toxicol Appl Pharmacol* 191: 22–39

121 Liu J, Lei D, Waalkes MP, Beliles RP, Morgan DL (2003) Genomic analysis of the rat lung following elemental mercury vapor exposure. *Toxicol Sci* 74: 174–181

122 Li GY, Kim M, Kim JH, Lee MO, Chung JH, Lee BH (2008) Gene expression profiling in human lung fibroblast following cadmium exposure. *Food Chem Toxicol* 46: 1131–1137

123 Chen HW, Su SF, Chien CT, Lin WH, Yu SL, Chou CC, Chen JJ, Yang PC (2006) Titanium dioxide nanoparticles induce emphysema-like lung injury in mice. *FASEB J* 20: 2393–2395

124 Katsuma S, Nishi K, Tanigawara K, Ikawa H, Shiojima S, Takagaki K, Kaminishi Y, Suzuki Y, Hirasawa A, Ohgi T et al (2001) Molecular monitoring of bleomycin-induced pulmonary fibrosis by cDNA microarray-based gene expression profiling. *Biochem Biophys Res Commun* 288: 747–751

125 Haider Y, Malizia AP, Keating DT, Birch M, Tomlinson A, Martin G, Ferguson MW, Doran PP, Egan JJ (2007) Host predisposition by endogenous transforming growth factor-β1 overexpression promotes pulmonary fibrosis following bleomycin injury. *J Inflamm* 4: 18

126 Satomi Y, Tsuchiya W, Mihara K, Ota M, Kasahara Y, Akahori F (2004) Gene expression analysis of the lung following paraquat administration in rats using DNA microarray. *J Toxicol Sci* 29: 91–100

127 Satomi Y, Tsuchiya W, Miura D, Kasahara Y, Akahori F (2006) DNA microarray analysis of pulmonary fibrosis three months after exposure to paraquat in rats. *J Toxicol Sci* 31: 345–355

128 Gate L, Langlais C, Micillino JC, Nunge H, Bottin MC, Wrobel R, Binet S (2006) Bitumen fume-induced gene expression profile in rat lung. *Toxicol Appl Pharmacol* 215: 83–92

129 Lee CT, Ylostalo J, Friedman M, Hoyle GW (2005) Gene expression profiling in mouse lung following polymeric hexamethylene diisocyanate exposure. *Toxicol Appl Pharmacol* 205: 53–64

130 Espinoza LA, Valikhani M, Cossio MJ, Carr T, Jung M, Hyde J, Witten ML, Smulson ME (2005) Altered expression of γ-synuclein and detoxification-related genes in lungs of rats exposed to JP-8. *Am J Respir Cell Mol Biol* 32: 192–200

131 Zhu L, Pi J, Wachi S, Andersen ME, Wu R, Chen Y (2008) Identification of Nrf2-dependent airway epithelial adaptive response to proinflammatory oxidant-hypochlorous acid challenge by transcription profiling. *Am J Physiol Lung Cell Mol Physiol* 294: L469–477

132 Kang JY (2008) Toxic responses of the heart and vascular system. In: C Klaassen (ed.): *Casarett & Doull's Toxicology: The Basic Sciences of Poisons*, 7th edn. McGraw-Hill, New York, 699–704

133 Yi X, Bekeredjian R, DeFilippis NJ, Siddiquee Z, Fernandez E, Shohet RV (2006) Transcriptional analysis of doxorubicin-induced cardiotoxicity. *Am J Physiol Heart Circ Physiol* 290: H1098–1102

134 Deng S, Kulle B, Hosseini M, Schluter G, Hasenfuss G, Wojnowski L, Schmidt A (2007) Dystrophin-deficiency increases the susceptibility to doxorubicin-induced cardiotoxicity. *Eur J Heart Fail* 9: 986–994

135 Lien YC, Noel T, Liu H, Stromberg AJ, Chen KC, St Clair DK (2006) Phospholipase c-δ1 is a critical target for tumor necrosis factor receptor-mediated protection against adriamycin-induced cardiac injury. *Cancer Res* 66: 4329–4338

136 Slamon DJ, Leyland-Jones B, Shak S, Fuchs H, Paton V, Bajamonde A, Fleming T, Eiermann W, Wolter J, Pegram M et al (2001) Use of chemotherapy plus a monoclonal antibody against HER2 for metastatic breast cancer that overexpresses HER2. *N Engl J Med* 344: 783–792

137 Pugatsch T, Abedat S, Lotan C, Beeri R (2006) Anti-erbB2 treatment induces cardiotoxicity by interfering with cell survival pathways. *Breast Cancer Res* 8: R35

138 Lattanzio FA Jr, Tiangco D, Osgood C, Beebe S, Kerry J, Hargrave BY (2005) Cocaine increases intracellular calcium and reactive oxygen species, depolarizes mitochondria, and activates genes associated with heart failure and remodeling. *Cardiovasc Toxicol* 5: 377–390

139 Thackaberry EA, Jiang Z, Johnson CD, Ramos KS, Walker MK (2005) Toxicogenomic profile of 2,3,7,8-tetrachlorodibenzo-*p*-dioxin in the murine fetal heart: Modulation of cell cycle and extracellular matrix genes. *Toxicol Sci* 88: 231–241

140 Aragon AC, Kopf PG, Campen MJ, Huwe JK, Walker MK (2008) *In utero* and lactational 2,3,7,8-tetrachlorodibenzo-*p*-dioxin exposure: Effects on fetal and adult cardiac gene expression and adult cardiac and renal morphology. *Toxicol Sci* 101: 321–330

141 Carney SA, Prasch AL, Heideman W, Peterson RE (2006) Understanding dioxin developmental toxicity using the zebrafish model. *Birth Defects Res A Clin Mol Teratol* 76: 7–18

142 Dunnick JK, Lieuallen W, Moyer C, Orzech D, Nyska A (2004) Cardiac damage in rodents after

exposure to bis(2-chloroethoxy)methane. *Toxicol Pathol* 32: 309–317

143 Dunnick J, Blackshear P, Kissling G, Cunningham M, Parker J, Nyska A (2006) Critical pathways in heart function: Bis(2-chloroethoxy)methane-induced heart gene transcript change in F344 rats. *Toxicol Pathol* 34: 348–356

144 Clerk A, Kemp TJ, Zoumpoulidou G, Sugden PH (2007) Cardiac myocyte gene expression profiling during H_2O_2-induced apoptosis. *Physiol Genomics* 29: 118–127

145 Slikker W Jr, Bowyer JF (2005) Biomarkers of adult and developmental neurotoxicity. *Toxicol Appl Pharmacol* 206: 255–260

146 Hardman JG, Gilman AG, Limbird LE (1996) *Goodman and Gilman's the Pharmacological Basis of Therapeutics*, 9th edn. McGraw-Hill, New York, 219–221

147 Bowyer JF, Harris AJ, Delongchamp RR, Jakab RL, Miller DB, Little AR, O'Callaghan JP (2004) Selective changes in gene expression in cortical regions sensitive to amphetamine during the neurodegenerative process. *Neurotoxicology* 25: 555–572

148 Thomas DM, Walker PD, Benjamins JA, Geddes TJ, Kuhn DM (2004) Methamphetamine neurotoxicity in dopamine nerve endings of the striatum is associated with microglial activation. *J Pharmacol Exp Ther* 311: 1–7

149 Barrett T, Xie T, Piao Y, Dillon-Carter O, Kargul GJ, Lim MK, Chrest FJ, Wersto R, Rowley DL, Juhaszova M et al (2001) A murine dopamine neuron-specific cDNA library and microarray: Increased COX1 expression during methamphetamine neurotoxicity. *Neurobiol Dis* 8: 822–833

150 Xie T, Tong L, Barrett T, Yuan J, Hatzidimitriou G, McCann UD, Becker KG, Donovan DM, Ricaurte GA (2002) Changes in gene expression linked to methamphetamine-induced dopaminergic neurotoxicity. *J Neurosci* 22: 274–283

151 Thomas DM, Francescutti-Verbeem DM, Liu X, Kuhn DM (2004) Identification of differentially regulated transcripts in mouse striatum following methamphetamine treatment – An oligonucleotide microarray approach. *J Neurochem* 88: 380–393

152 Thomas DM, Kuhn DM (2005) Cyclooxygenase-2 is an obligatory factor in methamphetamine-induced neurotoxicity. *J Pharmacol Exp Ther* 313: 870–876

153 Kuhn DM, Francescutti-Verbeem DM, Thomas DM (2006) Dopamine quinones activate microglia and induce a neurotoxic gene expression profile: Relationship to methamphetamine-induced nerve ending damage. *Ann NY Acad Sci* 1074: 31–41

154 Mayfield RD, Lewohl JM, Dodd PR, Herlihy A, Liu J, Harris RA (2002) Patterns of gene expression are altered in the frontal and motor cortices of human alcoholics. *J Neurochem* 81: 802–813

155 Lewohl JM, Wang L, Miles MF, Zhang L, Dodd PR, Harris RA (2000) Gene expression in human alcoholism: Microarray analysis of frontal cortex. *Alcohol Clin Exp Res* 24: 1873–1882

156 Lewohl JM, Dodd PR, Mayfield RD, Harris RA (2001) Application of DNA microarrays to study human alcoholism. *J Biomed Sci* 8: 28–36

157 Liu J, Lewohl JM, Dodd PR, Randall PK, Harris RA, Mayfield RD (2004) Gene expression profiling of individual cases reveals consistent transcriptional changes in alcoholic human brain. *J Neurochem* 90: 1050–1058

158 Liu J, Lewohl JM, Harris RA, Iyer VR, Dodd PR, Randall PK, Mayfield RD (2006) Patterns of gene expression in the frontal cortex discriminate alcoholic from nonalcoholic individuals. *Neuropsychopharmacology* 31: 1574–1582

159 Liu J, Lewohl JM, Harris RA, Dodd PR, Mayfield RD (2007) Altered gene expression profiles in the frontal cortex of cirrhotic alcoholics. *Alcohol Clin Exp Res* 31: 1460–1466

160 Saito M, Smiley J, Toth R, Vadasz C (2002) Microarray analysis of gene expression in rat hippocampus after chronic ethanol treatment. *Neurochem Res* 27: 1221–1229

161 Treadwell JA, Singh SM (2004) Microarray analysis of mouse brain gene expression following acute ethanol treatment. *Neurochem Res* 29: 357–369

162 Hashimoto JG, Wiren KM (2008) Neurotoxic consequences of chronic alcohol withdrawal: Expression profiling reveals importance of gender over withdrawal severity. *Neuropsychopharmacology* 33: 1084–1096

163 Mandel S, Grunblatt E, Youdim M (2000) cDNA microarray to study gene expression of dopaminergic neurodegeneration and neuroprotection in MPTP and 6-hydroxydopamine models: Implications for idiopathic Parkinson's disease. *J Neural Transm Suppl*: 117–124

164 Youdim MB (2003) What have we learnt from cDNA microarray gene expression studies about the role of iron in MPTP induced neurodegeneration and Parkinson's disease? *J Neural Transm Suppl*: 73–88

165 Holtz WA, Turetzky JM, O'Malley KL (2005) Microarray expression profiling identifies early

signaling transcripts associated with 6-OHDA-induced dopaminergic cell death. *Antioxid Redox Signal* 7: 639–648

166 Sakai R, Irie Y, Murata T, Ishige A, Anjiki N, Watanabe K (2007) Toki-to protects dopaminergic neurons in the substantia nigra from neurotoxicity of MPTP in mice. *Phytother Res* 21: 868–873

167 Mense SM, Sengupta A, Lan C, Zhou M, Bentsman G, Volsky DJ, Whyatt RM, Perera FP, Zhang L (2006) The common insecticides cyfluthrin and chlorpyrifos alter the expression of a subset of genes with diverse functions in primary human astrocytes. *Toxicol Sci* 93: 125–135

168 Slotkin TA, Seidler FJ (2007) Comparative developmental neurotoxicity of organophosphates *in vivo*: Transcriptional responses of pathways for brain cell development, cell signaling, cytotoxicity and neurotransmitter systems. *Brain Res Bull* 72: 232–274

169 Hossain MA, Bouton CM, Pevsner J, Laterra J (2000) Induction of vascular endothelial growth factor in human astrocytes by lead. Involvement of a protein kinase C/activator protein-1 complex-dependent and hypoxia-inducible factor 1-independent signaling pathway. *J Biol Chem* 275: 27874–27882

170 Cheng JP, Yuan T, Ji XL, Zheng M, Wang Y, Wang WH, Zhang QH (2006) [Analyzing differentially expressed genes in the brain of rat exposed to mercury chloride using cDNA microarray.] *Huan Jing Ke Xue* 27: 779–782

171 Sengupta A, Mense SM, Lan C, Zhou M, Mauro RE, Kellerman L, Bentsman G, Volsky DJ, Louis ED, Graziano JH, Zhang L (2007) Gene expression profiling of human primary astrocytes exposed to manganese chloride indicates selective effects on several functions of the cells. *Neurotoxicology* 28: 478–489

172 Rice RH, Mauro TM (2008) Toxic responses of the skin. In: C Klaassen (ed.): *Casarett & Doull's Toxicology: The Basic Sciences of Poisons*, 7th edn. McGraw-Hill, New York, 741–760

173 Cluzel-Tailhardat M, Bonnet-Duquennoy M, de Queral DP, Vocanson M, Kurfurst R, Courtellemont P, Le Varlet B, Nicolas JF (2007) Chemicals with weak skin sensitizing properties can be identified using low-density microarrays on immature dendritic cells. *Toxicol Lett* 174: 98–109

174 McDougal JN, Garrett CM, Amato CM, Berberich SJ (2007) Effects of brief cutaneous JP-8 jet fuel exposures on time course of gene expression in the epidermis. *Toxicol Sci* 95: 495–510

175 McDougal JN, Garrett CM (2007) Gene expression and target tissue dose in the rat epidermis after brief JP-8 and JP-8 aromatic and aliphatic component exposures. *Toxicol Sci* 97: 569–581

176 Ridd K, Zhang SD, Edwards RE, Davies R, Greaves P, Wolfreys A, Smith AG, Gant TW (2006) Association of gene expression with sequential proliferation, differentiation and tumor formation in murine skin. *Carcinogenesis* 27: 1556–1566

177 Bloom JC, Brandt JT (2008) Toxic responses of the blood. In: C Klaassen (ed.): *Casarett & Doull's Toxicology: The Basic Sciences of Poisons*, 7th edn. McGraw-Hill, New York, 455–484

178 Rokushima M, Omi K, Araki A, Kyokawa Y, Furukawa N, Itoh F, Imura K, Takeuchi K, Okada M, Kato I, Ishizaki J 007) A toxicogenomic approach revealed hepatic gene expression changes mechanistically linked to drug-induced hemolytic anemia. *Toxicol Sci* 95: 474–484

179 Rokushima M, Omi K, Imura K, Araki A, Furukawa N, Itoh F, Miyazaki M, Yamamoto J, Rokushima M, Okada M et al (2007) Toxicogenomics of drug-induced hemolytic anemia by analyzing gene expression profiles in the spleen. *Toxicol Sci* 100: 290–302

180 Forrest MS, Lan Q, Hubbard AE, Zhang L, Vermeulen R, Zhao X, Li G, Wu YY, Shen M, Yin S et al (2005) Discovery of novel biomarkers by microarray analysis of peripheral blood mononuclear cell gene expression in benzene-exposed workers. *Environ Health Perspect* 113: 801–807

181 Smith MT, Vermeulen R, Li G, Zhang L, Lan Q, Hubbard AE, Forrest MS, McHale C, Zhao X, Gunn L et al (2005) Use of 'Omic' technologies to study humans exposed to benzene. *Chem Biol Interact* 153–154: 123–127

182 Faiola B, Fuller ES, Wong VA, Pluta L, Abernethy DJ, Rose J, Recio L (2004) Exposure of hematopoietic stem cells to benzene or 1,4-benzoquinone induces gender-specific gene expression. *Stem Cells* 22: 750–758

183 Faiola B, Fuller ES, Wong VA, Recio L (2004) Gene expression profile in bone marrow and hematopoietic stem cells in mice exposed to inhaled benzene. *Mutat Res* 549: 195–212

184 Yoon BI, Li GX, Kitada K, Kawasaki Y, Igarashi K, Kodama Y, Inoue T, Kobayashi K, Kanno J, Kim DY et al (2003) Mechanisms of benzene-induced hematotoxicity and leukemogenicity: cDNA microarray analyses using mouse bone marrow tissue. *Environ Health Perspect* 111: 1411–1420

185 Hirabayashi Y, Yoon BI, Li GX, Kanno J, Inoue T (2004) Mechanism of benzene-induced hema-

totoxicity and leukemogenicity: Current review with implication of microarray analyses. *Toxicol Pathol* 32 Suppl 2: 12–16

186 Kaminski NE, Faubert Kaplan BL, Holsapple MP (2008) Toxic responses of the immune system. In: C Klaassen (ed.): *Casarett & Doull's Toxicology: The Basic Sciences of Poisons*, 7th edn. McGraw-Hill, New York, 485–556

187 Luebke RW, Holsapple MP, Ladics GS, Luster MI, Selgrade M, Smialowicz RJ, Woolhiser MR, Germolec DR (2006) Immunotoxicogenomics: The potential of genomics technology in the immunotoxicity risk assessment process. *Toxicol Sci* 94: 22–27

188 Baken KA, Vandebriel RJ, Pennings JL, Kleinjans JC, van Loveren H (2007) Toxicogenomics in the assessment of immunotoxicity. *Methods* 41: 132–141

189 Ezendam J, Staedtler F, Pennings J, Vandebriel RJ, Pieters R, Harleman JH, Vos JG (2004) Toxicogenomics of subchronic hexachlorobenzene exposure in Brown Norway rats. *Environ Health Perspect* 112: 782–791

190 Inadera H (2006) The immune system as a target for environmental chemicals: Xenoestrogens and other compounds. *Toxicol Lett* 164: 191–206

191 Zeytun A, McKallip RJ, Fisher M, Camacho I, Nagarkatti M, Nagarkatti PS (2002) Analysis of 2,3,7,8-tetrachlorodibenzo-*p*-dioxin-induced gene expression profile *in vivo* using pathway-specific cDNA arrays. *Toxicology* 178: 241–260

192 Nagai H, Takei T, Tohyama C, Kubo M, Abe R, Nohara K (2005) Search for the target genes involved in the suppression of antibody production by TCDD in C57BL/6 mice. *Int Immunopharmacol* 5: 331–343

193 He B, Munson AE, Meade BJ (2001) Analysis of gene expression induced by irritant and sensitizing chemicals using oligonucleotide arrays. *Int Immunopharmacol* 1: 867–879

194 Betts CJ, Moggs JG, Caddick HT, Cumberbatch M, Orphanides G, Dearman RJ, Ryan CA, Hulette BC, Frank Gerberick G, Kimber I (2003) Assessment of glycosylation-dependent cell adhesion molecule 1 as a correlate of allergen-stimulated lymph node activation. *Toxicology* 185: 103–117

195 Ryan CA, Gildea LA, Hulette BC, Dearman RJ, Kimber I, Gerberick GF (2004) Gene expression changes in peripheral blood-derived dendritic cells following exposure to a contact allergen. *Toxicol Lett* 150: 301–316

196 Gildea LA, Ryan CA, Foertsch LM, Kennedy JM, Dearman RJ, Kimber I, Gerberick GF (2006) Identification of gene expression changes induced by chemical allergens in dendritic cells: Opportunities for skin sensitization testing. *J Invest Dermatol* 126: 1813–1822

197 Hansen MB, Skov L, Menne T, Olsen J (2005) Gene transcripts as potential diagnostic markers for allergic contact dermatitis. *Contact Dermatitis* 53: 100–106

198 Kinser S, Jia Q, Li M, Laughter A, Cornwell P, Corton JC, Pestka J (2004) Gene expression profiling in spleens of deoxynivalenol-exposed mice: Immediate early genes as primary targets. *J Toxicol Environ Health A* 67: 1423–1441

199 Royaee AR, Hammamieh R, Mendis C, Das R, Jett M, DC HY (2006) Induction of immunomodulator transcriptional responses by cholera toxin. *Mol Immunol* 43: 1020–1028

200 Royaee AR, Jong L, Mendis C, Das R, Jett M, Yang DC (2006) Cholera toxin induced novel genes in human lymphocytes and monocytes. *Mol Immunol* 43: 1267–1274

201 Royaee AR, Mendis C, Das R, Jett M, Yang DC (2006) Cholera toxin induced gene expression alterations. *Mol Immunol* 43: 702–709

202 Ueno K, Yamaura K, Nakamura T, Satoh T, Yano S (2000) Acetaminophen-induced immunosuppression associated with hepatotoxicity in mice. *Res Commun Mol Pathol Pharmacol* 108: 237–251

203 Yamaura K, Ogawa K, Yonekawa T, Nakamura T, Yano S, Ueno K (2002) Inhibition of the antibody production by acetaminophen independent of liver injury in mice. *Biol Pharm Bull* 25: 201–205

204 Baken KA, Pennings JL, Jonker MJ, Schaap MM, de Vries A, van Steeg H, Breit TM, van Loveren H (2008) Overlapping gene expression profiles of model compounds provide opportunities for immunotoxicity screening. *Toxicol Appl Pharmacol* 226: 46–59

205 Gant TW (2007) Novel and future applications of microarrays in toxicological research. *Expert Opin Drug Metab Toxicol* 3: 599–608

206 Barrett MT, Scheffer A, Ben-Dor A, Sampas N, Lipson D, Kincaid R, Tsang P, Curry B, Baird K, Meltzer PS et al (2004) Comparative genomic hybridization using oligonucleotide microarrays and total genomic DNA. *Proc Natl Acad Sci USA* 101: 17765–17770

207 Pinkel D, Albertson DG (2005) Comparative genomic hybridization. *Annu Rev Genomics Hum Genet* 6: 331–354

208 Bastian BC, Olshen AB, LeBoit PE, Pinkel D (2003) Classifying melanocytic tumors based on DNA copy number changes. *Am J Pathol* 163: 1765–1770

209 Redon R, Ishikawa S, Fitch KR, Feuk L, Perry GH, Andrews TD, Fiegler H, Shapero MH, Carson AR, Chen W et al (2006) Global variation in copy number in the human genome. *Nature* 444: 444–454

210 Watson RE, Goodman JI (2002) Epigenetics and DNA methylation come of age in toxicology. *Toxicol Sci* 67: 11–16

211 van Steensel B (2005) Mapping of genetic and epigenetic regulatory networks using microarrays. *Nat Genet* 37 Suppl: S18–24

212 Barber RC, Dubrova YE (2006) The offspring of irradiated parents, are they stable? *Mutat Res* 598: 50–60

213 Vilarino-Guell C, Smith AG, Dubrova YE (2003) Germline mutation induction at mouse repeat DNA loci by chemical mutagens. *Mutat Res* 526: 63–73

214 Barber RC, Hickenbotham P, Hatch T, Kelly D, Topchiy N, Almeida GM, Jones GD, Johnson GE, Parry JM, Rothkamm K, Dubrova YE (2006) Radiation-induced transgenerational alterations in genome stability and DNA damage. *Oncogene* 25: 7336–7342

215 Li S, Washburn KA, Moore R, Uno T, Teng C, Newbold RR, McLachlan JA, Negishi M (1997) Developmental exposure to diethylstilbestrol elicits demethylation of estrogen-responsive lactoferrin gene in mouse uterus. *Cancer Res* 57: 4356–4359

216 Li S, Hansman R, Newbold R, Davis B, McLachlan JA, Barrett JC (2003) Neonatal diethylstilbestrol exposure induces persistent elevation of c-fos expression and hypomethylation in its exon-4 in mouse uterus. *Mol Carcinog* 38: 78–84

217 Pollack JR, Iyer VR (2002) Characterizing the physical genome. *Nat Genet* 32 Suppl: 515–521

218 Wang X, Tomso DJ, Chorley BN, Cho HY, Cheung VG, Kleeberger SR, Bell DA (2007) Identification of polymorphic antioxidant response elements in the human genome. *Hum Mol Genet* 16: 1188–1200

219 Kim VN, Nam JW (2006) Genomics of microRNA. *Trends Genet* 22: 165–173

220 Lee Y, Kim M, Han J, Yeom KH, Lee S, Baek SH, Kim VN (2004) MicroRNA genes are transcribed by RNA polymerase II. *EMBO J* 23: 4051–4060

221 Castoldi M, Schmidt S, Benes V, Noerholm M, Kulozik AE, Hentze MW, Muckenthaler MU (2006) A sensitive array for microRNA expression profiling (miChip) based on locked nucleic acids (LNA). *RNA* 12: 913–920

Molecular, Clinical and Environmental Toxicology. Volume 1: Molecular Toxicology
Edited by A. Luch

The role of toxicoproteomics in assessing organ specific toxicity

B. Alex Merrick[1] and Frank A. Witzmann[2]

[1] *National Institute of Environmental Health Sciences (NIEHS), Research Triangle Park, NC, USA*
[2] *Indiana University School of Medicine, Indianapolis, IN, USA*

Abstract. Aims of this chapter on the role of toxicoproteomics in assessing organ-specific toxicity are to define the field of toxicoproteomics, describe its development among global technologies, and show potential uses in experimental toxicological research, preclinical testing and mechanistic biological research. Disciplines within proteomics deployed in preclinical research are described as Tier I analysis, involving global protein mapping and protein profiling for differential expression, and Tier II proteomic analysis, including global methods for description of function, structure, interactions and post-translational modification of proteins. Proteomic platforms used in toxicoproteomics research are briefly reviewed. Preclinical toxicoproteomic studies with model liver and kidney toxicants are critically assessed for their contributions toward understanding pathophysiology and in biomarker discovery. Toxicoproteomics research conducted in other organs and tissues are briefly discussed as well. The final section suggests several key developments involving new approaches and research focus areas for the field of toxicoproteomics as a new tool for toxicological pathology.

Introduction

Toxicoproteomics applies global protein measurement technologies to toxicology testing and research. Aims of the field are the discovery of mechanisms governing key proteins in critical biological pathways creating adverse drug effects, development of biomarkers, and eventual prediction of toxicity based upon pharmacogenomic knowledge [1–4].

An increasing number of proteomic applications to many established scientific disciplines have generated great interest and enthusiasm in basic biology and medicine, as well as toxicology. There are well over 10 000 publications relating to some aspects or applications of proteomics in the biosciences. However, the numbers of published proteomic studies is quite limited in reporting primary data for drug-mediated adverse reactions and biochemical toxicities that lead to undesirable phenotypes [5–7]. Toxicoproteomics was initially developed under the auspices of toxicogenomics [8] and proteomics [9], but it has emerged as its own discipline. Toxicoproteomics is defined by goals of furthering mechanistic understanding of how specific exposures alter protein expression, protein behavior and response to cause injury and disease, but has also been greatly influenced by a growing body of research focusing upon key organs such as liver and kidney. The field has been augmented by tools from proteomics, bioinformatics and other enabling high throughput technologies. Interestingly, it might be argued that the overt pursuit of defining biomarkers as a major objective in toxicoproteomics research may not be appropriate or right. Biomarkers should be a natural progression of excellence in research from elucidating toxic mechanisms or modes of chemical actions in response to acute exposure to toxicants or during the long-term development of diseases caused or influenced by these exposures. Compared with such an important mission, the identification of biomarkers might be or should be a comparatively smaller part of the whole picture of toxicoproteomics. However, as a motivating factor, biomarker and toxicity signature discovery is very high in the minds of those who use proteomics in toxicology [10, 11]. Major drivers in toxicoproteomics are the commercial need to discover markers associated with drug exposure, efficacy or toxicity in the pharmaceutical arena, and also the urgencies of environmental hazard evaluation for the protection of public health. Finally, an overarching principle among all discovery technologies is that eventual placement of protein changes within biochemical pathways and processes will result from a mechanistic understanding of larger biochemical systems and signaling networks. Systems biology has come to represent this wider integration of functional genomics disciplines such as transcriptomics, proteomics, interactomics and metabolomics among organisms [12, 13].

Issues for toxicoproteomics and toxicogenomics in pharmaceutical data submission have been recently reviewed regarding non-clinical safety testing to regulatory agencies [14]. Several issues are still in development for data submission of genomics and proteomics studies to regulatory agencies. These would include data quality standards, wide differences in platforms and data formats, accepted criteria for data validation, relationships to traditional toxicological endpoints, added-value to established biochemical and molecular methods, animal-to-human extrapolation, mechanism of action, impact upon the NOAEL ("no observed adverse effect level"), early compared to adaptive or non-pharmacological responses, limits of bioinformatics algorithms, tools and available databases, and defined metrics of how and when genomic and proteomic data would influence regulatory decisions.

Toxicoproteomics and metabonomics have sometimes been called to task for their seemingly meager contributions to biomarker discovery compared to more well-established clinical chemistry and histopathology indicators. For example, a review of 13 toxicoproteomic and metabonomic studies with various nephrotoxic agents examined them for their respective abilities to determine specificity and sensitivity of renal toxicity. The review concluded that proteomics (and metabonomics) data compared very poorly with traditional methods of blood and urine chemistries and histopathology without significant improvements [15]. However, it is the potential for discovery and new insights into pathobiology and therapeutics that fuels interest in Omics technologies. In defense of these new fields, the same criticism could be levied upon pathology, histology and clinical chemistry for the many years and countless studies required for them to develop from the beginning of the twentieth century to present day capabilities. Although the discovery potential is high for new biomarkers from toxicoproteomics studies, the strategies for conducting proteomic analysis and using such data in drug development, preclinical safety and regulatory submission are far from standardized. The complexity of protein expression, multiple technology platforms, and emerging technical standards are major challenges for continued growth of the field. Researchers are finding that no one platform is best suited for toxicoproteomics research, and that more than one platform may be required for suitable proteome coverage.

If a primary goal for toxicoproteomics is to translate identified protein changes into improved biomarkers and signatures of chemical toxicity [7], then care must be exercised in designating any protein change observed during toxicoproteomic studies as "new biomarkers". Part of the challenge arises from an imprecise meaning of the term 'biomarker', accounting for its wide variation in use (and misuse) in scientific and regulatory communities [16, 17]. At a biochemical and molecular level, biomarkers can be narrowed down to "singular biological measures with reproducible evidence of a clear association with health, disease, adverse effect or toxicity". This is a necessarily limited definition for quantitative biochemical or molecular measures. Historical and more current examples of biomarkers are the detection of a single protein such as C-reactive protein in cardiovascular disease [18], an enzyme activity like alanine aminotransferase activity in liver injury [19], gene transcription products such as Her2/neu [20] in breast malignancies, gene mutations/polymorphisms like slow acetylators that affect xenobiotic metabolism [21] or small molecules/metabolites such as serum glucose, insulin and urinary ketone bodies in pathological or drug-induced diabetes. Many subcategories of biomarkers are also in popular usage, including biological, surrogate, prognostic, diagnostic and bridging biomarkers [16, 22]. Importantly, a major development of the large datasets derived from Omics technologies is the possibility of greater molecular topography compared to a singular biomarker. One of the major tenets of toxicoproteomics and other Omics analyses are that specific patterns of protein changes can comprise a consistent "signature" of toxicity [10] or a "combinatorial biomarker" [23] that is robust enough to be observed in spite of variations in biology, experimental design, or technology platforms. This is a critical assumption first, because there is great potential for including nonspecific or indirect protein changes in such a signature, and, secondly, because of the inherent challenges in establishing a causal linkage of multiple protein changes to a toxic or adverse phenotype. Specific descriptions of such toxicity signatures and biomarkers are at an early stage in the field of toxicoproteomics [6, 10].

Disciplines and platforms for toxicoproteomics research

Proteomics in a global protein analysis mode generally links separation and identification technologies to create a protein profile or differential protein display. Although the focus of proteomics has been grouped in various ways, Figure 1 shows representative subdisciplines of proteomic analysis that provide a means to categorize toxicoproteomics research.

Four factors often shape the manner in which researchers pursue their activities in toxicoproteomics; these include (1) the complex nature of proteins; (2) the particular portion of the proteome targeted for study; (3) the integrative relationship of toxicoproteomics studies with other Omics technologies; and (4) the driving forces behind specific toxicoproteomics projects. Each of these four factors should be considered. First, a primary objective in proteomics is the isolation and identification of individual proteins from complex biological matrixes. In toxicoproteomics analysis, the Tier I of proteomic analysis is to determine individual protein identities (fingerprint, amino acid sequence), their relative (or absolute) quantities and their spatial location within cell(s), tissues and biofluids of interest. Tier II of analysis globally screens for protein functions, protein interactions, three-dimensional structure and specific post-translational modifications. Tiers I and II of proteomic analysis encompass the seven intrinsic attributes of proteins that play a role in toxicoproteomic analysis [24] as shown in Figure 1. Proteomic platforms (Fig. 2) each vary greatly in their respective abilities to deliver data on all protein attributes simultaneously during one analysis.

Proteome mapping is the most descriptive of proteomic inquiries and usually focuses upon identifying all proteins in the sample or at a cellular location at hand. Profiling experiments necessarily require quantitation (relative to control, or absolute) to be comparative among samples. Implicit in protein mapping and profiling are considerations about the spatial "origin" of the sample. Often, sample origins are the same in profiling experiments for comparability; for example, serums are most comparable to serums, livers to livers and so on. Structural proteomics is usually defined as high-throughput determination of protein structures in three-dimensional space and is often determined by X-ray crystallography and NMR spectroscopy. This definition has been expanded in Figure 1 to included spatial location of proteins within the organism rather than continue to divide proteomic fields by specific levels of protein organization that might range from subcellular, cellular, organ, tissue, organism to species proteomics. A second factor to consider in toxicoproteomics is that the proteomes of most cells, tissues and organs are so vast that, unlike whole genome queries, proteomes cannot be completely analyzed by existing proteomic platforms. By default, toxicoproteomic studies most often analyze only a portion of the proteome contained in typical biological samples. A frequent strategy to broaden protein coverage is to take steps prior to analysis to reduce sample complexity (analyze a portion of the proteome or 'subproteome') by such procedures as subcellular fractionation, affinity or adsorptive chromatography or electrophoretic separation. Third, toxicoproteomic analysis may be conducted as an independent activity or alternately as a component of a large, formalized gene expression project for which the study design, type of experimental subjects and the availability or amount of biological specimens may

Tier I Proteomics

Protein Mapping

mhrnfrkwif yvflcfgvly vklgalssvv
alganiicnk ipglaprqra icqsrpdaii
vigegaqmgi necqyqfrfg rwncsalgek
tvfgqelrvg sreaaftyai taagvahavt
aacsqgnlsn cgcdrekqgy ynqaegwkwg
gcsadvrygi dfsrrfvdar eikknarrlm
nlhnneagrk vledrmqlec kchgvsgsct
tktcwttlpk frevghllke kynaavqvev

Protein Profiling

Differential Protein Expression

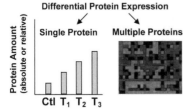

Single Protein   Multiple Proteins

Ctl T_1 T_2 T_3

Cell or Tissue Location

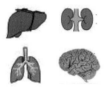

Tier II Proteomics

Posttranslational Modification

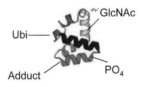

GlcNAc

Ubi—

Adduct PO_4

Structure

Protein Function

$$S \xrightarrow{E} P$$

Protein Interactions

Protein — Protein   DNA — Protein

Figure 1. Disciplines of toxicoproteomics to study effects of drug, chemical, disease or environmental stressor exposure. Proteomic analysis attempts to describe various protein attributes in a global manner. Tier I proteomic analysis involves protein mapping or profiling. Protein mapping for identification reflects the property of primary amino acid sequence; quantitations of all proteins from a defined space are inherent in protein profiling; and isolation or enrichment of proteins from a particular spatial location within cells or tissues help to characterize the organism's phenotype. Tier II proteomic analysis involves global determination of individual protein attributes (behavior and structure) regarding their three-dimensional structures, post-translational modifications, functional capabilities, and interactions and complexation with other biomolecules. In protein mapping, the underlined portions of an individual protein represent tryptic peptides for amino acid sequencing for identification by matrix-assisted laser desorption/ionization (MALDI) or tandem mass spectrometry (MS/MS). In protein profiling, changing levels of individual proteins (bar graph) or groups of proteins (cluster analysis) are measured over treatment (T1, T2, T3) or time. A proteome of interest occupies a specific spatial location for analysis and may comprise a subcellular organelle, tissue or organ. Protein structure may represent the β-pleated sheet or α-helix to form tertiary or quaternary protein folding. Specific post-translational moieties, such as ubiquitin (Ubi), phosphorylation (PO_4), glycosylation (GlcNAc), or chemical adduct, are covalently bound to specific amino acid residues on the protein that impart important functional and biophysical properties. Protein function may be: enzymatic, such as enzymatic (E) conversion of substrate (S) to product (P); structural, providing form and shape; translocational, across cells or tissues; signaling and transduction; or many other utilities to be carried out within cells and tissues. Protein interactions may occur between other proteins, between DNA and proteins, or between other biomolecules.

Proteomic Platforms

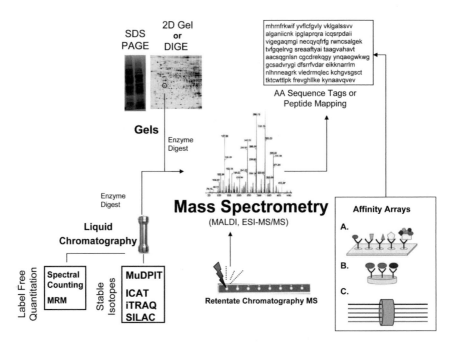

Figure 2. Proteomic platforms for toxicoproteomics studies. Proteomic platforms represent strategies for global separation and identification of proteins. Separations are generally accomplished by gel electrophoresis in toxicoproteomic studies, although more recent studies incorporate liquid chromatography (LC)-based platforms, such as linear column gradients or multidimensional chromatography (MuDPIT). Use of stable isotopes greatly facilitates protein quantitation (ICAT, isotope coded affinity tags; iTRAQ, isobaric tags for relative and absolute quantitation; SILAC, stable isotope labeling by amino acids in cell culture). Label-free methods such as multiple reaction monitoring (MRM) or spectral counting provide protein quantitation without stable isotopes (see text for further explanation). Mass spectrometry (MS) is the primary tool as a means of protein identification in proteomic analysis. Identification occurs by peptide mapping or amino acid (AA) sequencing. Retentate chromatography MS has been used for rapid profiling of biofluid samples using chemically reactive surfaces for separation and MALDI for generating protein mass spectra (i.e., SELDI technology). However, alternatives to MS-based identification in proteomic analysis exist in platforms based upon affinity arrays such as (A) antibody arrays, (B) antibody multiplexing and (C) fluorescently tagged antibody bound to bead suspensions such as the Luminex technology.

greatly impact sample preparation procedures and proteomic platform selection [25]. Fourth, the forces and individuals driving toxicoproteomics studies such as drug discovery, biophysical and chemical analyses, safety assessment, drug efficacy, absorption-distribution-metabolism-excretion (ADME) properties and clinical trials will greatly influence the study design, analysis and, importantly, biological interpretation of toxicoproteomic data.

The complexity of a "proteome" contained in a biological sample presents numerous challenges for comprehensively describing the seven attributes of protein expression during any single proteomic analysis [7]. The primary aims of proteomic analysis are to (1) achieve maximal coverage of the proteome (i.e., Tier I analysis) in each sample; (2) complete analysis at high throughput; (3) produce an accurate quantitative protein measurement; (4) deliver data and interpretable results in a timely period; and (5) use of discovery-oriented, open platforms.

All proteomic platforms typically share two common capabilities: a means of global separation and a technology for identification of proteins. Identification usually means assignment to international gene (i.e., NCBI) or protein (i.e., Uniprot or Swiss-Prot) identification numbers. The following proteomic platforms represent a brief description of the principal technologies used for separating and identifying proteins during toxicoproteomic studies as summarized in Figure 2.

Gel-based proteomics: Two-dimensional or difference gel electrophoresis with mass spectrometry

Two-dimensional (2D) gel electrophoresis systems have been combined with mass spectrometry (MS) in an established and adaptable platform since 1975, and it is the most commonly used proteomic platform to separate and comparatively quantitate protein samples [26]. Current state-of-the-art 2D gels use immobilized pH gradient (IPG) gels to separate proteins first by charge and then subsequently by mass using SDS-polyacrylamide gel electrophoresis (PAGE) for effective separation of complex protein samples. Proteins are separated to sufficiently homogeneity on 2D gels to permit MS identification. Typical IPG gels of 18–24 cm fitted with similarly sized SDS-PAGE gels can separate between 2000 and 3000 proteins. Each spot does not represent a unique protein (i.e., gene product) but often occur as post-translationally modified forms of the same protein. Fluorescent staining is often the most sensitive means of protein detection (nano- to microgram range). After electronic alignment (registration) of stained proteins in 2D gels by image analysis software, intensities of identical protein spots are compared among treatment groups and a ratio (fold change) is calculated for each protein using specialized software. In 2D-difference gel electrophoresis (DIGE)-MS, protein samples to be compared are labeled with either Cy2-, Cy3- and Cy5-based linkers. Labeled samples are mixed together and electrophoresed on the same gel. This procedure minimizes image analysis errors from trying to electronically register different gels since each dye (sample) is read at a different wavelength on the same gel [27]. Up to three or four samples can be run on the same 2D gel.

The combination of 2D-gel separation of proteins with MS provides a ready means of protein identification after protein excision, enzymatic digestion and MS analysis. The 2D-MS platform forms a versatile and discovery-oriented standardized approach for use in toxicoproteomic studies [28]. A downside to this platform is the limited coverage of a proteome that can be realized on 2D gels by even the most sensitive fluorescent stains.

One-dimensional gel-based proteomics platforms, 1D-gel liquid chromato-graphy-tandem MS (LC-MS/MS), may also be extremely effective for protein separation and identification using SDS-PAGE only (i.e., mass separation) with specially pre-processed samples such as immuno-depleted plasma [29] or cell secretomes [30]. Such pre-processing sufficiently reduces the original pro-tein complexity to allow small amounts of sample protein (micrograms) or serum (microliters) to be resolved to near protein homogeneity in stained pro-tein bands. Bands are enzymatically digested to obtain diagnostic peptides for protein identification after amino acid sequencing by LC-MS/MS.

Multi-dimensional, quantitative LC-MS/MS: MuDPIT, ICAT, iTRAQ and SILAC, and label-free quantitation

Multidimensional liquid chromatography (LC-LC) is used to separate protein digests (nano- to mirograms peptides) by charge (strong anion exchange) and hydrophobicity (C18) immediately prior to entry into a tandem mass spec-trometer for protein identification [31]. A premier representative of LC-MS/MS proteomics is the multidimensional protein identification technology or "MuDPIT" platform. This approach has also been called "shotgun pro-teomics" since entire protein lysates are trypsin-digested into thousands of peptide fragments without the need for any fractionation or processing prior to LC-LC separation and MS/MS identification. Advantages of this newer plat-form are the potential for detection and identification of low abundance pro-teins that may not be observed in gel-based protein separations. However, the MuDPIT platform is only semi-quantitative. The platform is very effective in proteomics mapping and discovery studies and should find great utility in tox-icoproteomics.

Other variations on the LC-MS/MS approach, closely linking LC separation to MS/MS instruments, have incorporated isotopic labeling strategies for pro-tein quantitation and in-depth proteomic profiling of samples. Examples of such platforms are isotope coded affinity tags (ICAT), isobaric tag for relative and absolute quantitation (iTRAQ) and stable isotope labeling with amino acids in cell culture (SILAC). These methods use "light" and "heavy" forms of isotopes in linkers that bind to functional groups of proteins (i.e., cysteines or amino groups) in lysates. SILAC and iTRAQ are particularly effective for metabolic incorporation of "light" and "heavy" forms of amino acids (i.e., $^1H{:}^2H/^{12}C{:}^{13}C/^{14}N{:}^{15}N$) into cellular proteins during cell culture incubations. Although sample throughput is slow and analysis time is lengthy, the protein coverage has been greatly expanded with the development of these new multi-dimensional proteomic platforms. Careful sample, dose and time selection to a few samples appears to be a successful strategy in achieving the most value from multi-dimensional, quantitative LC-MS/MS platforms.

Mass spectral data derived from shotgun proteomics approaches (e.g., 1D-LC-MS/MS) can also be used for relative or absolute protein quantitation and sample comparison. This can be accomplished without stable-isotope labeling or tagging, in any of several ways that include integration of ion chromatogram intensities [32, 33], spectral counting [34–37], or selected reaction monitoring (SRM) [38, 39]. Comparison of label-free techniques suggests these approaches, like isotope labeling, have their own strengths and weaknesses [40].

Higgs et al. [33] developed a comprehensive, fully automated, and label-free approach to relative protein quantification using data from LC-MS/MS analysis of proteolytic protein digests. The platform includes de-noising, mass and charge state estimation, chromatographic alignment, and peptide quantification *via* integration of extracted ion chromatograms. One important advantage of this technique lies in its ability to identify large numbers of proteins and provide abundance information for all of them, in a statistically robust manner. This approach has been applied to the development of protein biomarkers of cisplatin resistance in human ovarian cancer [41], and recently to evaluate the effect in the rat nucleus accumbens of ethanol self-administration in the posterior ventral tegmental area of the brain [42] where 1120 proteins where identified and comparatively quantified. The same technique was used to assess the toxic effect of JP-8 jet fuel exposure on rat alveolar type II epithelial cells, at sublethal levels that are occupationally relevant [43]. In that study 1135 unique proteins were identified with high confidence and quantified. *Post hoc* bioinformatic analysis of differentially expressed proteins suggested that the decreased cell viability of jet fuel-exposed cells corresponded to significant down-regulation of proteins involved in all manners of cell activity, but predominantly by declines in translational and protein synthetic machinery.

Spectral counting is a method for relative protein quantitation in MS-based experiments, and is based on the observation that the total number of detected peptides identifying a specific protein correlates strongly with the abundance of that protein. Simply put, one counts the total number of proteolytic peptide ions identified by MS/MS for a specific protein and on that basis relative comparisons can be made between samples. It is assumed that the more abundant a particular peptide, the more likely it will be selected by the mass spectrometer's operating software for MS/MS analysis, and the more it will be counted. Although this useful approach has yet to be exploited by toxicologists, it has found successful application in the analysis of protein complexes that yield comparatively small-scale datasets [44, 45] and in studying the effects of lipopolysaccharide treatment in initiating the cellular immune response [46].

SRM and its plural multiple reaction monitoring [47] enable "hypothesis-driven" or "candidate-based" analyses of protein expression, in contrast to the "discovery" orientation of most shotgun proteomics efforts. In SRM, only the current of ions with preselected mass to charge ratio (*m/z*) values are monitored. This improves detection sensitivity by decreasing the detector's response to other ions, thus decreasing the background noise. Inclusion of internal standards, specific isotopically ($^{13}C/^{15}N$) labeled peptides that are otherwise identical to the candidate peptides, enables absolute quantitation [38]. Although its application in toxicoproteomics studies is just now emerging,

SRM has been used toxicologically to detect and quantify the acrolein metabolite (3-hydroxypropyl)mercapturic acid in urine as a biomarker of cigarette-smoke-induced disease [48] and its potential utility in hypothesis-driven toxicoproteomics applications is outstanding.

Retentate chromatography MS: SELDI

Retentate chromatography-MS (RC-MS) is a high-throughput proteomic platform that creates a laser-based mass spectrum from a chemically absorptive surface. The principle of this approach is the adsorptive retention (pico- to nanograms protein) of a subset of sample proteins on a thin chromatographic support (i.e., hydrophobic, normal phase, weak cation exchange, strong anion exchange or immobilized metal affinity supports). The absorptive surfaces are placed on thin metal chips, which can be inserted into a specially modified matrix-assisted laser desorption/ionization (MALDI)-type mass spectrometer. The laser rapidly desorbs proteins from each sample on a metal chip to create a mass spectrum profile.

RC-MS can be performed upon any protein sample but thus far this platform has found greatest utility in the analysis of serum and plasma for disease biomarker discovery [6, 49]. The lead commercial platform of RC-MS proteomic platform is the surface-enhanced laser desorption/ionization time-of-flight (SELDI-TOF)-MS instrument [50]. Analysis of samples is relatively rapid. Only a few microliters of biofluid sample are necessary and hundreds of samples can be screened in a few days. Downsides are that only a fraction of the proteome can be analyzed (i.e., that adsorbs to the particular chemical surface), there are sample reproducibility issues, and protein identification of peaks is not readily achieved without additional analysis [51]. However, the RC-MS approach fits many problem areas as a proteomics discovery tool for defining drug or chemical exposure when rapid screening is needed for hundreds or thousands of pre-clinical or clinical samples.

Protein capture arrays: Antibody arrays

Protein capture arrays (any mass parallel array of proteins, peptides, capture ligands or adsorptive surfaces for protein analysis) represent a promising new proteomic tool that closely emulates the design for parallel analysis of DNA microarray technology [52]. Many different types of capture molecules can be arrayed (recombinant proteins, aptamers, peptides, drug libraries) but the most prevalent are antibody arrays that directly separate proteins from each other by affinity binding to specific protein targets. Generally, commercial antibody array platforms have widely varying sensitivities (pico- to microgram peptides) that fall into three classes based on targeted proteins: cytokine/chemokine arrays, cellular function protein arrays, and cell signaling arrays. However, antibody arrays are not presently available for any given cell type, biofluid or species. This platform provides a rapid screen for limited sets of proteins that may fit some applications in toxicoproteomics.

Toxicoproteomics studies in liver injury

The liver is the major organ for biotransformation and elimination of pharmaceutics from the body [53]. As a result, initial toxicoproteomics studies sought insights into drug-induced liver injury using rodent models of toxicity. Animal models of liver toxicity are often selected for prevalence of one phenotype such as necrosis, hepatitis, cholestasis, steatosis, fibrosis, cirrhosis or malignancy, but in fact many of these molecular processes occur simultaneously [54]. The removal from the marketplace of several widely prescribed drugs due to hepatotoxicity has attracted considerable attention that highlight underlying susceptibility factors to drug-induced injury including age, sex, drug-drug interactions, and genetic polymorphism in metabolic pathways involved in activation or disposition of therapeutic drugs [55]. Reactive intermediates produced during metabolism can be toxic or some compounds may dysregulate critical biochemical pathways or functions of the liver [53].

Chemical and drug-induced hepatic necroses produce reproducible phenotypes representative of compound families and common metabolic activation pathways in preclinical species. For example, a comprehensive determination of bioactivation pathways of organic functional groups on xenobiotics and pharmaceutical reagents has been extensively cataloged in an effort to guide drug design and avoid toxicity [56]. Such considerations provide a rationale for exploring and testing the capabilities of emerging Omics technologies like proteomics and transcriptomics upon acute hepatic injury. Which agents might be worthwhile for toxicoproteomics studies? A recent toxicogenomics study for classifying hepatotoxicants evaluated a representative list of 25 well-known model compounds or substances showing hepatotoxicity during testing [57]. The aim of this preclinical research report was to determine if biological samples from rats treated with various compounds could be classified based on gene expression profiles. Such model agents causing acute hepatonecrotic injury included acetaminophen, bromobenzene, carbon tetrachloride, hydrazine and others. Hepatic gene expression profiles were analyzed using a supervised learning method (support vector machines; SVMs) to generate classification rules. The SVM method was combined with recursive feature elimination to improve classification performance. The goal was to identify a compact subset of probe sets (transcripts) with potential use as biomarkers. DNA microarray data have been generated for each substance in this study [57]. Their list of representative hepatotoxic agents for preclinical testing served as a basis for examining the literature for corresponding toxicoproteomics studies.

Table 1 summarizes available primary data from toxicoproteomics studies. Generally, these studies have been conducted upon representative, model liver

Table 1. Toxicoproteomic analysis of liver and kidney toxicants.[a]

Reference	Chemical	Platform	Tissue	Results
[65]	APAP, AMAP	2D-MS	mouse liver	35 proteins IDed; altered proteins are known APAP adducts
[66]	APAP	2D-DIGE-MS	mouse liver	optimization study for 2D DIGE separation; several proteins altered with APAP
[67]	APAP	2D-DIGE-MS	mouse liver	DNA array/proteomics; ↓ Hsps; protein changes in 15 min
[68]	APAP, others	NEPHGE 2D-MS	rat liver, HepG2	113 proteins IDed in rat liver and 194 proteins in HepG2 cells; catalase, carbamoylphosphate synthetase-1, aldo-keto-reductase, altered
[72]	APAP	ICAT	mouse hepatocytes	optimization study for ICAT analysis
[69]	APAP, CCl₄, amiodarone	2D-MS	rat hepatocytes	31 proteins IDed of 113 proteins altered by APAP
[71]	APAP	ICAT	mouse liver	1632 protein IDs with 247 proteins diffentially expressed; ↑ hepatoprotective proteins in SJL resistant strain; ↑ loss mitochondrial proteins from C57BL/6 susceptible strain
[93]	APAP, ANIT, WY14643, PB	2D-DIGE-MS	rat liver, serum	liver (124), serum (101) proteomics (proteins IDed). 5 serum biomarkers liver toxicity: PNP, MDH, Gc, PON1, RBP
[70]	APAP, TCN, amiodarone	2D-MS	rat hepatocytes	for APAP, ↑15 and ↓25 protein IDed; ↓GPx; ↑PRx1,2
[137]	APAP	2D-MS	rat liver, serum	cluster analysis of transcriptomic, proteomic, clinical chemistry data
[138]	CCl₄	2D-MS	rat liver stellate cells	150 protein IDs, ↑ calcyclin, calgizzarin, galectin-1
[74]	CCl₄	2D-MS	rat liver	30 proteins IDed; proliferation and apoptosis proteins altered
[76]	BB	2D-MS	rat liver	DNA array/proteomics; IDed proteins infer degradation, oxidative stress from toxicity

(continued on next page)

Table 1. (continued)

Reference	Chemical	Platform	Tissue	Results
[30]	aflatoxin B_1	1D-gel LC-MS/MS	rat hepatocyte	rat secretome of 200 proteins IDed; ↓ α1-antitrypsin and 2-macroglobulin
[79]	WY14643, rosiglitizone	2D-MS	ob/ob mouse liver	↑ FA oxidation, lipogenesis in ob/ob with both PPAR activators; gluconeogenesis, glycolysis, AA metabolism affected with both
[78]	WY14643, oxazepam	2D-MS	mouse liver	DNA array/proteomics; subcellular fractions; protein IDs unique to each chemical
[139]	hydrazine, hypoxia	2D-MS	liver mitochondria	detection of carbonylated proteins after hypoxia
[82]	hydrazine	2D-DIGE-MS	rat liver	lipid, Ca^{2+}, thyroid, stress pathways activated by toxicity
[81]	hydrazine	2D-DIGE-MS	rat liver	↑10, ↓10 proteins IDed in lipid, Ca^{2+}, thyroid, stress pathways
[86]	thioacetamide	2D-MS	rat liver	STAP protein, stellate cells related to TAA-cirrhosis model
[85]	thioacetamide	SELDI, MALDI-TOF/TOF	rat liver	His-rich glycoprotein in serum related to TAA-cirrhosis model
[84]	thioacetamide	2D-MS	rat liver	liver cirrhosis model found ↓ FA β-oxidation, branched chain AA and methionine breakdown; ↑ oxidative stress, lipid peroxidation pathways
[98]	cyclosporine A	2D	rat liver, kidney	multiple 2D gel spots changed with cyclosporine in kidney (19 spots) and liver (29 spots)
[99]	cyclosporine A	2D AA sequencing	rat kidney	specific ↓ of calbindin in kidney by cyclosporine; IDed by AA sequencing; ELISA validation
[102]	cyclosporine A	2D gel, ^{35}S labeling	murine T cells	cyclosporine A induces >100 proteins not found in resting or activated murine T cells; proteins, like CSTAD, identified in later study [103]

(continued on next page)

Table 1. (continued)

Reference	Chemical	Platform	Tissue	Results
[110]	puromycin	2D-MS	rat kidney	proteomics/metabolomics to study renal glomerular toxicity
[140]	PbAc	2D-MS	rat kidney, medulla cortex	subacute PbAc altered 76 cortex proteins and 13 medullar proteins; ↓ calbindin, calcineurin, arginino- succinate synthetase; ↑ GSTM1
[108]	gentamicin	2D-MS	rat urine	20 proteins IDed; mitochondrial dysfunction in renal cortex
[115]	4-aminophenol, D-serine	2D-MS	rat plasma	FAH IDed as kidney damage biomarker
[114]	4-aminophenol, D-serine, cis-Pt	2D-MS	rat plasma	T-kininogen, inter-α-inhibitor H4P, complement c3 IDed as kidney damage biomarkers
[118]	DCVC	2D-DIGE-MS	porcine renal LLC-PK1 line	↑14, ↓9 proteins with DCVC; Hsp27 phosphorylation isoform found as marker of prosurvival by maintaining cell adhesion
[119]	DCVC	2D-DE Western blotting	rat kidney, proximal tubular cells	changes in the tyrosine phosphorylation of actin- and lamellipodia-related proteins
[120]	DCVC	2D-DE MS/MS	mouse kidney	autoprotection via low-dose alters protein expression response to lethal dose

[a] Summaries of liver and kidney toxicoproteomic studies are intended to briefly overview study details and results. Only abbreviations that are helpful in interpreting the summary notes are included. For further explanation, please refer to the citation. Abbreviations are listed in order of appearance in the table: APAP, acetaminophen (N-acetyl-p-aminophenol); AMAP, 3'-isomer of acetaminophen (3'-hydroxyacetanilide); 2D-MS, 2D gel-mass spectrometry; IDed, identified; 2D-DIGE-MS, 2D gel differential gel expression (Cy3,5 dyes) mass spectrometry; ↑,↓, increase, decrease; Hsp, heat shock protein; NEPHGE, non-equilibrium pH gel electrophoresis; ICAT, isotope coded affinity tags; CCl₄, carbon tetrachloride; ANIT, α-naphthylisothiocyanate; WY14643, Wyeth 14643 compound; PB, phenobarbital; PNP, purine nucleotide phosphorylase; MDH, malic dehydrogenase; Gc, vitamin D-binding protein; PON1, paraoxonase; RBP, retinol binding protein; TCN, tetracycline; GPx, glutathione peroxidase; PRx1,2, peroxiredoxin 1 and 2; BB, bromobenzene; LC-MS/MS, liquid chromatography tandem mass spectrometry; PPAR, peroxisome proliferator-activated receptor; Ca²⁺, calcium; STAP, stellate cell activation-associated protein; TAA, thioacetamide; SELDI, surface enhanced laser dissociated ionization; MALDI TOF/TOF, MALDI-based MS/MS; His, histidine; FA, fatty acid; AA, amino acid; CSTAD, cyclosporine A-conditional, T cell activation-dependent gene; PbAc, lead acetate; GSTM1, glutathione-S-transferase μ-1; FAH, fumarylacetoacetate hydrolase; cis-Pt, cisplatin; H4P, inter-α inhibitor H4P heavy chain; DCVC, dichlorovinyl-L-cysteine.

and kidney damaging agents relevant to preclinical assessment of toxicity. The agent, proteomic analysis platform, tissue or preparation and brief results for each study are summarized in Table 1. Liver toxicants will be addressed first.

Acetaminophen

Acetaminophen (*N*-acetyl-*p*-aminophenol, APAP) has been one of the most commonly tested agents for inducing hepatic injury in toxicoproteomics studies of the liver. It produces centrilobular hepatic necrosis in most preclinical species. Acute hepatocellular injury from acetaminophen exposure is primarily initiated by CYP2E1 bioactivation to form reactive intermediates such as *N*-acetyl-*p*-benzoquinone imine (NAPQI) that deplete glutathione (GSH) and then bind to critical cellular macromolecules [53]. Mitochondria are thought to be primary targets in acetaminophen toxicity with particular attention on the mitochondrial permeability transition [58]. It is worth noting that mitochondrial dysfunction underlies the pathogenesis of several toxicities in preclinical species especially in liver, skeletal and cardiac muscle, and the central nervous system (CNS) [59]. Evidence has also been accumulating for the contribution of non-parenchymal cells such as Kupffer cells, NK cells, neutrophils, and endothelial cells that secrete cytokines and chemokines during acetaminophen-induced liver injury [60–64].

Some of the earliest toxicoproteomics studies using 2D-MS platforms were conducted using standard 2D-MS [65] analysis as well as the 2D-DIGE-MS platform alone [66] or in combination with DNA microarrays [67]. Proteomic analysis of livers from these studies in mice identified altered proteins that are known targets for adduct formation such as mitochondrial proteins, heat shock proteins (HSPs), and other structural and intermediary metabolism proteins. A different type of 2D gel separation using a non-equilibrium approach to charge separation of proteins (NEPHGE) found 100–200 differentially expressed proteins in rat liver and HepG2 cells, especially in enzymes involved in intermediate metabolism [68].

Studies using rat hepatocytes exposed to acetaminophen and analyzed by 2D-MS have found it helpful to concurrently evaluate other cytotoxic pharmaceutical agents such as tetracycline, amiodarone and carbon tetrachloride [69, 70]. These studies found alterations in several metabolic enzymes and identified GSH peroxidase, peroxiredoxins 1 and 2 (PRX1, PRX2), which serve as cellular responsive antioxidative enzymes during toxicant exposure.

One of the first LC-MS/MS studies using ICAT technology that involved acetaminophen toxicity in mouse liver was published in 2005 [71]. It was preceded by an earlier optimization study for ICAT in mouse hepatocytes [72]. This study combined the more comprehensive ICAT analysis procedure with an adept choice of resistant (SJL) and susceptible (C57BL/6) mouse strains to investigate potential susceptibility factors (proteins and pathways) in acetaminophen toxicity [71]. Inherent differences in liver homogenate protein expression levels between resistant SJL and susceptible C57BL/6 mice were found by comparison of hepatic proteomics after vehicle (saline) treatment for 6 h. Of the 1236 proteins identified, 121 were differentially expressed between the two mouse strains. At 6 h after treatment with 300 mg/kg acetaminophen given intraperitoneally, 1632 proteins were identified from which 247 were different between the two strains and 161 proteins were more abundant in the SJL strain. Some of these naturally more abundant proteins (in the absence of toxicant) may have protective roles against toxicity including two- to fourfold increases in lactoferrin, galectin-1, tripeptidyl-peptidase II, proteasomal subunit β-type1 and DnaJ homolog A1. Upon administration of acetaminophen, comparative expression showed that SJL mice expressed from three- to tenfold higher levels of ubiquitin-like 2 (SUMO1) activating enzyme E1B, complement c5, cyclooxygenase 1 (COX-1), peroxiredoxin 1, glucose-regulated protein 170 (Grp170), heat shock protein 70 (Hsp70), glutathione-S-transferase μ-2 (GSTμ-2) and regucalcin. In addition to antioxidant enzyme functions, many of these up-regulated proteins may have a reparative role in degrading denatured and damaged proteins, cell proliferation and regeneration, and cellular stress response. A selective loss of several mitochondrial proteins from susceptible C57BL/6 mice suggested this organelle is particularly vulnerable to acetaminophen-induced hepatic injury.

Carbon tetrachloride

Carbon tetrachloride produces acute centrilobular hepatic necrosis but has been frequently used in a repeated exposure regimen over several weeks to produce an animal model of liver fibrosis [73]. Activation of hepatic stellate cells from a quiescent vitamin A-storing cell to a myofibroblast-like cell is a key event in excessive accumulation of fibril-forming extracellular matrix proteins and development of liver fibrosis. Proteomic analysis was performed on cellular and secreted proteins of normal and activated rat hepatic stellate cells either *in vitro* or *in vivo* after carbon tetrachloride for 8 weeks. Of the 43 altered proteins identified, 27 showed similar changes *in vivo* and *in vitro* including up-regulation of calcyclin, calgizzarin and galectin-1 as well as down regulation of liver carboxylesterase 10. These changes were confirmed in fibrotic liver tissues. A compendium of 150 stellate cellular and secreted proteins was identified.

Another carbon tetrachloride fibrosis study conducted a 2D-MS proteomic analysis upon liver tissues from rats exposed to carbon tetrachloride for a period of 4–10 weeks [74]. During this exposure period, collagen deposition and hydroxyproline content of fibrotic livers increased continuously. Differentially expressed proteins from proteomic analysis were categorized as proliferation-related proteins/enzymes (proliferating cell nuclear antigen p120, p40 and cyclin F ubiquitin-conjugating enzyme 7 UBC7), and apoptosis-related proteins, mainly caspase 12, which was absent in the control rats. These researchers found that proliferation- and apoptosis-related proteins are dynamically expressed during different stages of rat liver fibrosis induced by carbon tetrachloride.

Bromobenzene

Bromobenzene is another model liver toxicant whose metabolism, reactive intermediates, protein adducts and liver toxicity phenotype (centrilobular necrosis) have been well characterized [75]. A transcriptomic and proteomic comparison of bromobenzene conducted after 24-h exposure to a single dose of bromobenzene showed alterations in transcripts and genes involved in drug metabolism, oxidative stress, sulfhydryl metabolism and acute-phase response [76]. Of the 1124 proteins resolved from liver homogenates, 24 proteins were differentially expressed and identified as intermediary or drug metabolism enzymes.

Wyeth 14643

The peroxisome proliferator-activated receptors (PPARs) are ligand-activated transcription factors that modulate lipid and glucose homeostasis [77]. Wyeth 14643 (WY14643) is a hepatic metabolic enzyme inducer and acts as a potent agonist of PPARα, a member of the nuclear hormone receptor superfamily and a key transcriptional regulator of many genes involved in free fatty acid oxidation systems in liver. Global gene and protein expression changes were compared by cDNA microarray of mouse liver and 2D-MS of mouse liver subcellular fractions from B6C3F1 mice treated from 0.5 to 6 months with oxazepam and the peroxisome proliferator, WY14643 [78]. Each compound produces hepatocellular cancer after a 2-year bioassay of dietary exposure. The hypothesis was that each compound would produce cancer by different biochemical pathways and that transcript and protein changes measured prior to tumor formation (up to 6 months) would provide mechanistic insights into carcinogenesis. After 6 months, only 36 transcripts were altered after oxazepam compared to 220 transcripts with the Wyeth compound. Notable genes up-regulated in the signature profile for oxazepam were CYP2B20, Gadd45β, TNFα-induced protein 2 and Igfbp5. Up-regulated genes with WY14643 were cyclin D1, PCNA, Igfbp5, Gadd45β and CideA. Altered expression of over 100 proteins by proteomic analysis showed up-regulation of the cancer biomarker, α-fetoprotein in cytosol, and cell cycle-controlled p38-2G4 protein in microsomes during both treatments. Both transcriptomic and proteomic analyses were deemed complimentary in distinguishing between two chemical carcinogens that appear to proceed through different mechanisms and eventually lead to liver cancer as the common phenotype.

Insights into the therapeutic action of PPARα and PPARγ agonists, WY14643 and rosiglitazone, respectively, were reported in proteomic analysis of the ob/ob animal model of obesity disease [79]. Hepatic protein expression profiles were developed by 2D-MS analysis of lean and obese (ob/ob) mice, and obese mice treated with WY14643 or rosiglitazone. Livers from obese mice displayed higher levels of enzymes involved in fatty acid oxidation and lipogenesis compared to lean mice and these differences were further amplified by treatment with both PPAR activators. WY14643 normalized the expression levels of several enzymes involved in glycolysis, gluconeogenesis and amino acid metabolism in the obese mice to the levels of lean mice. Rosiglitazone only partially normalized levels of enzymes involved in amino acid metabolism. This study used an established mouse model of obesity disease to map metabolic pathways and discriminate between PPARα and PPARγ agonist effects by proteomic analysis.

Hydrazine

Hydrazine is a model, cross-species hepatotoxicant used as an industrial reagent and found as a drug metabolite of the structurally related pharmaceuticals, isoniazid (anti-tuberculosis drug) and the anti-hypertensive agent, hydralazine. Hydrazine typically causes initial steatosis, macrovesicular degeneration followed by marked hepatic necrosis. Transcriptomic studies suggest hydrazine initiates a process whereby the production and intracellular transport of hepatic lipids is favored over the removal of fatty acids and their metabolites [80].

Proteomics studies using 2D-DIGE-MS on the hepatotoxic effects of hydrazine were conducted in rats from 48 to 168 h [81, 82]. In one study, 2D gel patterns from liver were analyzed by principal component analysis (PCA) and partial least squares regression. PCA plots described the variation in protein expression related to dose and time. Regression analysis was used to select ten up-regulated proteins and ten down-regulated proteins that were identified by MS. Hydrazine treatment altered proteins in lipid metabolism, Ca^{2+} homeostasis, thyroid hormone pathways and stress response. In a second study, low-density cDNA microarrays and 2D-DIGE-MS proteomics of liver tissue and metabonomics analysis of serum was performed from hydrazine-treated rats at 48–168 h [81]. Their findings supported known effects of hydrazine toxicity and provided potential biomarkers of hydrazine-induced toxicity.

Thioacetamide

Thioacetamide is metabolically activated in liver to produce thioacetamide-*S,S*-dioxide as a reactive intermediate, which binds to liver macromolecules to initiate centrilobular necrosis [83]. Repeated administration of thioacetamide is an established technique for generating rat models of liver fibrosis and cirrhosis, depending upon dose and length of administration (weeks). A 2D-MS proteomic approach was used to profile liver protein changes in rat receiving thioacetamide for 3, 6 and 10 weeks to induce hepatic cirrhosis [84]. Expression of 59 proteins altered by thioacetamide were identified, including three novel, unannotated proteins. Down-regulation of enzymes were noted in pathways such as fatty acid β-oxidation, branched chain amino acids, and methionine breakdown, which may relate to succinyl-CoA depletion and affect heme and iron metabolism. Increased levels were found for enzymes responding to oxidative stress and lipid peroxidation such as GSTs. Finally, these proteomics data were integrated into a proposed overview model for thioacetamide-induced liver cirrhosis affecting succinyl-CoA and cytochrome P450 production combined with iron release and hydrogen peroxide generation.

In another model of thioacetamide-induced liver cirrhosis in rats, researchers searched for potential serum biomarkers using the SELDI proteomic approach [85]. A weak cation exchange surface was used to analyze serum by SELDI-MS from control (normal) rats, thioacetamide-induced liver cirrhosis rats and rats with bile duct ligation-induced liver fibrosis. A consistently down-regulated 3495-Da protein in cirrhosis samples was one of the selected significant biomarkers. This 3495-Da protein was purified on-chip and was trypsin digested on-chip for MS/MS identification, and was found to be a histidine-rich glycoprotein. This new protein was proposed as a novel preclinical biomarker for the rat cirrhosis model and might eventually prove useful for early clinical detection of liver cirrhosis and classification of liver diseases.

An innovative study involving stellate cell activation by 8-week treatment with thioacetamide utilized a proteomic approach that led to the discovery of a novel protein named STAP for "stellate cell activation-associated protein" [86]. Quiescent and thioacetamide-activated stellate cells were analyzed by 2D-MS [using electrospray ionization-(ESI)-MS/MS] to identify 43 proteins altered during the activation process. Up-regulation of collagen-α1 (I and III), γ-actin, neural cell adhesion molecule (N-CAM), calcyclin, calgizzarin and galectin-1 was detected. In particular, STAP was highly increased both in activated stellate cells and in fibrotic liver tissues induced by thioacetamide treatment. These researchers cloned the STAP gene and found it was a cytoplasmic protein, expressed only in stellate cells, with molecular mass of 21 496 Da and a 40% amino acid sequence homology to myoglobin. Biochemical characterization showed STAP is a heme protein exhibiting peroxidase activity toward hydrogen peroxide and linoleic acid hydroperoxide. These results indicate that STAP is a novel endogenous peroxidase catabolizing hydrogen peroxide and lipid hydroperoxides, both of which have been reported to trigger stellate cell activation and consequently promote progression of liver fibrosis. STAP was postulated to play a role as an anti-fibrotic scavenger of peroxides in the liver.

Detection of biomarkers in blood after liver injury

Blood is one of the most accessible and informative biofluids for specific organ pathology in preclinical studies. Biomarkers that can be assayed in biological fluids from preclinical species may hold relevance to human subjects [19]. The Human Proteome Organization (HUPO) is currently undertaking a comprehensive mapping of soluble human blood elements of the plasma proteome for an improved understanding of disease and toxicity [87]. Results from an international survey of soluble human blood proteins by chromatographic and electrophoretic separation have revealed several thousand resolvable proteins for which MS has provided evidence for over 1000 unique protein identifications [87, 88]. Researchers are also mapping the mouse [89] and rat [90] serum and plasma proteomes for use in preclinical and experimental studies. An excellent review has been published for 2D gel mapping of rat serum and rat tissue proteomic studies [90].

The sensitivity of 2D gel proteomic approaches to detect and measure alterations in the mouse or rat plasma proteomes has only recently been tested by various labs. Researchers have examined changes in the mouse plasma proteome focusing upon inflammation after cutaneous burn injury with superimposed *Pseudomonas aeruginosa* infection [91]. Up-regulations of inter-α-trypsin inhibitor heavy chain 4 and hemopexin were detected along with other mouse acute-phase proteins, including haptoglobin and serum amyloid A. In another inflammation study, reference maps of the mouse serum proteome were generated by 2D-MS from control animals and from mice injected with lipopolysaccharide (LPS) to induce systemic inflammation, and from mice transgenic for human apolipoproteins A-I and A-II [92]. The greatest changes were noted for haptoglobin and hemopexin.

Finally, a comparative plasma proteome analysis has been reported in which investigators used 1D-Gel LC-MS/MS analysis upon a few microliters of plasma from lymphoma-bearing SJL mice experiencing systemic inflammation [29]. After removal of albumin and immunoglobulins from plasma, these researchers identified a total of 1079 non-redundant mouse plasma proteins; more than 480 in normal and 790 in RcsX-tumor-bearing SJL mouse plasma. Of these, only 191 proteins were found in common. Many of the up-regulated proteins were identified as acute-phase proteins but several unique proteins, including haptoglobin, proteasome subunits, fetuin-B, 14-3-3ζ, and MAGE-B4 antigen, were found only in the tumor-bearing mouse plasma due to secretion or shedding by membrane vesicles, or externalized due to cell death. These results are very encouraging for the effectiveness of a proteomics approach for protein identification from small sample amounts, and for comparative proteomics in animal models of drug-induced toxicity or disease.

The application of serum or plasma protein maps in toxicoproteomics, such as those for serum profiling of liver injury, is just beginning to take shape. A recent study reported identification of serum proteins altered in rats treated with four liver-targeted compounds including acetaminophen, α-naphthylisothiocyanate (ANIT), phenobarbital and WY14643 at early, fulminant, and recovery periods of effect [93]. Nineteen serum proteins were identified as significantly altered from the four studies and among them, five serum proteins were of special interest as serum markers for early hepatic toxicity or functional alterations in rats, including vitamin D-binding protein (group-specific component, Gc-globulin), purine nucleotide phosphorylase (PNP), malic dehydrogenase (MDH), paraoxonase (PON1) and retinol-binding protein (RBP). Some of these proteins may serve as early predictive markers of hepatotoxicity for new drug candidates or may be more sensitive than other conventional methods.

The soluble portion of blood, serum or plasma, is regarded as a complex biofluid tissue. While many organs contribute various proteins as blood solutes, the liver is by far the most productive member of all organs and tissues. The liver parenchyma are often primary targets of drug-induced toxicity, and they also secrete many plasma proteins, which can be measured in preclinical species. Therefore, researchers have studied the secreted proteome of hepatocytes. Secreted proteins were separated and identified from primary rat hepatocytes using a collagen gel sandwich system. Proteomic analysis was conducted using a 1D gel LC-MS/MS procedure. More than 200 secreted proteins were identified; these included more than 50 plasma proteins, several structural extracellular matrix proteins and many proteins involved in liver regeneration. Secretion of two proteins, α1-antitrypsin and α2-macroglobulin, was greatly reduced in aflatoxin B_1-exposed hepatocytes. This study provides evidence that proteomic analysis of medium from hepatocyte sandwich culture might represent a new *in vitro* model and general approach for future discoveries of secreted biomarkers in drug-induced chemical toxicity.

Toxicoproteomic studies in kidney injury

Kidney is a primary organ for preclinical assessment in pharmaceutical development since its metabolic and excretory functions often render it susceptible to drug-induced toxicity [94]. The kidney is a major organ for filtration, reabsorption and secretion to maintain homeostasis of water-soluble salts and small molecules. The organ also has a considerable capacity for biotransformation of drugs and xenobiotics. Specific physiological characteristics of the kidney are localized to specific cell types (i.e., vascular endothelial and smooth muscle cells, mesangial cells, interstitial cells, podocytes, proximal and distal tubular epithelial cells), each of which demonstrates selective susceptibility to toxicity. Renal damage can be due to several different mechanisms affecting different segments of the nephron, renal microvasculature or interstitium. The nature of renal injury may be acute and recoverable. However, other drugs with repeated exposure can produce chronic renal changes that may lead to end-stage renal failure. The ability to perform kidney transplants and other organ replacements have saved many lives but relies on immunosuppressive drug treatment to prevent organ rejection. However, immunosuppressive drugs also run the risk of renal toxicity over time. New nephrotoxic markers amenable for multiple preclinical models and high-throughput screening is a major goal for toxicoproteomic and toxicogenomic technologies (Tab. 1) [94–96].

Cyclosporine A

Some of the groundbreaking studies that initiated the field of toxicoproteomics took place in the mid-1990s and involved investigating the side effects of the immunosuppressant drug, cyclosporine A. Cyclosporine A is a calcineurin inhibitor that has been a mainstay for immunosuppressive therapy following solid-organ transplantation. Cyclosporine A blocks immune responses by inhibiting the calcineurin-dependent dephosphorylation of the nuclear factor of activated T cells (NFAT). However, a dose-dependent nephrotoxicity occurs with high incidence that is characterized by non-histological functional deficits or functional decline, with calcium loss in urine (hypercalcinuria), vascular-interstitial lesions and calcification of renal tubules [97].

Initial 2D gel studies were conducted in rat liver and kidney samples that showed changes in 48 proteins in these tissues in rats treated with cyclosporine A. An unidentified protein present only in the kidney was uniquely down-regulated [98]. A subsequent 2D gel study of kidney homogenates identified a decrease in the 28-kDa kidney protein as calbindin-D using protein microsequencing. Importantly, this same study, using an ELISA, validated a time-dependent decrease in calbindin expression for up to 28 days of cyclosporine treatment [99]. These toxicoproteomic studies published a decade ago represented an important advance in understanding a part of cyclosporine-induced pathophysiology in kidney.

More recently, the contribution of calbindin-D28k has been clarified by the generation of genetically modified mice. Cyclosporine A-induced hypercalciuria represents two pathophysiological processes: a down-regulation of calbindin-D28k with subsequent impaired renal calcium reabsorption, and a cyclosporine A-induced high turnover bone disease [100]. In addition, there is evidence that one biochemical mechanism underlying cyclosporine A and other calcineurin inhibitors may be a drug-induced mitochondrial dysfunction [101].

The effects of cyclosporine A on gene up-regulation were advanced by a 2D gel proteomic analysis of newly synthesized [^{35}S]methionine-labeled proteins in murine T cells activated in the absence or presence of cyclosporine A [102]. Remarkably, these investigators found more than 100 proteins not present in resting or activated T cells that could be induced by cyclosporine A exposure. It is important to emphasize that the discovery nature of this proteomics study was capitalized upon (same researchers) with the identification of the corresponding genes under the same treatment conditions using a transcript enrichment technique called "representational difference analyses" [103]. Among the up-regulated transcripts, a new gene was found named *CSTAD*, for "cyclosporine A-conditional, T cell activation-dependent" gene. CSTAD encodes two proteins of 104 and 141 amino acids that are localized in mitochondria [103]. *CSTAD* up-regulation is observed in mice after cyclosporine A treatment, suggesting that up-regulation of CSTAD and perhaps many other genes are implicated in cyclosporine A toxicity. Thus, toxicoproteomics has played an important role in furthering the understanding of the critical proteins and biological pathways in cyclosporine A toxicity that should lead to better biomarkers for this important class of pharmaceutics.

Puromycin and gentamicin

The regionally specific structure and function of the kidney renders specialized areas more susceptible to toxicity from exposure to certain pharmaceutical agents. For example, puromycin aminonucleoside is an antibiotic that causes glomerular podocyte necrosis, nephrosis and proteinuria in rodent models [104]. Gentamicin is an aminoglycoside antibiotic that accumulates in proximal tubular epithelia and inhibits cell lysosomal function, producing phospholipidosis and tubular degeneration [105]. Some studies have begun to proteomically characterize specific regions such as the medulla and cortex [106] or subcellular structures of kidney cell types such as the nucleus [107]. In one study, the nephrotoxic effects of gentamicin on protein expression were studied in rat kidney. Results revealed the identities of more than 20 proteins involved the citric acid cycle, gluconeogenesis, fatty acid synthesis, and transport or cellular stress responses [108]. The authors believe that impairment of energy production and mitochondrial dysfunction were involved in gentamicin-induced nephrotoxicity.

Another approach to studying nephrotoxicity is by proteomic characterization of urine. Proteomic mapping of rat urine proteins studied by 2D-MS resolved 350 protein spots from which 111 protein components were identified including transporters, transport regulators, chaperones, enzymes, signaling proteins, cytoskeletal proteins, pheromone-binding proteins, receptors, and novel gene products [109]. One toxicoproteomics study examined urinary protein expression profiles to gain insight into puromycin-induced kidney toxicity [110]. Nephropathy and proteinuria caused by puromycin aminonucleoside in rats was studied by metabonomics and a 2D-MS proteomic analysis of urinary proteins from 8 to 672 h after dosing. Prior to exposure, major urinary protein (MUP), α2-microglobulin and glial fibrillary acid protein isoforms were the major urinary proteins found in addition to many other unidentified low-mass urinary proteins. Following puromycin treatment, a gradual increase in higher mass proteins was observed on 2D gels, particularly albumin, at 32 h after dosing. By 120 h, albumin, transthyretin and vitamin D-binding protein (Gc) were identified as major urinary proteins from puromycin-induced kidney damage. After 672 h, the urinary protein profile in 2D gels had largely returned to normal. Many of these plasma-derived proteins appearing in the urine over 0–672 h following puromycin were consistent with loss of glomerular integrity and major leakage of plasma protein in urine. This study suggests that urinary proteomics in conjunction with these other techniques, has the potential to provide significantly more mechanistic information than is readily provided by traditional clinical chemistries, and may be a productive means for biomarker discovery of nephrotoxic agents in preclinical species.

4-Aminophenol, D-serine and cisplatin

The nephrotoxin 4-aminophenol produces severe necrosis of the pars recta of the proximal tubules in the rat, which is thought to occur through formation of a toxic metabolite 1,4-benzoquinone imine [111]. D-Serine, an enantiomer of L-serine, is another model nephrotoxicant that selectively damages the pars recta of proximal tubules in the kidney, which may involve formation of toxic oxidative metabolites [112]. The chemotherapeutic agent, cisplatin is also a model nephrotoxicant and targets different portions of the kidney. It is metabolized to cytotoxic intermediates in proximal tubular epithelial cells and induces necrosis in distal tubules and collecting ducts along with causing mild glomerular toxicity [113]. Proteomic profiling using 2D-MS was used to investigate plasma protein changes in rats treated with 4-aminophenol, D-serine and cisplatin compared to saline controls [114]. Nontoxic isomers, L-serine, and transplatin, were also studied. Many plasma proteins were found that displayed dose- and temporal-dependent response to toxicants. Several isoforms of T-kininogen protein were identified as increasing in plasma at early time points and returning to baseline levels after 3 weeks with each nephrotoxicant but not with nontoxic compounds. In addition, inter-α inhibitor H4P heavy chain was increased in the 4-aminophenol and D-serine studies. A further set of proteins correlating with kidney damage was found to be a component of the complement cascade and other blood clotting factors, indicating a contribution of the immune system to the observed toxicity. It was proposed that T-kininogen may be required to counteract apoptosis in proximal tubular cells to minimize tissue damage following a toxic insult.

In a related study, plasma samples from 4-aminophenol and D-serine treated rats were profiled by 2D-MS, and showed dose- and time-dependent effects of various plasma proteins in response to these nephrotoxicants [115]. One toxicity-associated plasma protein was identified as the cellular enzyme, fumarylacetoacetate hydrolase (FAH), a key component of the tyrosine metabolism pathway. FAH was elevated in the plasma of animals treated with 4-aminophenol and D-serine at early time points and returned to baseline levels after 3 weeks. The protein was not elevated in the plasma of control animals or those treated with the non-toxic isomer, L-serine. The investigators raised the possibility that FAH might serve as a marker of kidney toxicity in preclinical species.

Dichlorovinyl-L-cysteine

Dichlorovinyl-L-cysteine (DCVC) is a model nephrotoxicant taken up by renal proximal tubular epithelia, where it is bioactivated by renal cysteine conjugate β-lyase to form reactive, cytotoxic intermediates [116]. DCVC is a metabolite of trichloroethylene but can be chemically synthesized for use in experimental studies [117].

A proteomic study of DCVC toxicity was conducted in LLC-PK1 porcine renal epithelial cells by 2D-DIGE-MS to determine early changes in stress-response pathways preceding focal adhesion disorganization linked to the onset of apoptosis [118]. DCVC treatment caused a greater than 1.5-fold up- and down-regulation of 14 and 9 proteins, respectively, prior to apoptosis. These included aconitase and pyruvate dehydrogenase, and those related to stress responses and cytoskeletal reorganization, such as cofilin, Hsp27, and αβ-crystallin. Most noticeable was a pI shift in Hsp27 on phosphorylation at Ser82. Only inhibition of p38 with SB203580 reduced Hsp27 phosphorylation, which was associated with accelerated reorganization of focal adhesions, cell detachment, and apoptosis. Inhibition of active JNK (JUN N-terminal kinase) localization at focal adhesions did not prevent DCVC-induced phosphorylation of Hsp27. Overexpression of a phosphorylation-defective mutant Hsp27 acted as a dominant negative form and accelerated DCVC-induced focal adhesion changes and onset of apoptosis. Early p38 activation appears to rapidly phosphorylate Hsp27, to maintain cell adhesion and to suppress renal epithelial cell apoptosis. This toxicoproteomics study combines both protein identification and post-translational modification to elucidate critical proteins (Hsp27) and protein attributes (phosphorylation) in critical pathways (p38 stress pathway) to gain insight into mechanisms of renal epithelial cell death.

More recently, de Graauw et al. [119] used phosphotyrosine proteomics (2-DE plus Western blotting) in cultured rat renal proximal tubule cells to demonstrate that DCVC-induced apoptosis is preceded by changes in the tyrosine phosphorylation status of actin-related protein 2 (Arp2), cytokeratin 8, t-complex protein 1 (TCP-1), chaperone containing TCP-1, and gelsolin precursor. It was concluded that the observed alterations are involved in the regulation of the F-actin reorganization and lamellipodia formation that precede renal cell apoptosis caused by DCVC.

Korrapati et al. [120] identified proteins indicative of DCVC-induced acute renal failure and autoprotection in mice using conventional, large-format 2-DE, Coomassie brilliant blue-based visualization, and MS/MS. Low-dose exposure (15 mg DCVC/kg i.p.) altered 30 proteins (9 up-regulated; 21 down-regulated) by 1.5-fold or more (at $p < 0.01$), while the lethal, high-dose exposure (75 mg DCVC/kg i.p.), altered the expression of 210 proteins (84 up-regulated; 126 down-regulated). As expected, when the low-dose exposure preceded the administration of the lethal dose by 72 h (autoprotection), the number and extent of differential protein expression was significantly reduced. The authors examined the 18 most radically altered proteins (>10-fold) and concluded that the DCVC-induced differential expression in proteins (involved in the biochemical mechanisms of renal injury and tissue repair) are implicated in the irreparable loss of renal structure and function after a high dose of DCVC, rather than being involved in recovery of structure and function in autoprotected mice.

Summary and future prospects

This chapter has focused upon the development of toxicoproteomics in liver and kidney injury because of their respective roles in xenobiotic biotransformation and excretion. However, protein profiling studies are being conducted in many other organs and tissues to profile adverse effects of therapeutics. Proteomic approaches are revealing new blood serum and tissue biomarkers in animal models of human neurodegenerative diseases like Parkinson's disease, Alzheimer's disease, and amyotrophic lateral sclerosis [121, 122]. Proteomic studies are being conducted in cardiotoxicity models with doxorubicin [123] and renin-angiotensin models of hypertension [124]. Comparative protein expression studies are systematically examining testicular toxicity in rats with several reproductive toxicants such as cyclophosphamide, sulfasalazine, 2,5-hexanedione and ethylene glycol monomethyl ether [125]. The effects of formaldehyde on rat lung [126] and protein adduct formation of 1-nitronaphthalene metabolites in rat lung [127, 128] are being examined by proteomic techniques to provide insights into pulmonary pathology by these agents. Thus, protein mapping and profiling studies are exploring a variety of preclinical animal assessment models of toxicity and disease.

The expectations of Omics technologies in pharmaceutical development are very high but the breakthroughs in drug discovery and improvements over traditional measures in preclinical assessment have not proceeded as quickly as anticipated. This situation is understandable since the platforms for proteomics continue to be in dynamic development. Furthermore, applications to toxicology settings are still being explored to match platform sensitivity for differential protein expression with preclinical biological samples. Many of the published toxicoproteomics reports reviewed here have served as proof-of-principle studies using Tier I proteomic analysis (Fig. 1). The approach has been to examine a well-characterized toxicant(s) and compare proteomics data output with known toxicological endpoints (i.e., serum and urine chemistries, histopathology). These efforts might be described as the "discovery phase" of toxicoproteomics where differential protein expressions are determined in response to compound exposure. However, many of these initial studies have often not been accompanied by any confirmation analysis using ELISA, Western blot, immunohistochemistry or functional assay (i.e., enzymatic activity). Two other areas show a slow progress in toxicoproteomics research. One is in the follow-up "hypothesis-driven research" that further characterizes discovery findings and establishes causal-linkage of toxicant exposure and effect. The other area is in "validation studies" of proposed biomarkers using independent and blinded study samples. However, the full cycle of discovery, focused confirmation analysis and hypothesis-testing for causality is achievable [86, 102, 103, 118]. Extensive validation studies of biomarkers represent a lengthier process.

Future trends in toxicoproteomics studies will see developments in several areas where special attributes of proteins can be exploited by proteomics in preclinical assessment. First, further refinements of MS/MS with intimately integrated multi-dimensional separation schemes will continue to dominate proteomic analysis for identification and quantification. MS instruments and software will become more user-friendly and accessible, such as the recently introduced orbitrap MS/MS instruments. Second, "reduction of sample complexity" or any pre-purification strategy prior to toxicoproteomics analysis will be very useful upon innovative application to appropriate biological samples and problem areas (i.e., immunodepletion of albumin, immunoglobulins in plasma) or research problem areas (i.e., phosphoprotein enrichment in protein signaling). Third, Tier II proteomics will begin to be applied to toxicoproteomics problem areas such as global and targeted protein phosphorylation [129–131] and chemoproteomics [132] using pharmaceutics or enzyme substrates like ATP [133] as mass capture-ligands for proteins. Fourth, toxicoproteomics is readily positioned to exploit accessible biofluids (i.e., serum/plasma, urine and cerebral spinal fluid) for biomarker development [134] and could be combined with transcriptomic analysis of blood leukocytes for a parallel approach in biomarker discovery [135]. Fifth, the astute use of genetically altered animals and cell models will enhance discovery of protein targets and mechanistic insights into adverse drug reactions. Finally, continued efforts for integration of proteomics, transcriptomics and toxicology data to derive mechanistic insight and biomarkers will be a continuing goal to maximize return on the investment in Omics technologies [25, 136, 137].

Challenges for toxicoproteomics in preclinical risk assessment are: use as a discovery tool for specific proteins affected by drug and toxicant action; better understanding of biochemistry and cell biology; and biomarker development. The discipline of proteome mapping will be a different and more complex enterprise from the high-throughput, linear-sequencing activities that have been so useful in mapping of the human genome. While the immensity of mapping and measuring the attributes in any one proteome is a large undertaking, biofluid proteomes such as serum/plasma, urine and cerebrospinal fluid hold the most immediate promise for preclinical assessment in terms of better biomarkers.

Although there are many challenges for toxicoproteomics in preclinical assessment, the opportunities are also close at hand for a greater understanding of toxicant action, the linkage to accompanying dysfunction and pathology, and the development of predictive biomarkers and signatures of toxicity.

Acknowledgments
This review was supported by the Intramural Research program of the NIH, National Institute of Environmental Health Sciences.

References

1 Chapal N, Molina L, Molina F, Laplanche M, Pau B, Petit B (2004) Pharmacoproteomic approach to the study of drug mode of action, toxicity, and resistance: Applications in diabetes and cancer. *Fundam Clin Pharmacol* 18: 413–422

2 Leighton JK (2005) Application of emerging technologies in toxicology and safety assessment: Regulatory perspectives. *Int J Toxicol* 24: 153–155

3 Ross JS, Symmans WF, Pusztai L, Hortobagyi GN (2005) Pharmacogenomics and clinical biomarkers in drug discovery and development. *Am J Clin Pathol* 124 Suppl: S29–41

4 Siest G, Marteau JB, Maumus S, Berrahmoune H, Jeannesson E, Samara A, Batt AM, Visvikis-Siest S (2005) Pharmacogenomics and cardiovascular drugs: Need for integrated biological system with phenotypes and proteomic markers. *Eur J Pharmacol* 527: 1–22

5 Bandara LR, Kennedy S (2002) Toxicoproteomics – A new preclinical tool. *Drug Discov Today* 7: 411–418

6 Petricoin EF, Rajapaske V, Herman EH, Arekani AM, Ross S, Johann D, Knapton A, Zhang J, Hitt BA, Conrads TP et al (2004) Toxicoproteomics: Serum proteomic pattern diagnostics for early detection of drug induced cardiac toxicities and cardioprotection. *Toxicol Pathol* 32 Suppl 1: 122–130

7 Wetmore BA, Merrick BA (2004) Toxicoproteomics: Proteomics applied to toxicology and pathology. *Toxicol Pathol* 32: 619–642

8 Waters MD, Fostel JM (2004) Toxicogenomics and systems toxicology: Aims and prospects. *Nat Rev Genet* 5: 936–948

9 Turner SM (2006) Stable isotopes, mass spectrometry, and molecular fluxes: Applications to toxicology. *J Pharmacol Toxicol Methods* 53: 75–85

10 Merrick BA, Bruno ME (2004) Genomic and proteomic profiling for biomarkers and signature profiles of toxicity. *Curr Opin Mol Ther* 6: 600–607

11 Silbergeld EK, Davis DL (1994) Role of biomarkers in identifying and characterizing environmentally induced disease. *Clin Chem* 40: 1363–1367

12 Hood L, Heath JR, Phelps ME, Lin B (2004) Systems biology and new technologies enable predictive and preventative medicine. *Science* 306: 640–643

13 Lin J, Qian J (2007) Systems biology approach to integrative comparative genomics. *Expert Rev Proteomics* 4: 107–119

14 Kasper P, Oliver G, Lima BS, Singer T, Tweats D (2005) Joint EFPIA/CHMP SWP workshop: The emerging use of omic technologies for regulatory non-clinical safety testing. *Pharmacogenomics* 6: 181–184

15 Gibbs A (2005) Comparison of the specificity and sensitivity of traditional methods for assessment of nephrotoxicity in the rat with metabonomic and proteomic methodologies. *J Appl Toxicol* 25: 277–295

16 MacGregor JT (2003) The future of regulatory toxicology: Impact of the biotechnology revolution. *Toxicol Sci* 75: 236–248

17 Hackett JL, Gutman SI (2005) Introduction to the Food and Drug Administration (FDA) regulatory process. *J Proteome Res* 4: 1110–1113

18 Yeh ET (2005) High-sensitivity C-reactive protein as a risk assessment tool for cardiovascular disease. *Clin Cardiol* 28: 408–412

19 Amacher DE (2002) A toxicologist's guide to biomarkers of hepatic response. *Hum Exp Toxicol* 21: 253–262

20 Ross JS, Fletcher JA, Linette GP, Stec J, Clark E, Ayers M, Symmans WF, Pusztai L, Bloom KJ (2003) The Her-2/neu gene and protein in breast cancer 2003: Biomarker and target of therapy. *Oncologist* 8: 307–325

21 Thier R, Bruning T, Roos PH, Rihs HP, Golka K, Ko Y, Bolt HM (2003) Markers of genetic susceptibility in human environmental hygiene and toxicology: The role of selected CYP, NAT and GST genes. *Int J Hyg Environ Health* 206: 149–171

22 Bilello JA (2005) The agony and ecstasy of "OMIC" technologies in drug development. *Curr Mol Med* 5: 39–52

23 Koop R (2005) Combinatorial biomarkers: From early toxicology assays to patient population profiling. *Drug Discov Today* 10: 781–788

24 Merrick BA (2004) Introduction to high-throughput protein expression. In: HK Hamadeh, CA Afshari (eds): *Toxicogenomics: Principles and Applications.* Wiley and Sons, New York, 263–281

25 Merrick BA, Madenspacher JH (2005) Complementary gene and protein expression studies and integrative approaches in toxicogenomics. *Toxicol Appl Pharmacol* 207: 189–194

26 Righetti PG, Castagna A, Antonucci F, Piubelli C, Cecconi D, Campostrini N, Antonioli P, Astner H, Hamdan M (2004) Critical survey of quantitative proteomics in two-dimensional electrophoretic approaches. *J Chromatogr A* 1051: 3–17

27 Freeman WM, Hemby SE (2004) Proteomics for protein expression profiling in neuroscience. *Neurochem Res* 29: 1065–1081

28 Yates JR (2004) Mass spectral analysis in proteomics. *Annu Rev Biophys Biomol Struct* 33: 297–316

29 Bhat VB, Choi MH, Wishnok JS, Tannenbaum SR (2005) Comparative plasma proteome analysis of lymphoma-bearing SJL mice. *J Proteome Res* 4: 1814–1825

30 Farkas D, Bhat VB, Mandapati S, Wishnok JS, Tannenbaum SR (2005) Characterization of the secreted proteome of rat hepatocytes cultured in collagen sandwiches. *Chem Res Toxicol* 18: 1132–1139

31 Macdonald N, Chevalier S, Tonge R, Davison M, Rowlinson R, Young J, Rayner S, Robert R (2001) Quantitative proteomic analysis of mouse liver response to the peroxisome proliferator diethylhexylphthalate (DEHP). *Arch Toxicol* 75: 415–424

32 Asara JM, Christofk HR, Freimark LM, Cantley LC (2008) A label-free quantification method by MS/MS TIC compared to SILAC and spectral counting in a proteomics screen. *Proteomics* 8: 994–999

33 Higgs RE, Knierman MD, Gelfanova V, Butler JP, Hale JE (2005) Comprehensive label-free method for the relative quantification of proteins from biological samples. *J Proteome Res* 4: 1442–1450

34 Colinge J, Chiappe D, Lagache S, Moniatte M, Bougueleret L (2005) Differential proteomics via probabilistic peptide identification scores. *Anal Chem* 77: 596–606

35 Liu H, Sadygov RG, Yates JR 3rd (2004) A model for random sampling and estimation of relative protein abundance in shotgun proteomics. *Anal Chem* 76: 4193–4201

36 Paoletti AC, Parmely TJ, Tomomori-Sato C, Sato S, Zhu D, Conaway RC, Connaway JW, Florens L, Washburn MP (2006) Quantitative proteomic analysis of distinct mammalian mediator complexes using normalized spectral abundance factors. *Proc Natl Acad Sci USA* 103: 18928–18933

37 Zybailov B, Mosley AL, Sardiu ME, Coleman MK, Florens L, Washburn MP (2006) Statistical analysis of membrane proteome expression changes in *Saccharomyces cerevisiae*. *J Proteome Res* 5: 2339–2347

38 Gerber SA, Rush J, Stemman O, Kirschner MW, Gygi SP (2003) Absolute quantification of proteins and phosphoproteins from cell lysates by tandem MS. *Proc Natl Acad Sci USA* 100: 6940–6945

39 Kirkpatrick DS, Gerber SA, Gygi SP (2005) The absolute quantification strategy: A general procedure for the quantification of proteins and post-translational modifications. *Methods* 35: 265–273

40 Old WM, Meyer-Arendt K, Aveline-Wolf L, Pierce KG, Mendoza A, Sevinsky JR, Resing KA, Ahn NG (2005) Comparison of label-free methods for quantifying human proteins by shotgun proteomics. *Mol Cell Proteomics* 4: 1487–1502

41 Fitzpatrick DPG, You JS, Bemis KG, Wery JP, Ludwig JR, Wang M (2007) Searching for potential biomarkers of cisplatin resistance in human ovarian cancer using a label-free LC/MS-based protein quantification method. *Proteomics Clin Appl* 1: 246–263

42 Witzmann FA, Hong D, Rodd ZA, Simon JR, Truitt WA, Wang M (2007) Synaptosomal protein expression in nucleus accumbens after EtOH self-administration in the posterior VTA. *FASEB J* 21: A477–A477

43 Witzmann FA, Lee K, Wang M, Yemane Y, Witten ML (2007) Pulmonary effects of JP-8 jet fuel exposure – Label-free quantitative analysis of protein expression in alveolar type II epithelial cells using LC/MS. *Toxicol Sci* 96: 102

44 Florens L, Carozza MJ, Swanson SK, Fournier M, Coleman MK, Workman JL, Washburn MP (2006) Analyzing chromatin remodeling complexes using shotgun proteomics and normalized spectral abundance factors. *Methods* 40: 303–311

45 Sardiu ME, Cai Y, Jin J, Swanson SK, Conaway RC, Florens L, Washburn MP (2008) Probabilistic assembly of human protein interaction networks from label-free quantitative proteomics. *Proc Natl Acad Sci USA* 105: 1454–1459

46 Ott LW, Resing KA, Sizemore AW, Heyen JW, Cocklin RR, Pedrick NM, Woods HC, Chen JY, Goebl MG, Witzmann FA, Harrington MA (2007) Tumor necrosis factor-α- and interleukin-1-induced cellular responses: Coupling proteomic and genomic information. *J Proteome Res* 6: 2176–2185

47 Janecki DJ, Bemis KG, Tegeler TJ, Sanghani PC, Zhai L, Hurley TD, Bosron WF, Wang M (2007) A multiple reaction monitoring method for absolute quantification of the human liver alcohol

dehydrogenase ADH1C1 isoenzyme. *Anal Biochem* 369: 18–26
48 Carmella SG, Chen M, Zhang Y, Zhang S, Hatsukami DK, Hecht SS (2007) Quantitation of acrolein-derived (3-hydroxypropyl)mercapturic acid in human urine by liquid chromatography-atmospheric pressure chemical ionization tandem mass spectrometry: Effects of cigarette smoking. *Chem Res Toxicol* 20: 986–990
49 Petricoin E, Wulfkuhle J, Espina V, Liotta LA (2004) Clinical proteomics: Revolutionizing disease detection and patient tailoring therapy. *J Proteome Res* 3: 209–217
50 Issaq HJ, Conrads TP, Prieto DA, Tirumalai R, Veenstra TD (2003) SELDI-ToF MS for diagnostic proteomics. *Anal Chem* 75: 148A–155A
51 Diamandis EP (2004) Mass spectrometry as a diagnostic and a cancer biomarker discovery tool: Opportunities and potential limitations. *Mol Cell Proteomics* 3: 367–378
52 Cutler P (2003) Protein arrays: The current state-of-the-art. *Proteomics* 3: 3–18
53 Park BK, Kitteringham NR, Maggs JL, Pirmohamed M, Williams DP (2005) The role of metabolic activation in drug-induced hepatotoxicity. *Annu Rev Pharmacol Toxicol* 45: 177–202
54 Kaplowitz N (2004) Drug-induced liver injury. *Clin Infect Dis* 38 Suppl 2: S44–48
55 Maddrey WC (2005) Drug-induced hepatotoxicity: 2005. *J Clin Gastroenterol* 39: S83–89
56 Kalgutkar AS, Gardner I, Obach RS, Shaffer CL, Callegari E, Henne KR, Mutlib AE, Dalvie DK, Lee JS, Nakai Y et al (2005) A comprehensive listing of bioactivation pathways of organic functional groups. *Curr Drug Metab* 6: 161–225
57 Steiner G, Suter L, Boess F, Gasser R, de Vera MC, Albertini S, Ruepp S (2004) Discriminating different classes of toxicants by transcript profiling. *Environ Health Perspect* 112: 1236–1248
58 Kon K, Kim JS, Jaeschke H, Lemasters JJ (2004) Mitochondrial permeability transition in acetaminophen-induced necrosis and apoptosis of cultured mouse hepatocytes. *Hepatology* 40: 1170–1179
59 Amacher DE (2005) Drug-associated mitochondrial toxicity and its detection. *Curr Med Chem* 12: 1829–1839
60 Liu ZX, Govindarajan S, Kaplowitz N (2004) Innate immune system plays a critical role in determining the progression and severity of acetaminophen hepatotoxicity. *Gastroenterology* 127: 1760–1774
61 Laskin DL, Laskin JD (2001) Role of macrophages and inflammatory mediators in chemically induced toxicity. *Toxicology* 160: 111–118
62 James LP, Simpson PM, Farrar HC, Kearns GL, Wasserman GS, Blumer JL, Reed MD, Sullivan JE, Hinson JA (2005) Cytokines and toxicity in acetaminophen overdose. *J Clin Pharmacol* 45: 1165–1171
63 Ishida Y, Kondo T, Tsuneyama K, Lu P, Takayasu T, Mukaida N (2004) The pathogenic roles of tumor necrosis factor receptor p55 in acetaminophen-induced liver injury in mice. *J Leukoc Biol* 75: 59–67
64 Ito Y, Bethea NW, Abril ER, McCuskey RS (2003) Early hepatic microvascular injury in response to acetaminophen toxicity. *Microcirculation* 10: 391–400
65 Fountoulakis M, Berndt P, Boelsterli UA, Crameri F, Winter M, Albertini S, Suter L (2000) Two-dimensional database of mouse liver proteins: Changes in hepatic protein levels following treatment with acetaminophen or its nontoxic regioisomer 3-acetamidophenol. *Electrophoresis* 21: 2148–2161
66 Tonge R, Shaw J, Middleton B, Rowlinson R, Rayner S, Young J, Pognan F, Hawkins E, Currie I, Davison M (2001) Validation and development of fluorescence two-dimensional differential gel electrophoresis proteomics technology. *Proteomics* 1: 377–396
67 Ruepp SU, Tonge RP, Shaw J, Wallis N, Pognan F (2002) Genomics and proteomics analysis of acetaminophen toxicity in mouse liver. *Toxicol Sci* 65: 135–150
68 Thome-Kromer B, Bonk I, Klatt M, Nebrich G, Taufmann M, Bryant S, Wacker U, Köpke A (2003) Toward the identification of liver toxicity markers: A proteome study in human cell culture and rats. *Proteomics* 3: 1835–1862
69 Kikkawa R, Yamamoto T, Fukushima T, Yamada H, Horii I (2005) Investigation of a hepatotoxicity screening system in primary cell cultures – "what biomarkers would need to be addressed to estimate toxicity in conventional and new approaches?" *J Toxicol Sci* 30: 61–72
70 Yamamoto T, Kikkawa R, Yamada H, Horii I (2005) Identification of oxidative stress-related proteins for predictive screening of hepatotoxicity using a proteomic approach. *J Toxicol Sci* 30: 213–227
71 Welch KD, Wen B, Goodlett DR, Yi EC, Lee H, Reilly TP, Nelson SD, Pohl LR (2005) Proteomic

identification of potential susceptibility factors in drug-induced liver disease. *Chem Res Toxicol* 18: 924–933

72 Lee H, Yi EC, Wen B, Reily TP, Pohl L, Nelson S, Aebersold R, Goodlett DR (2004) Optimization of reversed-phase microcapillary liquid chromatography for quantitative proteomics. *J Chromatogr B Analyt Technol Biomed Life Sci* 803: 101–110

73 Weber LW, Boll M, Stampfl A (2003) Hepatotoxicity and mechanism of action of haloalkanes: Carbon tetrachloride as a toxicological model. *Crit Rev Toxicol* 33: 105–136

74 Liu Y, Liu P, Liu CH, Hu YY, Xu LM, Mu YP, Du GL (2005) [Proteomic analysis of proliferation and apoptosis in carbon tetrachloride induced rat liver fibrosis]. *Zhonghua Gan Zang Bing Za Zhi* 13: 563–566

75 Nelson SD (1995) Mechanisms of the formation and disposition of reactive metabolites that can cause acute liver injury. *Drug Metab Rev* 27: 147–177

76 Heijne WH, Stierum RH, Slijper M, van Bladeren PJ, van Ommen B (2003) Toxicogenomics of bromobenzene hepatotoxicity: A combined transcriptomics and proteomics approach. *Biochem Pharmacol* 65: 857–875

77 Staels B, Fruchart JC (2005) Therapeutic roles of peroxisome proliferator-activated receptor agonists. *Diabetes* 54: 2460–2470

78 Iida M, Anna CH, Hartis J, Bruno M, Wetmore B, Dubin JR, Sieber S, Bennett L, Cunningham ML, Paules RS et al (2003) Changes in global gene and protein expression during early mouse liver carcinogenesis induced by non-genotoxic model carcinogens oxazepam and Wyeth-14,643. *Carcinogenesis* 24: 757–770

79 Edvardsson U, von Lowenhielm HB, Panfilov O, Nystrom AC, Nilsson F, Dahllöf B (2003) Hepatic protein expression of lean mice and obese diabetic mice treated with peroxisome proliferator-activated receptor activators. *Proteomics* 3: 468–478

80 Richards VE, Chau B, White MR, McQueen CA (2004) Hepatic gene expression and lipid homeostasis in C57BL/6 mice exposed to hydrazine or acetylhydrazine. *Toxicol Sci* 82: 318–332

81 Kleno TG, Kiehr B, Baunsgaard D, Sidelmann UG (2004) Combination of 'omics' data to investigate the mechanism(s) of hydrazine-induced hepatotoxicity in rats and to identify potential biomarkers. *Biomarkers* 9: 116–138

82 Kleno TG, Leonardsen LR, Kjeldal HO, Laursen SM, Jensen ON, Baunsgaard D (2004) Mechanisms of hydrazine toxicity in rat liver investigated by proteomics and multivariate data analysis. *Proteomics* 4: 868–880

83 Chilakapati J, Shankar K, Korrapati MC, Hill RA, Mehendale HM (2005) Saturation toxicokinetics of thioacetamide: Role in initiation of liver injury. *Drug Metab Dispos* 33: 1877–1885

84 Low TY, Leow CK, Salto-Tellez M, Chung MC (2004) A proteomic analysis of thioacetamide-induced hepatotoxicity and cirrhosis in rat livers. *Proteomics* 4: 3960–3974

85 Xu XQ, Leow CK, Lu X, Zhang X, Liu JS, Wong WH, Asperger A, Deininger S, Eastwood Leung HC (2004) Molecular classification of liver cirrhosis in a rat model by proteomics and bioinformatics. *Proteomics* 4: 3235–3245

86 Kawada N, Kristensen DB, Asahina K, Nakatani K, Minamiyama Y, Seki S, Yoshizato K (2001) Characterization of a stellate cell activation-associated protein (STAP) with peroxidase activity found in rat hepatic stellate cells. *J Biol Chem* 276: 25318–25323

87 Omenn GS, States DJ, Adamski M, Blackwell TW, Menon R, Hermjakob H, Apweiler R, Haab BB, Simpson RJ, Eddes JS et al (2005) Overview of the HUPO Plasma Proteome Project: Results from the pilot phase with 35 collaborating laboratories and multiple analytical groups, generating a core dataset of 3020 proteins and a publicly-available database. *Proteomics* 5: 3226–3245

88 Ping P, Vondriska TM, Creighton CJ, Gandhi TK, Yang Z, Menon R, Kwon MS, Cho SY, Drwal G, Kellmann M et al (2005) A functional annotation of subproteomes in human plasma. *Proteomics* 5: 3506–3519

89 Duan X, Yarmush DM, Berthiaume F, Jayaraman A, Yarmush ML (2004) A mouse serum two-dimensional gel map: Application to profiling burn injury and infection. *Electrophoresis* 25: 3055–3065

90 Gianazza E, Eberini I, Villa P, Fratelli M, Pinna C, Wait R, Gemeiner M, Miller I (2002) Monitoring the effects of drug treatment in rat models of disease by serum protein analysis. *J Chromatogr B Analyt Technol Biomed Life Sci* 771: 107–130

91 Duan X, Yarmush D, Berthiaume F, Jayaraman A, Yarmush ML (2005) Immunodepletion of albumin for two-dimensional gel detection of new mouse acute-phase protein and other plasma proteins. *Proteomics* 5: 3991–4000

92 Wait R, Chiesa G, Parolini C, Miller I, Begum S, Brambilla D, Galluccio L, Ballerio R, Eberini I, Gianazza E (2005) Reference maps of mouse serum acute-phase proteins: Changes with LPS-induced inflammation and apolipoprotein A-I and A-II transgenes. *Proteomics* 5: 4245–4253

93 Amacher DE, Adler R, Herath A, Townsend RR (2005) Use of proteomic methods to identify serum biomarkers associated with rat liver toxicity or hypertrophy. *Clin Chem* 51: 1796–1803

94 Witzmann FA, Li J (2004) Proteomics and nephrotoxicity. *Contrib Nephrol* 141: 104–123

95 Davis JW, Kramer JA (2006) Genomic-based biomarkers of drug-induced nephrotoxicity. *Expert Opin Drug Metab Toxicol* 2: 95–101

96 Janech MG, Raymond JR, Arthur JM (2007) Proteomics in renal research. *Am J Physiol Renal Physiol* 292: F501–512

97 Mihatsch MJ, Thiel G, Ryffel B (1989) Cyclosporin A: Action and side-effects. *Toxicol Lett* 46: 125–139

98 Benito B, Wahl D, Steudel N, Cordier A, Steiner S (1995) Effects of cyclosporine A on the rat liver and kidney protein pattern, and the influence of vitamin E and C coadministration. *Electrophoresis* 16: 1273–1283

99 Steiner S, Aicher L, Raymackers J, Meheus L, Esquer-Blasco R, Anderson NL, Cordier A (1996) Cyclosporine A decreases the protein level of the calcium-binding protein calbindin-D 28 kDa in rat kidney. *Biochem Pharmacol* 51: 253–258

100 Lee CT, Huynh VM, Lai LW, Lien YH (2002) Cyclosporine A-induced hypercalciuria in calbindin-D28k knockout and wild-type mice. *Kidney Int* 62: 2055–2061

101 Serkova N, Christians U (2003) Transplantation: Toxicokinetics and mechanisms of toxicity of cyclosporine and macrolides. *Curr Opin Investig Drugs* 4: 1287–1296

102 Mascarell L, Frey JR, Michel F, Lefkovits I, Truffa-Bachi P (2000) Increased protein synthesis after T cell activation in presence of cyclosporin A. *Transplantation* 70: 340–348

103 Mascarell L, Auger R, Alcover A, Ojcius DM, Jungas T, Cadet-Daniel V, Kanellopoulos JM, Truffa-Bachi P (2004) Characterization of a gene encoding two isoforms of a mitochondrial protein up-regulated by cyclosporin A in activated T cells. *J Biol Chem* 279: 10556–10563

104 Guan N, Ding J, Deng J, Zhang J, Yang J (2004) Key molecular events in puromycin aminonucleoside nephrosis rats. *Pathol Int* 54: 703–711

105 Sundin DP, Meyer C, Dahl R, Geerdes A, Sandoval R, Molitoris BA (1997) Cellular mechanism of aminoglycoside tolerance in long-term gentamicin treatment. *Am J Physiol* 272: C1309–1318

106 Witzmann FA, Fultz CD, Grant RA, Wright LS, Kornguth SE, Siegel FL (1998) Differential expression of cytosolic proteins in the rat kidney cortex and medulla: Preliminary proteomics. *Electrophoresis* 19: 2491–2497

107 Shakib K, Norman JT, Fine LG, Brown LR, Godovac-Zimmermann J (2005) Proteomics profiling of nuclear proteins for kidney fibroblasts suggests hypoxia, meiosis, and cancer may meet in the nucleus. *Proteomics* 5: 2819–2838

108 Charlwood J, Skehel JM, King N, Camilleri P, Lord P, Bugelski P, Atif U (2002) Proteomic analysis of rat kidney cortex following treatment with gentamicin. *J Proteome Res* 1: 73–82

109 Thongboonkerd V, Klein JB, Arthur JM (2003) Proteomic identification of a large complement of rat urinary proteins. *Nephron Exp Nephrol* 95: e69–78

110 Cutler P, Bell DJ, Birrell HC, Connelly JC, Connor SC, Holmes E, Mitchell BC, Monte SY, Neville BA, Pickford R et al (1999) An integrated proteomic approach to studying glomerular nephrotoxicity. *Electrophoresis* 20: 3647–3658

111 Crowe CA, Yong AC, Calder IC, Ham KN, Tange JD (1979) The nephrotoxicity of *p*-aminophenol. I. The effect on microsomal cytochromes, glutathione and covalent binding in kidney and liver. *Chem Biol Interact* 27: 235–243

112 Kaltenbach JP, Carone FA, Ganote CE (1982) Compounds protective against renal tubular necrosis induced by D-serine and D-2,3-diaminopropionic acid in the rat. *Exp Mol Pathol* 37: 225–234

113 Kuhlmann MK, Horsch E, Burkhardt G, Wagner M, Kohler H (1998) Reduction of cisplatin toxicity in cultured renal tubular cells by the bioflavonoid quercetin. *Arch Toxicol* 72: 536–540

114 Bandara LR, Kelly MD, Lock EA, Kennedy S (2003) A correlation between a proteomic evaluation and conventional measurements in the assessment of renal proximal tubular toxicity. *Toxicol Sci* 73: 195–206

115 Bandara LR, Kelly MD, Lock EA, Kennedy S (2003) A potential biomarker of kidney damage identified by proteomics: Preliminary findings. *Biomarkers* 8: 272–286

116 Chen JC, Stevens JL, Trifillis AL, Jones TW (1990) Renal cysteine conjugate β-lyase-mediated toxicity studied with primary cultures of human proximal tubular cells. *Toxicol Appl Pharmacol*

103: 463–473

117 Lash LH, Qian W, Putt DA, Hueni SE, Elfarra AA, Krause RJ, Parker JC (2001) Renal and hepatic toxicity of trichloroethylene and its glutathione-derived metabolites in rats and mice: Sex-, species-, and tissue-dependent differences. *J Pharmacol Exp Ther* 297: 155–164

118 de Graauw M, Tijdens I, Cramer R, Corless S, Timms JF, van de Water B (2005) Heat shock protein 27 is the major differentially phosphorylated protein involved in renal epithelial cellular stress response and controls focal adhesion organization and apoptosis. *J Biol Chem* 280: 29885–29898

119 de Graauw M, Le Devedec S, Tijdens I, Smeets MB, Deelder AM, van de Water B (2007) Proteomic analysis of alternative protein tyrosine phosphorylation in 1,2-dichlorovinyl-cysteine-induced cytotoxicity in primary cultured rat renal proximal tubular cells. *J Pharmacol Exp Ther* 322: 89–100

120 Korrapati MC, Chilakapati J, Witzmann FA, Rao C, Lock EA, Mehendale HM (2007) Proteomics of S-(1,2-dichlorovinyl)-L-cysteine-induced acute renal failure and autoprotection in mice. *Am J Physiol Renal Physiol* 293: F994–F1006

121 Vercauteren FG, Bergeron JJ, Vandesande F, Arckens L, Quirion R (2004) Proteomic approaches in brain research and neuropharmacology. *Eur J Pharmacol* 500: 385–398

122 Sheta EA, Appel SH, Goldknopf IL (2006) 2D gel blood serum biomarkers reveal differential clinical proteomics of the neurodegenerative diseases. *Expert Rev Proteomics* 3: 45–62

123 Merten KE, Feng W, Zhang L, Pierce W, Cai J, Klein JB, Kang YJ (2005) Modulation of cytochrome C oxidase-va is possibly involved in metallothionein protection from doxorubicin cardiotoxicity. *J Pharmacol Exp Ther* 315: 1314–1319

124 Elased KM, Cool DR, Morris M (2005) Novel mass spectrometric methods for evaluation of plasma angiotensin converting enzyme 1 and renin activity. *Hypertension* 46: 953–959

125 Yamamoto T, Fukushima T, Kikkawa R, Yamada H, Horii I (2005) Protein expression analysis of rat testes induced testicular toxicity with several reproductive toxicants. *J Toxicol Sci* 30: 111–126

126 Yang YH, Xi ZG, Chao FH, Yang DF (2005) Effects of formaldehyde inhalation on lung of rats. *Biomed Environ Sci* 18: 164–168

127 Wheelock AM, Boland BC, Isbell M, Morin D, Wegesser TC, Plopper CG, Buckpitt AR (2005) *In vivo* effects of ozone exposure on protein adduct formation by 1-nitronaphthalene in rat lung. *Am J Respir Cell Mol Biol* 33: 130–137

128 Wheelock AM, Zhang L, Tran MU, Morin D, Penn S, Buckpitt AR, Plopper CG (2004) Isolation of rodent airway epithelial cell proteins facilitates *in vivo* proteomics studies of lung toxicity. *Am J Physiol Lung Cell Mol Physiol* 286: L399–410

129 Collins MO, Yu L, Husi H, Blackstock WP, Choudhary JS, Grant SG (2005) Robust enrichment of phosphorylated species in complex mixtures by sequential protein and peptide metal-affinity chromatography and analysis by tandem mass spectrometry. *Sci STKE* 2005: pl6

130 Kim SY, Chudapongse N, Lee SM, Levin MC, Oh JT, Park HJ, Ho IK (2004) Proteomic analysis of phosphotyrosyl proteins in the rat brain: Effect of butorphanol dependence. *J Neurosci Res* 77: 867–877

131 Wang M, Xiao GG, Li N, Xie Y, Loo JA, Nel AE (2005) Use of a fluorescent phosphoprotein dye to characterize oxidative stress-induced signaling pathway components in macrophage and epithelial cultures exposed to diesel exhaust particle chemicals. *Electrophoresis* 26: 2092–2108

132 Gagna CE, Winokur D, Clark Lambert W (2004) Cell biology, chemogenomics and chemoproteomics. *Cell Biol Int* 28: 755–764

133 Beillard E, Witte ON (2005) Unraveling kinase signaling pathways with chemical genetic and chemical proteomic approaches. *Cell Cycle* 4: 434–437

134 Gao J, Garulacan LA, Storm SM, Opiteck GJ, Dubaquie Y, Hefta SA, Dambach DM, Dongre AR (2005) Biomarker discovery in biological fluids. *Methods* 35: 291–302

135 Merrick BA, Tomer KB (2003) Toxicoproteomics: A parallel approach to identifying biomarkers. *Environ Health Perspect* 111: A578–579

136 Quackenbush J (2005) Extracting meaning from functional genomics experiments. *Toxicol Appl Pharmacol* 207: 195–199

137 Fostel J, Choi D, Zwickl C, Morrison N, Rashid A, Hasan A, Bao W, Richard A, Tong W, Bushel PR et al (2005) Chemical effects in biological systems–data dictionary (CEBS-DD): A compendium of terms for the capture and integration of biological study design description, conventional phenotypes, and 'omics data. *Toxicol Sci* 88: 585–601

138 Kristensen DB, Kawada N, Imamura K, Miyamoto Y, Tateno C, Seki S, Kuroki T, Yoshizato K (2000) Proteome analysis of rat hepatic stellate cells. Hepatology 32: 268–277

139 Reinheckel T, Korn S, Mohring S, Augustin W, Halangk W, Schild L (2000) Adaptation of protein carbonyl detection to the requirements of proteome analysis demonstrated for hypoxia/reoxygenation in isolated rat liver mitochondria. *Arch Biochem Biophys* 376: 59–65

140 Witzmann FA, Fultz CD, Grant RA, Wright LS, Kornguth SE, Siegel FL (1999) Regional protein alterations in rat kidneys induced by lead exposure. *Electrophoresis* 20: 943–951

Molecular, Clinical and Environmental Toxicology. Volume 1: Molecular Toxicology
Edited by A. Luch
© 2009 Birkhäuser Verlag/Switzerland

High-throughput screening for analysis of *in vitro* toxicity

Willem G. E. J. Schoonen, Walter M. A. Westerink and G. Jean Horbach

Department of Pharmacology, NV Organon (Schering-Plough), Oss, The Netherlands

Abstract. The influence of combinatorial chemistry and high-throughput screening (HTS) technologies in the pharmaceutical industry during the last 10 years has been enormous. However, the attrition rate of drugs in the clinic due to toxicity during this period still remained 40–50%. The need for reduced toxicity failure led to the development of early toxicity screening assays. This chapter describes the state of the art for assays in the area of genotoxicity, cytotoxicity, carcinogenicity, induction of specific enzymes from phase I and II metabolism, competition assays for enzymes of phase I and II metabolism, embryotoxicity as well as endocrine disruption and reprotoxicity. With respect to genotoxicity, the full Ames, Ames II, Vitotox™, GreenScreen GC, RadarScreen, and non-genotoxic carcinogenicity assays are discussed. For cytotoxicity, cellular proliferation, calcein uptake, oxygen consumption, mitochondrial activity, radical formation, glutathione depletion as well as apoptosis are described. For high-content screening (HCS), the possibilities for analysis of cytotoxicity, micronuclei, centrosome formation and phospholipidosis are examined. For embryotoxicity, endocrine disruption and reprotoxicity alternative assays are reviewed for fast track analysis by means of nuclear receptors and membrane receptors. Moreover, solutions for analyzing enzyme induction by activation of nuclear receptors, like AhR, CAR, PXR, PPAR, FXR, LXR, TR and RAR are given.

Introduction

During the last 10 years the pharmaceutical industry has been overwhelmed by combinatorial chemistry and high-throughput screening (HTS) technologies. This not only resulted in huge libraries of compounds for screening on the biological targets, but also in a greater chemical diversity of potential drug candidates. Therefore, new, fast and predictive *in vitro* toxicology tests were needed to select the best drug candidates before the toxicological studies with animals had started. An additional reason for these *in vitro* toxicity assays is the current low overall success rate of newly developed drug candidates to reach the market. Only 1 out of 10 compounds reaches the market and toxicity is the main reason for this failure (40–50%) [1, 2]. Reduction of this attrition rate is one of the main challenges of the pharmaceutical industry leading to a sharp reduction of costs in drug development.

A better *in vitro* pre-screening strategy with improved selection criteria must certainly lead to a better selection of drug candidates and therefore to a reduction in the costs of clinical development. To approach this improvement, the previous observed unwanted pharmacological and toxicological side effects from compounds should be taken into account to improve the quality of the new medicines for further fine tuning and optimization. Between 1990 and 2007, there was a rather sharp variation in the cause of rejection in the development of new drugs [3]. In 1991, bad pharmacokinetics and bioavailability was, with 40%, the main justification to block further clinical development of a compound. In 2000, more predictive pre-clinical *in vitro* and *in vivo* assays reduced these phenomena towards only 10%. As a consequence of this decrease, clinical failure in drug development was shifted towards effectiveness, and toxicological and pharmacological safety. Nowadays, the combined failure rate on the toxic aspects even appears to be 40–50% [4–7].

Unfortunately, in 2006–2007 the failure values due to toxicity remained unchanged [4–7]. This only emphasizes the need for better and more predictive HTS *in vitro* toxicity tests. With the introduction of combinatorial chemistry, the number of synthesized compounds increased exponentially, while the amount of compound available was reduced simultaneously by at least 100- to 1000-fold towards 0.1 to 1 mg. This implied that for bioactivity measurements in animal models the new compounds should be extremely potent at 1–10 µg/kg and juvenile mice or rat studies were performed for the bioassays to minimize the amounts of compound needed. Another way to solve these quantitative problems was the development of *in vitro* bioactivity and toxicity assays with the highest priority. An additional and beneficial effect of these *in vitro* tests will be the reduction of animal studies that otherwise would have been needed in early pre-clinical development.

State of the art of HTS for *in vitro* bioassays within pharmacology

New molecular biological techniques developed in the eighties and nineties have made it possible to introduce specific DNA sequences by means of plasmids into cells and finally into the chromosomes of bacterial or mammalian cells. These so-called transient or stable transfections of cells led to the development of very sensitive assays for a diverse set of nuclear receptors [8–13] as well as for different classes of membrane receptors [9, 14, 15].

For the nuclear receptors (NR), it became apparent that specific DNA sequences in the promoter of a gene, the hormone responsive elements, could bind homodimeric or heterodimeric NR (Fig. 1A). For specific membrane receptors, the so-called G protein-coupled receptors (GPCR), a link could be made with the adenylate cyclase protein and the production of cyclic adenosine monophosphate (cAMP). This cAMP could be measured directly with ALPHAscreen® or dissociation-enhanced lanthanide fluorescence immunoassay (DELFIA) [16], or the cAMP could activate protein kinase A that phosphorylates the cAMP responsive element binding (CREB) protein. This activated CREB protein could also bind in a particular responsive element promoter setting of a gene [15, 17] (Fig. 1B). Moreover, if the original gene behind

A

Steroid receptor

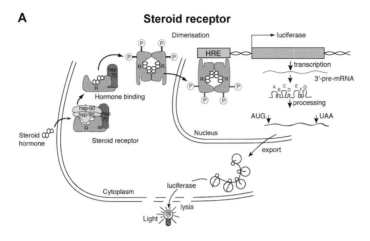

B

Membrane receptor

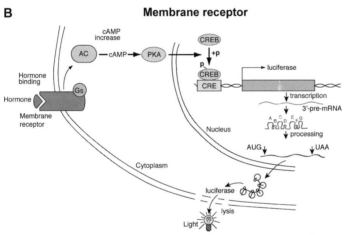

Figure 1. Mechanism of steroid receptor activation and membrane-coupled G protein receptor activation. A) The compound enters the cell through the cellular plasma membrane by diffusion and binds to the receptor (R), which leads first to the dissociation of heat shock protein (hsp) 90 and second of hsp70 during dimerization. During the dimerization process, the receptor also becomes activated by phosphorylation. This homodimeric steroid receptor or in some cases the heterodimeric nuclear receptor can pass the nuclear membrane. In the nucleus the dimeric receptor can bind to the hormone responsive element (HRE) in a particular promoter and the gene downstream to this promoter can be transcribed into mRNA and translated into protein. In case of steroid reporter based assays, this protein can be, e.g., luciferase, β-galactosidase or alkaline phosphatase. B) The compound binds to the membrane G protein receptor and induces the formation of cAMP *via* activation of adenylate cyclase (AC). cAMP then activates protein kinase A (PKA) and PKA phosphorylates the cAMP responsive element binding (CREB) protein, which in turn binds to the cAMP responsive element (CRE) in a particular promoter. The gene behind this promoter can be transcribed into mRNA and translated into protein. As above, this gene might be luciferase, β-galactosidase or alkaline phosphatase.

this promoter was replaced by either luciferase, β-galactosidase or alkaline phosphatase, these enzyme activities could be measured with luminometric assays at levels of 10^{-13}–10^{-18} M. These luminometric assays are much more sensitive than the spectrophotometric (10^{-2}–10^{-5} M), fluorometric (10^{-5}–10^{-8} M), and tritium (^{3}H) or iodine (^{125}I) radiolabeled (10^{-8}–10^{-12} M) measurements. Within the pharmaceutical industry this strategy has opened the possibility for fast screening devices with medium- and high-throughput screening (MTS and HTS). The analysis went from 96- into 384-well plates, which became common use in the nineties. Even 1536-well plates are in use since the start of the 21th century as the pipetting tasks with the appropriate robotic equipment became feasible. Drug discovery today heavily depends on the intensive and massive robotized screening devices in combination with large compound libraries and a great diversity of *in vitro* cell-based target assays. Especially in the pharmaceutical industry, *in vitro* screening for NR and GPCR, ion channels and enzymes, like tyrosine- or serine-dependent kinases with fluorescence resonance electron transfer (FRET), are really accepted and frequently performed on a large scale [18, 19]. These assays can all be performed in 96-, 384- and 1536-well plates at room temperature or in incubators at the appropriate temperature. The assays last for only 2–30 min for ion channel and enzyme analysis, 2–4 h for GPCR or 16–24 h for NR. The number of compounds measured per day can vary per assay, depending on the number of concentrations measured per compound, and the number of replicates. But one can say that, despite the diversification of measurements in between 28 000 and 156 000 compounds per day at one concentration can be analyzed.

State of the art of HTS for *in vitro* bioassays within toxicology

The first *in vitro* toxicological screening procedures were set-up in the seventies with the Ames test [20, 21] and the eighties with acute cytotoxicity testing with different kinds of cell lines, and chromosomal aberration as well as micronucleus tests with lymphocytes [22, 23]. The throughput in these assay systems was, with 2 up to 10 compounds per week, relatively low and inferior to HTS. Also, for genotoxicity assessment, molecular biological tools were introduced to develop assays for *Salmonella* and yeast strains. These assays examined the promoter of SOS repair genes in *Salmonella* or the promoter of genes involved in the restoration of chromosomal damage in yeast. In addition, these new techniques lead to the development of assays for cytotoxicity, cellular metabolic activation processes, cytochrome P450 competition, embryotoxicity and endocrine disruption. However, these relatively new assays still need in some cases the proper validation, recommendation, but most importantly the general acceptance from toxicologists. Only this will allow the use of these tools for the fast identification of unwanted pharmacological or toxicological side effects at an early stage of the drug development process. It may also help if the most relevant toxic chemicals are tested before analysis is started with animal studies. This is also relevant in the light of 'Registration, Evaluation, Authorization and restriction of Chemicals' (REACH), a new European Community Regulation on chemicals and their safe use. Here the demands are to test 30 000 compounds in animal studies within the next 10–15 years. These fast *in vitro* studies may help to classify the compound's toxicity in combination with quantitative structure activity relationship (QSAR) analysis for mutagenicity with DEREK, TopCat and MultiCASE [24–26].

Strategy for the development of non-toxic compounds

Within the pharmaceutical industry the drug developmental process can be divided into three main working groups (Fig. 2). A lead finding team analyzes new biological targets and new molecules for their cognate receptors or proteins. After the development of the particular assay(s) and the identification of the best lead molecule within a period of 2 years, this lead molecule is transferred into the lead optimization phase. This lead optimization phase takes 2–4 years to develop a compound that is biologically active in an animal model either after oral, subcutaneous or intravenous treatment, after which this developmental candidate is taken over by the proof-of-concept team. This proof-of-concept team is responsible for performing the pre-clinical animal toxicity studies and the first and second sets of clinical pharmacology and safety studies within 4 years. The venture team is then responsible for the further launch of the product, including registration and marketing, within the following 4 years. This can result in an overall development time of 10–16 years for a new drug. As one can imagine, the development of new drugs really demands special skills like team work.

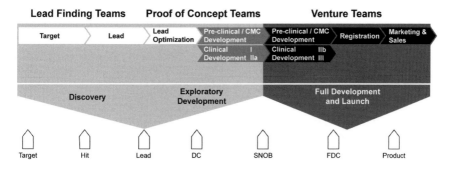

Figure 2. The research and development strategy. The three teams of lead finding, proof of concept and venture are together responsible for the marketing of a product. The time lines of approximately 16 years of development can be cut into smaller parts. The first phase of development concerns the target and lead finding with lead optimization and takes around 4 years. The second phase can be separated in (pre-)clinical development Phase I, which is performed in parallel with the clinical development Phase IIa. The third phase can be separated into (pre-)clinical development Phases II and III, which are followed by registration, marketing and sales. For the compounds the selection can be degraded into the selection of the hit, the lead, the developmental candidate (DC), statement of no objection (SNOB), full development candidate (FDC) and the product.

Within NV Organon, it was assessed from historical data analysis of compounds developed in between 1960 and 2000, that 50% of the candidate drugs were withdrawn from (pre-)clinical development due to toxicological problems. Identification, categorization and comparison of this historical set of compounds with a similar comparative analysis within Roche showed that the toxicity causes between the two companies were focused on completely different items (Tab. 1). For Roche's compounds, the most frequently identified failures were due to toxicity to the liver, the cardiovascular system, the skin or neuronal tissue. Since Organon is active in the field of contraceptives with estrogenic and progestagenic compounds, as well as in hormone-replacement therapy with estrogenic compounds, 40% of the toxic events at Organon were attributed to reproductive toxicity and genotoxicity. On the other hand, the more general toxic effects of (sub)acute, subchronic and chronic toxicity still accounted for 48% of the overall toxicity. These toxic effects could be categorized into 12% for each of the following four groups, i.e., hepatotoxicity, cardiotoxicity, nephrotoxicity or other organ toxicities (such as thymus, spleen, pituitary, lungs, pancreas, and gall bladder).

Table 1. Percentage of failure for toxicity reasons.

Toxicity failure	Organon	Roche
Genotoxicity/carcinogenicity	20	6
Reproductive toxicity	20	2
Hepatotoxicity	12	20
Cardiovascular safety	12	16
Skin toxicity	x	10
CNS side effects	x	10
Blood toxicity	x	6
Renal toxicity	12	4
Gastrointestinal toxicity	×	4

$\Sigma x = 12$

Thus, within NV Organon, a clear focus on these toxic aspects during lead optimization would be most beneficial for reducing the attrition rate. Still, only a small fraction of the causes of drug failure can easily be solved with a "yes-no" approach, such as genotoxicity and embryotoxicity, whereas the other types of toxicity might at best only lead to a ranking of compounds and a prioritization or pre-selection in the pre-clinical research phase. This apprehension of the toxicological concerns led, within Organon, to the development of a strategy to initiate HTS assays for the measurement of DNA and chromosomal damage, as well as for embryotoxicity. Subsequently, assays were developed to study (sub)acute cellular toxicity, followed by the up- or down-regulation of the activities of phase I and II enzymes by means of NR activation assays or enzyme competition assays. Since in nature endocrine disruption is mainly caused by estrogens, androgens, progestagens and/or glucocorticoids, these assays were developed as a spin-off from previous studies on the main endocrinological and pharmacological effects as well as on their main possible side effects. These endocrine studies were set-up in combination with the development programs of steroids for hormone-replacement therapy, female and male contraception as well as for immunosuppression.

Implementation of HTS *in vitro* toxicity analysis within the pharmaceutical industry

The implementation of *in vitro* toxicity assays is done within the pharmaceutical industry on different time points in the development route (Fig. 3). Simple QSAR analysis for mutagenicity is carried out *in silico* with the software program DEREK, TopCat, MultiCASE and an internally up-dated software program Mutalert. The chemical structures of the identified hits from each lead

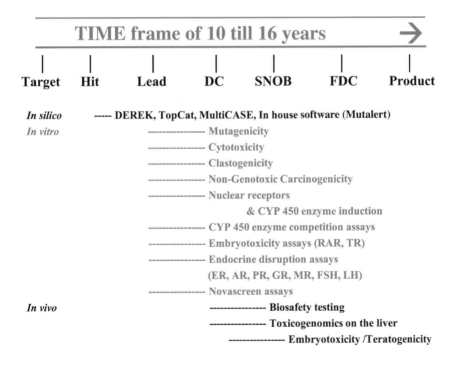

Figure 3. In relation to the time lines of the development strategy the *in vitro* toxicity studies can be implemented at several points in the time frame of development. Four points of implementation can be divided: *in silico* with DEREK, TopCat, MultiCASE and in house software (Mutalert); *in vitro* (light gray), and *in vivo* just at the start of the early toxicity profiling of 2 weeks; and later in the preclinical development Phase I or II. DC, developmental candidate; SNOB, statement of no objection; and FDC, full development candidate.

finding team are examined for mutagenic alerts. The first simple genotoxicity and cytotoxicity screens can take place at the start of the lead optimization phase, when a sufficient amount of compound is synthesized. Due to the relatively high amount of compound needed to prepare a stock solution of 10^{-1} M (approximately 10–20 mg high purity, and preferably crystallized compound), these assays are mostly postponed till the second or third year of the lead optimization phase. At this time point, the NR activation processes can also be studied as well as competition assays of particular cytochrome P450 enzymes (Fig. 3). Embryotoxicity and teratogenicity tests are performed during the clinical development phase. However, some assays are in development to measure the effects on vitamin A and thyroid hormone receptors somewhat earlier in the lead optimization process. This can also be done for endocrine disruption with the in-house available steroid hormone and membrane receptors or by outsourcing to Novascreen for binding analysis. In principle, with a stock solution of 300–500 µl of 10^{-1} M in DMSO, all the toxicity assay screens can be performed within 4 weeks. Therefore, parallel screening formats for a subset of assays are performed on a weekly basis, with a cycle time of once a month.

Screening procedures for genotoxicity

There are several ways to measure the induction of DNA damage. This can be done at the level of (1) the nucleotide, (2) the chromosome, and (3) the nucleus. For each of these levels *in vitro* genotoxicity tests have been developed [22, 23, 27] and the most common assays used are (1) the Ames test, (2) chromosomal sister chromatid exchange and chromosomal aberrations assays as well as the COMET test, and (3) the micronucleus test and analysis of centrosome segregation. For levels 1 and 2, HTS assays are or becoming available with Vitotox[TM], GreenScreen GC (yeast RAD54-GFP), RadarScreen and GreenScreen HC (human GADD45α-GFP) assays. These assays are described in relation to the current assay set-ups of Ames and chromosomal aberrations. For level 3, the feasibility of HTS has been arranged by means of high-content screening (HCS) or bioimaging techniques.

The nucleotide level

Simple restoration of a point or frameshift mutation in a gene involved in the production of histidine is used in the Ames mutagenicity test [20, 21]. The Ames test uses different *Salmonella typhimurium* strains, i.e., TA98 and TA1537 for frameshift mutations and TA100 and TA1535 for base pair substitutions [28]. After the incubation of the bacteria with a concentration range of the compound to be tested, the bacteria are grown in media without histidine, in which only revertants will survive. The number of these revertants is a measure for mutagenicity. Analysis of the cloning efficiency of these revertants can be performed by manual counting of the number of clones and/or using a dye for clone staining followed by spectrophotometric quantification. These incubations are mostly done in the absence and presence of a rat liver S9 fraction. This S9 fraction is a 9000 *g* supernatant of the liver homogenate of rats that were pre-treated for 24 h with Aroclor 1254 or a combination of compounds that are capable of activating the aryl hydrocarbon receptor (AhR), pregnane X receptor (PXR) and the constitutive androstane receptor (CAR). Commonly used agonistic compounds are β-naphthoflavone or 3-methylcholanthrene for AhR, rifampicin for PXR and phenobarbital for CAR. This treatment of the rat leads to the enhancement of liver metabolism by means of phase I enzymes and, in particular, the activation of cytochrome P450 enzymes. Thus, the S9 liver fraction stimulates the conversion of test compound into metabolites in the *in vitro* test system. The most commonly used reference compounds in the absence of S9 fraction are 4-nitroquinoline-1-oxide (4-NQO), methyl methanesulfonate (MMS), and nitropyrene, while in the presence of S9 fraction benzo[*a*]pyrene (B[*a*]P), 2-acetylaminofluorene (AAF), 7,12-dimethylbenz[*a*]anthracene, aflatoxin B_1, 2,7-dinitrofluorene, nitrofurantoin, and cyclophosphamide are used [29, 30]. Several labs have automated the full Ames assay or an assay of one of the particular strains into a 384-well plate method. Besides this full Ames test, a fast track analysis with the Ames II assay kit also exists [31, 32] using only the TA98 strain and TA mix, which is a mixture of strains TA7001, TA7002, TA7003, TA7004, TA7005, and TA7006 (Xenometrix, Osteomedical Group, Nijkerk, The Netherlands). This test is, however, according to in-house data sets less predictive than the full Ames and Vitotox[TM] tests with respect to less potent genotoxicants (data not shown).

A more modern version of the Ames test, uses the *S. typhimurium* TA104 strain, which is modified by molecular biological engineering (Vitotox[TM], Thermo Technologies, Finland) [33–35]. In TA104, DNA damage is shown more directly due to the activation of the SOS repair system. In these bacteria a luciferase gene is introduced by molecular design and this gene is under the transcriptional control of the recN promoter. Normally, this recN promoter is strongly repressed and prevents the expression of the luciferase gene. However, in the presence of a DNA-damaging genotoxic compound, the RecA regulator protein recognizes the resultant free ends or mismatches in DNA. This results in a cascade of biochemical interactions and reactions, leading to the derepression of the strong recN promoter, subsequent transcription of the luciferase gene and an increase in the production of the luciferase protein. With the proper amount of luciferin and ATP molecules, the synthesis of light can then be visualized. The light enhancement is an indication for genotoxicity. A second *S. typhimurium* TA104 strain constitutively expressing luciferase is used as a control cell line for measuring cytotoxicity. A ratio score of 1.5 between the genotoxicity and cytotoxicity strains is essential in this Vitotox[TM] screen to assess real genotoxicity. Compared to the Ames test, which needs for an endpoint measurement at least 3–5 days, the format of the Vitotox[TM] assay with

15 measurements in 3 h in the range of 10^{-6}–10^{-4} M at 5 dosages in a 384-well white culture plate is much faster and more in line with HTS procedures. The predictivity of the Vitotox^{TM} test in relation to the full Ames test was, with 90%, rather high. The sensitivity score was 90% for the true positives and the specificity score was 90% for the true negatives (Tab. 2). For this test, overall 156 reference and specific Organon compounds were investigated [36].

Table 2. Predictive values of Vitotox^{TM}, GreenScreen GC and RadarScreen assays in comparison with scores for full Ames tests and a set of 156, 81, and 154 compounds, respectively. The compounds are references, a specific set of pre-clinical and clinical developmental compounds of Organon and a specific clastogenic set of steroidal compounds [50].

	Vitotox^{TM}		GreenScreen GC		RadarScreen	
		n		n		n
Sensitivity	0.90	48	0.39	33	0.55	47
Specificity	0.90	108	0.98	48	0.52	107
Predictivity	0.90	156	0.74	81	0.53	154

The chromosomal level

The above-mentioned bacterial mutagenicity (Ames) and genotoxicity (Vitotox^{TM}) tests underestimate mitotic and meiotic segregation failures as common to eukaryotic cells. For the recognition of these meiotic segregation or clastogenic failures several cell types, such as mice, rat and human lymphocytes, Chinese hamster ovary (CHO) or lung (V79) cells are commonly used. In these cells, treatment with genotoxic compounds may lead in the metaphase into chromosomal aberrations, aneuploidy, sister chromatid exchange or the formation of micronuclei [22, 23]. Also with these clastogenic assays, the phase I enzymes of the liver are of crucial importance for the identification of the activity of metabolites of the tested compound. In these clastogenic assays, as with the Ames and Vitotox^{TM} screening, 3-h incubations are also performed in the absence or presence of rat liver S9 fractions. Although the overall assay times for clastogenicity are relatively short (24–48 h), the microscopical analysis of 200 metaphases per compound per dose is a long procedure. Skilled personal are occupied for up to 2 or 3 weeks with the overall quantification of one compound per assay [37]. Despite this thorough analysis, the *in vitro* clastogenicity scoring is hampered by a lot of false-positive and false-negative scores. Two items play a pivotal role in the imperfect profiling under these *in vitro* conditions. Cells are exposed during a very short period of 3–24 h towards an extensively high dose of compound (10 mM or 5 mg/ml), which is extremely high for *in vitro* treatment. It can sometimes be very difficult to dissolve compounds at such high concentrations in the cell media used. Another assumption is that the compound should be given at the highest dose at a concentration that induces at least 50% of cell death. This implies that the cells are struggling for their existence and usually ATP becomes depleted. Therefore, it is very plausible that the meiotic segregation will be hampered and will finally collapse. On this aspect, a project group of ECVAM (European Centre for the Validation of Alternative Methods) has taken the initiative to adapt the rules for *in vitro* clastogenicity testing [38, 39]. The ECVAM group asserts that the final dose levels should not exceed 1 mM or 0.5 mg/ml, and the maximal toxicity should not exceed the level of 20–50%, which can best be measured by mitochondrial activity (see below, section on cytotoxicity). Thus, new molecular biology techniques may also help here to simplify the overall analysis and read-outs.

In this respect, attention was drawn by yeast (*Saccharomyces cerevisiae*), as chromosomal segregation in the meiotic phase in yeast resembles the eukaryotic system. In yeast the focus was at the activation of the RAD54 promoter, whereas in vertebrates the RAD51 promoter plays a crucial role [40]. Both RAD54 and RAD51 are recombinational repair genes belonging to the class of RAD52 genes, which are involved in the repair of chromosomal double strand break damage [41–44]. Deficiencies of these genes lead to increased chromosomal damage before cell death occurs. Especially clastogenic, but also mutagenic, compounds can activate the RAD54 promoter in yeast or RAD51 promotor in vertebrates. Therefore, this assay may have an additive value in identifying genotoxic compounds that only disturb the chromosomal segregation. Two different commercially available yeast strains have been prepared. In one strain the RAD54 gene behind the RAD54 promoter has been replaced by green fluorescent protein (GFP) (GreenScreen GC, Gentronix, UK) [35, 45–47], while in the other strain β-galactosidase was introduced (Radar-Screen, reMYND, Belgium) [48, 49]. The end-point measurement is assessed with GFP after 24 h and with β-galactosidase after 6 h. The control analysis for cytotoxicity can, in principle, be done in the same sample by adsorption measurement with a spectrophotometer. The whole assay can be performed in 96- or 384-well white (view) culture plates at seven dosages between 10^{-6} and 10^{-3} M. In case of luminescence measurements, white culture plates deliver a 100-fold higher signal than black culture plates. A drawback of the system with GFP is that some of the compounds tested gave autofluorescence at the wavelength of GFP. With the specific set of Organon compounds this occurred with 15% of the compounds, while other compounds (also 15%) required activation by the rat liver S9 fraction. However, with the GFP yeast strain this was very difficult as the S9 fraction caused quenching of the GFP signal. These problems could be overcome with the β-galactosidase enzyme. For this enzyme a luciferin derivative is available at Promega (Madison, WI). This 6-*O*-β-galactopyranosyl-luciferin can be cleaved into free luciferin by β-galactosidase, which in turn can be quantified with firefly luciferase. At the wavelength of 650 nm, no quenching occurred with the rat liver S9 fraction and no interference of test compounds was found. Due to luminometry, this assay can be very sensitive, and genotoxic and/or clastogenic compounds can be identified within a dose range of $10^{-6}-10^{-3}$ M. With the GreenScreen GC assay, the predictivity with the full Ames score was lower in comparison with the Vitotox^{TM} evaluation, leading to a true positive score of 39% and a true negative score of 98%. With RadarScreen, the obtained values for true positives and negatives were 55% and 52%, respectively (Tab. 2). However, this RadarScreen assay can also be used for identification of real clastogenic and/or carcinogenic compounds. Thus, the large difference between the true negative scores of GreenScreen GC and RadarScreen assays was due to the large number of false negatives scored with RadarScreen assay as a consequence of the identification of clastogenic compounds. Within the Organon steroidal portfolio [50], we identified 33 out of 40 compounds, as being real positive or negative compounds, leading to a score of 82.5% true positives and negatives, respectively. With respect to the compounds used in the full Ames test, 12 Ames negative compounds were identified with clastogenic human lymphoblast or CHO assays. From these 12 compounds, 8 were identified with the RadarScreen assay. Overall analysis of 132 compounds with this RadarScreen assay, in comparison with the available *in vitro* clastogenicity/aneuploidy assays, showed that the sensitivity for the positive compounds was 80%, the specificity for the negative compounds was 77% and the predictivity was 78% (Tab. 3). Analysis for the GreenScreen GC assay on 44 compounds showed a sensitivity of 22%, a specificity of 95% and a predictivity of 57% (Tab. 3).

In analogy with the RAD54-GFP GreenScreen GC assay in yeast, an assay known as GreenScreen HC has been developed with the human lymphoblastoid TK6 cell line incorporating the human promoter of the growth arrest and DNA damage gene (GADD45α) in front of the GFP gene [51]. The early validation data for GreenScreen HC indicates a much higher sensitivity (>80%) for genotoxic carcinogens than for the yeast-based assays, with better identification of aneugens, topoisomerase inhibitors, and nucleoside analogues, while retaining a very high specificity (>95%). Greenscreen HC also has an S9 protocol available that allows the identification of pro-genotoxins.

Table 3. Predictive values for Vitotox^{TM}, GreenScreen GC and RadarScreen assays in comparison with scores for *in vitro* clastogenicity/aneuploidy for a set of 132, 44, and 130 compounds respectively. The compounds are references, a specific set of pre-clinical and clinical developmental compounds of Organon and a specific clastogenic set of steroidal compounds [50].

| | Vitotox^{TM} | | GreenScreen GC | | RadarScreen | |
		n		*n*		*n*
Sensitivity	0.29	85	0.22	23	0.80	83
Specificity	0.89	47	0.95	21	0.77	47
Predictivity	0.51	132	0.57	44	0.78	130

Selection procedure on the nucleotide and chromosomal level

A positive score in a mutagenic, genotoxic, clastogenic or carcinogenic assay is seen as a strong negative factor for a compound, which leads most times to deselection of that compound. At present the Ames II and Vitotox[TM] tests are used in the pre-selection phase within our company. A low throughput of 2 to 3 compounds per week for Ames II is feasible, while analysis of 16 to 160 compounds per week for Vitotox[TM] assays is possible. In the late developmental phase, clastogenicity is still measured using an *in vitro* chromosomal aberration test and an *in vitro* or *in vivo* micronucleus test. Both tests are time consuming and not suitable for incorporation in the early research phase. Implementation of the RadarScreen assay and final acceptance may help in this aspect. Furthermore, efforts are being made to develop a real eukaryotic hepatocyte HepG2 test in the early research phase. Introduction and the combined usage of the Vitotox[TM], RadarScreen and hepatocyte tests may lead to a better deselection and/or ranking of the lead compound in the early phase of lead optimization. At this moment, the sensitivity of the Vitotox[TM] and RadarScreen assays together is in comparison with Ames 96%, with clastogenicity 78% and with carcinogenicity 82% (Tab. 4). Further fine tuning of the data sets and analysis may lead to a further optimization. Chemical optimization may than finally lead to a better drug.

The nucleus level

For testing on this level the formation of micronuclei is commonly used. Using specific staining and microscopy, the formation of micronuclei can be scored manually. Similarly, the segregation of centrosomes can be analyzed with microscopy techniques and antibody coloration of the centrosomes. However, a lot of cells have to be visually inspected, which is very time consuming. HCS is a new bioimaging technique, in which the counting of the micronuclei and

Table 4. Sensitivity values of Vitotox[TM] and RadarScreen assays separately and combined according to scores for *in vitro* mutagenicity in the Ames test for a set of 156 compounds, for *in vitro* clastogenicity assays for a set of 85 compounds, and for *in vitro* carcinogenicity assays for a set of 50 compounds, comprising references, a specific set of pre-clinical and clinical developmental compounds of Organon and a specific clastogenic set of steroidal compounds [50].

| | Sensitivity (+ and −) | | | | | |
	Mutagenicity		Clastogenicity		Carcinogenicity	
Vitotox[TM]	43/48	0.90	25/85	0.29	15/50	0.30
RadarScreen	26/47	0.55	66/83	0.80	40/49	0.82
Vitotox[TM] + RadarScreen	46/48	0.96	66/85	0.78	41/50	0.82

of the centrosomes can be taken over by specific software. HCS is discussed in a later paragraph.

Screening procedures for cytotoxicity

A simple disturbance or unbalance in energy metabolism, cytoskeletal organization, membrane integrity or a cell-specific function, such as protein production and excretion, uptake and secretion of metabolic waste products, or glycogen storage, may form the initiation phase of cellular toxicity. Knowledge of the mechanism of action involved in these cellular cytotoxicity processes may lead to the identification of the biological cause. Although this sentence seems to make things simple, they are more complicated than that. First of all, general cytotoxicity can be determined in almost any cell type, but in some cases specific tissue functions may be needed to identify the real toxic effect. Therefore, the most appropriate tissue-specific and relevant cell lines, primary cells or tissue slices may be needed. To give a few examples, digoxin is a human-specific potassium channel inhibitor [52, 53]. Thus cytotoxicity can only be demonstrated in human cell(s) or cell lines. Doxorubicin, also known as adriamycin, on the other hand, is a cytostatic compound that intercalates with the DNA, but more specifically with mitochondrial DNA [54, 55]. Moreover, this compound is a topoisomerase inhibitor. This implies that every cell that divides will be affected by doxorubicin. Nevertheless, cells that are more dependent on their mitochondrial capacity, like muscle cells, might suffer much more from such a compound. In particular, the largest congregate of muscle cells in our body, the heart, is the primary target organ for doxorubicin toxicity. The real course of the cardiotoxicity of doxorubicin is the formation of the doxorubicinol [56, 57]. Beside these tissue- or species-specific compounds, compounds that are active on many cell types also exist. For instance, the cytostatics, 5-fluorouracil, a thymidine kinase inhibitor, methotrexate or amethopterin, an inhibitor of the enzyme dihydrofolate reductase, which produces tetrahydrofolate an important cofactor for the enzyme reaction with thymidylate synthetase, as well as cytarabin, a cytosine derivative, all prevent proper DNA replication. Since this principle is so essential for cell division, these compounds cause cytotoxicity at high concentrations in bacteria, yeast and all eukaryotic cell types.

These general phenomena of cytotoxicity can be studied in different ways. Various laboratories have developed "easy-to-interpret" end-point assays. The most direct one measures the DNA content of cells using [^{3}H]thymidine [58–60] or bromodeoxyuridine incorporation into the DNA [60], crystal violet [61] or Hoechst 33342 coloration [62–64]. The membrane integrity can simply be measured by calcein-acetoxymethyl (Calcein-AM) uptake [65, 66]. For more defined metabolic degradation or oxygen stress, the depletion of glutathione or the formation of reactive oxygen species (ROS) may indicate the mechanism of action of toxicity. Glutathione depletion can be measured with monochlorobimane [67–69] and the production of ROS with dichlorofluorescein [70–72]. The influence on mitochondrial activity can be measured in many ways; the spectrophotometric tetrazolium dyes 3-(4,5-dimethylthiazol-2-yl)-2,5-diphenyltetrazolium bromide (MTT) [59, 60], nitro blue tetrazolium (NBT), XTT and WST-1 [73] assays as well as the fluorometric Alamar Blue[TM] assay [74–77] and the oxygen consumption FRET assay (probe A65N-2, Luxcel Biosciences) [78] are the most commonly used ones. These mitochondrial activity assays have been extended with the luminometric assays, such as CytoLite[TM] [79], ATP-Lite[TM], and CellTiter Glo[TM] [80, 81], measuring NADH production, ATP synthesis and oxygen consumption, respectively. As there are many pathways leading to cell death, ideally a combination of these assays should be used for the evaluation of the "mechanistic" parameters. The *status quo* within our company is that, nowadays, the following assay markers are assessed to obtain a clear picture on cytotoxicity: mitochondrial cellular activity (ATP-Lite[TM], Cyto-Lyte[TM], Alamar Blue[TM], oxygen consumption), DNA proliferation (Hoechst 33342), calcein uptake (Calcein-AM), and glutathione depletion (monochlorobimane).

Previously, a validation was performed with these luminometric and fluorometric assays on four different cell lines, i.e., liver hepatocyte HepG2 cells, endometrium ECC-1 cells and cervix HeLa cells (all three of human origin) and CHO cells [82, 83]. Besides these cell lines, mouse L929 and/or primary human fibroblast cells were also studied in a collaboration with Débiton (Inserm, Clermond-Ferrand, France). The overall validation in these assays with a set of 110 compounds, selected with respect to their *in vivo* toxicity in rats and/or humans, showed that there is a very high resemblance in toxic effects between the different cells or cell lines used. An example is given for Calcein-AM with HeLa cells, and for monochlorobimane, ATP-Lite[TM], Cyto-Lite[TM], Alamar Blue[TM] and Hoechst 33342 assays with HepG2 cells (Fig. 4). It is shown that both doxorubicin and oligomycin B are very strong cytotoxicants even at 10^{-7} M. Oligomycin B is an uncoupler of the electron transport chain in mitochondria. Moreover, *N*-ethylmaleimide and perphenazine are relatively strong toxicants at 10^{-5} M, followed by the genotoxicants nitropyrene and nitrofurantoin at 3.16×10^{-5} M. Paracetamol and quinidine sulfate appeared still non-toxic up to a dose of 3.16×10^{-5} M in these assays. However, paracetamol is toxic in these assays at a dose of 10^{-2} M. The glutathione depletion and calcein uptake measurements had a large overlap with the mitochondrial activity and DNA proliferation assays. The ROS assay, on the other hand, was very insensitive and only a few compounds were identified in this assay as being toxic. For this reason, this assay was left out of the testing program and a more sensitive Nrf2-RE luciferase-based luminescent assay for ROS has been developed in HepG2 cells.

For mitochondrial activity and DNA proliferation assays, incubations of 72 h are needed. To avoid confluency of the wells, incubations were carried out with 10^4 cells/well in 96-well white view plates. For calcein uptake and glutathione depletion, the incubations only lasted for 24 h and were performed in

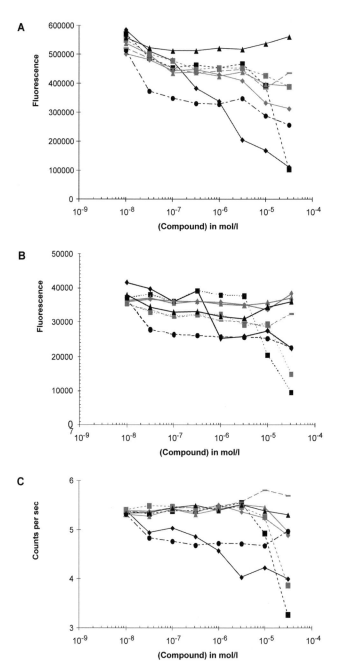

Figure 4. Cytotoxic effects of doxorubicin (—◆—), *N*-ethylmaleimide (- -■- -), nitrofurantoin (—▲—), nitropyrene (—◆—), oligomycin B (—●—), paracetamol (——), perphenazine (- -■- -), and quinidine sulfate (—▲—) on calcein uptake in HeLa cells (A) and glutathione depletion in HepG2 cells (B) after 24-h incubations, and effects on mitochondrial activity status with ATP-Lite[TM]

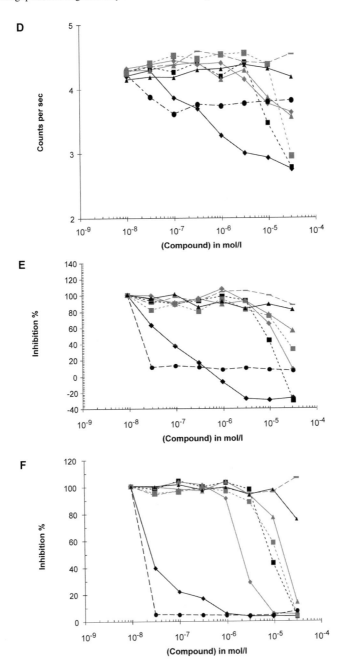

(C), CytoLite™ (D), Alamar Blue™ (E), as well as DNA proliferation with Hoechst 33342 (F) in HepG2 cells after 72-h incubations. The control effect with only 0.1% DMSO and no compound is given at the 10^{-8} M concentration.

96-well plates with 3×10^4 cells/well. In this screening device, 44–220 compounds at seven dosages between 3.16×10^{-8} M and 3.16×10^{-5} M can be screened on a weekly basis for a set of tests, but with the focus on just one assay 4400 compounds can be screened. With the above-mentioned techniques only 70% of the compounds were characterized as being toxic.

These levels found in-house are in line with the Multicenter Evaluation of *In Vitro* Cytotoxicity (MEIC) studies [84–88] as well as with the ACuteTox program [89, 90]. These MEIC and ACuteTox studies also used human HepG2 cells [89–91] and HeLa cells [89, 90, 92]. Approximately 250 toxic and nontoxic compounds were evaluated on these cell lines and showed a very high correlation between the *in vitro* test results and the *in vivo* data [52, 93–100]. For a small number of these compounds, in these studies a very high correlation was found between the concentrations of *in vitro* cellular genotoxicity and *in vivo* serum levels. These data imply that *in vitro* cellular toxicity assays can predict the *in vivo* toxicity for at least acute and more chronic toxicity, as serum levels measured *in vivo* at toxicity endpoints correspond to the *in vitro* toxicity concentrations [100–102]. Taking into account all the efforts made to improve the sensitivity of the cytotoxicity assays over the last 25 years, a breakthrough with luminometric assays was expected, but has nevertheless not been made. This contradiction can be explained by the fact that toxicity occurs at high concentration levels. Thus, if the control activity level of normals cells (at 100%) can be clearly distinguished from the complete toxicity level (at 0%) with a large enough difference in signal, no gain will be made whatsoever with more sensitive assay techniques. This will only occur when the mechanism of action leads to a positive activation signal, which can be identified at lower dosage levels. This kind of principle has been found in the Vitotox™ and RadarScreen assays used for genotoxicity and/or clastogenicity as mentioned above. Moreover, this principle is also applicable to the assays for apoptosis in which luciferin analogues are used for the cleavage by caspase 3/7 and/or 8/9 [103, 104]. These specific Promega substrates, i.e., Z-DEVD-aminoluciferin and LETD-aminoluciferin, may help to identify apoptosis induction, if induced directly by the compound *via* caspase 3 due to mitochondrial impairment or *via* caspase 8 due to inflammation and activation of the interleukin-1β and TNFα secretion. An example is given for caspase 3 with both HeLa and HepG2 cells after 4 h and 24 h (Fig. 5), showing that the caspase 3 signaling may differ for different cell types. Staurosporine and TNFα show a clear dose-dependent increase in caspase 3 activity after 4 h in HeLa cells and after 24 h in HepG2 cells. TNFα is given in combination with actinomycin D, which inhibits proteases that might breakdown caspase 3. Without the addition of actinomycin D, the signal activation of caspase 3 may remain unnotified. Dexamethasone is an anti-inflammatory compound and causes a reduction in caspase 3 production. The inflammatory compound lipopolysaccharide (LPS) does not act on either caspase 3 or caspase 8. LPS normally activates the secretion of interleukin-1β and TNFα *in vivo* by means of, for instance, intermediate liver Kupffer or blood T-helper cells. Such an activation is missing in the *in vitro* setting due to the

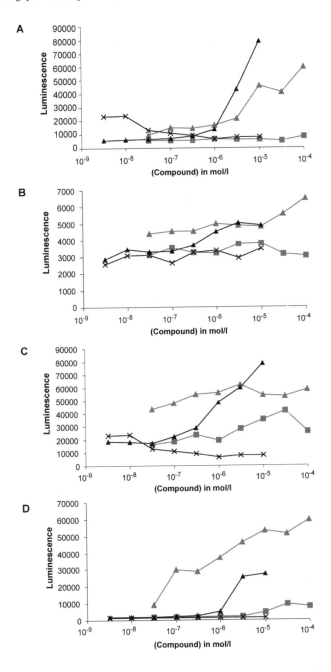

Figure 5. Dose-dependent apoptotic effects of dexamethasone (—✕—), doxorubicin (—■—), staurosporin (—▲—) in mol/l, and of TNFα (—▲—) in ng/ml on the activation of caspase 3 in HeLa cells (A, C) or HepG2 cells (B, D) after incubation of 4 h (A, B) and 24 h (C, D). Actinomycin D was added to all incubations at 3×10^{-7} mol/l to prevent protease activity, while TNFα was used from 0.039 up to 125 ng/ml.

absence of cross-talk between different cell types. On the other hand, TNFα in combination with actinomycin D also enhances the production of caspase 3 in both HeLa and HepG2 cells. For a more physiological setting and normal cellular hierarchy, precision-cut tissue slices might provide a better tool. Following a similar line of reasoning, in the cell lines used, the metabolism of the compounds into reactive intermediates may not have been sufficiently adequate, and precision-cut tissue slices or primary cell lines may again be preferential.

The importance of cellular metabolism by phase I and II enzymes

In liver cells two categories of enzymes, phase I and II enzymes, are involved in the metabolism of drugs and xenobiotic compounds [105]. The main role of phase I enzymes is the introduction of a functional group by means of hydroxylation or the modification of an existing group by O-, S-, N-dealkylation or demethylation to facilitate phase II reactions. Thus, phase I enzymes lead to more polar compounds, while phase II enzymes are involved in reactions like glucuronidation, sulfation, phosphorylation, methylation, N-acetylation or glutathione conjugation. The most important phase I enzymes are the cytochrome P450 (CYP) enzymes. Of these, CYP1A1, CYP1A2, CYP2B6, CYP2C8, CYP2C9, CYP2C19, CYP2D6, CYP2E1 and CYP3A4 are most common in human metabolism in intestine and liver cells. For phase II enzymes there is a large diversification of subtypes for UDP-glucuronosyltransferase (UGT), sulfotransferase (SULT), glutathione-S-transferase (GST), and N-acetyltransferase (NAT). Besides subtyping, enzyme polymorphisms exist in humans, e.g., for CYP2C9, CYP2D6, UGT1A1 and NAT-2. Some of these variations can even clarify why certain drugs are more or less affective or more or less harmful for certain ethnic groups [106]. Both phase I and II enzymes can be very important for the toxification or detoxification mechanisms of compounds. An example of a toxification mechanism through the phase I enzyme CYP1A1 and the phase II enzyme epoxide hydroxylase is the biochemical transformation of B[a]P into the carcinogenic metabolite B[a]P-7,8-diol-9,10-epoxide [107, 108]. The absence of CYP1A1 in knockout mice prevents B[a]P from becoming carcinogenic [109]. Other examples of the relevance of the presence of certain phase II enzymes are Gilbert's disease and the Crigler-Najjar syndrome Type I and II [110, 111]. In these diseases the phase II enzyme UGT1A1 is affected. With Gilbert's disease a patient has one functional and one dysfunctional gene encoding this enzyme and expresses therefore 50% of the normal level. This leads to a reduction in the degradation of bilirubin, a degradation product of heme, and a yellow coloration of the skin (jaundice) and of the white of the eye. Ultraviolet light as well as sunlight will degrade this bilirubin. Extremely high levels of bilirubin can lead to cholestasis and, in combination with an extremely high intake of paracetamol, estrogens or androgens, a similar phenomenon may occur. With Crigler-Najjar syndrome Type II only 10% of the normal enzyme level remains and with this inborn error a child will need a liver transplantation at the age of 6 or 7 years, as complete liver failure will occur around that age. With Crigler-Najjar syndrome Type I and 0% of UGT1A1 the infant will die within 3 weeks after birth. These examples are just to show that for metabolism both phase I and II enzymes can be of equipotent importance. The main point is that, if a toxic event occurs in certain individuals, the evolving accumulation of a toxic product may be due to either enhanced activity of phase I enzymes or reduced activity of phase II enzymes. Localization of the cause of this imbalance is becoming most urgent and represents a main point of concern.

The wrong classification for 30% of the toxic compounds as being nontoxic (see above) when using permanent tumor-like cell lines, in comparison with the *in vivo* situation, makes this method seem inappropriate. In many cases this discrepancy can be explained by the low expression levels of CYP enzymes and/or phase II enzymes, resulting in the complete absence of endogenous metabolism in these cell lines [112–114]. Therefore, primary liver cells are mainly considered for this kind of studies, as they give a better representation of the *in vivo* situation [115]. In a large number of comparative studies on the metabolizing activity of human HepG2 or rat H4IIE cells with freshly prepared primary human or rat hepatocytes, comparing the mRNA level by quantitative PCR techniques and by measuring the conversion of specific substrates for the main CYP enzymes with HPLC-MS, the primary hepatocytes showed higher levels of metabolism. However, what is often not taken into account, is that permanent cell lines are cultured under minimal essential culture conditions. This means that the cells are not exposed to xenobiotic compounds, as normal liver cells are by means of food exposure. A direct comparison for human HepG2 cells without pre-treatment with xenobiotics showed that, on the mRNA expression level, primary hepatocyte cells produce 10–100-fold more mRNA for the above-mentioned enzymes [116]. With the new luminescence techniques with luciferin analogues Luc-CEE, Luc-ME, Luc-HE and Luc-BE (P450-GLO substrates Promega), which are very specific for CYP1A1, CYP1A2, CYP2C9 and CYP3A4, respectively, it became very easy to measure the HepG2 CYP enzyme activities. Consequently, direct comparison on the enzyme level could also be performed for HepG2 cells and primary hepatocytes. Remarkably, the differences in enzyme activity between the cell line and fresh hepatocytes was much smaller and only in the range of 5.6-fold for CYP1A1, 80-fold for CYP1A2, and 15-fold for CYP3A4. A similar conclusion was found with the less specific fluorophore benzyloxyresorufin, which is mainly converted by the CYP1A1, CYP1A2, CYP2B6 and CYP 3A4 enzymes (Fig. 6A). With this substrate the biological conversion between the primary hepatocytes appeared equal to that of HepG2 cells (Fig. 6B). On the other hand, CYP2C9 appeared to be completely absent in HepG2 cells, and could only be measured at high levels in primary hepatocytes. With respect to CYP2C9, HepG2 cells were clearly lacking activity, but with respect to CYP1A1, CYP1A2 and CYP3A4 activity differences were much smaller than expected. Specific

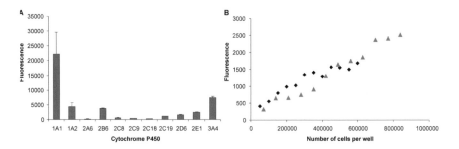

Figure 6. The conversion of benzyloxyresorufin is indicated for specific cytochrome P450 enzyme, i.e., CYP1A1, 1A2, 2A6, 2B6, 2C8, 2C9, 2C19, 2C19, 2D6, 2E1 and 3A4, in a supersome fraction of baculovirus in insect cells (A). The conversion of BROD is given as fluorescence change per number of human primary hepatocytes (◆) or HepG2 cells (▲) (B).

induction of the enzyme levels in HepG2 cells might even make it possible to obtain similar levels of these enzymes as compared to hepatocytes. In this respect, the nuclear receptors AhR, CAR and PXR are of high importance and the next section is devoted to CYP enzyme activations mediated through these NR.

Activation of cellular metabolism by nuclear receptors

Several NR appear heavily involved in the up-regulation of phase I and II enzymes and therefore in drug disposition and metabolism. Some of these NR are involved in the induction of only very specific CYP enzymes. This mechanism is due to the presence of very specific responsive elements for one particular NR in the promoter site of each CYP enzyme. With respect to the phase II conjugation enzymes, it appears that in some promoters very specific responsive elements for only one particular receptor are available, while in other promoters more diverse responsive elements for several of these NR are available. This makes synergistic effects between different NR possible. For AhR, a specific activation was identified for several genes, i.e., CYP1A1, CYP1A2, CYP1B1, UGT1A1, UGT1A6 and SULT1A1 [116–119]. The AhR agonists 2,3,7,8-tetrachlorodibenzo-p-dioxin (TCDD), 3-methylcholanthrene, indirubin, indigo and β-naphthoflavone [120–122] could induce these genes in HepG2 cells [116, 117]. Indirubin is a tryptophan derivative and a structural isomer of indigo, a color dye. Indirubin is a compound that can be identified as a physiological substrate at concentrations of 10^{-8} M in the human serum [120]. Induction on the mRNA level was 100–1000-fold for CYP1A1, 10–100-fold for 1A2 and 4–20-fold for UGT1A1 and 1A6 (Fig. 7A), while on the enzyme activity level the induction was 10–100-fold for CYP1A1 and 10-fold for 1A2, as shown for indirubin (Fig. 7B). For PXR, a specific activation was identified for CYP3A4 [123–126], and to a lesser extent for UGT1A1 and

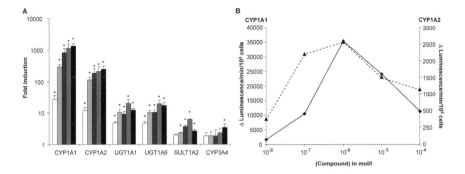

Figure 7. Dose-dependent stimulation at 10^{-8} (☐), 10^{-7} (▨), 10^{-6} (■), 10^{-5} (■) and 10^{-4} (■) mol/l of indirubin on the mRNA expression of CYP1A1, CYP1A2, UGT1A1, UGT1A6, SULT1A2 and CYP3A4 (A). Enzyme activity of CYP1A1 (—◆—) and CYP1A2 (--▲--) measured with P450-Glo-specific substrates (Promega) (B).

SULT1A2. The PXR agonists rifampicin and T0901317 showed a 2- to maximally 4-fold increase in the mRNA expression of CYP3A4, UGT1A1 and SULT1A2 [116, 117, 127–129]. For the enzyme levels of CYP3A4 a 2-fold increase was also observed [116, 117]. CAR is a constitutive inhibitor of CYP3A4 and an activator of CYP2B6. The CAR agonists 6-(4-chloro-phenyl)imidazo[2,1-*b*][1,3]thiazole-5-carbaldehyde *O*-(3,4-dichlorobenzyl)-oxime (CITCO) and phenobarbital can activate the mRNA expression of CYP2B6, SULT1A2 and SULT2A1 in HepG2 cells [116, 117, 123–126]. A specific mechanism occurs for PXR and CAR, in which they can operate synergistically in certain promoters and inhibitory in others. Moreover, both receptors activate or inactivate the promoter sites with a specific partner, the retinoic X receptor (RXR). The heterodimeric PXR/RXR or CAR/RXR receptors bind to their respective responsive elements in the promoter.

The metabolic activation mechanism *via* NR can become even more complicated with the introduction of the more pharmacologically active receptors, like the liver X receptor (LXRα), the farnesoid X receptor (FXR), and the peroxisome proliferator-activated receptors (PPARα, γ, δ), all of which can also heterodimerize with RXR [130]. This implies that if more of this kind of receptors become activated, competition for the RXR will be increased. The high binding potency of these receptor types with RXR, will then define which pathways become more activated, more inhibited or suppressed. The FXR/RXR inhibits the mRNA expression of CYP7A1 in an indirect manner *via* the small heterodimer protein-1 (SHP-1) [131], while the LXRα/RXR activates the mRNA expression of CYP7A1 [132, 133]. Fatty acid metabolism is regulated by FXR and LXRα in a counteractive way. The PPARα/RXR is involved in the transcription of enzymes involved in fatty acid catabolism, such as acetyl-coenzyme A acetyltransferase (Acaa1) and acyl-coenzyme A oxidase-1 (Acox1), cytosolic acyl-coenzyme A thioesterase 1 (Cte1), fatty acid binding protein 4 (Fabp4), and more particularly in activation of CYP4A10 and CYP4A14 [131], while PPARγ/RXR resulted in a marked decrease in the expression of the liver enzymes phosphoenolpyruvate carboxykinase, pyruvate carboxylase and glucose-6-phosphatase, leading to a reduced gluconeogenesis [134].

The presence of AhR, PXR and CAR has been demonstrated in HepG2 cells and these receptors are clearly implicated in the activation of phase I and II enzymes. It has been shown that activation of these receptors can lead to mRNA expression and enzyme activities at levels comparable to those in human primary hepatocytes [116, 117]. Since the permanent cell lines are always cultured under conditions with minimal essential media and 10% bovine calf serum, one idea was to enhance the phase I and II enzyme activities by adding xenobiotic compounds in the media and thereby mimicking more physiological conditions. Therefore, a cocktail of indirubin (10^{-8} M), CITCO (10^{-7} M) and T0901317 (10^{-7} M) was used to induce AhR, PXR and CAR, respectively. Indirubin was also used alone at 10^{-6} M as, at this particular dose, it also activates PXR and CAR. In the absence or presence of these two test combinations the cytotoxicity experiments for Alamar Blue^{TM} and Hoechst 33342 were repeated with the 110 toxic compounds (see above). The idea was that the remaining 30% of these toxic compounds, which were misclassified as non-toxic, might be classified as toxic in the novel set-up. Remarkably, the toxicity of the test set was not, as expected, increased after addition of these cocktails but decreased. This decrease was found at incubation periods of 3 days, as well as after 5 and 7 days. Thus, elongation of the incubation periods in this respect did not improve the toxicity findings.

For B[a]P, a compound that is activated into a carcinogenic compound *via* metabolism of CYP1A1 and CYP1A2, we examined this strange finding more accurately. First of all we were able to demonstrate that B[a]P was cytotoxic in HepG2 cells and that this cytotoxicity could be reduced specifically by the addition of the AhR antagonist α-naphthoflavone, but also by addition of ketoconazole an inhibitor of CYP1A1 and CYP3A4, or of furafyllin an inhibitor of CYP1A2 (Fig. 8A). This demonstrates that HepG2 cells are able to produce B[a]P metabolites. Culturing in combination with indirubin at 10^{-6} M led to a sharp increase in the phase I enzyme activities of CYP1A1 and CYP1A2 by 375- and 30-fold, respectively (Fig. 8B). Despite this tremendous increase, the cytotoxicity of B[a]P was reduced at least by 2-fold (Fig. 8C). Since the phase II enzymes UGT1A1, UGT1A6, SULT1A2 and SULT2A1 are also highly induced by indirubin, the most logical explanation is that enhancement of these phase II enzymes leads to a better inactivation or degradation of the reactive B[a]P intermediates. Thus, with these results, we want to stress that, besides phase I enzymes, phase II enzymes can also be important in the flow of a compound from a non-toxic state, through a reactive toxicant intermediate, to a non-toxic end product. Using this particular cocktail of agonists, one simply activates the metabolic defense system towards its maximal capacity.

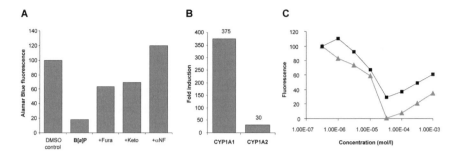

Figure 8. (A) Inhibition of the cytotoxic effects induced by benzo[*a*]pyrene (B[*a*]P) with either furafyllin (Fura, 10^{-6} M), an inhibitor of CYP1A2, ketoconazole (Keto, 10^{-6} M), an inhibitor of CYP1A1 and CYP3A4, or α-naphthoflavone (αNF, 10^{-5} M), an arylhydrocarbon receptor antagonist. (B) The induction level of indirubin (10^{-6} M) on CYP1A1 and CYP1A2 activity in HepG2 cells with P450-Glo-specific substrates (Promega). (C) The dose-dependent effect on cytotoxicity of B[*a*]P in the absence (—▲—) or presence of indirubin (10^{-6} M) (—■—) (C).

From these cocktail treatments we learned that metabolic activation of CYP1A1, CYP1A2, CYP2B6 and CYP3A4 did not lead to a higher identification of toxic compounds. Moreover, it was also impossible to influence the sensitivity of cytotoxicity assays with luminometric assays, as seen with genotoxicity and carcinogenicity assays. So, despite all the effort put into optimization of these *in vitro* cytotoxicity assays we were fated to a final level of identification of 70% of the toxic compounds. Fortunately, in this area of cytotoxicity analysis new techniques are being developed. Initially, bioimaging techniques were hampered by poor software programs, but nowadays a fast 96-well analysis can be performed in 10 min. The advantages of this HCS are discussed below.

High content screening or bioimaging techniques

At the moment HCS equipment is available from several suppliers, such as BD Biosciences (BD Pathway™), Molecular Devices (Discovery-1), GE Healthcare (In cell analyzer 1000/3000), and Thermo (Cellomics). The advantage of bioimaging is that it is no longer necessary that 100 000 cells are examined for one single biological marker; individual cells can be studied with five different biological biomarkers. Using HepG2 cells, O'Brien and his colleagues [135] showed that toxicity predictions can be extremely good with fluorometric markers for DNA coloration, membrane permeability, mitochondrial membrane potential, nuclear area and apoptotic induction. In their approach they examined 230 marketed drugs that were either hepatotoxic, cardiotoxic, nephrotoxic or non-toxic. From 97 hepatotoxic drugs, 13 compounds produced reactive metabolites and 12 were inducing idiosyncrasy. Remarkably, 97% of these drugs were predicted quite well with this technique. From 46 drugs that are also toxic for other organs, including heart and kidney, the prediction score with liver tissue was still 80% with these HepG2 cells. For another 48 benchmark toxic products, and 38 non-toxic drugs (including 12 benchmark non-toxic products), the score values were 96% and 100%, respectively. Thus, the *in vitro* prediction of clinically identified toxic compounds with this 'Cellomics' technique was 87% [135]. The overall score for non-toxic compounds with 97% was also extremely good. As shown with the idiosyncratic and reactive metabolite compounds, metabolic activity should have been present in these HepG2 cells on an acceptable level to lead to such high prediction scores. The same technique was used for the prediction of toxicity with primary human hepatocytes. Surprisingly, the overall score with these physiological more relevant cells was only 67% with 400 toxic compounds [136]. This novel assay technique has an additional advantage that a medium- to high-throughput screening potential can be reached. Since the same cells can be used for five measurements, the number of plates can be reduced with respect to independent population plate assays.

The article of O'Brien et al. [135] was the first to describe the positive predictive outcome with so many toxic compounds in HepG2 cells. In addition, the predictive value of this technique was also confirmed for eight compounds of Johnson & Johnson, all being toxic in the clinic with humans, but non-toxic with rats [137]. Bioimaging techniques have already been in use for a while for the observation of cellular kinetic changes or specific toxicological analysis [138–141]. For instance, the analysis of micronuclei can be simplified tremendously using this technique. Good results have also been obtained for CHO-K1 cells in the presence and absence of S9 fraction with a set of 46 compounds, leading to a sensitivity of 88% and a selectivity of 100% [142]. Moreover, phospholipidosis can be measured microscopically in hepatocytes with specific fluorophores like Nile Red [143], *N*-(7-nitrobenz-2-oxa-1,3-diazol-4-yl)-1,2-dihexadecanoyl-*sn*-glycero-3-phosphatidylcholine (NBD-PC) [144], NBD-phosphoethanolamine (NBD-PE) [145] or LipidTox [146]. For these assays different cell lines can be used, like CHO-K1 and V79, HeLa, HepG2 and U-937 cells. Furthermore, the sensitivity of the measurement of bioimaging makes it feasible to measure the difference between control and affected cells with a 60-fold increased sensitivity for NBD-PE and CHO cells (Fig. 9). Consequently, bioimaging will rapidly gain acceptance in the field of toxicology with respect to cytotoxicity and clastogenicity.

Screening for non-genotoxic carcinogenicity with nuclear receptors

The activation of CYP enzymes in HepG2 cells by NR has already been mentioned above. For two of the NR, AhR and PPARα, a clear correlation between the receptor activation with certain ligands and carcinogenicity is seen. Activation of the AhR with dioxin (TCDD) leads to an increased incidence of liver tumors in both rats and humans [147–149], whereas activation of PPARα with the compound Wyeth 14643 (WY14643) leads to liver tumor induction in

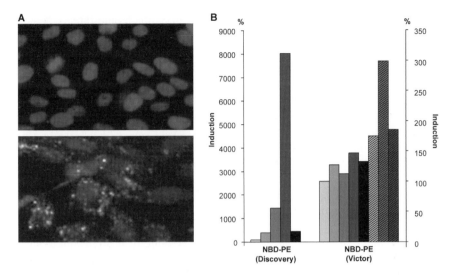

Figure 9. Accumulation of the fluorophore *N*-(7-nitrobenz-2-oxa-1,3-diazol-4-yl)-1,2-dihexadecanoyl-*sn*-glycero-3-phosphoethanolamine (NBD-PE) in CHO-K1 cells in the absence (A, top) and presence (A, bottom) of 10^{-5} M amiodarone by means of bioimaging (Discovery-1, Molecular Devices). Quantification of the fluorophore NBD-PE with Discovery-1 software or a fluorometer (Victor II, PerkinElmer) (B) at 10^{-8} (▨), 3.16×10^{-7} (▨), 10^{-7} (■), 3.16×10^{-6} (■), 10^{-6} (■), 3.16×10^{-6} (▨), 10^{-5} (▨), and 3.16×10^{-5} (▨) mol/l. Control level with 0.1% DMSO vehicle is set at 100%.

rats and mice, but not in humans [150]. The difficulty in this aspect of identification lies in the fact that not all inducers are necessarily non-genotoxic carcinogens. For instance, grapes, fruits and vegetables are mentioned as very healthy, although some of the ingredients enhance the AhR activity, and thus induce AhR-driven pathways. These fruits and vegetables are recommended to protect individuals against tumor development. Resveratrol is a major constituent of the grape skin and known as an AhR antagonist [151]. Moreover, many flavonoids are described as agonists and/or antagonists of AhR, of substrates and/or inhibitors of several CYP enzymes and phase II enzymes [152, 153]. This appears somewhat similar to the reduction of cytotoxicity described for B[*a*]P above. The main question now remains of whether these protective mechanisms increase or decrease the tumor incidence. In rats, indigo is a more potent AhR activator than indirubin, while in humans the effects of induribin prevail. In rats, the beneficial effects of indole derivatives on the intervention and prevention of tumors have been shown [154, 155]. In humans, indirubin is commonly used in Chinese health care for treatment of chronic myelotic leukemia. Among the interaction with AhR, other cellular mechanisms, such as the interaction with cyclin-dependent kinases (e.g., p27^{Kip1}) and glycogen synthase kinase 3, are suggested [156]. Thus, a compound that activates the AhR can either be beneficial or carcinogenic. Since it is difficult to predict in which class an AhR activator will fall, it is advisable to steer away from AhR activation during drug development.

With respect to AhR activation, a division can be made into three kinds of toxic compounds. The compounds can be divided into the following three classes: (1) genotoxic compounds, like 3-methylcholanthrene; (2) compounds that need bioactivation into a genotoxicant either by CYP1A1 (e.g., B[a]P), by CYP1A2 (e.g., AAF), or by CYP3A4 (e.g., aflatoxin B_1) and (3) non-geno-toxic compounds, like TCDD and polychlorobiphenyls (PCBs). All these compounds induce tumors in rats and humans and are all directly responsible for the activation of the AhR and thus CYP1A1, CYP1A2 and CYP1B1. The genotoxic compound 3-methylcholanthrene also directly interacts with the DNA for intercalation, while B[a]P needs to be metabolized by CYP1A1 and epoxide hydroxylase into the genotoxic metabolite B[a]P-7,8-dihydro-diol-9,10-epoxide. TCDD, on the other hand, activates the AhR, but neither this compound nor one of its metabolites intercalate with the DNA. The exact mechanism of action is still unknown.

For the identification of non-genotoxic carcinogenicity, it is relevant to analyze whether a compound activates both the rat and human AhR or whether species selectivity exists. For the AhR, a simple cellular assay is available, in both rat H4IIE cells and human HepG2 cells, that makes use of the metabolism of 3-cyano-7-ethoxycoumarin (CEC) or specific P450-GLO substrates (Promega) by CYP1A1 and CYP1A2 into fluorescent or luminescent products. That species differences between human and rat in activation of AhR do occur is shown by potency differences of compounds between the rat and human cell lines as well as by compounds that are only active in one of those cell lines [157]. For TCDD and 3-methylcholanthrene the activity is similar in both cell lines (Fig. 10). However, indigo activity is dominant in the rat cell line, while indirubin is much stronger in the human cell line. On the other hand, compounds like menadione, Org A, B, and C only activated the human AhR receptor, while flutamide, Org D, PCB 156 and PCB157 were very specific for the rat receptor (data not shown).

The throughput of the AhR activation assay is relatively high, since it can be performed in 96- and 384-well plates. In each plate a concentration range of $10^{-9}-10^{-4}$ M is used, leading to 8 compounds per plate. A throughput of 800 compounds per day and 4000 per week is feasible.

Another mechanism of action for non-genotoxic carcinogens in mice and rats is by PPARα activation. PPARα activation is, for example, involved in the tumor induction by WY14643 and appears to be murine PPARα dependent [158]. Transgenic mice with the humanized PPARα were dramatically less sensitive for tumor induction. These effects are more or less similar to the activation responses with AhR, but seem to be restricted to rodents and not to humans. Consequently, species differences should also be taken into account for PPARα, as for AhR. In addition, the levels of PPARα in human HepG2 cells are low and, therefore activation of the PPARα pathway needs to be studied in H4IIE cells. Tests for PPARα can be set-up in a manner similar to that for AhR.

Separation of non-genotoxic compounds from genotoxic compounds and/or genotoxic intermediates can be carried out with the genotoxicity and

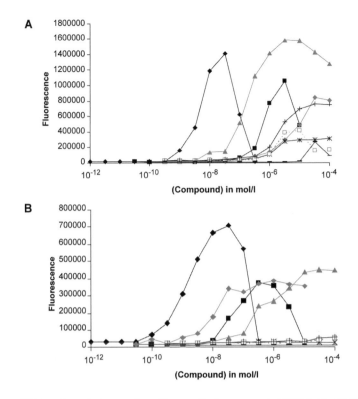

Figure 10. Dose-dependent conversion of 3-cyano-7-ethoxycoumarin (CEC) in HepG2 (A) and H4IIE cells (B) upon exposure to TCDD (—◆—), 3-methylcholanthrene (—■—), indirubin (—▲—), indigo (—◆—), Org A (—✖—), Org B (--□--), Org C (—+—), and menadione (—▬—).

clastogenicity screening procedures. The Ames II, Vitotox[TM] and RadarScreen assays can be used in either the presence or absence of S9 fraction. This results in a sound strategy to identify compounds with a risk potential for non-genotoxic carcinogenicity. Since rat studies for carcinogenicity will not identify the human receptor-specific compounds, this *in vitro* analysis helps to identify potential risk candidates at an early stage of the developmental process. This leads to the feasibility to optimize the chemical structure of the compound in such a way that the potency for the physiological target remains, while the potency for AhR and/or PPARα can be reduced or minimized to negligible.

Screening for competition of CYP and UGT enzymes

For many human CYP and UGT enzymes, molecular biology has led to the preparation of baculovirus or insect cell-derived supersomes that express specific enzymes in high abundance [159]. The development of protocols for CYP enzyme assays has been performed by the GenTest corporation (BD Biosciences) [160–162]. The inhibition of CYP-mediated metabolism is often the mechanism for drug-drug interaction. To identify which of the CYP enzymes are involved in the metabolism of the compound, competition assays using cDNA-expressed enzymes or human liver microsomes are introduced as a routine part of the drug screening. The fluorometric assays for CYP1A1, CYP1A2, CYP2C8, CYP2C9, CYP2C19, CYP2D6, CYP2E1 and CYP3A4 were developed for 96-well plates initially and have been optimized with slight changes of the protocols for 384-well plate assays (Tab. 5). For each CYP assay, specific fluorophores are used as substrate and specific inhibitors are used as reference compound (Tab. 5, Fig. 11A, B). The throughput of these CYP-inhibition assays is 40 compounds per day at different concentrations in the dose range of 10^{-7}–10^{-4} M. Dose-dependent inhibition curves for each enzyme are shown in Figure 11C and D. In a similar manner the competition of compounds can be measured and the findings can be expressed as potency against the reference compound. By performing these incubation assays for all compounds at the early toxicity screening, we can obtain information about the likelihood of drug-drug interactions in the clinic and about the contribution of each specific CYP enzyme. With this knowledge we can look at the molecular structure again and may propose modifications to the molecule to improve bioactivity, bioavailability and/or potency.

UGT enzymes can be studied similarly to CYP enzymes. Gentest Corporation has also developed 96-well assays for these determinations. The assays can be done in three different ways, using liquid chromatography-mass spectrometry (LC-MS), with radioactivity, or with fluorometry. In case of LC-MS, identification of the glucuronidated products should be feasible and the LC-MS capacity should be sufficient [163]. With radiometric assays, [^{14}C]UDP-glucuronic acid is used as cofactor for conjugating the glucuronic acid to the different substrates [164]. Separation can be examined by LC or thin layer chromatography (TLC). In case of fluorometry, 7-methyl-4-methylcoumarin is most commonly used as fluorometric substrate in combination with UDP-glucuronic acid as cofactor. In this way it is possible to perform competition assays with the same strategy as with the above-described CYP enzymes. Unfortunately, 7-methyl-4-methylcoumarin is not a very potent substrate for UGTs; nevertheless, the first assays were developed with this fluorophore [165, 166]. Recently a better fluorophore, scopoletin (7-hydroxy-6-methyl-coumarin), has been identified. This has made the assay more sensitive and feasible for HTS as described for UGT1A1 [167, 168]. The inhibition of phase II enzymes, like UGT-mediated metabolism can also be relevant for drug-drug interaction. To identify which of the UGT enzymes is involved in the metabolism of the compound, it is also of importance to introduce these competition assays using cDNA-expressed enzymes or human liver microsomes to a routine part of the drug screening. The fluorometric assays for UGT1A1, UGT1A3, UGT1A4, UGT1A6, UGT1A7, UGT1A8, UGT1A9,

Table 5. Human cytochrome P450 (CYP) enzyme assays, according to Gentest Corporation (BD Biosciences) with slight modifications due to parallel 384-well plates *in vitro* testing. The final concentrations in the well are given, while additions are combined as follows: 5 µl of a √10 dose range of 10^{-7}–10^{-4} M compound or inhibitor in buffer with 0.1% DMSO, 5 µl substrate, followed by 10 µl enzyme in combination with the cofactors. Exception was for CYP2D6 with quinidine and CYP3A4 with ketoconazole where the inhibitor dose range was in between 10^{-8} and 10^{-5} M.

CYP§	Substrates	(µM)	Inhibitors	(app. IC$_{50}$ nM)	Buffer* (mM)	Cofactors Solution#	Enzyme dilutions (Gentest) (nM)	Excitation wavelength (nm)	Emission wavelength (nm)
1A1	3-cyano-7-ethoxycoumarin	5	ketoconazole	153	100	A	2.5	409	460
1A2	3-cyano-7-ethoxycoumarin	5	furafylline	476	100	A	2.5	409	460
2C8	dibenzylfluoresceine	2	quercetin	14500	100	A	2.5	485	538
2C9	dibenzylfluoresceine	1	sulfaphenazole	439	25	A	10	485	538
2C19	dibenzylfluoresceine	2	tranylcypromine	7810	50	A	5	485	538
2D6	3-[2-(N,N-diethyl-N-methylammonium)ethyl]-7-methoxy-4-methylcoumarin	1.5	quinidine	65	100	B	7.5	390	460
2E1	7-methoxy-4-trifluoromethylcoumarin	133	diethyldithiocarbamic acid	13800	44	A	8	409	530
3A4	7-benzyloxyquinoline	40	ketoconazole	245	100	A	15	409	530
3A4	dibenzylfluoresceine	1	ketoconazole	86	100	A	2.5	485	538
3A4	7-benzyloxyresorufine	50	ketoconazole	80	100	A	15	530	590

§ Catalogue numbers for CYP enzymes of BD Biosciences were: 1A1 (P211), 1A2 (P203), 2C8 (P456262), 2C9 (P456258), 2C19 (P456259), 2D6 (P217), 2E1 (P206), and 3A4 (P202).

* Buffer constitution: 0.5 M potassium phosphate buffer pH 7.4

Cofactor solution A: 25 mM 0.5 M potassium phosphate buffer pH 7.4, 2.6 mM NADP$^+$, 6.6 mM glucose-6-phosphate, 6.6 mM glucose-6-phosphate dehydrogenase. Cofactor solution B: 25 mM 0.5 M Tris buffer pH 7.4, 16.4 µM NADP$^+$, 0.82 mM glucose-6-phosphate, 0.82 mM MgCl$_2$ and 0.8 U/ml glucose-6-phosphate dehydrogenase.

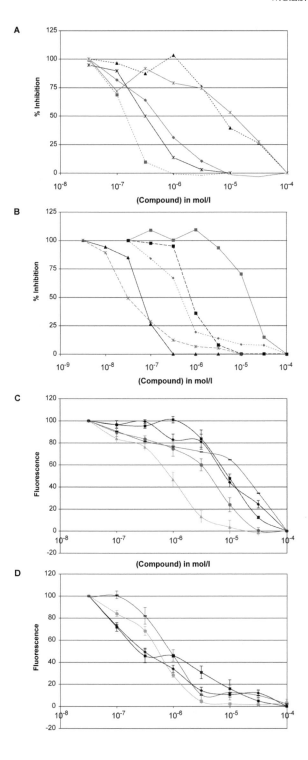

Table 6. Human UDP-glucuronosyltransferase (UGT) enzyme assays, according to Gentest corporation (BD Biosciences) with slight modifications due to parallel 384-well plates *in vitro* testing. The final concentrations in the well are given, while additions are combined as follows: 5 µl of a √10 dose range of 10^{-7}–10^{-4} M compound or inhibitor in buffer with 0.1% DMSO, 5 µl substrate, followed by 10 µl enzyme in combination with the cofactors.[a]

UGT[$]	Substrate	(µM)	Inhibitors	(app. IC$_{50}$ µM)	Enzyme dilutions (Gentest) (mg/ml)
1A1	scopoletin	50	bilirubin	80.7	0.4
			raloxifen	6.6	
1A3	scopoletin	100	naringenin	88.7	1.0
1A6	scopoletin	150	1-naphtol	9.9	0.2
1A7	scopoletin	100	2-naphtol	50.2	0.16
1A8	scopoletin	150	1-naphtol	216	0.1
1A9	scopoletin	25	2-naphtol	3.4	0.08
1A10	scopoletin	200	eugenol	5.3	0.6
2B15	scopoletin	100	eugenol	5.0	0.6
2B17	scopoletin	100	testosterone	7.8	1.0

[$] Catalogue numbers for UGT enzymes of BD Biosciences were: 1A1 (456411), 1A3 (456413), 1A6 (456416), 1A7 (456407), 1A8 (456418), 1A9 (456419), 1A10 (456410), 2B15 (456435), and 1B17 (456437).

[a] UGT1A4 (456414), UGT2B4 (456424) and UGT2B7 (456427) could not be measured with scopoletin (7-hydroxy-6-methylcoumarin) or 4-methylumbelliferin; buffer constitution: 0.2 M potassium phosphate buffer pH 7.4; cofactor solution: 25 mM 0.5 M potassium phosphate buffer pH 7.4, 1 mM MgCl$_2$, 4 mM UDP-glucuronic acid and 0.025 mg/ml alamethicin; excitation wavelength is 380 nm and emission wavelength is 510 nm.

UGT1A10 and UGT2B15 were first developed for 96-well plates and have been optimized with slight changes of the protocols for 384-well plate assays (Tab. 6). The throughput of these UGT assays is 40 compounds per day at different concentrations in a dose range of 10^{-7}–10^{-4} M. The dose-dependent inhibition curves for these enzymes are shown for the specific inhibitors of these enzymes (Fig. 11C, D). In a similar way, the competition of compounds can be measured and the findings can be expressed as potency against the reference compound. Again, by doing these incubation assays for all compounds during the early toxicity screening, we can obtain information about the like-

Figure 11. Competition curves of cytochrome P450 (A, B) and UDP-glucuronosyltransferase (C, D) enzymes with their respective inhibitors. Representative dose-dependent inhibition curves are given in (A) for ketoconazole on CYP1A1 (--■--), furafylline on CYP1A2 (—♦—), tranylcypromine on CYP2C19 (--▲--), diethyldithiocarbamic acid on CYP2E1 (—✳—), and ketocanozole on CYP3A4 (with 7-BQ, —✳—); and in (B) for quercetin on CYP2C8 (—■—), sulfaphenazole on 2C9 (--♦--), quinidine on CYP2D6 (--■--), ketoconazole on CYP3A4 (with BzRes, —▲— and DBF, --✳--). For each CYP enzyme the substrate was used as indicated in Table 5. Representative dose-dependent inhibition curves with scopoletin as substrate are given in (C) for bilirubine on UGT1A1 (—♦—), naringenin on UGT1A3 (—■—), 1-naphtol on UGT1A6 (—●—), 2-naphtol on UGT1A7 (—♦—), and 1-naphtol on UGT1A8 (—▬—); and in (D) for 2-naphtol on UGT1A9 (—♦—), eugenol on UGT1A10 (—■—), eugenol on UGT2B15 (—●—), and testosterone on UGT2B17 (—▬—). 7-BQ, 7-benzyloxyquinoline; BzRes, 7-benzyloxyresorufine; DBF, dibenzylfluoresceine.

lihood of drug metabolism and disposition in the human environment and about the contribution of each specific UGT. The bioactivity, bioavailability and/or potency of a compound might be improved by small molecular modifications.

Screening for embryotoxicity

Next to genotoxicity, one of the main issues in drug development within our company is embryotoxicity. For every potential drug intended for use in women of child-bearing potential, costly *in vivo* tests in rabbits and rats are regulatory requirements. Especially the teratogenic effects found with thalidomide in the sixties have led to these regulatory procedures [169]. These pivotal tests are scheduled quite late in development. So, high investments have already been made in (pre-)clinical studies before the decision is made to drop the compound on basis of embryotoxicity.

Many research groups have put effort in developing easy and cheap *in vitro* screening assays to predict embryotoxicity. The primary goal was to develop assays that could replace the classical *in vivo* tests. However, embryonic development is rather complex and not all facets can be identified in one or several *in vitro* models. On the other hand, for the purpose of HTS the predictive value for embryotoxicity does not need to be above 95%. A predictive value of 70–80% might suffice for *in vitro* model systems for a preselection.

Recently, under coordination of ECVAM, the Institute for Health and Consumer Protection, an extensive inter-laboratory validation of the three mostly used *in vitro* embryotoxicity tests has been performed [170]. The three evaluated assays are: whole embryo culture, a micromass assay and an embryonic stem cell test. A report on this validation has been released and the variety of papers published on these assays demonstrate [171, 172] that the systems have a reasonable to good predictive value, sensitivity, specificity and selectivity. In particular, the micromass assay and the embryonic stem cell test have the potency to be transferred into a high- to medium-throughput format. However, with the validation of these assays, compounds like 5-fluorouracil, 5-bromo-2-deoxyuridine and methotrexate, all inhibitors of thymidine incorporation into the DNA, as well as all-*trans*-retinoic acid (vitamin A) were used [173]. The first three compounds are genotoxicants and cytostatics, and vitamin A is an agonist for the retinoic acid receptor (RARα, β, γ). Besides these compounds, the embryotoxic compounds valproic acid and methoxyacetic acid were also used in these studies. Remarkably, these two compounds appear to increase the sensitivity for estrogen, progesterone, androgen and thyroid hormone receptors from 2- to 8-fold [174, 175]. As the effects of vitamin A, thyroid hormone, diethylstilbestrol (an estrogen) and androgens for teratogenicity are well known, NR assays may become of interest in this respect.

Vitamin A plays an important role during neural development. Exogenous administration of vitamin A can cause an open spinal cord during embryonic development [176]. An excess of vitamin A may also lead to other malformations in face, limbs, heart and central nervous system [177, 178]. For the thyroid hormone receptor (TR), the thyroid hormones, 3,5,3'-triiodothyronine (T_3) and 3,5,3',5'-tetraiodothyronine (thyroxine, T_4) are of crucial importance. Thyroid hormones are involved in the development of extremities or the disappearance of, for instance, the tail in tadpoles and frogs [179, 180]. Very low levels of thyroid hormone result in stillbirths, abortions, a decrease in the litter size and congenital malformations [181]. Very high levels of thyroid hormone, on the other hand, lead to additional fingers, cleft lip and palate, an increase in skeletal malformations and ear deformities [182]. Thiouracil, propyl thiouracil, gold thioglucose, and iodoacetate can prevent the formation of T_3 and T_4 on the iodide oxidation level by inhibiting the enzyme deiodonase [183, 184], whereas resorcinol, carbimazole, *p*-aminobenzoic acid and sulfaguanidine can be active as goitrogenic substances [185]. Overexpression of iodonase type III in amphibians can prevent the tail regeneration [186], and this type of iodonase is largely expressed in the placenta to prevent maternal thyroid hormone influencing the fetus [187]. For both RARα and TR luciferase read-out assays can be used.

Besides these NR, the membrane receptors for dopamine and serotonin can also play an important role in the neural development of the brain. Several subtypes for both membrane receptors have been identified in the brain [188, 189]. It is not difficult to imagine that production of dopamine and serotonin at a too-early stage or at the wrong location in the brain will activate the wrong receptors [190]. Thus, exposure to dopaminergic and/or serotonergic compounds or anti-dopaminergic and/or anti-serotonergic compounds at a too-early or late time point in the development will be devastating for the proper outgrowth of nervous tissue [191, 192]. Not only compounds that influence these receptors, but also compounds influencing the turnover of dopamine or serotonin may lead to malformations. Consequently, compounds used for anti-depression, psychosis, Parkinson's disease or prolactonemia may be hazardous. The question is whether in this case prevention is possible by making better compounds with a much more selective profile, if the specific serotonin or dopamine receptor is involved in this process, or whether contraception during the time of drug treatment is a better option. Thus, although we can manipulate nature in certain ways, we have to realize that the influences of compounds with endocrine effectivity can manifest themselves completely differently on embryos, pups/babies and/or children from that after puberty in adulthood, when all developmental processes have been finalized.

Nevertheless, the increase in knowledge of the developmental processes in embryos and pups/babies means that new screening methods may be identified. The development of zebrafish eggs into small larvae and fish is very promising in this aspect. First of all, these experiments with zebra fish eggs and larvae are not part of the legislation of animal experimentation. The egg and larval development can be studied in a 96-well plate, which makes embryological development studies relatively easy. Failures in neural development can be followed quite well in the transparent eggs, while if hatched the locomotor activity can be studied [193, 194]. Since the eggs and larvae grow in normal fresh water, compound treatment can be easily performed by adding compounds to the water. The compounds can be taken up by diffusion through the skin and in later age through the gills. The progress in this field within the last five years has been gigantic and one can only dream about the possibilities in this area for the future.

Endocrine disruption and reprotoxicity

Reproductive failure in females or males can depend on a variety of problems. If gonadotropins like luteinizing hormone (LH) and follicular stimulating hormone (FSH) are not secreted in sufficiently high levels from the pituitary, this will hamper the androgen and estrogen production as well as the progestagen production. The levels of testosterone and 5α-dihydrotestosterone (DHT), as well as of progesterone will remain low if LH excretion is reduced to very low levels. Estradiol-17β (E$_2$), estriol (E$_3$) and/or estrone (E$_1$) levels will be untraceable if FSH secretion is not induced in combination with LH. LH is needed for the surge of the precursor molecules dehydroepiandrosterone, androstenedione and/or testosterone. A blockade of LH and FSH secretion, on the other hand, can be caused in rats and rabbits by oral or subcutaneous treatment with androgenic, estrogenic and/or progestagenic compounds. These compounds cause a negative feedback mechanism *via* the hypothalamus or pituitary on the LH and/or FSH secretion.

The biological activity of androgenic compounds can be measured in orchidectomized male rats and mice with the Hershberger test [195]. Anti-androgenic compounds, like flutamide, hydroxy-flutamide, finasteride [196] and casodex can be identified in a so-called Anti-Hershberger test. Estrogenic activity of compounds can be analyzed in ovariectomized female rats and mice by means of uterine growth or vaginal smears. The latter test is better known as the Allen-Doisy test [197]. Anti-estrogenic compounds, like tamoxifen, 4-hydroxy-tamoxifen, raloxifen and ICI 164.384 [198] can be studied in a so-called Anti-Allen-Doisy test. Progestagen treatment in female rats leads to ovulation inhibition [199], while progestagen treatment in female ovariectomized rabbits leads to growth of the endometrium, also known as the McPhail test [200]. Anti-ovulation inhibition and anti-McPhail tests can be carried out to identify antiprogestagens, like RU486 (mifepristone), onapristone or Org 31710. Moreover, a more general hormone screening test in young males can be used to identify estrogenic, androgenic and/or glucocorticoid activities in compounds [199]. With these *in vivo* assays, the androgenic, estrogenic, progestagenic or glucocorticoid activities of a compound and its metabolites as well as its counterpart antihormonal activities can be measured. For the direct analysis of a compound several *in vitro* assays have been developed and examples are given below.

For a fast track assay of endocrine drug activities, estrogenic, androgenic, progestagenic and glucocorticoid receptor analysis can be performed using either the 'old-fashioned' radiolabeled binding assays [201, 202] or the sensitive luminometric assays. Several laboratories have succeeded in the preparation of assays with CHO [10, 203, 204], human uterotrophic osteoblast U2OS [205], human breast tumor MCF-7 [11], T47D [12, 205–208] or MDA-MB-453 [209], human cervix HeLa [10, 210] or yeast [211–213] cells. Yeast cells might have an advantage, since a combination with S9 mixtures is possible. These cells were prepared by transient or stable transfections with the human cognate receptor in combination with the hormone responsive elements for the respective receptors. These responsive elements could be of synthetic origin, combining three consecutive estrogen responsive elements (EREs) or glucocorticoid responsive elements (GREs). The sequence of androgen, progesterone and mineralocorticoid responsive elements are identical or very similar to that of GREs. These synthetic responsive elements are also present in the natural promoter of certain genes. For instance for the estrogenic response we used a rat oxytoxin promoter [214] as this leads to higher activation levels than that of the very low density lipoprotein (VLDL) promoter [215]. Others used the vitellogenin [216] or progesterone promoter [217]. For GRE, commonly used promoters are the mouse mamma tumor virus (MMTV) promoter [218], the tyrosine aminotransferase (TAT) promoter [219], or phosphoenolpyruvate carboxykinase (PEPCK) promoter [220]. Stable cell lines can be prepared by double transfections with plasmids containing either the neomycine or hygromycine resistancy genes in combination with either the cognate receptor and/or the promoter with the luciferase gene construct. Selection of the most active pools and finally single cell cloning will deliver the appropriate testing device.

Validation of the androgen, estrogen, progesterone and glucocorticoid receptors has been performed in-house in CHO cells with luciferase as the read-out system [10, 203, 204, 221]. The overall in-house comparison of *in vitro* data with *in vivo* data showed that the predictions of the activities identified *in vitro* were in close proximity with the minimal active dose levels of the subcutaneous *in vivo* treatments. This was found with all four of the endocrine assays. Examples of the sensitivity of the *in vitro* agonistic and antagonistic assays are given for reference compounds (Fig. 12). In collaboration with Biodetection Systems (Amsterdam, The Netherlands), the specific data set was compared with the U2OS cell line for estrogenic and androgenic activities [205]. This analysis revealed that the findings with both cellular systems show a very high correlation between the most active and most inactive compounds. Also, the correct prediction between the *in vitro* and *in vivo* activity appeared to be relatively high. These assays have a very high potential for HTS, since analysis at three concentrations of 10^{-7}, 10^{-6} and 10^{-5} M leads to a fast identification of endocrine activity of a compound. In such an analysis approximately 10 800 compounds can be analyzed per assay per day. Thus, in the case of the REACH project with 30 000 compounds, this would be 3 days per assay

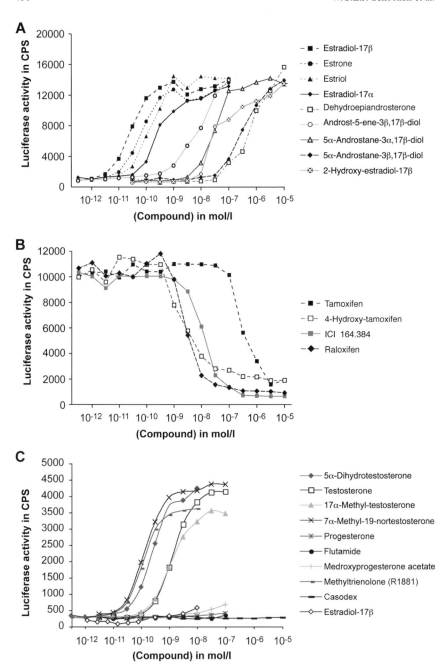

Figure 12. (continued on pages 439 and 440) Dose-dependent stimulation with agonists (A, C, E, G) and dose-dependent inhibition with antagonists (B, D, F, H) for the estrogen receptor α (ERα) (A, B), androgen receptor (AR) (C, D), progesterone receptor (PR) (E, F), and glucocorticoid receptor (GR) (G, H) in Chinese hamster ovary CHO cells, stably transfected with the human cognate receptor and a luciferase reporter-based read-out system.

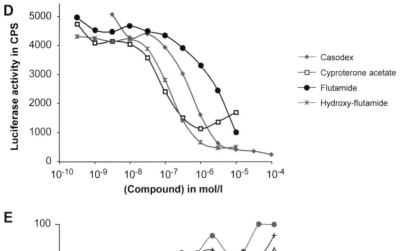

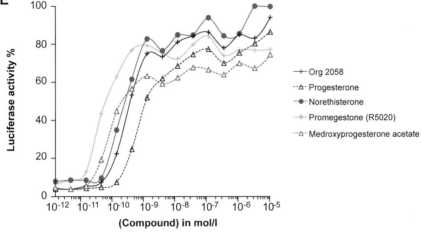

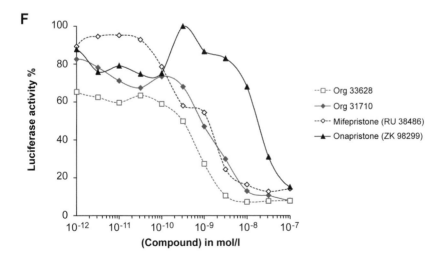

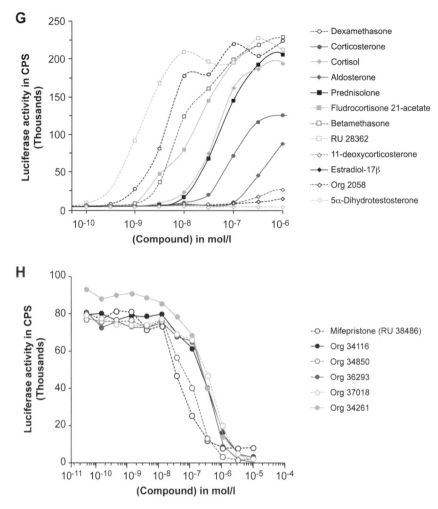

Figure 12. (continued; legend see p. 438)

and 24 days for all agonistic and antagonistic assays. Therefore, these assays might become of use for the fast selection and prioritization of compounds for endocrine disruption. Nevertheless, one should be careful as pro-hormones still need an activation step and might be missed with these screening procedures [222]. On the other hand the positive scores are definitely true positives.

Besides at the level of the steroid hormones, a disturbance can also be induced by overstimulation of the gonadotropin releasing hormone (GnRH) secretion or by a blockade of this secretion. GnRH influences the secretion of both LH and FSH. A GnRH antagonist can block the whole reproductive process by blocking GnRH secretion, while a GnRH agonist can block the process by depleting the GnRH pool, preventing the further secretion of LH and FSH. Moreover, a dopaminergic compound can cause a blockade for the final surge of LH secretion just before ovulation to enhance the production of progesterone by blocking the activity of the enzyme C17,20-lyase. An absence of such a surge will also lead to infertility. Thus, there are various ways to interrupt the reproductive cycle. This implies that assays for the identification of GnRH, dopamine, serotonin, FSH and LH receptors will lead to a better fine-tuning of the cause of incidence.

So, a low level of estrogens in females or androgens in males may tell us whether endocrine disruption is present. However, in fish, the testis can develop into a sort of ovotestis, containing both parts of an ovarium and parts of a testis. Such a development may already start at an early stage or even before puberty or just during adulthood. An adult male fish may also produce more vitellogenin from the liver due to the presence of estrogenic compounds in the waste water. If these aspects are seen, the conclusion is more or less focused on compounds with estrogenic activity. There are, however, other examples in which the development of a testis does simply not occur in fish or snails. In this case the cause may lay in the presence of anti-androgens that block the formation of the testicular development at the embryonic stage.

Other assay types

Besides cell death, changes in cellular proliferation can have serious consequences *in vivo*. Proliferative potencies of compounds can be tested in conjunction with cytotoxicity assays, measuring the DNA content. Depending on the chemical and pharmacological class of the compound, tailor-made *in vitro* systems can be developed to study the most often encountered toxicities or adverse pharmacological side effect of these drugs. Examples are the development of tests to rank compounds as being specific for one ion channel or as modulators of potassium channels (QT prolongation) or hepatic heme production.

Ion channel measurement is becoming more feasible with the new molecular techniques. After overexpression of the ion channels of interest, membrane fractions can be obtained for specific interactive analysis. Active uptake can be measured since these ion channels are usually operational *via* an ATPase reaction. The ATP equivalents can be measured with ATP-Lite™ or CellTiter-GLO™ kits. These assays are becoming available for P-glycoprotein pumps, multi drug resistance pumps (MDR1 and MDR2), as well as for the organic anion transporting pumps (OATP). Solvo and Promega are preparing such assay types. For the hERG-channel, special assays have also been developed with patch clamp techniques or very fast ion channel measurements with Vipor equipment. So, within the field of toxicology and ADME, great progress in preparing HTS assays for regular testing has been made during the last 10 years.

Conclusions

For toxicological screening a huge battery of *in vitro* tests is available or under development. Still, implementation of these tests within toxicology has only just started within the pharmaceutical industry. In the regulatory world for drug registration these new *in vitro* assays are not well accepted yet. The currently used *in vivo* methods are still recognized as the gold standard. Although this acceptance is based on a wide experience with these *in vivo* tests and the availability of large databases, the prediction for human toxicity is still inadequate.

Nevertheless, alternative methods will only be able to replace the current regulatory required studies in an effective way when they can mimic the *in vivo* biological complexity. In this aspect, *in vitro* methods may play an important role in (pre-)clinical development. If specific mechanistic issues are involved in the toxicity of a class of compounds, these *in vitro* assays may help to recognize the toxicological finding in one of the toxic species and to interpret the human relevance of this finding. Thus, in the future, the role of *in vitro* toxicology in drug development will become more prominent. This inevitably will lead into a reduction of the attrition rate and the use of experimental animals.

Acknowledgments
We would like to thank Drs Gerard Griffioen and Annick Lauwers of reMYND, Leuven, Belgium for the use of the yeast RAD54-lacZ strain and the specific culture media, Drs Richard Walmsley and Andrew Knight for the use of the ingredients of the GreenScreen GC assay and Mrs Nelleke Bisseling for the preparation of the drawings.

References

1 Brown D, Superti-Furga G (2003) Rediscovering the sweet spot in drug discovery. *Drug Discov Today* 8: 1067–1077
2 DiMasi JA (2001) Risks in new drug development: Approval success rates for investigational drugs. *Clin Pharmacol Ther* 69: 297–307
3 Kola I, Landis J (2004) Can the pharmaceutical industry reduce attrition rates? *Nat Rev Drug Discov* 3: 711–715
4 Caldwell GW, Ritchie DM, Masucci JA, Hageman W, Yan Z (2001) The new pre-clinical paradigm: Compound optimization in early and late phase drug discovery. *Curr Top Med Chem* 1: 353–366
5 Schuster D, Laggner C, Langer T (2005) Why drugs fail – A study on side effects in new chemical entities. *Curr Pharm Des* 11: 3545–3559
6 Smith DA, Schmid EF (2006) Drug withdrawals and the lessons within. *Curr Opin Drug Discov Dev* 9: 38–46
7 Cuatrecasas P (2006) Drug discovery in jeopardy. *J Clin Invest* 116: 2837–2842
8 Gagne D, Balaguer P, Demirpence E, Chabret C, Trousse F, Nicolas JC, Pons M (1994) Stable luciferase transfected cells for studying steroid receptor biological activity. *J Biolumin Chemilumin* 9: 201–209
9 Naylor LH (1999) Reporter gene technology: The future looks bright. *Biochem Pharmacol* 58: 749–757
10 Dijkema R, Schoonen WGEJ, Teuwen R, Struik van der E, Ries de RJH, Kar SAT, Olijve W (1998) Human progesterone receptor A and B isoforms in CHO cells. I. Stable transfection of receptor and receptor-responsive reporter genes: Transcription modulation by (anti)progestagens.

J Steroid Biochem Mol Biol 64: 140–156

11 Balaguer P, Boussioux A-M, Demirpence E, Nicolas JC (2001) Reporter cell lines are useful tools for monitoring biological activity of nuclear receptor ligands. *Luminescence* 16: 153–158

12 Willemsen P, Scippo ML, Kausel G, Figueroa J, Maghuin-Rogister G, Martial JA, Muller M (2004) Use of reporter cell lines for detection of endocrine-disrupter activity. *Anal Bioanal Chem* 378: 655–663

13 Lai C, Jiang X, Li X (2006) Development of luciferase reporter-based cell assays. *Assay Drug Dev Technol* 4: 307–315

14 Stratowa C, Himmler A, Czernilofsky AP (1995) Use of a luciferase reporter system for characterizing G-protein-linked receptors. *Curr Opin Biotech* 6: 574–581

15 Durocher Y, Perret S, Thibaudeau E, Gaumond MH, Kamen A, Stocco R, Abramovitz M (2000) A reporter gene assay for high-throughput screening of G-protein-coupled receptors stably or transiently expressed in HEK293 EBNA cells grown in suspension culture. *Anal Biochem* 284: 316–326

16 Gabriel D, Vernier M, Pfeifer MJ, Dasen B, Tenaillon L, Bouhelal R (2003) High throughput technologies for direct cyclic AMP measurement. *Assay Drug Dev Technol* 1: 291–303

17 Kunapuli P, Ransom R, Murphy KL, Pettibone D, Kerby J, Grimwood S, Zuck P, Hodder P, Lacson R, Hoffman I et al (2003) Development of an intact cell reporter gene β-lactamase assay for G protein-coupled receptors for high-throughput screening. *Anal Biochem* 314: 16–29

18 Hemmilä IA, Hurskainen P (2002) Novel detection strategies for drug discovery. *Drug Discov Today* 7: 150–156

19 Cómez-Henz A, Aguilar-Caballos MP (2007) Modern analytical approaches to high throughput screening. *Trends Anal Chem* 26: 171–182

20 Ames BN, Lee FD, Durston WE (1973) An improved bacterial test system for the detection and classification of mutagens and carcinogens. *Proc Natl Acad Sci USA* 70: 782–786

21 McCann J, Choi E, Yamasaki E, Ames BN (1975) Detection of carcinogens as mutagens in the *Salmonella*/microsome test: Assay of 300 chemicals. *Proc Natl Acad Sci USA* 72: 5135–5139

22 Miller B, Pötter-Locher F, Seelbach A, Stopper H, Utesch D, Madle S (1998) Evaluation of the *in vitro* micronuclei test as an alternative to the *in vitro* chromosomal aberration assay: Position of the GUM working group on the *in vitro* micronucleus test. *Mutat Res* 410: 81–116

23 McGregor JT, Casciano D, Müller L (2000) Strategies and testing methods for identifying mutagenic risks. *Mutat Res* 455: 3–20

24 Richard AM (1998) Structure-based methods for predicting mutagenicity and carcinogenicity: Are we there yet? *Mutat Res* 400: 493–507

25 Cariello NE, Wilson JD, Britt BH, Wedd DJ, Burlinson B, Gombar V (2002) Comparison of the computer programs DEREK and TOPKAT to predict bacteria mutagenicity. *Mutagenesis* 17: 321–329

26 Greene N (2002) Computer systems for the prediction of toxicity: An update. *Adv Drug Deliv Rev* 54: 417–431

27 Mitchell AD (2000) *In vitro* genetic toxicity testing. In: SC Gad (eds): *In vitro Toxicology*, 2nd edn. CRC Press, Boca Raton, 94–127

28 Mortelmans K, Zeiger E (2000) The Ames *Salmonella*/microsome mutagenicity assay. *Mutat Res* 455: 29–60

29 Musatov, SA, Anisimov VN, André V, Vigreux C, Godard T, Gauduchon P, Sichel F (1998) Modulatory effects of melatonin on genotoxic response of reference mutagens in the Ames test and the comet assay. *Mutat Res* 417: 75–84

30 Jemnitz K, Veres Z, Torok G, Toth E, Vereckey L (2004) Comparative study in the Ames test of benzo[*a*]pyrene and 2-aminoanthracene activation using rat hepatic S9 and hepatocytes following *in vivo* or *in vitro* activation. *Mutagenesis* 19: 245–250

31 Flückiger-Isler S, Baumeister M, Braun K, Gervais V, Nasler-Nguyen N, Reimann R, Van Gompel J, Wunderlich HG, Engerhardt G (2004) Assessment of the performance of the Ames II™ assay: A collaborative study with 19 coded compounds. *Mutat Res* 558: 181–189

32 Piegorsch WW, Simmons SJ, Margolin BH, Zeiger E, Gidrol XM, Gee P (2000) Statistical modeling and analyses of a base-specific *Salmonella* mutagenicity assay. *Mutat Res* 467: 11–19

33 Lelie van der D, Regniers L, Borremans B, Provoost A, Verschaeve L (1997) The VITOTOX® test, an SOS bioluminescence *Salmonella typhimurium* test to measure genotoxicity kinetics. *Mutagenesis* 20: 449–454

34 Verschaeve L, Van Gompel J, Thilemans L, Regniers L, Vanparys P, van der Lelie D (1999) VITO-

TOX® bacterial genotoxicity and toxicity test for the rapid screening of chemicals. *Environ Mol Mutagen* 33: 240–248

35 Van Gompel J, Woestenborghs F, Beerens D, Mackie C, Cahill PA, Knight AW, Billinton N, Tweats DJ, Walmsley RM (2005) An assessment of the utility of the yeast GreenScreen assay in pharmaceutical screening. *Mutagenesis* 20: 449–454

36 Westerink WMA, Stevenson JCR, Schoonen WGEJ (2009) Evaluation of the Vitotox and Radarscreen assay for the rapid early assessment of genotoxicity in the research phase of drug development. *Mutat Res (submitted)*

37 Jacobson-Kram D, Contrera JF (2007) Genetic toxicity assessment: Employing the best science for human safety evaluation part I: Early screening for potential human mutagens. *Toxicol Sci* 96: 16–20

38 Kirkland D, Pfuhler S, Tweats D, Aardema M, Corvi R, Darroudi F, Elhajouji A, Glatt H, Hastwell P, Hayashi M et al (2007) How to reduce false positive results when undertaking *in vitro* genotoxicity testing and thus avoid unnecessary follow-up animal tests: Report of an ECVAM Workshop. *Mutat Res* 628: 31–55

39 Tweats DJ, Scott AD, Westmoreland C, Carmichael PL (2007) Determination of genetic toxicity and potential carcinogenicity *in vitro* – Challenges post the seventh amendment to the European cosmetics directive. *Mutagenesis* 22: 5–13

40 Pastink A, Eeken JCJ, Lohman PHM (2001) Genomic integrity and the repair of double-strand DNA breaks. *Mutat Res* 480–481: 37–50

41 Clever B, Interthal H, Schmuckli-Maurer J, King J, Sigrist M, Heyer WF (1997) Recombinational repair in yeast: Functional interactions between Rad51 and Rad54 proteins. *EMBO J* 16: 2535–2544

42 Sonoda E, Sasaki MS, Buerstedde J-M, Bezzubova O, Shinohara A, Ogawa H, Takata M, Yamaguchi-Iwai Y, Takeda S (1998) Rad51-deficient vertebrate cells accumulate chromosomal breaks prior to cell death. *EMBO J* 17: 598–608

43 Arbel A, Zenvirth D, Simchen G (1999) Sister chromatid-based DNA repair is mediated by RAD54, not by DMC1 or TID1. *EMBO J* 18: 2648–2658

44 Dronkert MLG, Beverloo HB, Johnson RD, Hoeijmakers JHJ, Jasin M, Kanaar R (2000) Mouse RAD54 affects DNA double-strand break repair and sister chromatid exchange. *Mol Cell Biol* 20: 3147–3156

45 Billinton N, Barker MG, Michel CE, Knight AW, Heyer W-D, Goddard NJ, Fielden PR, Walmsley RM (1998) Development of a green fluorescent protein reporter for a yeast genotoxicity biosensor. *Biosens Bioelectr* 13: 831–838

46 Cahill PA, Knight AW, Billington N, Barker MG, Walsh L, Keenan PO, Williams CV, Tweats DJ, Walmsley RM (2004) The GreenScreen® genotoxicity assays: A screening validation programme. *Mutagenesis* 19: 105–119

47 Knight AW, Billinton N, Cahill PA, Scott A, Harvey JS, Roberts KJ, Tweats DJ, Keenan PO, Walmsley RM (2007) An analysis of results from 305 compounds tested with the yeast *RAD54-GFP* genotoxicity assay (GreenScreen GC) – including relative predictivity of regulatory tests and rodent carcinogenesis and performance with autofluorescent and coloured compounds. *Mutagenesis* 22: 409–416

48 Cole GM, Schild D, Lovett ST, Mortimer RK (1987) Regulation of RAD54- and RAD52-lacZ gene fusions in *Saccharomyces cerevisiae* in response to DNA damage. *Mol Cell Biol* 7: 1078–1084

49 Averbeck D, Averbeck S (1994) Induction of the genes RAD54 and RNR2 by various damaging agents in *Saccharomyces cerevisiae*. *Mutat Res* 315: 123–138

50 Joosten HFP, Acker FAA, Dobbelsteen van den DJ, Horbach GJMJ, Krajnc EI (2004) Genotoxicity of hormonal steroids. *Toxicol Lett* 151: 113–134

51 Hastwell PW, Chai LL, Roberts KJ, Webster ThW, Harvey JS, Rees RW, Walmsley RM (2006) High-specificity and high-sensititvity genotoxicity assessment in a human cell line: Validation of the GreenScreen HC GADD45α-GFP genotoxicity assay. *Mutat Res* 607: 160–175

52 Jover R, Pondosa X, Castell JV, Gómez-Lechón MJ (1992) Evaluation of the cytotoxicity of ten chemicals on human cultured hepatocytes: Predictability of human toxicity and comparison with rodent cell culture systems. *Toxicol In Vitro* 6: 47–52

53 Okey AB, Roberts EA, Harper PA, Denison MS (1986) Induction of drug-metabolizing enzymes: Mechanisms and consequences. *Clin Biochem* 19: 132–141

54 Marcillat O, Zhang Y, Davies KJ (1989) Oxidative and non-oxidative mechanisms in the inactiva-

tion of cardiac mitochondrial electron transport chain components by doxorubicin. *Biochem J* 259: 181–189

55 Zhou S, Starkov A, Froberg MK, Leino RL, Wallace KB (2001) Cumulative and irreversible cardiac mitochondrial dysfunction induced by doxorubicin. *Cancer Res* 61: 771–777

56 Olson RD, Mushlin PS, Brenner DE, Fleischer S, Cusak BJ, Chang BK, Boucek RJ (1988) Doxorubicin cardiotoxicity may be caused by its metabolite doxorubicinol. *Proc Natl Acad Sci USA* 85: 3585–3589

57 Mordente A, Minotti G, Martorana GE, Silvestrini A, Giardina B, Meucci E (2003) Anthracycline secondary alcohol metabolite formation in human or rabbit heart: Biochemical aspects and pharmacologic implications. *Biochem Pharmacol* 66: 989–998

58 Fries de R, Mitsuhashi M (1995) Quantification of mitogen induced human lymphocyte proliferation: Comparison of Alamar Blue[TM] assay to ^{3}H-thymidine incorporation assay. *J Clin Lab Anal* 9: 89–95

59 Gieni RS, Li Y, HayGlass KT (1995) Comparison of ^{3}H-thymidine incorporation with MTT- and MTS-based bioassays for human and murine IL-2 and IL-4 analysis. Tetrazolium assays provide markedly enhanced sensitivity. *J Immunol Methods* 187: 85–93

60 Wagner U, Burkhardt E, Failing K (1999) Evaluation of canine lymphocyte proliferation: Comparison of three different colorometric methods with the ^{3}H-thymidine incorporation assay. *Vet Immunol Immunopathol* 70: 151–159

61 Nagahama T, Sawada M, Gonzalez FJ, Yokoi T, Kamataki T (1996) Stable expression of human CYP2E1 in Chinese hamster cells: High sensitivity to *N,N*-dimethylnitrosamine in cytotoxicity testing. *Mutat Res* 360: 181–186

62 Blaheta RA, Franz M, Auth MKH, Wenisch HJC, Markus BH (1991) A rapid non-radioactive fluorescence assay for the measurement of both cell number and proliferation. *J Immunol Methods* 142: 199–206

63 Lydon MJ, Keeler KD, Thomas DB (1980) Vital DNA coloring and cell sorting by flow microfluorometry. *J Cell Physiol* 102: 175–181

64 Richards WL, Song MK, Krutsch H, Everts RP, Marsden E, Thorgeirsson SS (1985) Measurement of cell proliferation using Hoechst 33342 for the rapid semiautomated microfluorometric determination of chromatin DNA. *Exp Cell Res* 159: 235–246

65 Dhar S, Nygren P, Liminga G, Sundstrom G, de la Torre M, Nilsson K, Larsson R (1998) Relationship between cytotoxic drug response patterns and activity of drug efflux transporters mediating multidrug resistance. *Eur J Pharmacol* 346: 315–322

66 Marbeuf-Gueye G, Salerno M, Quidu P, Garnier-Suillerot A (2000) Inhibition of the P-glycoprotein and multidrug resistance protein efflux of anthracyclines and calceinacetoxymethyl ester by PAK-104P. *Eur J Pharmacol* 391: 207–216

67 Fernandez-Checa JC, Kaplowitz N (1990) The use of monochlorobimane to determine hepatic GSH levels and synthesis. *Anal Biochem* 190: 212–219

68 Stevenson D, Wokosin D, Girkin J, Grant MH (2002) Measurement of the intracellular distribution of reduced glutathione in cultured rat hepatocytes using monochlorobimane and confocal laser scanning microscopy. *Toxicol In Vitro* 16: 609–619

69 Young PR, ConnorsWhite AL, Dzido GA (1994) Kinetic analysis of the intracellular conjugation of monochlorbimane by IC-21 murine macrophage glutathione-*S*-transferase. *Biochim Biophys Acta* 1201: 461–465

70 Gabriel C, Camins A, Sureda FX, Aquirre L, Escubedo E, Pallàs M, Camarasa J (1997) Determination of nitric oxide generation in mammalian neurons using dichlorofluorescin diacatate and flow cytometry. *J Pharmacol Toxicol Methods* 38: 93–98

71 Sanner BM, Meder U, Zidek W, Tepel M (2002) Effects of glucocorticoids on generation of reactive oxygen species in platelets. *Steroids* 67: 715–719

72 Soliman MK, Mazzio E, Soliman KFA (2002) Levodopa modulating effects of inducible nitric oxide synthase and reactive oxygen species in glioma cells. *Life Sci* 72: 185–198

73 Berridge MV, Tan AS, McCoy KD, Wang R (1996) The biochemical and cellular basis of cell proliferation assays that use tetrazolium salts. *Biochemica* 4: 14–19

74 Andrews MJ, Garle MJ, Clothier RH (1997) Reduction of the new tetrazolium dye, Alamar Blue[TM], in cultured rat hepatocytes and liver fractions. *ATLA* 25: 641–653

75 Nakayama GR, Caton MC, Nova MP, Parandoosh Z (1997) Assessment of the Alamar Blue assay for cellular growth and viability *in vitro*. *J Immunol Methods* 240: 205–208

76 O'Brien J, Wilson I, Orton T, Pognan F (2000) Investigation of the Alamar Blue (resazurin) fluo-

rescent dye for the assessment of mammalian cell cytotoxicity. *Eur J Biochem* 267: 5421–5426

77 Slaughter MR, Bugelski PJ, O'Brien PJ (1999) Evaluation of Alamar Blue reduction for *in vitro* assay of hepatocyte toxicity. *Toxicol In Vitro* 13: 567–569

78 Hynes J, Hill R, Papkovsky DB (2006) The use of a fluorescence-based oxygen uptake assay in the analysis of cytotoxicity. *Toxicol In Vitro* 20: 785–792

79 Chan JH, Harminder SD, Powell-Richards A, Jones DRE, Harris IM (2001) Effect of ABO blood group mismatching on corneal epithelial cells: An *in vitro* study. *Br J Ophthalmol* 85: 1104–1109

80 Hannah R, Beck M, Moravec R, Riss T (2001) CellTiter-GloTM luminescent cell viability assay: A sensitive and rapid method for determining cell viability. *Promega Cell Notes* 2: 11–13

81 Germain MA, Hatton A, Williams S, Matthews JB, Stone MH, Fisher J, Ingham E, (2003) Comparison of the cytotoxicity of clinically relevant cobalt-chromium and alumina ceramic wear particles *in vitro*. *Biomaterials* 24: 469–479

82 Schoonen WG, de Roos JA, Westerink WM, Debiton E (2005) Cytotoxic effects of 110 reference compounds on HepG2 cells and for 60 compounds on HeLa, ECC-1 and CHO cells. II. Mechanistic assays on NAD(P)H, ATP and DNA contents. *Toxicol In Vitro* 19: 491–503

83 Schoonen WG, Westerink WM, de Roos JA, Debiton E (2005) Cytotoxic effects of 100 reference compounds on Hep G2 and HeLa cells and of 60 compounds on ECC-1 and CHO cells. I. Mechanistic assays on ROS, glutathione depletion and calcein uptake. *Toxicol In Vitro* 19: 505–516

84 Ekwall B, Sandström B (1978) Combined toxicity to HeLa cells of 30 drug pairs, studied by a two-dimensional microtitre method. *Toxicol Lett* 2: 285–292

85 Ekwall B, Sandström B (1978) Improved use of the metabolic inhibition test to screen combined drug toxicity to HeLa cells-preliminary study of 61 drug pairs. *Toxicol Lett* 2: 293–298

86 Ekwall B, Johansson A (1980) Preliminary studies on the validity of *in vitro* measurement of drug toxicity using HeLa cells. I. Comparative *in vitro* cytotoxicity of 27 drugs. *Toxicol Lett* 5: 299–307

87 Clemedson C, McFarlane-Abdulla E, Andersson M, Barile B, Calleja MC, Chesné C, Clothier R, Cottin M, Curren R, Dierickx PJ et al (1996) MEIC evaluation of acute systemic toxicity. Part II. *In vitro* results from 68 toxicity assays used to test the first 30 reference chemicals and a comparative cytotoxicity analysis. *ATLA* 24: 273–311

88 Clemedson C, Ekwall B (1999) Overview of the MEIC results: I. The *in vitro–in vitro* evaluation. *Toxicol In Vitro* 13: 657–663

89 Clemedson C, Blaauboer B, Castell J, Prieto P, Ristyelli L, Vericat JA, Wendel A (2006) ACuteTox – Optimation and pre-validation of an *in vitro* test strategy for predicting human acute toxicity. *ALTEX* 23 Suppl: 254–258

90 Clemedson C, Kolman A, Forsby A (2007) The integrated acute systemic toxicity traject (ACuteTox) for the optimisation and validation of alternative *in vitro* tests. *Altern Lab Anim* 35: 33–38

91 Thabrew MI, Hughes RD, McFarlane IG (1997) Screening of hepatoprotective plant components using a HepG2 cell cytotoxicity assay. *J Pharmacol* 49: 1132–1135

92 Ekwall B (1980) Toxicity to HeLa cells of 205 drugs as determined by the metabolic inhibition test supplemented by microscopy. *Toxicology* 17: 273–295

93 Ekwall B (1980) Preliminary studies on the validity of *in vitro* measurement of drug toxicity using HeLa cells. III. Lethal action to man of 43 drugs related to the HeLa cell toxicity of the lethal drug concentrations. *Toxicol Lett* 5: 319–331

94 Ekwall B (1980) Preliminary studies on the validity of *in vitro* measurement of drug toxicity using HeLa cells. II. Drug toxicity in the MIT-24 system compared with mouse and human lethal dosage of 52 drugs. *Toxicol Lett* 5: 309–317

95 Ekwall B, Bondesson I, Castell JV, Gómez-Lechón MJ, Hellberg S, Högberg J, Jover R, Pondosa X, Romert L, Stenberg K, Walum E (1989) Cytotoxicity evaluation of the first ten MEIC chemicals: Acute lethal toxicity in man predicted by cytotoxicity in five cellular assays and by oral LD50 tests in rodents. *ATLA* 17: 83–100

96 Clothier RH, Hulme LM, Smith M, Balls M, (1987) Comparison of the *in vitro* cytotoxicities and acute *in vivo* toxicities of 59 chemicals. *Mol Toxicol* 1: 571–577

97 Bondesson I, Ekwall B, Hellberg S, Romert L, Stenberg K, Walum E (1989) MEIC – A new international multicenter project to evaluate the relevance to human toxicity of *in vitro* cytotoxicity tests. *Cell Biol Toxicol* 5: 331–347

98 Barile FA, Dierickx PJ, Kristen U (1994) *In vitro* toxicity testing for prediction of acute human toxicity. *Cell Biol Toxicol* 10: 155–162

99 Pondosa X, Núñez C, Castell JV, Gómez-Lechón MJ (1997) Evaluation of the cytotoxic effects of MEIC chemicals 31–50 on primary culture of rat hepatocytes and hepatic and non-hepatic cell lines. *ATLA* 25: 423–436

100 Scheers EM, Ekwall B, Dierickx PJ (2001) *In vitro* long-term cytotoxicity testing of 27 MEIC chemicals on Hep G2 cells and comparison with acute human toxicity data. *Toxicol In Vitro* 15: 153–161

101 Ekwall B (1999) Overview of the final MEIC results: II The *in vitro-in vivo* evaluation, including the selection of a practical battery of cell tests for prediction of acute lethal blood concentrations in humans. *Toxicol in Vitro* 13: 665–673

102 Rosenkranz HS, Cunningham AR (2000) The high production volume chemical challenge program: The rodent LD50 and its possible replacement. *ATLA* 28: 271–277

103 O'Brien MA, Daily WJ, Hesselberth E, Moravec RA, Scurria MA, Klaubert DH, Bulleit RF, Wood KV (2005) Homogeneous, bioluminescent protease assay: Caspase 3 as a model. *J Biomol Screen* 10: 137–148

104 Liu D, Li C, Chen Y, Burnett C, Liu XY, Downs S, Colllins RD, Hawegir J (2004) Nuclear transport of proinflammatory transcription factors is required for massive liver apoptosis induced by bacterial lipopolysaccharide. *J Biol Chem* 279: 48434–48442

105 Parkinson A (2003) Biotransformation and xenobiotics. In: CD Klaassen and JB Watkins (eds): *Casarett & Doull's Essentials of Toxicology*. McGraw-Hill, New York, 71–97.

106 Meyer UA (2000) Pharmacogenetics and adverse drug interactions. *Lancet* 356: 1667–1671

107 Bao H, Vepakomma M, Sarkar MA (2002) Benzo[*a*]pyrene exposure induces CYP1A1 activity and expression in human endometrial cells. *J Steroid Biochem Mol Biol* 81: 37–45

108 Huang WY, Chatterjee N, Chanock S, Dean M, Yeager M, Schoen RE, Hou LF, Berndt SI, Yadavalli S, Johnson CC, Hayes RB (2005) Microsomal epoxide hydrolase polymorphisms and risk for advanced colorectal adenoma. *Cancer Epidemiol Biomarkers Prev* 14: 152–157

109 Uno S, Dalton TP, Shertzer HG, Genter MB, Warshawsky D, Talaska G, Nebert DW (2001) Benzo[*a*]pyrene-induced toxicity: Paradoxical protection in Cyp1a1$^{(-/-)}$ knockout mice having increased hepatic BaP-DNA adduct levels. *Biochem Biophys Res Commun* 289: 1049–1056

110 Seppen J, Bosma PJ, Goldhoorn BG, Bakker CTM, Chowdhury JR, Chowdhury NR, Jansen PLM, Oude Elferink RPJ (1994) Discrimination between Crigler najjar Type I and II by expression of mutant bilirubin uridine phosphate-glucuronosyltransferase. *J Clin Invest* 94: 2385–2391

111 Aono S, Adachi Y, Uyama E, Yamada Y, Keino H, Nanno T, Koiwai O, Sato H (1995) Analysis of genes for bilirubin UDP-glucuronosyltransferase in Gilbert's syndrome. *Lancet* 345: 958–959

112 Hewitt NJ, Hewitt P (2004) Phase I and II enzyme characterization of two sources of HepG2 cell lines. *Xenobiotica* 34: 243–256

113 Rodriguez-Antona C, Donato MT, Boobis A, Edwards RJ, Watts PS, Castell JV, Gomez-Lechon MJ (2002) Cytochrome P450 expression in human hepatocytes and hepatoma cell lines: Molecular mechanisms that determine lower expression in cultured cells. *Xenobiotica* 32: 505–520

114 Wilkening S, Stahl F, Bader A (2003) Comparison of primary human hepatocytes and hepatoma cell line HepG2 with regard to their biotransformation properties. *Drug Metab Disp* 31: 1035–1042

115 Gomez-Lechon MJ, Donato MT, Castell JV, Jover R (2004) Human hepatocytes in primary culture: The choice to investigate drug metabolism in man. *Curr Drug Metab* 5: 443–462

116 Westerink WMA, Schoonen WGEJ (2007) Phase I enzyme levels in cryopreserved primary human hepatocytes and HepG2 cells and their induction in HepG2 cells. *Toxicol In Vitro* 21: 1581–1591.

117 Westerink WMA, Schoonen WGEJ (2007) Phase II enzyme levels in HepG2 cells and cryopreserved human primary human hepatocytes and their induction in HepG2 cells. *Toxicol In Vitro* 21: 1592–1602

118 Köhle C, Bock KW (2007) Coordinate regulation of phase I and II xenobiotic metabolism by the Ah receptor and Nrf2. *Biochem Pharmacol* 73: 1853–1862

119 Fletcher N, Wahlström D, Lundberg R, Nilsson KC, Stockling K, Hellmold H, Håkansson H (2005) 2,3,7,8-Tetrachlorodibenzo-*p*-dioxin (TCDD) alters the mRNA expression of critical genes associated with cholesterol metabolism, bile acid biosynthesis, and bile transport in rat liver: A microarray study. *Toxicol Appl Pharmacol* 207: 1–24

120 Adachi J, Mori Y, Matsui S, Takigami H, Fujino J, Kitagawa H, Miller CA III, Kato T, Saeki K, Matsuda T (2001) Indirubin and indigo are potent aryl hydrocarbon receptor ligands present in

human urine. *J Biol Chem* 276: 31475–31478

121 Runge D, Kohler C, Kostrubsky VE, Jager D, Lehmann T, Runge DM, May U, Stolz DB, Strom SC, Fleig WE, Michalopoulos GK (2000) Induction of cytochrome P450 (CYP)1A1, CYP1A2, and CYP3A4 but not of CYP2C9, CYP2C19, multidrug resistance (MDR-1) and multidrug resistance associated protein (MRP-1) by prototypical inducers in human hepatocytes. *Biochem Biophys Res Commun* 273: 333–341

122 Yueh MF, Kawahara M, Raucy J (2005) Cell-based high-throughput bioassays to assess induction and inhibition of CYP1A enzymes. *Toxicol In Vitro* 19: 275–287

123 Bock KW, Kohle C (2004) Coordinate regulation of drug metabolism by xenobiotic nuclear receptors: UGTs acting together with CYPs and glucuronide transporters. *Drug Metab Rev* 36: 595–615

124 Maglich JM, Stoltz CM, Goodwin B, Hawkins-Brown D, Moore JT, Kliewer SA (2002) Nuclear pregnane X receptor and constitutive androstane receptor regulate overlapping but distinct sets of genes involved in xenobiotic detoxification. *Mol Pharmacol* 62: 638–646

125 Mankowski DC, Ekins S (2003) Prediction of human drug metabolizing enzyme induction. *Curr Drug Metab* 4: 381–391

126 Xie, W, Yeuh MF, Radominska-Pandya A, Saini SP, Negishi Y, Bottroff BS, Cabrera GY, Tukey RH, Evans RM (2003) Control of steroid, heme, and carcinogen metabolism by nuclear pregnane X receptor and constitutive androstane receptor. *Proc Natl Acad Sci USA* 100: 4150–4155

127 Moore JT, Moore LB, Maglich JM, Kliewer SA (2003) Functional and structural comparison of PXR and CAR. *Biochim Biophys Acta* 1619: 235–238

128 Waxman DJ (1999) P450 gene induction by structurally diverse xenochemicals: Central role of nuclear receptors CAR, PXR, and PPAR. *Arch Biochem Biophys* 369: 11–23

129 Wei P, Zhang J, Dowhan DH, Han Y, Moore DD (2002) Specific and overlapping functions of the nuclear hormone receptors CAR and PXR in xenobiotic response. *Pharmacogenomics J* 2: 117–126

130 Repa JJ, Turley SD, Lobaccaro JMA, Medina J, Li L, Lustig K, Shan B, Heyman RA, Dietschy JM, Mangelsdorf DJ (2000) Regulation of absorption and ABC1-mediated efflux of cholesterol by RXR heterodimers. *Science* 289: 1524–1529

131 Gupta S, Stravitz RT, Dent P, Hylemon PB (2001) Down-regulation of cholesterol 7α-hydroxylase (CYP7A1) gene expression by bile acids in primary rat hepatocytes is mediated by the c-jun N-terminal kinase pathway. *J Biol Chem* 276: 15816–15822

132 Anderson SP, Dunn C, Laughter A, Yoon L, Swanson C, Stulnig TM, Steffensen KR, Chandraratna RAS, Gustafsson JÅ, Corton JC (2004) Overlapping transcriptional programs regulated by the nuclear receptors peroxisome proliferator-activated receptor α, retinoic X receptor and liver X receptor in mouse liver. *Mol Pharmacol* 66: 1440–1452

133 Goodwin B, Watson MA, Kim H, Miao J, Kemper JK, Kliewer SA (2003) Differential regulation of rat and human CYP7A1 by the nuclear orphan receptor liver X receptor α. *Mol Endocrinol* 17: 386–394

134 Wilson TM, Lambert MH, Kliewer SA (2001) Peroxisome proliferator-activated receptor γ and metabolic disease. *Annu Rev Biochem* 70: 341–367

135 O'Brien PJ, Irwin W, Diaz D, Howard-Cofield E, Krejsa CM, Slaughter MR, Gao B, Kaludercic N, Angeline A, Bernardi P, Brain P, Hougham C (2006) High concordance of drug-induced human hepatotoxicity with *in vitro* cytotoxicity measured in a novel cell-based model using high content screening. *Arch Toxicol* 80: 580–604

136 Xu JJ (2007) High content screening with primary rat hepatocytes. Lecture Symposium on: *High Content Screening for Toxicity* (Informa). Le Meridian, Vienna, Austria, 28 June 2007

137 Vanparys P (2007) New models for toxicity testing. 2nd Conference of *Predictive Human Toxicity and ADME Toxicity Studies*. Brussels, Belgium, 25–26 January 2007

138 Herrera G, Diaz L, Martinez-Romero A, Gomes A, Villamon E, Callaghan RC, O'Connor JE (2007) Cytomics: A multiparametric, dynamic approach to cell research. *Toxicol In Vitro* 21: 176–182

139 Giuliano KA, Haskins JR, Taylor DL (2003) Advances in high content screening for drug discovery. *Assay Drug Dev Technol* 1: 565–577

140 Haney SA, LaPan P, Pan J, Zhang J (2006) High-content screening moves to the front of the line. *Drug Discov Today* 11: 889–894

141 Houck KA, Kavlock RJ (2008) Understanding mechanisms of toxicity: Insights from drug discovery research. *Toxicol Appl Pharmacol* 277: 163–178

142 Diaz D, Scott A, Carmichael P, Shi W, Costales C (2007) Evaluation of an automated *in vitro* micronucleus assay in CHO-K1 cells. *Mutat Res* 630: 1–13

143 McMillian MK, Grant ER, Zhong Z, Parker JB, Li L, Zivin RA, Burczynski ME, Johnson MD (2001) Nile red binding to HepG2 cells: An improved assay for *in vitro* studies of hepatosteatosis. *In Vitro Mol Tox* 14: 177–190

144 Reasor MJ, Hastings KL, Ulrich RG (2006) Drug-induced phospholipidosis: Issues and future directions. *Exp Opin Drug Safety* 5: 567–583

145 Fujimara H, Dekura E, Kurabe M, Shimazu N, Koitabashi M, Toriumi W (2007) Cell-based fluorescence assay for evaluation of new-drugs potential for phospholipidosis in an early stage of drug development. *Exp Toxicol Pathol* 58: 375–382

146 Nioi P, Perry BK, Wang EJ, Gu YZ, Snyder RD (2007) *In vitro* detection of drug-induced phospholipidosis using gene expression and fluorescent phospholipid based methodologies. *Toxicol Sci* 99: 162–173

147 Fingerhut MA, Halperin WE, Marlow DA, Piacitelli LA, Honchar PA, Sweeney MH, Greife AL, Dill PA, Steenland K, Suruda AJ (1991) Cancer mortality in workers exposed to 2,3,7,8-tetrachlorodibenzo-*p*-dioxin. *N Engl J Med* 324: 212–218

148 Manz A, Berger J, Dwyer JH, Flesch-Janys D, Nagel S, Waltsgott H (1991) Cancer mortality among workers in chemical plant contaminated with dioxin. *Lancet* 338: 959–964

149 Zober A, Messerer P, Huber P (1990) Thirty-four-year mortality follow-up of BASF employees exposed to 2,3,7,8-TCDD after the 1953 accident. *Int Arch Occup Environ Health* 62: 139–157

150 Vanden Heuvel PJ (1999) Peroxisome proliferator-activated receptors (PPARS) and carcinogenesis. *Toxicol Sci* 47: 1–8

151 Stewart JR, Artime MC, O'Brian CA (2003) Resveratrol: A candidate nutritional substance for prostate cancer prevention. *J Nutr* 133: 2440S–2443S

152 Galati G, O'Brien PJ (2004) Potential toxicity of flavonoids and other dietary phenolics: Significance for their chemopreventive and anticancer properties. *Free Radic Biol Med* 37: 287–303

153 Moon YJ, Wang X, Morris ME (2006) Dietary flavanoids: Effects on xenobiotic and carcinogen metabolism. *Toxicol In Vitro* 20: 187–210

154 Chen I, McDougal A, Wang F, Safe S (1998) Aryl hydrocarbon receptor-mediated antiestrogenic and antitumorigenic activity of diindolylmethane. *Carcinogenesis* 19: 1631–1639

155 Kim YS, Milner JA (2005) Targets for indole-3-carbinol in cancer prevention. *J Nutr Biochem* 16: 65–73

156 Meijer L, Sheare J, Bettayeb K, Ferandin Y (2006) Diversity of intracellular mechanisms underlying the anti-tumor properties of indirubins. In: L Meijer, N Guyard, L Skaltsounis, G Eisenbrand (eds): *Indirubin, The Red Shade of Indigo*. Life in Progress Editions, Roscoff, 235–246

157 Westerink WMA, Stevenson JCR, Schoonen WGEJ (2007) CYP1A induction and species differences between human HepG2 and rat H4IIE cells. *Toxicol Lett* 172 Suppl 1: S98

158 Morimura K, Cheung C, Ward JM, Reddy JK, Gonzalez FJ (2005) Differential susceptibility of mice humanized for peroxisome proliferator-activated receptor α to Wy-14,643-induced liver tumorigenesis. *Carcinogenesis* 27: 1074–1080

159 Crespi CL, Miller VP (1999) The use of heterologously expressed drug metabolizing enzymes – State of the art and prospects for the future. *Pharmacol Ther* 84: 121–131

160 Crespi CL, Miller VP, Penman BW (1997) Microtiter plate assays for inhibition of human, drug-metabolizing cytochromes P450. *Anal Biochem* 248: 188–190

161 Miller VP, Stresser DM, Blanchard AP, Turner S, Crespi CL (2000) Fluorometric high-throughput screening for inhibitors of cytochrome P450. *Ann NY Acad Sci* 919: 26–32

162 Stresser DM, Turner SD, Blanchard AP, Miller VP, Crespi CL (2002) Cytochrome P450 fluorometric substrates: Identification of isoform-selective probes for rat CYP2D2 and human CYP3A4. *Drug Metab Disp* 30: 845–852

163 Nakajima M, Sakata N, Ohashi N, Kume T, Yokoi T (2002) Involvement of multiple UDP-glucuronosyltransferase 1A isoforms in glucuronidation of 5-(4'-hydroxyphenyl)-5-phenylhydantoin in human liver microsomes. *Drug Metab Disp* 30: 1250–1256

164 Kurkela M, Garcia-Horsman JA, Luukkanen L, Mörsky S, Taskinen J, Baumann M, Kostianen R, Hirvonen J, Finel M (2003) Expression and characterization of recombinant human UDP-glucuronosyltransferases (UGTs). *J Biol Chem* 278: 3536–3544

165 Collier AC, Tingle MD, Keelan JA, Paxton JW, Mitchell MD (2000) A highly sensitive fluores-

cent microplate method for the determination of UDP-glucuronosyl transferase activity in tissues and placental cell lines. *Drug Metab Disp* 28: 1184–1186

166 Uchaipichat V, Mackenzie PI, Guo X-H, Gardner-Stephen D, Galetin A, Houston JB, Miners JO (2004) Human UDP-glucuronosyltransferases: Isoform selectivity and kinetics of 4-methylumbelliferone and 1-naphthol glucuronidation, effects of organic solvents, and inhibition by diclofenac and probecenid. *Drug Metab Disp* 32: 413–423

167 Broudy MI, Crespi CL, Patten CJ (2001) *A sensitive fluorometric high throughput inhibition assay for human UDP glucuronosyl transferase (UGT) 1A1*. Poster presentation: BD Biosciences, Property of Becton, Dickinson and Company

168 Trubetskoy OV, Finel M, Kurkela M, Fitzgerald M, Peters NR, Hoffman FM, Trubetskoy VS (2007) High throughput screening assay for UDP-glucuronosyltransferase 1A1 glucuronidation profiling. *Assay Drug Dev Technol* 5: 343–354

169 Wu JJ, Huang DB, Pang KR, Hsu S, Tyring SK (2005) Thalidomide: Dermatological indications, mechanisms of action and side-effects. *Br J Dermatol* 153: 254–273

170 Genschow E, Spielmann, H, Scholz G, Seiler A, Brown N, Piersma A, Brady M, Clemann N, Huuskonen H, Paillard F et al (2002) The ECVAM international validation study on *in vitro* embryotoxicity tests: Results of the definitive phase and evaluation of prediction models. *ATLA* 30: 151–176

171 Piersma AH (2004) Validation of alternative methods for developmental toxicity testing. *Toxicol Lett* 149: 147–153

172 Piersma AH (2006) Alternative methods for developmental toxicity testing. *Basics Clin Pharmacol Toxicol* 98: 427–431

173 Piersma AH, Janer G, Wolterink G, Bessems JGM, Hakkert BC, Slob W (2008) Quantitative extrapolation of *in vitro* whole embryo culture embryotoxicity data to developmental toxicity *in vivo* using the benchmark dose approach. *Toxicol Sci* 101: 91–100

174 Jansen MS, Nagel SC, Miranda PJ, Lobenhofer EK, Afshari CA, McDonnell DP (2004) Short-chain fatty acids enhance nuclear receptor activity through mitogen activated protein kinase activation and histone deacetylase inhibition. *Proc Natl Acad Sci USA* 101: 7199–7204

175 Tabb MM, Blumberg B (2006) New modes of action of endocrine-disrupting chemicals. *Mol Endocrinol* 20: 475–482

176 Campbell LR, Dayton DH, Sohal GS (1986) Neural tube defects: A review of human and animal studies on the etiology of neural tube defects. *Teratology* 34: 171–187

177 Sulik KK, Dehart DB, Rogers JM, Chernoff N (1995) Teratogenicity of low doses of all-trans retinoic acid in presomite mouse embryos. *Teratology* 51: 398–403

178 Arnhold T, Elmazar MMA, Nau H (2002) Prevention of vitamin A teratogenesis by phytol or phytanic acid results from reduced metabolism of retinol to the teratogenic metabolite, all-trans retinoic acid. *Toxicol Sci* 66: 274–282

179 Wang Z, Brown DD (1993) Thyroid hormone-induced gene expression program for amphibian tail resorption. *J Biol Chem* 268: 16270–16278

180 Brown DD, Wang Z, Furlow JD, Kanamori A, Schwartman RA, Remo BF, Pinder A (1996) The thyroid hormone-induced tail resorption program during *Xenopus laevis* metamorphosis. *Proc Natl Acad Sci USA* 93: 1924–1929

181 Hollowell JG Jr, Hannon WH (1997) Teratogen update: Iodine deficiency, a community teratogen. *Teratology* 55: 389–405

182 Yang XF, Xu J, Hou XH, Guo HL, Hao LP, Yao P, Liu LG, Sun XF (2006) Developmental toxic effects of chronic exposure to high doses of iodine in the mouse. *Reprod Toxicol* 22: 725–730

183 Sanders JP, van der Geyten S, Kaptein E, Darras VM, Kühn ER, Leonard JL, Visser TJ (1997) Characterization of a propylthiouracil-insensitive type 1 iodothyronine deiodinase. *Endocrinology* 138: 5153–5160

184 Köhrle J (2000) The deiodinase family: Selenoenzymes regulating thyroid hormone availability and action. *Cell Mol Life Sci* 57: 1853–1863

185 Barrington EJW (1975) *An Introduction to General Comparative Endocrinology*, 2nd edn. Clarendon Press, Oxford, 147–184

186 Huang H, Marsh-Armstrong N, Brown DD (1999) Metamorphosis is inhibited in transgenic *Xenopus laevis* tadpoles that overexpress type III deiodinase. *Proc Natl Acad Sci USA* 96: 962–967

187 Glincer D (1984) Thyroid dysfunction in the pregnant patients. In: LJ DeGroot, PR Larsen, G Hennemann (eds): *The Thyroid and Its Diseases*. Wiley, New York

188 Jaber M, Robinson SW, Missale C, Caron MG (1996) Dopamine receptors and brain function. *Neuropharmacology* 35: 1503–1519

189 Hoyer D, Clarke DE, Fozard JR, Hartig PR, Martin GR, Mylecharane EJ, Saxena PR, Humphrey PPA (1994) VII. International union of pharmacology classification of receptors for 5-hydroxttryptamine (serotonin). *Pharmacol Rev* 46: 157–203

190 Shuey DL, Sadler TW, Lauder JM (1992) Serotonin as a regulator of craniofacial morphogenesis site specific malformations following exposure to serotonin uptake inhibitors. *Teratology* 46: 367–378

191 Jurand A, Martin LVH (1990) Teratogenic potential of two neurotropic drugs, haloperidol and dextromoramide, tested on mouse embryos. *Teratology* 42: 45–54

192 Van Cauteren H, Vandenberghe J, Marsboom R (1986) Protective activity of ketanserin against serotonin-induced embryotoxicity and teratogenicity in rats. *Drug Dev Res* 8: 179–185

193 Huuskonen H (2005) New models and molecular markers in evaluation of developmental toxicity. *Toxicol Appl Pharmacol* 207 Suppl: 495–500

194 Hill AJ, Teraoka H, Heideman W, Peterson RE (2005) Zebrafish as a model vertebrate for investigating chemical toxicity. *Toxicol Sci* 86: 6–19

195 Hershberger LG, Shipley, EG, Meyer RK (1953) Myometric activity of 19-nortestosterone and other steroids determined by modified levator and muscle method. *Proc Soc Exp Biol Med* 83: 175–180

196 Ashley J, Lefevre PA (2000) Preliminary evaluation of the major protocol variables for the Hershberger castrated male rat assay for the detection of androgens, antiandrogens, and metabolic modulators. *Regul Toxicol Pharmacol* 31: 92–105

197 Allen A, Doisy EA (1923) An ovarian hormone. Preliminary report on its location, extraction and partial purification and action in test animals. *J Am Med Assoc* 81: 819–821

198 Wakeling AE (1995) Use of pure antioestrogens to elucidate the mode of action of oestrogens. *Biochem Pharmacol* 49: 1545–1549

199 Van der Vies J, de Visser J (1983) Endocrinological studies with desogestrel. *Arzneimittelforschung* 33: 231–236

200 McPhail MK (1934) The assay of progestin. *J Physiol* 83: 145–156

201 Bergink EW, van Meel F, Turpijn EW, van der Vies J (1983) Binding of progestagens to receptor proteins in MCF-7 cells. *J Steroid Biochem* 19: 1563–1570

202 Schoonen WGEJ, Joosten JWH, Kloosterboer HJ (1995) Effects of two classes of progestagens, pregnane and 19-nortestosterone derivatives, on cell growth of human breast tumor cells. I. MCF-7 cell lines. *J Steroid Biochem Mol Biol* 55: 423–437

203 Schoonen WGEJ, Dijkema R, Ries de RJH, Wagenaars JL, Joosten JWH, de Gooyer ME, Deckers GH, Kloosterboer HJ (1998) Human progesterone receptor A and B isoforms in CHO cells. II. Comparison of binding, transactivation and ED_{50} values of several synthetic (anti)progestagens *in vitro* in CHO and MCF-7 cells and *in vivo* in rabbits and rats. *J Steroid Biochem Mol Biol* 64: 157–170

204 Schoonen WGEJ, Ries de RJH, Joosten JWH, Mathijssen-Mommers GJW, Kloosterboer HJ (1998) Development of a high throughput *in vitro* bioassay for the assessment of potencies of steroids for the progesterone receptor with a luciferase reporter system in CHO cells. *Anal Biochem* 261: 222–224

205 Sonneveld E, Riteco JAC, Jansen HJ, Pieterse B, Brouwer A, Schoonen WG, van der Burg B (2006) Comparison of *in vitro* and *in vivo* screening models for androgenic and estrogenic activities. *Toxicol Sci* 89: 173–187

206 Legler J, van den Brink CE, Brouwer A, Murk AJ, van der Saag PT, Vethaak AD, van der Burg B (1999) Development of a stably transfected estrogen receptor-mediated luciferase reporter gene assay in the human T47D breast cancer cell line. *Toxicol Sci* 48: 55–66

207 Blankvoort BMG, de Groene EM, van Meeteren-Kreikamp AP, Witkamp RF, Rodenburg RJT, Aarts JMMJG (2001) Development of an androgen reporter gene assay (AR-LUX) utilizing a human cell line with an endogenously regulated androgen receptor. *Anal Biochem* 298: 93–102

208 Wilson VS, Bobseine K, Gray LE Jr, (2004) Development and characterization of a cell line that stably expresses an estrogen-responsive luciferase reporter for the detection of estrogen receptor agonist and antagonists. *Toxicol Sci* 81: 69–77

209 Wilson VS, Bobseine K, Lambright CR, Gray LE Jr, (2002) A novel cell line, MDA-kb2, that stably expresses an androgen- and glucocorticoid-responsive reporter for the detection of hormone receptor agonists and antagonists. *Toxicol Sci* 66: 69–81

210 Wong SP, Li J, Shen P, Gong Y, Yap SP, Yong EL (2007) Ultrasensitive cell-based bioassay for the measurement of global estrogenic activity of flavonoid mixtures revealing additive, restrictive, and enhanced actions in binary and higher order combinations. *Assay Drug Dev Technol* 5: 355–362

211 Bovee TFH, Helsdingen RJR, Rietjens IMCM, Keijer J, Hoogenboom RLAP (2004) Rapid yeast estrogen bioassays stably expressing human estrogen receptors α and β, and green fluorescent protein: A comparison of different compounds with both receptor types. *J Steroid Biochem Mol Biol* 91: 99–109

212 Bovee TFH, Helsdingen RJR, Hamers ARM, Van Duursen MBM, Nielen MWF, Hoogenboom RLAP (2007) A new highly specific and robust yeast androgen bioassay for the detection of agonists and antagonists. *Anal Bioanal Chem* 389: 1549–1558

213 Bovee TFH, Schoonen WGEJ, Hamers ARM, Bento MJ, Peijnenburg AACM (2008) Screening of synthetic and plant-derived compounds for (anti-)estrogenic and (anti-)androgenic activities. *Anal Bioanal Chem* 390: 1111–1119

214 Richard S, Zingg HH (1990) The human oxytocin gene promoter is regulated by estrogens. *J Biol Chem* 265: 6098–6103

215 Schippers IJ, Kloppenburg M, Waardenburg R, Ab G (1994) Cis-acting elements reinforcing the activity of the estrogen-response element in the very-low-density apolipoprotein II gene promoter. *Eur J Biochem* 221: 43–51

216 Chang TC, Nardulli AM, Lew D, Shapiro DJ (1992) The role of estrogen response elements in expression of the *Xenopus laevis* vitellogenin B1 gene. *Mol Endocrinol* 6: 346–354

217 Kastner P, Krust A, Turcotte B, Stropp U, Tora, L, Gronemeyer H, Chambon P (1990) Two distinct estrogen regulated promoters generate transcripts encoding the two functionally different human progesterone receptor forms A and B. *EMBO J* 9: 1603–1614

218 Perlmann T, Eriksson P, Wrange Ö (1990) Quantitative analysis of the glucocoticoid receptor-DNA interaction at the mouse mammary tumor virus glucocorticoid response element. *J Biol Chem* 265: 17222–17229

219 Sun YN, DuBois DC, Almon RR, Pyszczynski NA, Jusko WJ (1998) Dose-dependent and repeat-dose studies for receptor/gene-mediated pharmacodynamics of methylprednisolone on glucocorticoid receptor down-regulation and tyrosine aminotransferase induction in rat liver. *J Pharmcokinet Biopharm* 26: 619–648

220 Scott DK, Strömstedt PE, Wang JC, Granner DK (1998) Further characterisation of the glucocorticoid response unit in the phosphoenolpyruvate carboxykinase gene. The role of the glucocorticoid receptor-binding sites. *Mol Endocrinol* 12: 482–491

221 Deckers GH, Schoonen WGEJ, Kloosterboer HJ (2000) Influence of the substitution of 11-methylene, Δ^{15}, and/or 18-methyl groups in norethisterone on receptor binding, transactivation assays and biological activities in animals. *J Steroid Biochem Mol Biol* 72: 83–92

222 Soto AM, Maffini MV, Schaeberle CM, Sonnenschein C (2006) Strengths and weaknesses of *in vitro* assays for estrogenic and androgenic activity. *Best Pract Res Clin Endocr Metab* 20: 15–33

Glossary

AA	aristolochic acid
AAF	2-acetylaminofluorene
AAN	aristolochic acid (induced) nephropathy
ABT	American Board of Toxicology
AC	adenylate cyclase
ACF	aberrant crypt foci
ADME	absorption, distribution, metabolism, excretion
AGT	alkylguanine transferase
AhR	aryl hydrocarbon receptor
AIMS	amplification of intermethylated sites
AIP	apoptosis-inducing protein 1
ALARA	'as low as reasonably achievable'
AMPH	amphetamine
AMPK	5'AMP-activated protein kinase
AMS	accelerator mass spectrometry
ANIT	α-naphthylisothiocyanate
AP	apurinic (or) apyrimidinic
APAP	acetaminophen (N-acetyl-p-aminophenol)
APE1	AP endonuclease-1
AR	androgen receptor
ARE	androgen response element
ARNT	Ah receptor nuclear translocator
ASDP	Asian sand dust particles
ATM	ataxia telangiectasia-mutated (kinase)
ATR	ataxia telangiectasia-related (kinase)
ATRIP	ATR-interacting protein
ATS	Academy of Toxicological Sciences
B[a]P	benzo[a]pyrene
BAT	biological tolerance values (at the workplace)
BEI	Biological Exposure Indices
BEN	Balkan endemic nephropathy
BER	base excision repair
BHD1	β-hairpin domain
BLC	B lymphocyte chemoattractant
BP	benzo[a]pyrene
BPA	bisphenol A
BPDE	benzo[a]pyrene (BP) 7,8 diol 9,10 epoxide
CA	chromosome aberrations

CAR	constitutive androstane receptor
CAR	constitutively active/androstane receptor
CBP	CREB (cAMP-responsive element-binding) protein
CCL	chemokine ligand
CCL1	CC-chemokine ligand 1
CCR	chemokine receptor
CCRP	cytoplasmic CAR retention protein
Cdk	cyclin-dependent protein kinase
CE-LIF	capillary electrophoresis-laser-induced fluorescence
CFC	chlorofluorocarbon
ChIP	chromatin immunoprecipitation
CIA	chemiluminescence immunoassay
CITCO	6-(4-chlorophenyl)imidazo[2-1-*b*][1,3]thiazole-5-carbaldehyde *O*-(3,4-dichlorobenzyl)oxime
CLICK	cluster identification *via* connectivity kernels
CNS	central nervous system
CNV	copy number variants
COX	cyclooxygenase
CRE	cAMP responsive element
CREB	cAMP responsive element binding
CS	Cockayne syndrome
CSTAD	cyclosporine A-conditional, T cell activation-dependent (gene)
CTN2	centrin-2
CYP	cytochrome P450-dependent monooxygenase
DAPK	promoter hypermethylation
DAQ	dopamine-quinones
DBP	dibenzo[*a*,*l*]pyrene
DBPDE	dibenzo[*a*,*l*]pyrene (DBP) 11,12-diol 13,14 epoxide
DC	developmental candidate
DC	dendritic cell
DCVC	dichlorovinyl-L-cysteine
DDB2	DNA damage-binding protein 2
DDE	dichlorodiphenyldichloroethylene
DDT	dichlorodiphenyltrichloroethane
DELFIA	dissociation-enhanced lanthanide fluoroimmunoassay
DEN	diethylnitrosamine
DEP	diesel exhaust particles
DES	diethylstilbestrol
DHT	dihydrotestosterone
DIGE	2D-difference gel electrophoresis
DILI	drug-induced liver injury
DMBA	7,12-dimethylbenz[*a*]anthracene
DNA-PK	DNA-dependent protein kinase
DNBS	2,4-dinitrobenzenesulfonic acid
DNFB	2,4-dinitro-1-fluorobenzene

DNMT	DNA methyltransferases
DR	dioxin receptor
DSB	double strand break
DYRK-2	dual-specificity tyrosine-(Y)-phosphorylation regulated kinase 2
EAF	enzyme altered foci
ECD	electrochemical detection
ECVAM	European Centre for the Validation of Alternative Methods
EDC	endocrine disrupting chemicals
EE	17α-ethinyl estradiol
EGFR	epidermal growth factor receptor
EH	epoxide hydrolase.
ELISA	enzyme-linked immunosorbent assay
EPA	Environmental Protection Agency
EPHX1	(microsomal) epoxide hydrolase
EPIG	Extracting microarray (gene) expression Patterns and Identifying coexpressed Genes
ER	estrogen receptor
ERCC1	excision repair cross complementing-1 (protein)
ERE	estrogen responsive element
ERK	extracellular signal-regulated kinase
ESI	electrospray ionisation spectrometry
FA	Fanconi anemia
FAH	fumarylacetoacetate hydrolase
FDA	U.S. Food and Drug Administration
FDC	full development candidate
FDR	false discovery rate
FDVE	fluoromethyl-2,2-difluoro-1-(trifluoromethyl)vinyl ether
FICZ	6-formylindolo[3,2-*b*]carbazole
FPG	formamidopyrimidine-DNA-glycosylase
FRET	fluorescence resonance electron transfer
FSH	follicular stimulating hormone
FXR	farnesoid X receptor
Gc	group-specific component (vitamin D-binding protein)
GFP	green fluorescent protein
GGR	global genome repair
GI	gastrointestinal
GINA	Genetic Information Nondiscrimination Act
GnRH	gonadotropin releasing hormone
GO	Gene Ontology
GPCR	G protein-coupled receptor(s)
GR	glucocorticoid receptor
GRE	glucocorticoid responsive element
GST	glutathione-*S*-transferase
HAH	halogenated aromatic hydrocarbon(s)
HCB	hexachlorobenzene

HCC	hepatocellular carcinoma
HCS	high-content screening
HEP	Human Epigenome Project
HIF-1	hypoxia-inducible factor 1
HIPK-2	homeodomain interacting protein kinase 2
HMG	high-mobility group (protein)
HMG-CoA	3-hydroxy-3-methyl-glutaryl-CoA
hOGG1	human 8-oxoguanine DNA glycosylase 1
HPCE	high-performance capillary electrophoresis
HPLC	high-performance liquid chromatography
HR	homologous recombination
HRE	hormone responsive element
HTS	high-throughput screening
HUPO	Human Proteome Organization
IARC	International Agency for Research on Cancer
ICAT	isotope coded affinity tag
ICH	International Conference on Harmonisation of Technical
ICI	Imperial Chemical Industries
ICL	interstrand cross-link
iDC	immature dendritic cell
IDM	methyisobutylxanthine
IFR3	interferon-γ responsive factor
IGF	insulin-like growth factor
IHC	immunohistochemistry
IKK	inhibitor of κB kinases
IPCS	International Programme on Chemical Safety
IPG	immobilized pH gradient
IQ	2-amino-3-methylimidazo[4,5-f]quinoline
IR	ionizing radiation
ISQ	1,5-dihydroxyisoquinoline
ISSX	International Society for the Study of Xenobiotics
iTRAQ	isobaric tags for relative and absolute quantitation
IUTOX	International Union of Toxicology
IWGT	International Workshop on Genotoxicity Testing
JNK	c-JUN N-terminal kinase
JNK/SAPK	c-Jun N-terminal/stress-activated protein kinase
JP-8	Jet propulsion fuel
JUND	JunD (protein)
KIM-1	kidney injury molecule-1
LC	liquid chromatography
LCES/MS	liquid chromatography-electrospray mass spectrometry
LH	luteinizing hormone
LIF	laser-induced fluorescence
L-NAME	N^{G}-nitro- arginine methyl ester
LOH	loss of heterozygosity

LPS	lipopolysaccharide
LSD	lysergic acid diethylamide
LTBP	latent TGFβ binding protein
LXR	liver X receptor
MALDI	matrix-assisted laser desorption/ionization
MAPK	mitogen-activated protein kinase
MBDE	mass-balance differential equation
3-MC	3-methylcholanthrene
Mdm2	murine double minute 2 (protein)
MDR	multidrug resistance
MeDIP	methylated DNA immunoprecipitation
MEF	mouse embryo fibroblasts
MeHg	methylmercury
MEIC	Multicenter Evaluation of *In vitro* Cytotoxicity
MEKK1	MAPK/ERK kinase kinase 1
METH	methamphetamine
MGMT	O^6-methylguanine-DNA methyltransferase
MLA	mouse lymphoma assay
MLN	mesenteric lymph node(s)
MMR	mismatch repair
MMS	methyl methanesulfonate
MMTV	mouse mamma tumor virus
MN	micronuclei
MPG	methylpurine-DNA glycosylase
MPTP	1-methyl-4-phenyl-1,2,3,6-tetrahydropyridine
MR	mineralocorticoid receptor
Mrp2	MDR-associated protein 2
MS	mass spectrometry
MSI	microsatellite instability
MTS	medium-throughput screening
MTT	3-(4,5-dimethylthiazol-2-yl)-2,5-diphenyltetrazolium bromide
MuDPIT	multidimensional protein identification technology
NAPQI	*N*-acetyl-*p*-benzoquinone imine
NAT	*N*-acetyltransferase
NBD-PC	*N*-(7-nitrobenz-2-oxa-1,3-diazol-4-yl)-1,2-dihexadecanoyl-*sn*-glycero-3-phosphatidyl choline
NBD-PE	NBD-phosphoethanolamine
NBS	Nijmegen breakage syndrome
NBT	nitro blue tetrazolium
NCoR	nuclear receptor corepressor
NDP	nedaplatin
NEPHGE	non-equilibrium pH gradient electrophoresis
NER	nucleotide excision repair
NES	nuclear export signal
NHEJ	non-homologous end-joining

NIEHS	National Institute for Environmental Health Sciences
NLS	nuclear localization signal
NNN	N'-nitrosonornicotine
NQO1	NAD(P)H-dependent quinone oxidoreductase-1
4-NQO	4-nitroquinoline-1-oxide
NR	nuclear receptor
NRE	nuclear receptor response element
Nrf2	nuclear factor erythroid-2 p45-related factor 2
OAT	ornithine aminotransferase
OATP	organic anionic transporting pump
OCT	organic cation transporter
OGG1	8-oxoguanine DNA glycosylase 1
6-OH-DA	6-hydroxydopamine
OP	octylphenol
8-oxo-dGuo	7,8-dihydro-8-oxo-2'-deoxyguanosine
PAH	polycyclic aromatic hydrocarbon(s)
PA1	plasminogen activator 1
PAPS	3'-phosphoadenosine-5'-phosphosulfate
PARP	Poly(ADP-ribose) polymerase
PB	phenobarbital
PBDE	polybrominated bipehenyl ethers
PBP	PPAR-binding protein
PBREM	phenobarbital (PB)-responsive enhancer module
PBTK	physiologically based toxicokinetic (models)
PCA	principal component analysis
PCB	polychlorinated biphenyl(s)
PCB126	3,3',4,4',5-pentachlorobiphenyl
PCDD	polychlorinated dibenzodioxins
PCDF	polychlorinated dibenzofurans
PCNA	proliferating cell nuclear antigen
PEPCK	phosphoenolpyruvate carboxykinase
PepT	peptide transporter
P-gp	P-glycoprotein
PhIP	2-amino-1-methyl-6-phenylimidazo[4, 5-b]pyridine
PIKK	phosphatidylinositol 3-kinase-like kinase
PMS2	postmeiotic segregation increased 2
POLH	polymerase eta
POP	persistent organic pollutant(s)
PPAR	peroxisome proliferator-activated receptor
PR	progesterone receptor
PSTC	Predictive Safety Testing Consortium
PXR	pregnane X receptor
PXR/SXR	pregnane X receptor/steroid X receptor
QSAR	quantitative structure-activity relationship
RAR	retinoic acid receptor

RB	retinoblastoma (protein)
RC-MS	retentate chromatography-mass spectrometry
RDA	representational difference analysis
REACH	Registration, Evaluation, Authorization and restriction of Chemicals
RHD	Rel homology domain
RLGS	restriction landmark genomic scanning
ROFA	residual oil fly ash
ROS	reactive oxygen species
RPA	replication protein A
RPT	rapidly perfused tissues
RXR	retinoic X receptor
SCE	sister chromatid exchange
SELDI-TOF	surface-enhanced laser desorption/ionization time-of-flight
SFS	synchronous fluorescence spectroscopy
SHP-1	small heterodimer protein-1
SILAC	stable isotope labeling by amino acids in cell culture
SMRT	silencing mediator for retinoic acid receptor and thyroid hormone receptor
SNOB	statement of no objection
SNP	single nucleotide polymorphism
SNPC	substantia nigra pars compacta
SOFT	Society of Forensic Toxicology
SOT	Society of Toxicology
SRM	selected reaction monitoring
SSB	single strand break
SSR	structure-specific recognition (protein)
STAP	stellate cell activation-associated protein
SULT	sulfotransferase
SVM	support vector machine
SXR	steroid X receptor
TBA	tumor-bearing animal
TBTO	bis(tri-n-butyltin)oxide
TCDD	2,3,7,8-tetrachlorodibenzo-p-dioxin
TCPOBOP	1,4-bis-[2-(3,5-dichloropyridyloxy)]benzene
TCR	transcription-coupled repair
TEL	tetraethyl lead
TEPP	tetraethyl pyrophosphate
TFF1	trefoil factor 1
TFIIH	transcription factor IIH
TFT	trifluorothymidine
TGD	transglutaminase domain
TGF	transforming growth factor
TGIF	5'-TG-3'-interacting factor
TLC	thin layer chromatography

TLS	translesion synthesis
TNF	tumor necrosis factor
TPA	12-*O*-tetradecanoylphorbol 13-acetate
TR	thyroid hormone receptor
TRE	thyroid hormone response element
TSH	thyroid-stimulating hormone
TSP1	thrombospondin-1
UDP	uridine diphosphate
UDPGA	uridine diphosphate glucuronic acid
UGT	UDP-glucuronosyltransferase
VDAC2	voltage-dependent anion channel-selective protein-2
vGCKR	glucokinase regulatory protein
VLDL	very low density lipoprotein
VOC	volatile organic compounds
X/NRE	xenobiotic/nuclear receptor response elements
XAP	X-associated protein
XLF	XRCC4-like factor
XP	xeroderma pigmentosum
XPA-XPF	xeroderma pigmentosum complementation group A-F (proteins)
XPC	xeroderma pigmentosum protein C
XRCC4	X-ray repair cross-complementing-4
XREM	xenobiotic-responsive enhancer module
XS	xenosensor

Index

The EXS-Series
Experientia Supplementum

Experientia Supplementum (EXS) is a multidisciplinary book series originally created as supplement to the journal *Experientia* which appears now under the cover of *Cellular and Molecular Life Sciences*. The multi-authored volumes focus on selected topics of biological or biomedical research, discussing current methodologies, technological innovations, novel tools and applications, new developments and recent findings.
The series is a valuable source of information not only for scientists and graduate students in medical, pharmacological and biological research, but also for physicians as well as practitioners in industry.

Forthcoming titles:

Molecular, Clinical and Environmental Toxicology, Volume 2, EXS 100, A. Luch (Editor), 2009
Molecular, Clinical and Environmental Toxicology, Volume 3, EXS 101, A. Luch (Editor), 2009

Published volumes:

Proteomics in Functional Genomics, EXS 88, P. Jollès, H. Jörnvall (Editors), 2000
New Approaches to Drug Development, EXS 89, P. Jollès (Editor), 2000
The Carbonic Anhydrases, EXS 90, Y.H. Edwards, W.R. Chegwidden, N.D. Carter (Editors), 2000
Genes and Mechanisms in Vertebrate Sex Determination, EXS 91, G. Scherer, M. Schmid (Editors), 2001
Molecular Systematics and Evolution: Theory and Practice, EXS 92, R. DeSalle, G. Giribet, W. Wheeler (Editors), 2002
Modern Methods of Drug Discovery, EXS 93, A. Hillisch, R. Hilgenfeld (Editors), 2003
Mechanisms of Angiogenesis, EXS 94, M. Clauss, G. Breier (Editors), 2004
NPY Family of Peptides in Neurobiology, Cardiovascular and Metabolic Disorders: from Genes to Therapeutics, EXS 95, Z. Zukowska, G.Z. Feuerstein (Editors), 2006
Cancer: Cell Structures, Carcinogens and Genomic Instability, EXS 96, L.P. Bignold (Editor), 2006
Plant Systems Biology, EXS 97, S. Baginsky, A.R. Fernie (Editors), 2006
Neurotransmitter Interactions and Cognitive Function, EXS 98, E.D. Levin (Editor), 2006